MX 9711138 4

AF443666

Sensors

Volume 3

Chemical and Biochemical Sensors

Part II

Sensors

A Comprehensive Survey

Edited by
W. Göpel   (Universität Tübingen, FRG)
J. Hesse   (Zeiss, Oberkochen, FRG)
J. N. Zemel (University of Pennsylvania,
                    Philadelphia, PA, USA)

Published:
Vol. 1   Fundamentals and General Aspects
             (Volume Editors: T. Grandke, W. H. Ko)
Vol. 2/3 Chemical and Biochemical Sensors, Part I/II
             (Volume Editors: W. Göpel, T. A. Jones †, M. Kleitz,
             I. Lundström, T. Seiyama)
Vol. 4   Thermal Sensors
             (Volume Editors: T. Ricolfi, J. Scholz)
Vol. 5   Magnetic Sensors
             (Volume Editors: R. Boll, K. J. Overshott)
Vol. 6   Optical Sensors
             (Volume Editors: E. Wagner, R. Dändliker, K. Spenner)

Remaining volumes of this closed-end series:

Vol. 7   Mechanical Sensors (scheduled for 1992)
Vol. 8   Cumulative Index and Selected Topics (scheduled for 1993)

Distribution

VCH, P. O. Box 101161, D-6940 Weinheim (Federal Republic of Germany)

Switzerland: VCH, P. O. Box, CH-4020 Basel (Switzerland)

United Kingdom and Ireland: VCH (UK) Ltd., 8 Wellington Court, Wellington Street,
      Cambridge CB1 1HZ (England)

USA and Canada: VCH, Suite 909, 220 East 23rd Street, New York, NY 10010-4606 (USA)

ISBN 3-527-26769-7 (VCH, Weinheim)   ISBN 0-89573-675-6 (VCH, New York)

Sensors

A Comprehensive Survey

Edited by
W. Göpel, J. Hesse, J. N. Zemel

Volume 3

Chemical and Biochemical Sensors
Part II
Edited by
W. Göpel, T. A. Jones †, M. Kleitz,
I. Lundström, and T. Seiyama

Weinheim · New York · Basel · Cambridge

Series Editors:
Prof. Dr. W. Göpel
Institut für Physikalische und
Theoretische Chemie der Universität
Auf der Morgenstelle 8
D-7400 Tübingen, FRG

Prof. Dr. J. Hesse
Carl Zeiss,
ZB „Entwicklung"
Postfach 1380
D-7082 Oberkochen, FRG

Prof. Dr. J. N. Zemel
Center for Sensor Technology
University of Pennsylvania
Philadelphia, PA 19104-6390, USA

Volume Editors:
Prof. Dr. W. Göpel
see above

Dr. T. A. Jones †
Health and Safety
Executive
Sheffield, UK

Dr. M. Kleitz
L.I.E.S.G./
E.N.S.E.E.G.
Domaine Universitaire,
B.P. 75
F-38402 Saint-Martin
d'Hères, France

Prof. I. Lundström
Linköping Institute of
Technology
Dept. of Physics and
Measurement Technology
S-58183 Linköping, Sweden

Prof. T. Seiyama
Tokuyama Soda Co., Ltd.
Tenjin 1-10-24
Chuo-ku, Fukuoka-shi,
Japan 810

Published jointly by
VCH Verlagsgesellschaft mbH, Weinheim (Federal Republic of Germany)
VCH Publishers Inc., New York, NY (USA)

Editorial Directors: Dipl.-Phys. W. Greulich, Dipl.-Chem. Dr. M. Weller, N. Banerjea-Schultz
Production Manager: Dipl.-Wirt.-Ing. (FH) H.-J. Schmitt
Indexing: Borkowski, Schauernheim

Library of Congress Card No.: applied for

British Library Cataloguing-in-Publication Data:
Sensors: a comprehensive survey: Vol 3. Chemical and biochemical sensors, Part II. – (Sensors)
   I. Goepel, W.   II. Jones, T. A.   III. Kleitz, M.
   IV. Series
   502.8
   ISBN 3-527-26769-7

Deutsche Bibliothek Cataloguing-in-Publication Data:
Sensors: a comprehensive survey / ed. by W. Göpel … –
Weinheim ; Basel (Switzerland) ; Cambridge ; New York, NY
VCH.
NE: Göpel, Wolfgang [Hrsg.]
Vol. 3. Chemical and biochemical sensors. – Part 2. Ed. by W.
   Göpel … – 1991
   ISBN 3-527-26769-7 (Weinheim …)
   ISBN 0-89573-675-6 (New York)

Printed on acid-free paper

Composition: Filmsatz Unger + Sommer GmbH, D-6940 Weinheim.
Printing: DiesbachMedien, D-6940 Weinheim.
Bookbinding: Großbuchbinderei J. Schäffer, D-6718 Grünstadt.
Printed in the Federal Republic of Germany

Preface to the Series

The economic realities of productivity, quality, and reliability for the industrial societies of the 21st century are placing major demands on existing manufacturing technlogies. To meet both present and anticipated requirements, new and improved methods are needed. It is now recognized that these methods must be based on the powerful techniques employing computer-assisted information systems and production methods. To be effective, the measurement, electronics and control components, and sub-systems, in particular sensors and sensor systems, have to be developed in parallel as part of computer-controlled manufacturing systems. Full computer compatibility of all components and systems must be aimed for. This strategy will, however, not be easy to implement, as seen from previous experience. One major aspect of meeting future requirements will be to systematize sensor research and development.

Intensive efforts to develop sensors with computer-compatible output signals began in the mid 1970's; relatively late compared to computer and electronic measurement peripherals. The rapidity of the development in recent years has been quite remarkable but its dynamism is affected by the many positive and negative aspects of any rapidly emerging technology. The positive aspect is that the field is advancing as a result of the infusion of inventive and financial capital. The downside is that these investments are distributed over the broad field of measurement technology consisting of many individual topics, a wide range of devices, and a short period of development. As a consequence, it is not surprising that sensor science and technology still lacks systematics. For these reasons, it is not only the user who has difficulties in classifying the flood of emerging technolgical developments and solutions, but also the research and development scientists and engineers.

The aim of "Sensors" is to give a survey of the latest state of technolgy and to prepare the ground for a future systematics of sensor research and technology. For these reasons the publishers and the editors have decided that the division of the handbook into several volumes should be based on physical and technical principles.

Volume 1 (editors: T. Grandke/Siemens (FRG) and W. H. Ko/Case Western Reserve University (USA)) deals with general aspects and fundamentals: physical principles, basic technologies, and general applications.

Volume 2 and 3 (editors: W. Göpel/Tübingen University (FRG), T. A. Jones †/Health and Safety Executive (UK), M. Kleitz/LIESG-ENSEEG (France), I. Lundström/Linköping University (Sweden) and T. Seiyama/Tokuyama Soda Co. (Japan)) concentrate on chemical and biochemical sensors.

Volume 4 (editors: J. Scholz/Sensycon (FRG) and T. Ricolfi/Consiglio Nazionale Delle Ricerche (Italy)) refers to thermal sensors.

Volume 5 (editors: R. Boll/Vacuumschmelze (FRG) and K. J. Overshott/Gwent College (UK) deals with magnetic sensors.

Volume 6 (editors: E. Wagner and K. Spenner/Fraunhofer-Gesellschaft (FRG), and R. Dändliker/Neuchâtel University (Switzerland)) treats optical sensors.

Volume 7 (editors: N. F. de Rooij/Neuchâtel University (Switzerland), B. Kloeck/Hitachi (Japan), and H. H. Bau/University of Pennsylvania (USA)) presents mechanical sensors.

Each volume is, in general, divided into the following three parts: specific physical and technological fundamentals and relevant measuring parameters; types of sensor and their technologies; most important applications and discussion of emerging trends.

It is planned to close the series with a volume containing a cumulated index and selected topics.

The series editors wish to thank their colleagues who have contributed to this important enterprise wheter in editing or writing articles. Thank is also due to Dipl.-Phys. W. Greulich, Dr. M. Weller, and Mrs. N. Banerjea-Schultz of VCH for their support in bringing this series into existence.

W. Göpel, Tübingen   J. Hesse, Oberkochen   J.N. Zemel, Philadelphia, PA

August 1991

Preface to the Volumes "Chemical and Biochemical Sensors"

Planning "Sensors", it soon became clear that chemical and biochemical sensors would have to be treated in two volumes to appropriately present the wealth of material.

Thus, these volumes present for the first time a comprehensive description of chemical and biochemical sensors with emphasis placed upon both, technical and scientific fundamentals and applications. The aim is to offer well-founded knowledge to scientists and technicians and to show todays technical capabilities in this sensor field. Furthermore, both volumes together are intended to foster the future developments and applications of sensors and at the same time serve as a useful reference work.

The arrangement of the material presented here deviates in some way from that of the other volumes in the series "Sensors", which are devoted to physical sensors (mechanical, thermal, magnetic, optical sensors). With those sensors the internal structure of each single volume follows a classification by the input signal of the first transduction principle, which in most cases is identical with a classification according to the measurand. (For a complete discussion of sensor definitions and classifications see Volume 1, Chapter 1). The number of measurands in chemical and biochemical systems, however, is many orders of magnitude larger than in physical systems because of the huge number of different compounds which can occur in gaseous, liquid, and solid media. Therefore another structuring criterion had to be used. As described in detail in Chapter 1 of the present volumes, different types of (bio-)chemical sensors may be classified according to the different sensor properties used for the detection of chemical state, ie, of concentrations, partial pressures, or activities of particles. We adopted this classification to organize the "core" of the present volumes which consits of a description of "basic sensors" (like liquid and solid electrolyte sensors, etc.). For the sake of completeness, the core should be surrounded by a "shell" of articles devoted to other important aspects of the field:

- physical and chemical parameters and measurands;
- the theoretical physical or physico-chemical background underlying the sensing mechanisms (selected textbook knowledge on mechanics, optics, thermodynamics, kinetics, statistics, etc);
- the technology to produce sensor elements or components (thin-film, thick-film, ceramics technologies, etc.);
- applications (car engine regulation, environmental control, etc.).

We tried to cover all aspects by organizing the books in the following way:

First volume:

- Definitions, typical examples for chemical and biochemical sensors, and some historical remarks are given in Chapters 1 and 2.
- Chemical sensor technologies and interdisciplinary tasks to design chemical sensors are described in Chapter 3.

- Physical and physical chemistry basics of different detection principles and also pattern recognition approaches for multicomponent analysis with sensor arrays are described in Chapters 4–6.
- The major part of the volumes consists of a careful description of basic sensors in Chapters 7–13. They include liquid electrolyte sensors, solid electrolyte sensors, electronic conductivity and capacitance sensors, field effect sensors, calorimetric sensors, optochemical sensors, and mass sensitive sensors.

Second volume:

- Biosensors often make use of transducer properties of the basic sensors mentioned above and usually have additional biological components. They are therefore described in a separate Chapter 14.
- Application aspects are dealt with in Chapters 15–25. Here, the possibilities and limitations of sensors if compared with the conventional instrumentation in analytical chemistry and calibration aspects are described first. Specific facettes of certain fields of applications are then presented by specialists from different fields including environmental, biotechnological, medical, or chemical process control.

A major input to the present books originally came from Dr. T. A. Jones from the National Health and Safety Executive in Sheffield, U. K. He was a distinguished scientist in chemical sensor basic research and an expert in the particular field of combustible gas sensors. In 1989 he passed away. We could like to take the opportunity here to thank him for his input and enthusiastic support in the planning phase of the volumes on chemical and biochemical sensors.

After all this effort, the editors would now like to thank all participating authors for their contributions and their help in structuring the book by coordinating their individual chapters to the overall guidelines. The editors would also like to thank Dr. Klaus Schierbaum and the staff of VCH, particularly the editorial staff Mrs. N. Banerjea-Schultz and Dipl.-Phys. W. Greulich for their professional input and patience.

Wolfgang Göpel, Michel Kleitz, Ingemar Lundström, Tetsuro Seiyama
Tübingen   Grenoble   Linköping   Fukuoka

August 1991

Contents
Volume 3: Chemical and Biochemical Sensors, Part II

Volume 2: Chemical and Biochemical Sensors, Part I

List of Contributors

Volume 3: Chemical and Biochemical Sensors, Part II

Dr. Hansjörg Albrecht
Laser-Medizin-Zentrum GmbH
Krahmerstraße 6–10
W-1000 Berlin 45, FRG
Tel.: (0049-30) 8344002
Tfx: (0049-30) 8344004

Dr. Hiromichi Arai
Kyushu University
Materials Science & Technology
6-1 Kasugakoen Kasuga-shi
Fukuoka 816, Japan
Tel.: (0081-92) 5739611 x 310
Tfx: (0081-92) 5752318

Dr. Friedrich G. K. Baucke
Schott Glaswerke
Hattenbergstr./Postfach 2480
D-6500 Mainz 1, FRG
Tel.: (0049-6131) 333239
Tfx: (0049-6131) 333341

Dr. Karen Colbow
Prof. Konrad Colbow
Simon Fraser University
Dept. of Physics
Barnaby, British Columbia V 5 A 156,
Canada
Tel.: (001-604) 2913162
Tfx: (001-604) 2913592

Dr. Martin Gerber
Boehringer Mannheim Co.
Biochemistry R & D Division
9115 Hague Road/P. O. Box 50100
Indianapolis, IN 46250-0100, USA
Tel.: (001-317) 5767589
Tfx: (001-317) 5767525

Dr. Michael Hofer
Joanneum Research
The Optical Sensor Institute
Steyrergasse 17
A-8010 Graz, Austria
Tel.: (0043-316) 8020222
Tfx: (0043-316) 8020181

Dr. Klaus Kaltenmaier
Dr.-Ing. Häfele Umweltverfahrens-
technik GmbH & Co.
Gerwigstraße 69
D-7500 Karslruhe, FRG
Tel.: (0049-721) 616093
Tfx: (0049-721) 621599

Dr. Petra Krämer
University of California
Dept. of Entomology
Davis, CA 95616, USA
Tel.: (001-916) 7525109
Tfx: (001-916) 7521537

Prof. Dr. Hans-Heinrich Möbius
Ernst-Moritz-Arndt-Universität
Inst. f. Physikalische Chemie
Soldtmannstraße 16
DO-2200 Greifswald, FRG
Tel.: (0037-822) 75479
Tfx: (0037-822) 63260

Prof. Dr. Michael Oehme
Norwegian Inst. f. Air Research
Dept. Organic Analyt. Chemistry
Postboks 64
N-2001 Lillestrom, Norway
Tel.: (0047-6) 814170
Tfx: (0047-6) 819247

Dr. Kenneth F. Reardon
Colorado State University
Dept. of Agricultural and
Chemical Engineering
Fort Collins, CO 80523, USA
Tel.: (001-303) 491 65 05
Tfx: (001-303) 491 73 69

Prof. Dr. Friedrich Scheller
Zentralinsitut für Molekularbiologie
Robert-Rössle-Straße 10
DO-1115 Berlin-Buch
Tel.: (0037-2) 346 36 81/-29 18
Tfx: (0037-2) 349 41 61

Dr. Thomas Scheper
Universität Hannover
Inst. f. Technische Chemie
Callinstraße 3
D-3000 Hannover, FRG
Tel.: (0049-511) 762 25 09
Tfx: (0049-511) 762 34 56

Prof. Dr. Rolf D. Schmid
GBF
Mascheroder Weg 1
D-3300 Braunschweig, FRG
Tel.: (0049-531) 618 13 00
Tfx: (0049-531) 618 13 03

Prof. Dr. Hanns-Ludwig Schmidt
TU München
Allgemeine Chemie und Biochemie
D-8050 Freising-Weihenstephan, FRG
Tel.: (0049-8161) 7135 53/-54
Tfx: (0049-8161) 7135 83

Dr. Florian Schubert
Physikalisch-Technische Bundesanstalt
Abbestraße 2-12
DW-1000 Berlin 10, FRG
Tel.: (0049-30) 348 12 35
Tfx: (0049-30) 348 14 90

Dr. Wolfgang Schuhmann
TU München
Allgemeine Chemie und Biochemie
D-8050 Freising-Weihenstephan, FRG
Tel.: (0049-8161) 7135 19
Tfx: (0049-8161) 7135 83

Dr. Wolfgang Trettnak
Joanneum Research
The Optical Sensor Institute
Steyrergasse 17
A-8010 Graz, Austria
Tel.: (0043-316) 802 02 22
Tfx: (0043-316) 802 01 81

Dr. Karl Wulff
Boehringer Mannheim GmbH
Abtlg. GT-GL
Sandhofer Straße 116
D-6800 Mannheim, FRG
Tel.: (0049-621) 759 40 19
Tfx: (0049-621) 759 30 87/7 50 28 90

Volume 2: Chemical and Biochemical Sensors, Part I

Dr. Mårten Armgarth
Sensistor AB
Box 76
S-58 102 Linköping, Sweden
Tel.: (0046-13) 1134 22
Tfx: (0046-13) 1234 22

Dr. Gilbert E. Boisdé
CEA-CEN-Saclay
DEIN-SAI
F-91 191 Gif-sur-Yvette Cedex, France
Tel.: (0033-1) 6908 8543
Tfx: (0033-1) 6908 7819

Prof. Dr. Karl Cammann
Westfälische Wilhelms-Universität
Anorg. Chem. Insitut
Lehrstuhl für Analyt. Chemie
Wilhelm-Klemm-Straße 8
D-4400 Münster, FRG
Tel.: (0049-251) 8333141
Tfx: (0049-251) 833169

Dr. Pierre Fabry
L.I.E.S.G./E.N.S.E.E.G
Domaine Universitaire, B. P. 75
F-38 402 St. Martin d'Heres, France
Tel.: (0033-76) 8265 57
Tfx: (0033-76) 8266 30

Dr. Jaques Fouletier
L.I.E.S.G./E.N.S.E.E.G.
Domaine Universitaire, B. P. 75
F-38 402 St. Martin d'Heres, France
Tel.: (0033-76) 8265 57
Tfx: (0033-76) 8266 30

Prof. Dr. Günter Gauglitz
Universität Tübingen
Inst. f. Physik u. Theor. Chemie
Auf der Morgenstelle 8
D-7400 Tübingen, FRG
Tel.: (0049-7071) 2969 27
Tfx: (0049-7071) 2969 10

Prof. Dr. Wolfgang Göpel
Universität Tübingen
Inst. f. Physik u. Theor. Chemie
Auf der Morgenstelle 8
D-7400 Tübingen, FRG
Tel.: (0049-7071) 2969 04
Tfx: (0049-7071) 2969 10

Dr. T. Alwyn Jones †
Health and Safety Executive
Broad Lane
Sheffield S3 7HQ, UK

Dr. Michel Kleitz
L.I.E.S.G./E.N.S.E.E.G.
Domaine Universitaire, B. P. 75
F-38 402 St. Martin d'Heres, France
Tel.: (0033-76) 8265 57
Tfx: (0033-76) 8266 30

Prof. Ingemar Lundström
Linköping Inst. of Technology
Dept. of Physics and Measurement
Technology
S-58 183 Linköping, Sweden
Tel.: (0046-13) 2812 00
Tfx: (0046-13) 1375 68

Dr. Maarten S. Nieuwenhuizen
Prins Maurits Laboratory TNO
P. O. Box 45
NL-2280 AA Rijkswijk,
The Netherlands
Tel.: (0031-15) 843519
Tfx: (0031-15) 843991

Dr. Claes I. Nylander
Sensistor AB
Box 76
S-58102 Linköping, Sweden
Tel.: (0046-13) 113422
Tfx: (0046-13) 123422

Dipl.-Chem. Friedrich Oehme
Hühnerbühl 34
D-7883 Görwihl, FRG
Tel.: (0049-7754) 7358

Dr. Klaus-Dieter Schierbaum
Universität Tübingen
Inst. f. Physik u. Theoret. Chemie
Auf der Morgenstelle 8
D-7400 Tübingen, FRG
Tel.: (0049-7071) 295282
Tfx: (0049-7071) 296910

Prof. Tetsuro Seiyama
Tokuama Soda Co., Ltd.
Tenjin 1-10-24
Chuo-ku, Fukuoka-shi, Japan 810
Tel.: (0081-92) 7516566
Tfx: (0081-92) 7111089

Dr. Elisabeth Seibert
L.I.E.S.G./E.N.S.E.E.G.
Domaine Universitaire, B. P. 75
F-38402 St. Martin d'Heres, France
Tel.: (0033-76) 826557
Tfx: (0033-76) 826630

Stefan Vaihinger
Universität Tübingen
Inst. f. Physik. u. Theoret. Chemie
Auf der Morgenstelle 8
D-7400 Tübingen, FRG
Tel.: (0049-7071) 296933
Tfx: (0049-7071) 296910

Dr. Albert van den Berg
CSEM SA
Rue de la Maladiere 71
CH-2007 Neuchâtel, Switzerland
Tel.: (0041-38) 205387
Tfx: (0041-38) 254078

Dr. Hendrik H. van der Vlekkert
Priva B.V.
Zijlweg 3/P. O. Box 18
2678 LC/2678 ZG De Lier,
The Netherlands
Tel.: (0031-1745) 13921
Tfx: (0031-1745) 17195

Dr. Bartholomeus H. van der Schoot
Université de Neuchâtel
Institut de Microtechnique
Rue A.-L.-Brequet 2
CH-2007 Neuchâtel, Switzerland
Tel.: (0041-38) 205387
Tfx: (0041-38) 254078

Prof. Dr. Adrian Venema
Delft University of Technology
Electrical Engineering Faculty
4 Mekelweg
NL-2629 CD Delft, The Netherlands
Tel.: (0031-15) 786466
Tfx: (0031-15) 785755

Dr. Peter Walsh
Health and Safety Executive
Res. and Lab. Service Division
Broad Lane
Sheffield S3 7 HQ, UK
Tel.: (0044-742) 768141 x 3179
Tfx: (0044-742) 755792

Dr. habil. Hans-Dieter Wiemhöfer
Universität Tübingen
Inst. f. Physik u. Theoret. Chemie
Auf der Morgenstelle 8
D-7400 Tübingen, FRG
Tel.: (0049-7071) 296753
Tfx: (0049-7071) 296910

Prof. Dr. Otto Wolfbeis
Joanneum Research
The Optical Sensor Institute
Steyrergasse 17
A-8010 Graz, Austria
Tel.: (0043-316) 8020222
Tfx: (0043-316) 8020181

Prof. Jay N. Zemel
University of Pennsylvania
Center for Sensor Technologies
Philadelphia, PA 19104-6390, USA
Tel.: (001-215) 8988545
Tfx: (001-215) 8981130

14 Specific Features of Biosensors

Hanns-Ludwig Schmidt, Wolfgang Schuhmann, TU München,
Freising-Weihenstephan, FRG
Friedrich W. Scheller, Zentralinstitut für Molekularbiologie, Berlin-
Buch, FRG
Florian Schubert, PTB Berlin, FRG

Contents

14.1   Introduction and Definitions

The selectivity and specificity of a special group of chemical sensors originate from biological recognition systems connected to a suitable transducer. Specific molecular recognition is a fundamental prerequisite and expression of life, based on affinity between complementary structures (enzyme-substrate, antibody-antigen, receptor-hormone), and in biosensors this property is used for the production of concentration-proportional signals. According to Turner [1], a biosensor is "a device incorporating a biological sensing element either intimately connected to or integrated within a transducer" (Figure 14-1).

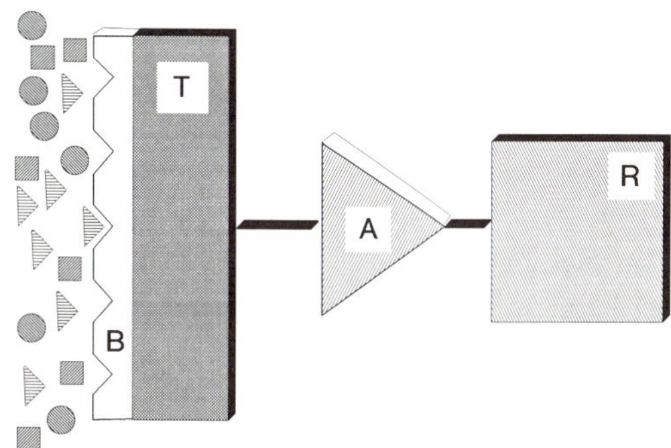

Figure 14-1.
Principle of setup of a biosensor. B = bioactive layer, containing "recognition molcules" (enzymes, antibodies, receptor proteins); T = transducer, a probe sensitive to the primary signal produced by the recognition process (potentiometric or amperometric electrode, FET (Field-effect transistor), piezoelectric crystal); A = amplifier, R = recorder.

Usually, the biological part of a biosensor is a macromolecule, often submitted to a conformational change in context with the "recognition" (binding) of its partner. In nature this effect may immediately be used for transduction and amplification, eg, in the "ion channels" of the nerve tissue, and also in some biosensors belonging to the group of affinity sensors this principle is tried to be imitated. "Affinity sensors" use lectins, antibodies, or receptors as the recognition element, and they transfer changes of properties provoked by the binding of the partners to optoelectronic devices, potentiometric electrodes, or field effect transistors (FETs). In another group of biosensors, the "metabolism sensors", the specificity and catalytic power of enzymes, organelles, microorganisms, or tissues to convert the analyte and to produce secondary chemical or thermal signals, suitable for the recognition by potentiometric or amperometric electrodes, field effect transistors, or thermistors, is used. The nature of the biological part of biosensors, usually a protein, also implies the disadvantage of thermal and chemical instability, and hence among the greatest efforts to optimize biosensors is the development of methods for the stabilization of the biological part.

Biosensor development started from the physical entrapment of soluble enzymes or the fixation of enzyme membranes to a transducer (first generation), was continued by applying bulk immobilizations of the biocompound onto the transducer (second generation) and is now at the stage of covalent binding of monolayers of orientated recognition molecules directly to the surface of semiconductor devices, leading to biosensors with very short response times. This technical progress in the last 10 years is characterized by successes with techniques for the immobilization and stabilization of biomolecules on the one hand, and the miniaturization and functionalization of more and more sophisticated transducers, mainly of the semiconductor type, on the other. Biochemists and engineers have learned from each other, and are still learning, in order to develop reliable sensors, not only using and increasingly approaching, but also imitating natural principles of recognition, transduction, and amplification.

14.2 Biological Fundamentals of Chemical Sensing and Transduction

While most reviews on biosensors are organized according to physical transducers (thermistors, potentiometric and amperometric electrodes, fiber optics, field-effect transistors), in this study the biological aspects will be in the foreground as the principle of order, and we shall look at the recognition systems and transduction possibilities that nature has developed, with the aim to check which of its inventions have already been used and which could perhaps be used in the future as a part of a biosensor. In this respect, biological and corresponding physical transduction systems are compared (Table 14-1), and the chapter is organized according to the recognition systems with the aim of optimizing existing biosensors and of conceiving new types with regard to how sensing and transducing take place in nature.

Table 14-1. Complementary molecules used in biological recognition, biological principles of transduction and amplification, and (possible) artificial equivalents (FET = field-effect transistor).

Substance to be recognized	Complementary macromolecule	Recognition effect /biol. transduction	Equivalent physical transducer
Substrate, inhibitor	Enzyme	Change of concentration or (redox) potential	Potentiometric or amperometric electrode, FET, thermistor
Ion, molecule	Carrier Protein	Change of concentration or (redox) potential	Ion selective electrode or FET, basing on ionophores
Antigen, hapten	Antibody	Conformation change, weight change	Electrode, FET (indirect also by indicator reaction), piezoelectric sensor
Hormone, drug, toxin, neuro-transmitter	Receptor	Conformation change, membrane permeability change, production of secondary messenger	Potentiometric or amperometric electrode or piezoelectric sensor with orientated receptor
DNA, RNA	DNA, RNA	Formation of hydrogen bridges	Excitation transfer (fluorescence quenching), indicator reactions in labelled molecules
Carbohydrate	Lectin or other specific protein	Conformation change	Excitation transfer (fluorescence quenching), indicator reactions in labelled molecules

14.2.1 Molecular Recognition as a Basis of Biological Communication and Organization

In biology, sensing is a fundamental principle for the communication of a living system with its environment. Nature has developed sensors for light, pressure, temperature, humidity, osmotic pressure, sound, taste, and odor, and only the last two are by definition chemical sensors, using molecular recognition for communication. Molecular recognition, however, is also an integrated part of the internal communication between the cells and organs of an organism, eg, by means of hormones and neurotransmitters. In any case, this function of external and internal chemical communication is performed by the binding between complementary molecules, which is thus a general process of vital importance, responsible for the organization and the protection of organisms and the regulation of their metabolism.

How do biological molecules recognize and bind to each other? The binding between enzyme and substrate, antigen and antibody, or hormone and receptor is due to the existence of complementary structures and conformations, providing corresponding distributions of polar and non-polar groups, which interact through Coulomb and Van der Waals forces or by hydrogen bridges. In the case of "affinity sensors" a reversible binding between the partners (R = "receptor"[1]; S = substrate) is considered to lead to an equilibrium. Binding affinity constants between 10^3 L/mol and 10^{15} L/mol have been observed [2], providing a large range for determination limits. At a given receptor concentration the complex concentration and hence the signal should be proportional to the analyte concentration. However, in practice this is not always so, eg, in the case of immunosensors competition between the analyte and a labeled analyte is often used, and a correlation between signal and analyte concentration results, which has to be checked empirically. Another complication with affinity sensors is their regeneration. Often, especially in cases of high affinity, a very slow dissociation is observed, which in practice conflicts with the long-term use of the sensor.

$$S + R \underset{k_{-1}}{\overset{k_1}{\rightleftharpoons}} SR \qquad\qquad S + E \underset{k_{-1}}{\overset{k_1}{\rightleftharpoons}} SE \overset{k_2}{\rightarrow} P + E \qquad (14\text{-}1)$$

$$\frac{[SR]}{[S] \cdot [R]} = K = k_1/k_{-1} \qquad\qquad K_M = (k_{-1} + k_2)/k_1 \qquad (14\text{-}2)$$

$$K_M = (v_{max}/v - 1) \cdot [S] \,.$$

In contrast, "metabolic sensors" have the advantage to regenerate their affinity structures by themselves, because the substrate S is converted to the product P, which has a different affinity to the receptor, in this case an enzyme E. The response of enzyme-based sensors is induced by the production of P, and follows the Michaelis-Menten kinetics. The simplest case is the reversible formation of an intermediate complex SE, which then decomposes with a velocity constant k_2 to the product. The "affinity constant" in this case, called the Michaelis constant K_M, is an expression of the substrate concentration, at which half the maximum

---

1 "Receptor" in this part is generally used for a "recognition molecule", not in the specific meaning as in Section 14.2.2.

reaction velocity is observed (for $v = v_{max}/2 \rightarrow K_M = [S]$). This also indicates half-saturation of the enzyme and, in practice, only concentrations below this value lead to proportional signals, because with saturation of the enzyme the production of P and hence the related signal become independent of [S]. This is true for a given amount of active enzyme, which can decrease by denaturation or inhibition. In most practical cases, the response and response time of metabolic sensors are determined by diffusion processes, namely diffusion of S through a membrane to the enzyme and diffusion of P to the transducer or back into the solution, but the complicated equilibrium resulting from these processes implies the possibility of enlarging the range of the biosensor by diffusion control.

The specificity of biological recognition systems can be extremely high, as is known from nucleic acid interactions. Also in enzyme-substrate recognition we can find highly expressed specificities. However, on the other hand many enzymes can bind or convert similar substrates such as homologues, but nevertheless in general they strictly distinguish between chiral centers. In the context of the development of biosensors, an important fact is that many of these systems can bind in addition other substances by which they can be modulated. The resulting activating or inhibitory effect may be used for the indirect determination of these substances. Sometimes the "docking" of the partners is accompanied by a conformational change of the macromolecule. This conformational change can be or induce a molecular switching and start a biological transduction.

The most direct principle of biological transduction is the opening of an "ion channel" (Figure 14-2), a protein responsible for the selective permeation of ions through a membrane, as a consequence of the conformational change implied by the binding of a partner. This is followed by depolarization of the membrane (most biological membranes exert chemical potentials, maintained by the "active transport" of ions against a gradient; as a consequence many membranes also develop electrical potentials, eg, in the nerves of higher animals -70 mV for the interior towards the exterior), a process which takes place within the range of milliseconds. Another general principle of biological transduction is the formation of a "secondary messenger" substance within the cell, which can initiate the permeability change of ion channels, or start a cascade of chemical reactions, mostly the activation of enzymes, as a signal amplification process. The response time of these processes can be in the range of seconds to minutes.

The best known biological transduction process for an external signal is that of light perception in the eye of mammals [3–5]. The absorption of a light quantum by rhodopsin, the visual purple, results in a conformational change of this protein, which induces, in a cascade of reactions, the degradation of a substance responsible for the opening of Na^+ channels. They close, and a hyperpolarization of the cell membrane is stimulated, starting the nerve pulse. Different color perceptions are basing on the computation of the relative intensities of signals at only three different primary light perceptors ("sensor array"). Similarily, the mechanism of odor or taste perception [6] starts with the activation of an enzyme producing a second messenger which allows the change of channel permeabilities. The most sensitive (bio)chemical sensor is the antenna of the male butterfly, which is capable of recognizing only a few hundred molecules of a pheromone per milliliter of air, and which can even identify concentration gradients in this range. Bacterial chemotaxis is also based on chemical sensors, but with a different transduction (see Section 14.2.3).

The chemical and electrical amplification factors of some natural sensing processes have been determined. They are well known for the transduction of light [7]: the absorption of a

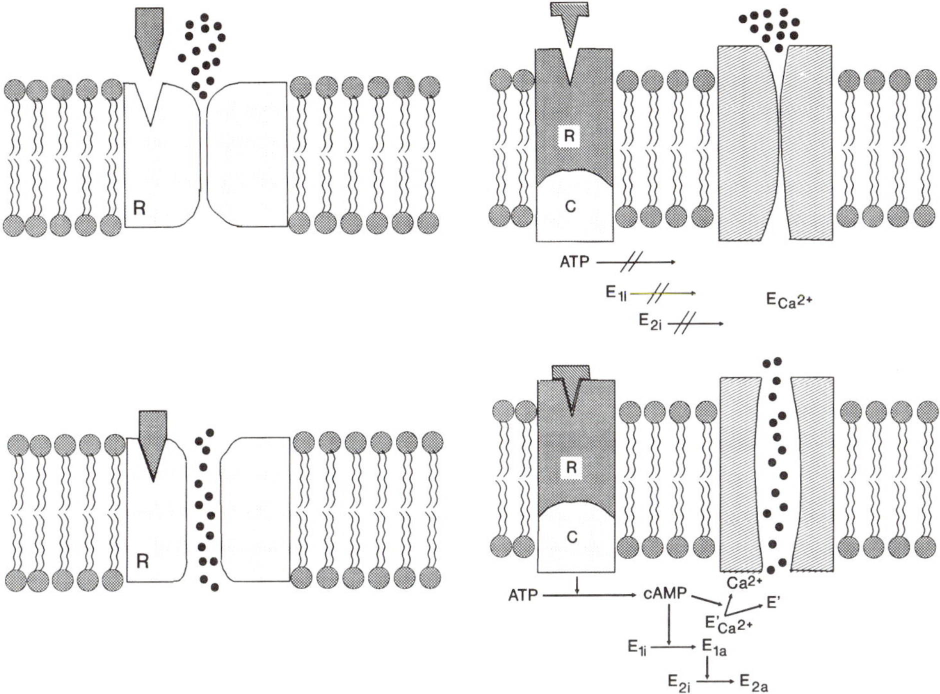

Figure 14-2. Principles of biological transduction and amplification. Binding of an effector at a receptor protein R induces a conformation change which can directly have for consequence the opening of an ion channel (left) with electrolyte influx and membrane depolarization, or initiate the formation of a "second messenger" (here cyclic adenosine monophosphate, cAMP) by adenylate cyclase, which can then, in catalytic amplification steps, lead to electrochemical or chemical amplification and signal spreading.

light quantum of wavelength 500 nm implies an energy uptake of $5 \cdot 10^{-19}$ J; the energy equivalent of the resulting electric signal in the nerve is between 10^{-16} and 10^{-13} J, hence we have an amplification factor of 10^3–10^6. In the case of tasting, the binding energy between a flavor molecule and a corresponding receptor is below 10^{-20} J, and the energy implied in the current induced is near 10^{-17} J; again, and amplification of 10^3 results. The chemical amplification through a cascade of catalytic processes can reach factors of up to 10^9. An example is the hormonal regulation of the blood glucose concentration: whereas the concentrations of adrenalin or glucagon are in the range 10^{-11}–10^{-9} M, the concentration of the second messenger in the cell interior is of the order of 10^{-6} M, by which concentration changes for glucose in the mM-range are produced.

14.2.2 Transduction and Amplification of the Chemical Signals from Hormones and Neurotransmitters

The tertiary structure of hormone or neurotransmitter receptor proteins [8, 9] and their integration and orientation within the cell membrane are implied in their amino acid sequence.

This consists of helix segments with hydrophobic side-chains sticking within the membrane bilayer, while hydrophilic loops inside or outside the membrane provide sites for the binding of the signal molecule or of modulators, and serve as contact to subunits or other proteins. Normally receptor proteins possess one, four, or seven membrane-integrated segments. In the latter cases they may form ion channels, the permeability of which is controlled by the protein's conformation.

Receptors with one membrane integrated segment (eg, insulin receptor) normally perform purely chemical transduction. It starts, mediated by so-called G-proteins (from GTP binding), with the activation of an enzyme which produces "secondary messenger" molecules in the cell interior [10]; the latter molecules modulate enzymes or enzyme cascades. This process implies not only an amplification of the signal by a factor of 10^3 in each step, but also the possibility of tuning or modulation of the primary signal. As already mentioned, the response time of this purely chemical transduction is in the range of minutes.

Responses to other hormones or to neurotransmitters are much faster. They normally imply electrical or chemoelectrical transduction and amplification. A change in the membrane ion permeability induces a change in the local membrane potential. "Voltage-gated ion channels" in the neighborhood, as present in nerve axons, are opened by this potential change, and an "action potential" is initiated. The structure of voltage-gated Na^+ channels and their reactions with neurotoxins are well known [11]; they can be clogged by molecules bearing on one side a charged group of the size of the hydratized cation but being in fact a large organic molecule (tetrodotoxin of the Japanese fugu fish, saxitoxin of the mussel Saxidomos giganteus [12]), or they can be modulated by drugs [13]. This observation could be a starting point for the development of biosensors for these substances.

"Primary ligand-gated ion channels", preferably with four membrane permeating helix segments, exert the fastest response (millisecond range), and are mainly present in the neuromuscular junction or in synapses (eg, acetylcholine- or amino acid-gated channels). Many of these channels can be activated, blocked, or modulated by alkaloids, peptides, or synthetic drugs. "Secondary ligand-gated channels" change their conformation and permeability when they are modified by an intrinsic system at the end of an internal chemical amplification cascade; their response time is in the order of seconds. Finally we find, preferably at the end of axons, "voltage-gated Ca^{2+}-channels", which start a Ca^{2+}-ignited enzyme activation cascade [14]. In most cases the response of all these systems can be quenched or modulated by external and internal substances.

A new method, developed in neurophysiology, permits the study of the ion permeability and electric conductivity of single ion channels. In the "patch-clamp" technique, tiny parts of plasma membranes are tightly pressed to glass micropipettes by means of suction, and the ion flux through this membrane patch and influences of modulating agents are measured by means of electrical methods [15]. Owing to recombinant DNA techniques, intact receptors and channel proteins are also now available [7], and their potential for application as recognition and transduction systems is under discussion [16]. The aim envisaged, and realized in a few cases, is to embed them in an orientated way in artificial membranes, using Langmuir-Blodgett techniques [15], and to stabilize the systems obtained by polymerization [17]. The same principle may be useful for preparing sensors on the basis of piezoelectric crystals. Another approach is even to use intact tissues, in which chemoreceptors with transducing and amplifying elements are already integrated [18].

14.2.3 Recognition and Transduction in Bacterial Chemotaxis

Pheromone production and perception form a very common means of chemical communication between different individuals of a society, and it is especially developed for well defined molecules among insects. In this case the molecular mechanism of transduction is not very different from that of olfactory perception. On the other hand, chemical signals also exist between microorganism, and chemotaxis is a widespread phenomenon leading to cell motion. Especially bacterial chemotaxis is an important mechanism in the nutrition of mobile bacteria and their protection from toxic substances. Hence the effector substances are not as specialized as pheromones, but rather are molecules of more general availability [19].

These substances, nutritives (attractants) or toxic molecules (repellents), after being "recognized" by the bacteria, modulate the flagellum rotation and induce a movement of the cell uphill or downhill a concentration gradient. The molecular mechanism generally consists in the formation of a complex between the molecule in question and a specific receptor in the periplasmic space. The complex, sometimes directly the substance, then docks to a "methyl-accepting chemotaxis protein (MCP)", a transducer in the plasma membrane, stimulating a conformational change of this protein. As a consequence, it is activated by methylation and produces a secondary chemical signal that triggers the flagellum. So far, about 30 different chemoreceptors are known, mainly for various sugars and amino acids, for which they have very low K_M values in the range 0.1–1 µM. As these receptor proteins are soluble, they could possibly be used in electrode chambers. However, even if the reconstitution of the MCPs within membranes were possible, the use of their transduction process for sensors cannot be imagined at present. On the other hand, the reactions of microorganisms to changes in environmental quality (modification of movement, fluorescence, or luminescence) is becoming increasingly interesting as the basis of "bioprobes", indicators of environmental quality.

14.2.4 Enzyme Catalysis as a Basis of Chemical Transduction

Enzymes are biocatalysts involved in the performance of metabolic reactions. They have outstanding properties with regard to substrate and product recognition, but in nature they do no directly exert functions of transduction. However, as they can produce secondary chemical signals recognizable by physical transducers, and as even some enzyme-catalyzed reactions can be used for signal amplification in recycling systems, many enzymes have been introduced in the concept of biosensors. About 2000 have been characterized so far, a large number are commercially availabe, for many the tertiary structure is known, and methods for their immobilization have been developed. Therefore, it is not surprising that most biosensors described so far are based on enzymes. Like enzymes whole tissues, organelles, and bacteria can be used as chemical transducers, recognizing specifically a particular substrate and producing a general secondary chemical signal for a transducer. "Microbial sensors" have been used for the monitoring of more than 50 parameters in bioreactors and environmental control [20].

According to the reaction types catalyzed, enzymes are classified into six main groups; among these, oxidoreductases (transfering electrons or other redox equivalents), hydrolases (hydrolysing various bonds), and some lyases (adding small molecules to double bonds or the reverse) are of great interest in the context of biosensors. It is nearly exclusively the reactants of the reactions catalyzed by these enzymes for which transducers are available (Table 14-2).

Table 14-2. Examples for enzyme reactions producing secondary chemical signals and corresponding transducers. Trivial names of the enzymes and systematic numbers (EC-numbers) are given.

Enzyme [EC-number]	Reaction catalyzed (coupled auxiliary reaction)		Substrate indicated	Secondary chem. signal	Transducer
Glucose oxidase [EC 1.1.3.4]	β-D-Glucose + O_2	$\rightarrow$ β-D-Gluconolactone + H_2O_2	β-D-Glucose	O_2, H_2O_2	Amperometric electrode for O_2 or H_2O_2
Malate dehydrogenase [EC 1.1.1.37]	S-malate + NAD^+	$\rightarrow$ Oxaloacetate + NADH + H^+	S-malate, oxaloacetate	NADH	Amperometric electrode, optode
L-Amino acid oxidase [EC 1.4.3.2]	L-aminoacid + O_2 + H_2O	$\rightarrow$ α-ketoacid + H_2O_2 + NH_3	L-amino acid	NH_3, O_2, H_2O_2	Potentiometric electrode, FET
Urease [EC 3.5.1.5]	Urea + H_2O	$\rightarrow$ 2 NH_3 + CO_2	Urea	NH_3, CO_2, H^+	Potentiometric electrode, FET
Acetylcholine esterase [EC 3.1.1.7]	Acetylcholine + H_2O	$\rightarrow$ Acetic acid + choline	Acetylcholine	H^+	Potentiometric electrode, FET
β-Galactosidase [EC 3.2.1.23]	Lactose + H_2O (β-D-Glucose + O_2	$\rightarrow$ Galactose + glucose $\rightarrow$ β-D-Gluconolactone + H_2O_2)	Lactose	O_2, H_2O_2	Amperometric electrode for O_2 or H_2O_2
Citrate lyase [EC 4.1.3.6]	Citrate (oxaloacetate + NADH + H^+	$\rightarrow$ Acetate + oxaloacetate $\rightarrow$ L-malate + NAD^+)	Citrate	NADH	Amperometric electrode, optode
Amino acid decarboxylases [EC 4.1.1.n]	Aminoacid	$\rightarrow$ Amine + CO_2	Amino acid	CO_2	Potentiometric electrode, FET
Amino acid ammonia lyases [EC 4.3.1.n]	Aminoacid	$\rightarrow$ α, β-unsaturated acid + NH_3	Amino acid, unsaturated acid	NH_3	Potentiometric electrode, FET

These transducers include potentiometric electrodes and field effect transistors for H^+ (and indirectly CO_2 or NH_3), pH- and O_2-sensitive fiber-optic devices, and amperometric electrodes for O_2, H_2O_2, and redox mediators.

The K_M-values of enzymes for their substrate are between 10^{-3} to 10^{-5} M, and their turn-over number (number of substrate molecules converted per active site) is of the order of a few thousand per second, but can reach 600000 s^{-1} (carboanhydrase). Enzymes not only recognize and convert their substrates, but they can also be reversibly inhibited by substrate-simulating molecules and irreversibly by substances that modify functional groups. They can also be affected in their activity by substances that influence their conformation (special effectors, cations, or protons).

In biosensors, enzymes specifically recognize their substrates and produce secondary chemical signals, such as electrons (oxidoreductases) or protons (hydrolases), which can be recognized by suitable transducers. On the basis of the stoichiometry, any reactant of the reaction catalyzed can be determined. The signal obtained, however, must not always be directly proportional to the substrate concentration. Whereas hydrolyses in aqueous solutions are almost complete and irreversible, many oxidoreductions stop after partial turnover, attaining equilibria given by the redox potential of the partners in question. The most promising but also the most complicated systems for biosensors are based on oxidoreductases, a group of enzymes catalyzing the transfer of redox equivalents.

14.2.5 Direct and Mediated Electron Transfer Between Proteins

Many fundamental metabolic processes in biological systems involve the transfer of charges or redox equivalents. The transport of electrons between macromolecules must be effected without the formation of highly reactive and cytotoxic radicals. Nature has also achieved the separation of charges from the surrounding electrolyte and the transport of energy without dissipation by the ordered arrangement of redox proteins in membranes and by the construction of hydrophobic microenvironments within the proteins. The indispensably ordered electron transfer in transport chains present in chloroplasts (photosynthetic electron transport), mitochondria (respiratory chain), or bacterial membranes is only possible on the basis of orientation and cooperation of the different redox centres. In many cases, even with the optimum orientation of the proteins, an electron transfer can only be explained by charge-transfer or tunneling processes. In other cases it is performed by means of mediators entering or contacting the active site, eg, quinones and small proteins such as cytochrome c or plasto-cyanine. Substrate-converting oxidoreductases perform their task by hosting the reaction partners within their active site, either simultaneously or sequentially connected with an intermediate change in their proper redox state. The intrinsic mechanism by which an electron transfer occurs is strongly dependent on the structure of the different protein moieties involved in the reaction [21, 22]. Five main types of redox centers are found in proteins involved in electron and hydrogen transfer; these are, in order of increasing redox potential, iron-sulfur clusters, flavins, pyrroloquinolinequinones (PQQ), heme, and distorted tetragonal copper complexes. Embedding of these centers in the proteins provides wide ranges of redox potentials. In the communication between redox proteins and their substrates, two-electron transfers ($NAD(P)^+$-dependent dehydrogenases) or two one-electron steps (flavoproteins, quinoproteins) are dominant, whereas real one-electron transfers are the domain of small metal proteins (iron-sulfur, heme, and copper proteins), preferably for electron transfer between these proteins. Comparative kinetic studies, X-ray crystallographic measurements [23] and selective chemical modification studies [24, 25 and references cited therein] suggest that partner specificity, as exhibited in the most investigated case of horse heart cytochrome c [21, 26, 27], is associated with interactions between charged amino acid residues on the protein surfaces. The exposed heme edge of cytochrome c is below the protein surface and surrounded by a highly conservative ring of lysine residues. These positively charged residues are necessary for the formation of a protein-protein precursor complex with the complementary structure of the binding domains of the physiological partners, eg, cytochrome oxidase consisting of negatively charged glutamate and aspartate side-chains [28]. The interaction of the proteins favors a specific orientation and minimizes the distance between the redox centres, facilitating effective and fast electron transfer. Nevertheless, any electron transfer from and to cytochrome c has to overcome a considerable spatial separation of donor and acceptor redox centers. This seems to be possible through mixing the t_{2g} orbitals of the central iron atom with the π^* orbitals of the porphyrin ring, extending the d-electron density of the metal effectively to the exposed heme edge [20], from where direct electron transfer becomes possible, but tunneling may also be involved in the redox process [20, 29].

Altering the overall charge of cytochrome c by specific chemical modifications such as acylation of the lysine residues markedly decreases the rate of electron transfer to physiological redox partners [30]. On the other hand, substitution by guanido groups involves no decrease in the activity compared with the native protein. Protamines and polylysine as positively charged proteins exert a great competitive inhibition on the activity of the reaction between cytochrome c and cytochrome oxidase; this reaction is also sensitive to changes in the ionic strength of the reaction mixture. Possible applications of electron-transfer processe between small redox proteins in amperometric biosensors are shown in Section 14.3.3.1.

14.2.6   Biological Recognition Systems Without Transduction and Chemical Conversion

In addition to "signal molecules" and substrates, cells have to recognize many other substances, eg, transport metabolites, foreign substances, and molecules responsible for cell organization. The corresponding recognition process does not lead to any immediate transduction process but could be the basis for an artifical transduction system. Any cell is capable of performing a highly selective passive and active transport of ions, amino acids, sugars, and vitamins; this phenomenon is the prerequisite of any ordered metabolism, the function of nerve cells, of the kidney, and of the intestine. The proteins responsible for this selective transport are called "carriers" or carrier proteins. They are integrated proteins of biomembranes, and their binding of substrates follows, like that of enzymes, saturation kinetics, and they can also be inactivated by reagents for functional groups.

Whereas biological carrier molecules have not yet been used for the development of biosensors, carrier-like molecules, eg, natural ionophore antibiotics such as valinomycin or synthetic analogs, have been the basis of very specific ion selective electrodes [31], and the recent development of the corresponding chemistry [32] should lead to further progress in this area.

The most versatile biological recognition system developed by higher animals is the immune system. It is capable of producing complementary structures even to molecules and shapes with which no living matter has ever been confronted during evolution. Therefore, it is ideal for the development of specific detection systems, as demonstrated by the admirably selective and sensitive immune methods radio immuno assay (RIA) and enzyme-linked immunosorbent assay (ELISA), especially after the availability of monoclonal antibodies in large amounts. The antibody-antigen system has also been used for the construction of "immuno sensors" [33], in which for example an antigen-bound enzyme, as in ELISA, has an indicator function, the activity of which is measured by an electrochemical process. As antibodies against almost any biologically interesting substance (metabolites, drugs, toxins, pesticides) can be produced, the potential of their use seems to be unlimited. However, so far immuno electrodes are based on the same principle as other immunological methods, and need an indicator reaction.

Immuno sensors with direct electrochemical or optical triggering based on the antigen-antibody reaction have not yet been described (very recently a system for the detection of inflammatory proteins has been developed [34]) because they would demand an orientated immobilization of the antibody on a transducer. Corresponding systems should nevertheless be possible, because the binding is certainly accompanied by changes in net charge distribution and optical properties, and even the weight change should be measurable by means of piezoelectric crystals. In any case, however, problems should arise from the virtually kinetic irreversibility of the complex formation. The possibility of producing catalytic antibodies (synthetic enzymes [35]) by immunizations with transition-state analogs as antibodies is certainly of outstanding importance and will be discussed in Section 14.2.7.

Cell recognition and cell movement, and tissue organization and differentiation are directed by oligosaccharide structures from glycolipids and glycoproteins on the cell membrane surface. The complementary carbohydrate-binding proteins have high specificities and they may be subjected to conformational changes when binding their partners [36]. A special group are the lectins [37], ubiquitous proteins from plants, bacteria, invertebrates, and vertebrates. Although their physiological role is not yet completely understood, more than 100 representatives have been purified, and for about ten the complete sequence is known. Applications have been described for blood typing and as reagents for simple and complex carbohydrates in solution and on cell surfaces. For these purposes, labeled lectins and lectin conjugates (eg, with radioactive tracers, enzymes, biotin, fluorescent dyes, or colloidal gold) serve as specific and sensitive reagents. So far, only one example of the use of a lectin in a glucose biosensor has been described, the principle of which is a competition between a fluorescence-labeled polysaccharide and glucose for binding to the lectin concanavalin A. Nevertheless one can expect a promising potential among these substances. This may also be true for other carbohydrate-binding proteins, the structure elucidation of which is in progress [38].

The most specific and sensitive biological recognition is that between nucleic acids, owing to complementary sequences of hydrogen bridges provided by heteroaromatic bases. This recognition is the basis of the preservation and expression of the genetic code. Visualization of hybridization between complementary strains is possible by means of radioactive labeling, and by this method sensitivities have been attained that are capable of detecting the nucleic acids of only a few cells. Attempts to develop DNA probes with non-radioactive visualization have used fluorimetric or enzymatic detection systems [39]. The principle of fluorimetric detection is based on energy transfer between different fluorescent dyes; for enzymatic identification methods the probe DNA can be labeled with biotin and, after its hybridization with the (immobilized) analyte DNA, a suitable avidin-marked enzyme is added, which binds to the hybrid DNA and can then be used for the formation of a colored or luminescent product.

14.2.7   Towards Artificial Biomimetic Recognition Systems Suitable for Biosensors

Organic chemistry and biochemistry have reached the stage where they can not only describe the structure and function of most complex biological recognition systems [31, 40], but also imitate nature in constructing molecules of similar functions and properties, often using a suitable biochemical machinery for synthesis. In this context the production of receptors by recombinant DNA methods should be mentioned. However, this is even more true with relatively simple molecules such as ionophors and liquid ion exchangers. Natural ionophores such as the antibiotics valinomycin and nonactin have been surpassed in variability and versatility by new classes of soluble cation-complexing agents such as crown ethers, already in use in ion-selective membrane electrodes [41], and by host molecules for anions, even with catalytic properties [42], and for organic molecules [43]. Artificial receptors for nucleotides and peptides have been prepared which mimic the multi-point binding strategy of their hosts within flexible cavities by means of hydrophobic and hydrogen-bonding groups linked to a macrocyclic ring [44]. Even when the recognition and binding between these host and guest molecules are so far not accompanied by a transduction, the promotion of the chemistry would be promising in the context of biosensors, because the host molecules mentioned have the advantage over natural equivalents of being heat stable and more resistant towards chemicals.

New methods have been developed for the construction of artificial enzymes. A general method for obtaining new substrate-specific catalytic proteins is to develop antibodies towards transition states of the reaction in question by using molecules as antigens which simulate the transition state [33]. Although these catalytic antibodies would have the disadvantages implied for protein molecules, the host molecules mentioned before [41] and specifically functionalized host molecules [45] would imply the properties of a catalyst and hence be potentially suitable for chemical transduction in a sensor.

Even if the availability of natural hormone receptors and ion channels, due to genetic engineering methods, for the construction of sensors were within the range of possibility, they could not be used directly for this purpose. These proteins need intact membranes and an orientated integration into these membranes for their functioning. In fact, reconstitution of the nicotinic acetylcholine receptor and a few other receptors in (artificial) bilayer membranes has been achieved, and the electrical conductivity of this membrane could be modulated by means of the effector [46]. A sensor on the basis of the acetylcholine receptor would be of great interest for the monitoring of nerve gases.

The main prerequisite for a general adaptation of this possibility of building a sensor would be to produce monolayer or bilayer membranes on transducers and to charge them with receptors in an orientated way (Figure 14-3) [47]. A possibility capable of realizing this concept is the Langmuir-Blodgett technique, which enables artificial mono- or bilayer membranes to be produced and transfered [48]. Although we are actually far from the production of perfect and stable membranes, a method has been reported for stabilization, involving the secondary cross-linking of membrane-integrated polymerizable amphiphiles [16, 49]. The orientated integration of modified antibodies or parts of them has been possible [50]. Finally the chemistry developed for semiconductor substitution [51] provides possibilities for generating lipid monolayers on silica by means of silanization and to start from there towards orientated functionalization [52].

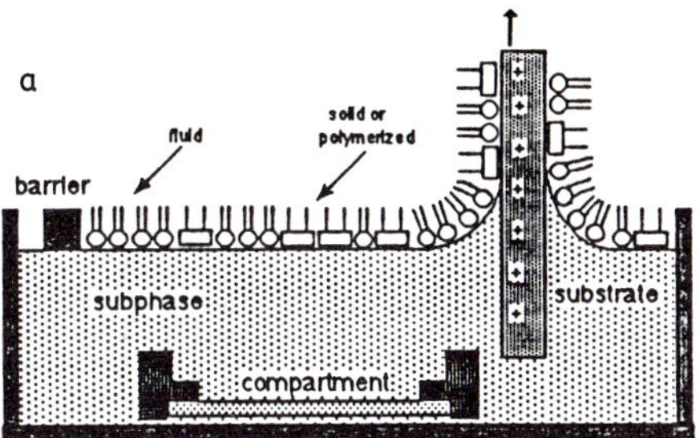
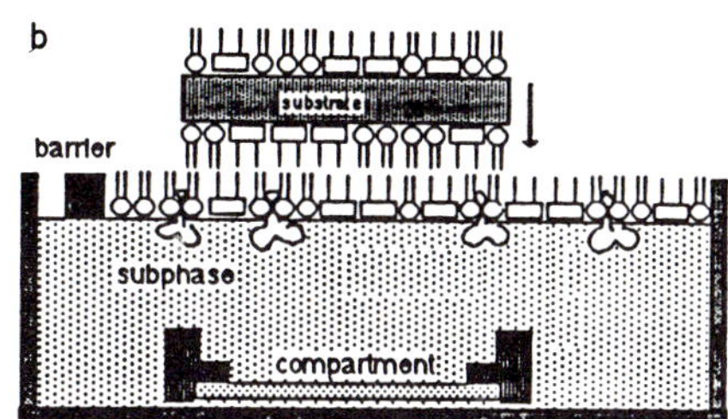

Figure 14-3. Schematic view of the deposition technique ((a) dip — (b) touch mode) of a heterogeneous lipid/protein biplayer. The biospecific ligand is incorporated into fluid patches composed of one lipid component whereas the bulk of the lipid layer (which is composed of a second amphiphile) is stabilized either by crystallization or by cross-linking, using a polymerizable lipid.

In conclusion, a better understanding of biological recognition, transduction, and amplification, and the translation of this knowledge into sensor concepts, in addition to the better adaptation of physical sensing and transduction to biological needs, will be a good basis for future developments of biosensors. Maybe forthcoming techniques will help to detect substances by conformational changes, inducing, for example, a modulation of the permeability of synthetic membranes, or by inducing changes in the polarizations of such membranes, which would trigger currents in semiconductor devices. The possible adaptation of the physical parts of biosensors to the biological is the topic of the following section.

14.3 Biosensors from Coupling Suitable Transducers to Biochemical Recognition Systems

As a biosensor, by definition, consists of a biological recognition system in intimate physical and functional contact with an artificial transducer, among the possible combinations the most efficient ones have to be conceived with regard to the demands of the actual analytical problem. The choice will not only be determined by the optimum sensitivity or the response time to be attained, but also by the properties of the matrix and the concentration range of the substance to be monitored. This may be explained by means of an example.

Glucose is the most common and most important substance to be determined, occurring in food, blood, and fermentation broths, and there is no other substrate for which so many kinds and variations of biosensors have been described. Most of them use glucose oxidase (GOD) as the biological recognition system. As will be derived from the following formula, many possibilities of transduction are possible, and all of them have been realized.

The stoichiometry of the reaction permits determination of glucose by (Figure 14-4):

1. Monitoring the O_2 consumption demands a constant and sufficient O_2 supply
    a) Reduction of O_2 at a platinum electrode (Clark electrode).
    b) Optical measurement of fluorescence quenching by the diradical O_2. Off-line measurement with fiber optics may be possible.

2. Monitoring of H_2O_2 production or, in the presence of an artifical electron acceptor, monitoring of the formation of its reduced form.
   a) Anodic oxidation of H_2O_2 and amperometric current measurement.
   b) Anodic reoxidation of a enzymatically reduced redox mediator. This has been performed successfully with various soluble and immobilized mediators.
3. Determination of changes in the local pH value (may be interfered with by pH shifts in the surroundings).
   a) Potentiometric pH determination with ion-selective electrodes or field-effect transistors.
   b) Optical pH monitoring with indicator dyes on optodes and in flow-injection systems.
4. Calorimetric measurement of the reaction enthalpy (sensitivity can be increased by coupling of H_2O_2 fission with catalase).

Instead of glucose oxidase, glucose dehydrogenase (GDH) can be used, and according to the reaction shown in Figure 14-4 glucose may be monitored by:

5. Fluorimetric, photometric, or electrocatalytic determination of the reduced coenzyme NADH.
6. Potentiometric determination of H_3O^+.
7. Optical or electrochemical determination of the PQQ/PQQH$_2$ (pyrroloquinoquinoline) ratio.
8. Anodic reoxidation of a reduced redox mediator formed by reaction with PQQH$_2$ within the active site of the enzyme.

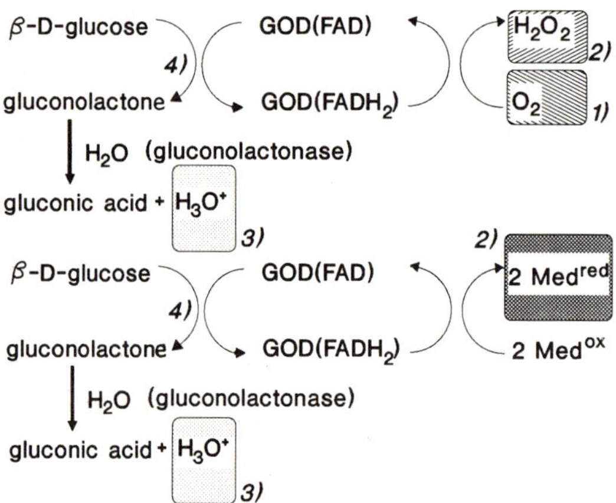

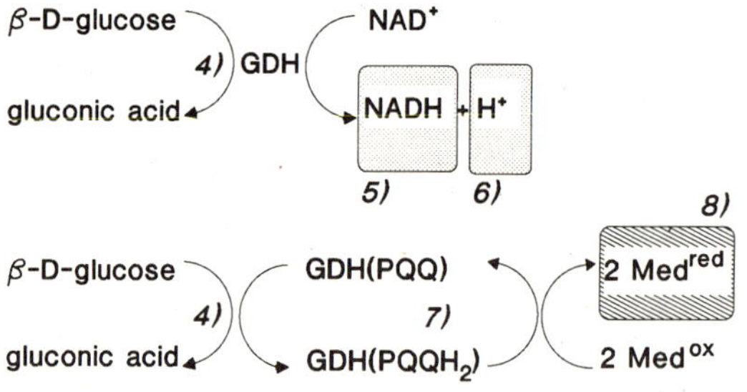

Figure 14-4.
Oxidation of glucose using different enzymes produces substances which may be detected by various techniques.

How should one select the optimum device for a given analytical problem and how should one adapt a transducer to a given recognition system or vice versa? Many aspects of biosensors combined with transducer principles have been extensively reviewed in the literature [53–57] and are also topics of previous chapters in this volume. Therefore, this chapter is focused on a biological approach; some transducer systems will be discussed in terms of the underlying biochemical recognition and/or catalytic process in the following, but major emphasis will be laid on the communication between biological redox systems and amperometric electrodes.

14.3.1 Fundamentals, Construction, and Application of Amperometric Biosensors

The most precise, versatile, and promising possibility for the transduction of a biological "recognition" into an electrical signal is the amplification of a current passing through an electrochemical cell. Electroanalytical techniques are fairly sensitive; to obtain currents of 10^{-9} A as little as 10^{-14} mol s^{-1} of one-electron redox reactions with subsequent electron transfer have to take place. For an amperometric measurement a defined potential is applied at a working electrode with respect to a reference electrode while the circuit is closed by means of a counter electrode. For low current densities, the error arising from the dependence of the potential from the cell current may be neglected and a two-electrode set-up is sufficient for amperometric investigations (Figure 14-5).

As has been pointed out in Section 14.2.5, the wide variety of substrates, mediators, and protein oxidoreductions in nature is realized by means of only five types of redox centers and by only a few types of mechanisms. Order is attained by precise fitting of substrates into the active sites of enzymes and by orientated docking of proteins with each other. A communication between biomolecules and non-natural transducers has to take this into account, and provide "biocompatible" electron carriers and electrode surfaces. In an amperometric biosensor, the function of a biomolecule, of a cell, or of cell layers, and also of a microorganism used as a selectivity element, is to generate redox-active charge carriers in a stoichiometric relationship with the substrate to be determined. These charge carriers, which may be biological or synthetic, should be compounds easily converted by oxidation or reduction at an electrode surface poised at the appropriate potential for the desired redox reaction. An electrolytic process occurs at the electrodes, leading to a finite current which is passed through the electrochemical cell. In fact, all redox reactions take place in the interphase region between the bulk solution and the electrode surface, and the potential gradient applied between two electrodes exists only in a narrow interphase region near the electrode surfaces (ca. 10^{-6} m), while in the bulk solution electroneutrality remains. Hence, molecules in this area do not "feel" the applied potential. The charge transport to the surfaces of the electrodes is bound to mass transport by diffusion processes, and current generation is only performed by molecules in the area of the potential gradient or which attain this region during the time of the experiment. Thus, in order to enhance the probability that a charge carrier will reach the electrode within a short time, the biochemical reaction has to take place close to its surface. For this aim, the biological compound in a biosensor has to be immobilized as close as possbile to the electrode surface (see Section 14.4). As a further prerequisite for the development of amperometric biosensors, mechanisms of communication among biomolecules have to be studied. In principle there are

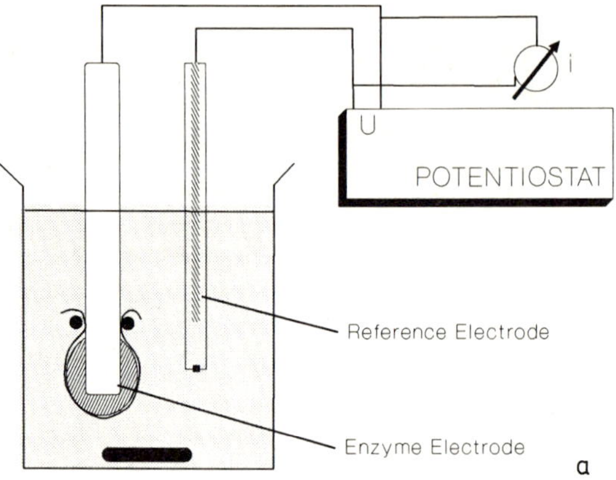

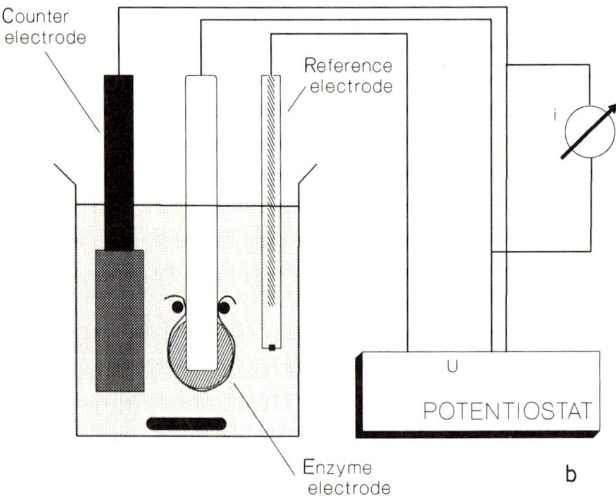

Figure 14-5. Schematic setup of a) a two-electrode and b) a three-electrode amperometric system with an enzyme electrode.

two main kinds of charge transfer between two redox species or a redox compound and an electrode surface. First, a direct electron-transfer process from the biomolecule can occur, requiring extremely short distances between the redox centers involved. Direct electron-transfer processes are bound to the existence of easily accessible redox sites, as can be demonstrated in the following for small electron-transferring proteins. Secondly, mediated electron transfer takes place between redox sites, the structure of which is incompatible with direct electron transfer. With a relatively small redox mediator which can communicate with both redox partners, efficient electron-transfer rates are possible. For highmolecular-weight enzymes, mediated electron transfer is more probable, and biosensors based on enzyme catalysis have

Figure 14-6.

a) Promoted electron transfer. The electrode surface is modified so as to obtain a correct orientation of a biological redox species and to permit direct electron transfer (see Section 14.3.1.1). b) Mediated electron transfer by means of mobile redox species (see Section 14.3.1.2). b1) A soluble enzyme communicates with the electrode by means of a soluble redox mediator. b2) An enzyme, cross-linked in the presence of an inert protein such as albumin forms an enzyme membrane on the electrode surface. Electron transfer can be achieved by means of a soluble redox mediator. b3) Entrapment of an enzyme within a polymer matrix with electron transfer by soluble redox compounds. b4) Covalent binding of enzymes to the electrode surface and electron transfer by means of low-molecular-mass weight mediators. c) Mediated electron transfer at electrode surfaces modified with a redox compound by adsorption or covalent binding. Electrocatalytic properties facilitate electron transfer from proteins, coenzymes, or redox mediators by decreasing possible overpotentials (see Section 14.3.1.3). d) Enzymes with covalently bound mediators possibly transfer electrons due to tunneling processes from the active site of the enzyme via the bound mediators to the electrode surface (see Section 14.3.1.4). e) Enzymes bound covalently to or been entrapped in organic conducting polymers can be coupled by soluble or polymerbound mediators, and also by direct electron transfer with the electrode (see Section 14.3.1.5).

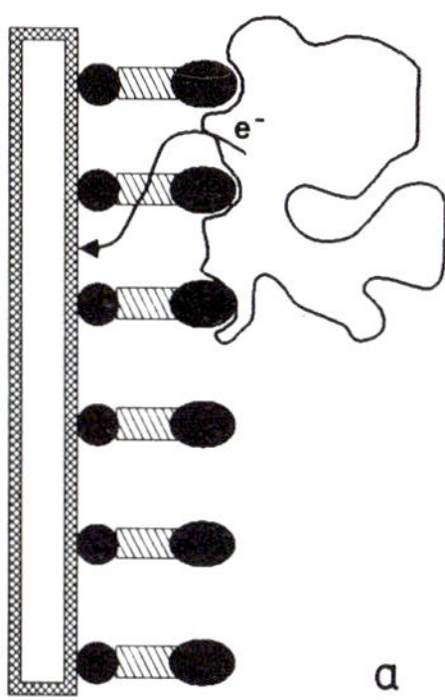
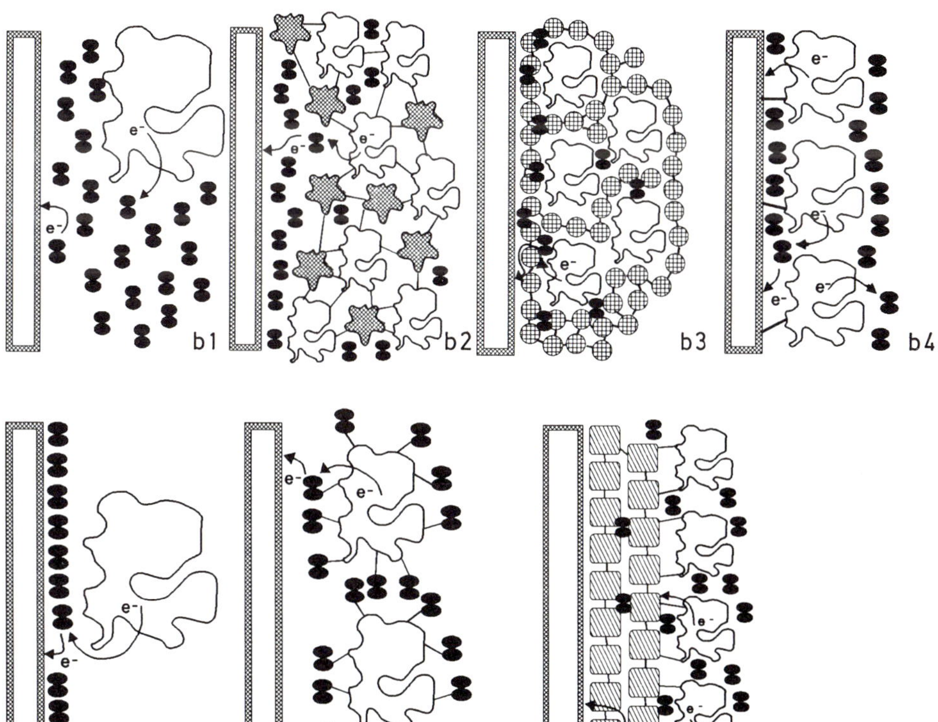

to be constructed with regard to this fact. The different possibilities of achieving electric communication between biological redox compounds and electrode surfaces, leading finally to a substrate-proportional current in the amperometric cell, are shown in Figure 14-6; they will be treated in detail in the following sections.

14.3.1.1 Promoted Electron Transfer Between Small Redox Proteins and Electrode Surfaces

The understanding of the factors that promote rapid redox reactions between electron transferring proteins such as cytochrome c and their physiological reaction partners is of fundamental importance for the investigation of the direct "electrochemistry" of these compounds (see Section 14.2.5). The behavior of these redox proteins, especially mitochondrial cytochrome c, on electrodes has been studied [58] using DC and AC cyclic voltammetry, impedance spectrometry [59], and rotating disk and ring-disk measurements. A theoretical treatment of these electrochemical methods and descriptions of suitable experimental setups has been given [60].

It is well known that redox proteins do not always exhibit "well-behaved" reversible electrochemical responses at bare metal electrodes. Adsorption of the protein on an unmodified surface may result in a unproductive binding owing to incorrect orientation [61] or might even have deleterious effects on the protein. In order to attain orientated binding, "promoters" can be used. A promoter is an organic molecule capable of modifying an electrode surface in order to produce a suitable interface for rapid electron-transfer reactions. By itself a promoter is not redox active in the potential region investigated, and thus is not able to act as a mediator. At metal electrodes treated with the promoters 4,4'-bipyridyl [59, 62, 63], 1,2-bis(4-pyridyl) ethylene [64], bis(4-pyridyl)disulphide, L-cysteine [65], purine [66], pyridine-n-aldehyde thiosemicarbazone ($n = 2, 3, 4$) [67], and other molecules [68, 69], "well-behaved" redox reactions of cytochrome c have been observed. The electrode reaction is then similar to the physiological electron transfer between cytochrome c and cytochrome c oxidase, for example the redox process can be inhibited by polylysine or by derivatization of cytochrome c [59, 70].

All of the promoter molecules indicated above have two different functionalities, one being suitable for its binding to the electrode surface and the other for its interactions with the protein. Electrochemical [59] and spectroelectrochemical methods [71], surface-enhanced Raman spectroscopy [66, 72], and ellipsometry indicate an "end-on" mode of adsorption of the promoter molecule on the electrode surface. An array of adsorbed promoter molecules displays an arrangement of weakly basic nitrogen groups on the electrode surface, which probably forms hydrogen bonds to protonated lysine side-chains of the protein cytochrome c. This leads to transient binding of the protein, with orientation of the exposed heme edge towards the electrode, thus facilitating rapid electron transfer. A subsequent conformation change, followed by a breakdown of the electrode-protein hydrogen bond network, allows diffusion of the product molecules away from the modified electrode. The surface site is left vacant, ready to bind another cytochrome c molecule. This anisotropic and reversible binding of the protein on the electrode surface decreases the overall free energy of activation [69, 73], hence increasing the electron-transfer rate. Nevertheless, the electron transfer must still run off over a distance of at least 1.2 nm, the minimum distance between the electrode and the plane of closest approach to the heme edge. Direct electrochemistry of proteins is possible not only at metal electrodes in the presence of promoter molecules, but also directly at specifically treated electrode surfaces such as pyrolytic graphite [74, 75], tin-doped indium oxide [76], ruthenium oxide [77], and extensively pretreated metal surfaces [73, 78]. In most cases, reversible electron transfer is only observed when the electrode surface has properties resembling those present in the natural redox partner. In addition to different cytochromes, a range of other metalloproteins, including ferredoxin, rubredoxin, azurin, and plastocyanin, show direct quasi-reversible electrochemistry at these modified electrode surfaces. Finally, the electrochemistry of plastocyanin [79, 80], rubredoxin [81], and bacterial ferredoxin [82], proteins with negatively charged interaction domains, is promoted and stabilized after electrode-surface protonation or interfacial binding of multivalent cations.

Possible applications of promoted electron transfer are biological fuel cells, microbial activity monitors, and electrochemical enzyme electrodes [83]. Terminal oxidases of both, prokaryotes and eukaryotes, contain at least two redox centers and are efficiently coupled, with one of them, to the redox enzymes which precede them in the physiological electron-transport chain. In an example of an enzyme-based electrocatalytic fuel cell, an electron transfer is achieved by coupling the reduction of dioxygen over cytochrome cd1 (Pseudomonas ferrocytochrome C_{551}-O_2 oxidoreductase, E.C. 1.9.3.2) to cytochrome C_{551} and horse-heart cytochrome c and finally to a gold electrode modified with 1,2-bis(4-pyridyl)ethene [84]. It was also possible to couple the respiratory chain of rat liver mitochondria or from *Paracoccus denitrificans* via exogenous cytochrome c to an electrode modified with bis(4-pyridyl) disulfid [85]. Analogously, the oxidation of L-lactate to pyruvate with flavocytochrome b2 could be monitored by using reoxidation of its natural electron acceptor cytochrome c at a promoter-modified electrode [86].

The application of direct electrochemistry of small redox proteins is not restricted to cytochrome c. For example, the hydroxylation of aromatic compounds was possible by promoted electron transfer from p-cresol methylhydroxylase (a monooxygenase from *Pseudomonas putida*) to a modified gold electrode [87] via the blue copper protein azurin. All these results prove that well-oriented non-covalent binding of redox proteins on appropriate electrode surfaces increases the probability of fast electron transfer, a prerequisite for unmediated biosensors. Although direct electron-transfer reactions based on small redox proteins and modified electrode surfaces are not extensively used in amperometric biosensors, the understanding of possible electron-transfer mechanisms is important for systems with proteins bearing catalytic activity.

14.3.1.2 Mediated Electron Transfer from Enzymes to Electrodes by Means of Soluble Redox Mediators

Direct electron exchange is rarely encountered with highmolecular-weight enzymes whose active centers lie deeply buried in the polypeptide structure. Nevertheless, even here a knowledge of the electron-transfer mechanism is a very fundamental presupposition for the development of amperometric enzyme electrodes. Especially, one has to distinguish between oxidoreductases with prosthetic groups bound tightly to the protein matrix (eg, flavoproteins, PQQ enzymes), and oxidoreductases without an integrated cofactor (eg, NAD^+-dependent dehydrogenases), whose cosubstrate can be used as a redox mediator. The steric conditions, which are known for some oxidoreductases, seriously decrease the accessibility of the active site. Hence, mechanisms involving natural or artificial electroactive compounds, which act as "electron shuttles", must provide redox coupling between the electrode and the redox center in the biological compound. An ideal mediator must bring along the following properties [88, 89]:

1. It must have a well-defined electron stoichiometry between its oxidized and reduced states.
2. Its formal redox potential must be appropriate for the desired redox reaction.
3. The mediator must exhibit fast heterogeneous and homogeneous electron-transfer rates with both the biomolecule and the electrode surface.
4. It must be sufficient soluble to permit a shuttle mechanism for electron transfer.
5. It must not interfere with optical monitoring of the biocomponent.
6. Its interaction with the biocomponent should not alter its redox potential.
7. It must be stable in both the oxidized and the reduced form.

For electron-transfer measurements it is very important that the formal potential of the mediator is close to that of the biomolecule in question. One can show from the Nernst equation that this formal potential for one-electron reactions should be within ± 118 mV of that of the biocomponent. Very often, the direct electron transfer from the biomolecule to the electrode will suffer from irreversibility and need high overpotentials. In this case, from a practical point of view, a good mediator should have a redox potential between the formal potential of the biomolecule and the observed overvoltage. Compilations of compounds that mediate biological redox systems have been published [88–90].

Electrochemical techniques, especially direct current cyclic voltametry, are very useful in evaluating the properties of new mediators, eg, redox potential, stability, electrochemical rate constant, and especially the rate of the reaction with the enzyme [91]. A first hint about the stability can be obtained from measuring the peak current in subsequent cycles of the voltammogram. A decrease in the charge transferred in a half scan can originate from side-reactions of the redox compound. In fact, efficient coupling of an enzymatic redox process in question with a non-physiological mediator should be achieved, and this will be indicated by a significant effect on the related cyclic voltammogram. When the mediator transports electrons by shuttling between a reduced enzyme and an anode, the effect seen on the cyclic voltammogram must be an increase in the anodic current concomitant with a decrease in the cathodic current (Figure 14-7). This effect will be the more significant the higher is the rate constant of the reaction between the reduced enzyme and the oxidized mediator, as can be proved by mathematical treatments [92]. Quantitative kinetic data on the rate constant of electron transfer between an enzyme and a redox mediator may also be obtained from cyclic voltammetric experiments [91, 93]. In addition, measurements with rotating electrodes, in which problems arising from poorly defined mass-transport characteristics are eliminated, can be used to obtain quantitative kinetic data, eg, diffusion coefficients and information about immobilized enzyme layers [93, 94].

In mediated electron transfer, the immobilization of the enzyme does not significantly influence the mechanism of current generation, but the mass transport, and hence the response time and linear range of the enzyme electrode. As has been shown in Figure 14-4, in glucose oxidase (GOD) β-D-glucose is oxidized to D-gluconolactone by reduction of the cofactor flavine adenine dinucleotide. In a second step the reduced cofactor is reoxidized by the natural acceptor O_2, yielding H_2O_2. The glucose determination can be performed either by cathodic

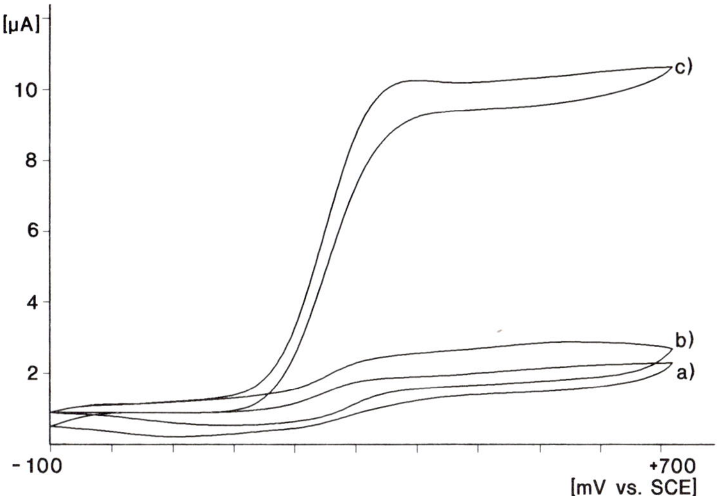

Figure 14-7. Cyclic voltammogram demonstrating mediated electron transfer from glucose oxidase to a platinum electrode with ferrocene carboxylic acid. a) 2mM ferrocene carboxylic acid; b) +10 mg glucose oxidase; c) +30 mg glucose. -100 to $+700$ mV vs. SCE; 5 mV s^{-1}; 0,1 M phosphate buffer (ph 7.4) with 0.5 M NaCl; platinum disk electrode, 1 mm diameter.

reduction of the dioxygen consumption or by anodic oxidation of the hydrogen peroxide produced [95, 96]. Corresponding devices are widely used, especially in clinical analysis and for process control of fermentation processes in biotechnology and the food industry [97–99]. The main problems with these biosensors are their dependence on the oxygen tension and the high overpotential needed for the oxidation of hydrogen peroxide (at platinum electrodes $+700$ mV vs. SCE). These restrictions are of special importance with sensors for *in-vivo* applications, because here the partial pressure of oxygen can be lower than the concentration of glucose, which leads to a stoichiometric limitation of the enzymatic reaction by oxygen.

One possiblity of overcoming these problems is to design sensors in which oxygen tension is no longer the limiting factor. In early attempts the glucose oxidase electrode was rinsed with air-saturated buffer prior to the measurement of blood glucose [100]. In another device, a "two-dimensional" enzyme electrode [101, 102], oxygen was additionally admitted to diffuse to the electrode from two sides while the mass transport of glucose into the enzyme membrane was restricted to only one. Interfering ascorbic acid can be completely eliminated by electroenzymatic oxidation [103]. Overpotentials for the oxidation of H_2O_2 could be successfully reduced with carbon electrodes modified by vapour deposition of palladium and gold to form a catalytic surface [104], on which glucose oxidase was adsorbed [105]. It even seems that such electrodes can work by direct electron transfer from the enzyme to the catalytic surface [106].

Many of these investigations were aimed at the development of electrodes for *invivo* application. Additional serious problems arising in this context are degradation of the enzyme by serum proteases, adsorption of platelets, proteins, or other blood components, and hence the risk of thrombosis and infection [107]. Nevertheless, needle-type glucose electrodes have been constructed and also used in short-term applications, working in the physiological range of blood glucose level avoiding the above-mentioned limitation from oxygen tension by applying sophisticated membrane-covering methods [108, 109]. Recently, even microelectrodes fabricated by means of standard silicon processing techniques have been used in planar amperometric microcells for the determination of glucose [110, 111]. Thus, cheap mass production can be extended to the design and construction of amperometric enzyme electrodes. This is not only true for glucose electrodes; even when the primary focus has been the determination of glucose, the list of extensions to substrates of other oxidases is considerable [112–114].

Another approach to achieve independence of O_2 with oxidase electrodes is the replacement of the natural electron acceptor by artificial mediators. This leads to amperometric enzyme electrodes for glucose with lower working potentials and with less cross-sensitivity to interfering substances. Electron acceptors for glucose oxidase include 2,6-dichlorophenolindophenol [115], hexacyanoferrate(III) [116], tetrathiafulvalene [117], tetracyano-*p*-quinodimethane [118], quinones [119–121], and ferrocene derivatives [122, 123]. The major problem with all these mediated oxidase electrodes is the kinetic competition of the mediator with oxygen in the active site of the enzyme [124]. Another drawback is the possible instability of the oxidized mediator in the presence of O_2. To date ferrocenes and organic conducting charge-transfer salts (see Section 14.3.1.5) seem to provide the best mediator characteristics in both respects. Especially in the case of ferrocene derivatives the formal potentials and the reactions rates with active sites of enzymes can be tailor-made by substitution at the cyclopentadienyl rings (Table 14-3).

Most of the ferrocene derivatives are insoluble in water, and therefore the mechanism of their action as mediators was not clear until recently. The fact that polyvinylferrocene did not show a catalytic current with reduced glucose oxidase gave a first hint on a shuttle mechanism, because this polymer would not be able to enter the active site and to diffuse between the electrode surface and the enzyme. On the other hand, with 1,1'-dimethylferrocene adsorbed on the surface of a graphite electrode [125] or co-immobilized into a cross-linked polycrylamide gel [126], an amperometric enzyme electrode for glucose with a working potential far below that for H_2O_2 oxidation was obtained. This must be due to the fact that the ferricinium ions produced at the graphite electrode are capable of replacing oxygen as electron acceptor within the reduced enzyme. As a consequence of the competition between oxygen and the artificial

Table 14-3. Formal potentials of ferrocene derivatives and electrochemically determined rate constants of the related ferricinium ions with reduced glucose oxidase.

Ferrocene derivative	$E_{1/2}$ in mV Vs. SCE	$k_s \cdot 10^{-5}$ in L/mol s
1,1'-dimethyl-3-(2-aminoethyl)ferrocene[a]	75	Not determined
1,1'-dimethylferrocene[b]	100	0.77
Ferrocene[b]	165	0.26
Hydroxymethylferrocene[c]	185	9.0
(2-aminoethyl)ferrocene[c]	200	44.0
Vinylferrocene[b]	250	0.3
Ferrocenemonocarboxylic acid[c]	275	1.8
Aminomethylferrocene[a]	309	Not determined
1,1'-ferrocenedicarboxylic acid[c]	395	0.26
Methyldimethylaminoferrocene[b]	400	5.25
Polyvinylferrocene[b]	450	–

a) from [53];   b) from [125];   c) from [138]

acceptor, the maximum current calculated for the oxidation of this mediator is not attained [127]. Assuming a shuttle mechanism and an electron transfer due to diffusion of the reduced virtually water-insoluble mediator or even additionally by an electron hopping between mediator molecules, and a real or sufficient solubility of the oxidized form of the mediator, the stability of these sensors must be limited by leaking of the ferricinium cations from the electrode. In fact, in cyclic voltammetric experiments a significant decrease in the peak current in subsequent scans is observed, and also in rotating ring-disk experiments with the graphite disk modified by 1,1'-dimethylferrocene losses of mediator could be demonstrated [127]. Additionally, the response of an exhausted electrode could be partially restored by a new "loading" with 1,1'-dimethylferrocene (Figure 14-8). These results explain findings of Brooks et al. [128], who, however, had attributed the decrease in the response of their electrode to leaking or denaturation of the enzyme. Improved immobilization methods, eg, after oxidation of the sugar residues of the glycoenzyme glucose oxidase [129] or by covalent binding to defined anchor groups on the electrode surface, showed definitely that long-term application is limited by the leaking of the 1,1'-dimethylferricinium cations [130]. As ferrocenes show chronic toxicity with dogs [131], subcutaneous implantations of ferrocene-mediated glucose electrodes have to be reconsidered [132, 133]. On the other hand these electrodes have found application in one-shot glucose determinations in home-monitoring of diabetes [134]. Analogously, other enzymes such as D-galactose oxidase, glycolate oxidase, and L-amino acid oxidase [135], cholesterol oxidase [136], NADH oxidase [137], pyruvate oxidase, xanthine oxidase, sarcosine oxidase, diaphorase, glutathione reductase, carbon monoxide oxidase, and flavocytochrome b_2 [138] can be coupled to ferrocenes as electron-transfer mediators leading to biosensors for the corresponding substrates. Other substrates can be determined on the basis of coupling or

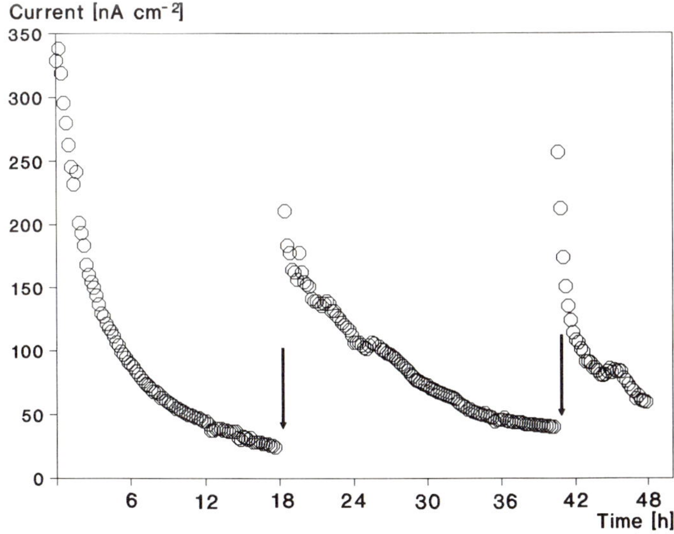

Figure 14-8. Response of a graphite-glucose oxidase electrode in an automatic flow-injection system. 100 µL of 5 mM glucose solution were injected in a buffer stream of flow-rate 2.5 mL min⁻¹. The response could be partially restored by addition of new 1,1'-dimethylferrocene to an exhausted electrode.

competing reactions with the ferrocene-mediated glucose oxidase electrode. For example, reactions producing ATP can be monitored by phosphorylation of glucose with hexokinase and determination of the concomitantly decreased glucose concentration [139].

Ferrocene and its derivatives even mediate the enzymatic redox reactions of PQQ-dependent enzymes. As an example, the mediation of the enzymatic glucose dehydrogenation by PQQ-glucose dehydrogenase has been reported [140]; the same enzyme also can use phenazines as electron acceptors [141].

14.3.1.3   Electrocatalytic Properties of Modified Electrodes for Redox Reactions of Biological Molecules

Over 250 oxidoreductases (dehydrogenases) use the cosubstrate β-nicotinamide adenine (phosphate) dinucleotide $(NAD(P)^+)$ to oxidize a substrate SH_2 with concomitant reduction of the cofactor to NAD(P)H. In the most cases the cofactor is bound simultaneously with the substrate in the active site of the enzyme, allowing transfer of a hydride ion from the substrate to $NAD(P)^+$.

$$SH_2 + NAD(P)^+ + H_2O \xrightleftharpoons{\text{dehydrogenase}} NAD(P)H + S + H_3O^+ \qquad (14\text{-}3)$$

$$NAD(P)H \xrightarrow{\text{electrode surface}} NAD(P)^+ + H_3O^+ + 2e^- \qquad (14\text{-}4)$$

Reoxidation of the cosubstrate at an appropriate electrode surface will lead to the generation of a current that is proportional to the concentration of the substrate, hence the coenzyme can be used as a kind of mediator. The formal potential of the $NADH/NAD^+$ couple is -560 mV vs. SCE (KCl-saturated calomel electrode) at pH 7, but for the oxidation of reduced nicotinamide adenine dinucleotide (NADH) at unmodified platinum electrodes potentials >750 mV vs. SCE have to be applied [142] and on carbon electrodes potentials of 550–700 mV vs. SCE [143]. Under these conditions the oxidation proceeds via radical intermediates facilitating dimerization of the coenzyme and forming side-products. In the anodic oxidation of NADH the initial step is an irreversible heterogeneous electron transfer. The resulting cation radical $NADH^{+\cdot}$ looses a proton in a first-order reaction to form the neutral radical $NAD\cdot$, which may participate in a second electron transfer (ECE mechanism) or may react with $NADH^{+\cdot}$ (disproportionation) to yield NAD^+ [144]. The irreversibility of the first electron transfer seems to be the reason for the high overpotential required in comparison with the enzymatically determined oxidation potential.

At chemically modified electrodes, NADH can be oxidized to enzymatically active NAD^+ at much lower potentials than at bare electrode surfaces. Specified functionalities have been introduced to the electrode surface by a number of different immobilization techniques (see Section 14.4). It is presumed that a redox compound able to oxidize NADH in solution may also act as a suitable mediator when it is fixed to an electrode surface. In cyclic voltammetric experiments, the current of the oxidation wave of the mediator must be significantly increased in the presence of NADH, while the corresponding reduction current is decreased (Figure 14-9).

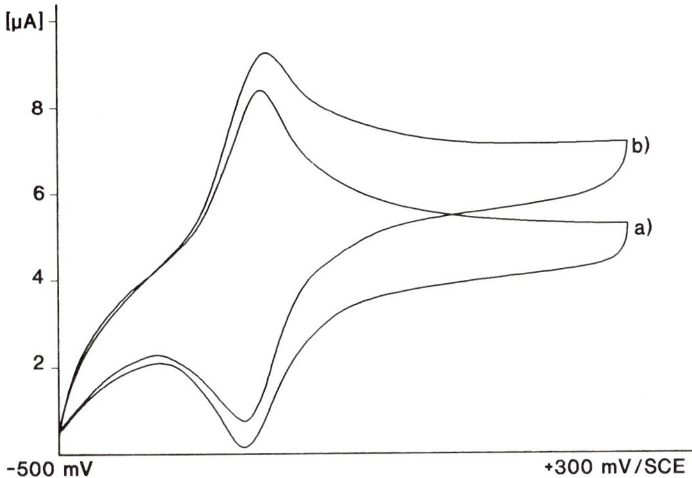

Figure 14-9. Cyclic voltammogram of a graphite electrode modified with β-naphthoyl Nile Blue. a) buffer (pH 8.0); b) after addition of NADH (10 mM). The increase in the anodic current is attributed to electrocatalytic oxidation of NADH at the mediator-modified electrode. −500 to +300 mV vs. SCE; 5 mV s^{-1}; 0.1 M phosphate buffer (pH 8.0) with 0.5 M NaCl; graphite disk electrode, 6.4 mm diameter.

Various redox compounds that fulfil catalyst characteristics have been investigated in systems with recycling of NAD$^+$ by electrocatalytic methods. Quinones, formed either by oxidation of carbon surfaces [143, 145] or adsorbed to the electrode surface [146, 147], phenazines [148, 149], phenoxazine derivatives such as Meldola Blue [182], β-naphthoyl-Nile Blue [151, 152] and 1,2-benzophenoxazine-7-one [153], and also the organic conducting salt N-methyl phenazinium tetracyanoquinodimethanide (TTF$^+$TCNQ$^-$) [154, 155], ferricinium ions [156, 157] and hexacyanoferrat(III) ions [158, 159] can act as catalysts for the electrochemical oxidation of NADH. It is assumed that in corresponding electron-transfer reactions a charge-transfer complex between the immobilized mediator and NADH is formed. The intermediate reduced redox mediator will be reoxidized electrochemically. Most systems mentioned, however, suffer from poor electrode stabilities.

14.3.1.4 Modification of Enzymes by Covalent Binding of Redox Mediators

A new approach to avoid dissolved redox mediators in amperometric enzyme electrodes was introduced by Degani and Heller [160, 161] who, after modifying the enzyme with covalently bound ferrocenecarboxylic acid, observed a direct electron transfer from glucose oxidase to a gold or platinum electrode. Glucose oxidase is a structurally rigid glycoprotein with two identical polypeptide chains and a hydrodynamic radius of 43 Å. The incorporation of about 12 ferrocenecarboxylic acid molecules between the two subunits of the enzyme drastically shortened the tunneling distance for electron transfer from the active site to the surface of the macromolecule. This was probably attained by an electron-hopping mechanism. One may imagine that the mechanism of mediation with soluble ferrocene derivatives (see Section 14.3.1.2) was thus "extended" from the solution into the active site of the enzyme without the necessity for real diffusion of mediator molecules. The increased flexibility of enzyme-immobilized ferrocene derivatives, attained by changing from ferrocenecarboxylic acid to ferroceneacetic acid or ferrocenebutanoic acid significantly enhanced the kinetics of electron transfer [162]. This "wiring" of enzymes is certainly a very interesting field of research, especially in connection with conducting polymers as electrode material (see Section 14.3.1.5).

14.3.1.5 Binding of Enzymes to Conducting Polymer Electrodes

The conducting organic polymer polypyrrole is being increasingly used as an electrode material in amperometric biosensors. The ramified network is assumed to increase the probability of fast electron-transfer reactions. One approach to bind enzymes to the polymer is to perform the electrochemical polymerization of pyrrole in the presence of the enzyme [163–165]. With glucose oxidase a dependence of the activity of the immobilized enzyme on the amount of applied enzyme and the pH value during the polymerization has been observed. Apart from an enclosure, an electrostatic binding of the negatively charged enzyme to the positive polymer chains has been assumed [163]. The electropolymerization reaction can be performed on tiny electrodes, and suitable materials are gold or platinum. Thus, on gold microelectrodes obtained by lithographic techniques on silicon, two different polypyrrole electrodes on a single chip have been realized. By Immobilizing glucose oxidase on one electrode and galactose oxidase on the other, the simultaneous determination of both sugars in a very small volume was permitted [166]. The derivatization of the N-position of polypyrrole with ferrocene moieties probably led to a polymer capable of direct electron transfer from the active site of glucose oxidase to the electrode material [167]. Even the covalent binding of enzymes to polypyrrole electrodes has been realized after introduction of functional groups in the β-position of the heterocycle by nitration and subsequent electrochemical reduction [168]; enzyme electrodes with fast response times and remarkable stability were obtained.

For electrodes based on conducting organic charge-transfer salts such as TTF^+TCNQ^- (a complex of the radical cation of tetrathiafulvalene and the radical anion tetracyano-p-quinodimethane) or NMP^+TCNQ^- (N-methylphenaziniumtetracyano-p-quinodimethane), direct [155, 169] and mediated [154] electron transfer mechanisms have been described. In analogy with the theory of outer-sphere electron transfer [170], Kulys and co-workers [118, 171] have developed a mathematical model which permits to evaluate the depth of the active site of some oxidoreductases from the steric requirements of inorganic redox couples (Table 14-4).

Table 14-4. Calculated depth of active site for some oxidoreductases [54].

Enzyme	Depth of active center in Å
Glucose oxidase	8.7
Flavocytochrome b_2	4.2
Peroxidase	4.1
Diaphorase	6.3
Lipoamidedehydrogenase	4.8
Laccase	4.3

In the case of a mediator-free mechanism [169], electron transfer must occur at the equilibrium potential of the substrate or the cofactor. On the other hand, in a mediated electron-transfer reaction [154], current generation is bound to the redox potential of the mediator redox couple. For glucose oxidase and xanthin oxidase on a charge-transfer salt electrode, a mediated electron transfer has been demonstrated, because the substrate oxidation proceeds at the mediator's conversion potential and the current generation is inhibited by oxygen at low substrate concentrations [154]. The mediator (TTF^+, NMP^+, $TCNQ^-$) is probably liberated from a layer near the electrode surface owing to a slight dissolution of the electrode material. In contrast, in the case flavocytochrome b_2 or peroxidase, the electron exchange between the active site and the conducting-salt electrode seems to be direct [172, 173], as could be concluded from the lack of the dependence of the current from the potential. In contradistinction, Albery and co-workers [155, 169], referring to results of kinetic measurements, assumed a direct electron transfer for glucose oxidase, xanthine oxidase, and amino acid oxidases. The existence of these different electron-transfer mechanisms is in good agreement with the estimated depth of the active sites of the enzymes investigated.

14.3.2 Use of Calorimetric Devices

Any chemical reaction is accompanied by an energy conversion, in the most cases heat production, and normally this heat is proportional to the amount of substance converted. It can therefore be a measure of its amount. In an insulated adiabatic system of defined heat capacity (calorimeter), the heat produced leads to a proportional temperature rise, and even in open "semi-adiabatic" systems proportional temperature changes are observed, however, these systems must be calibrated for substance determinations. Very sensitive devices for the measurement of temperature changes are thermistors, which are semiconductor resistances with high temperature coefficients, eg, $3-4\%\ °C^{-1}$.

Heat is a very unspecific expression of a chemical reaction, but it can become indicative for a given substrate when the latter is selectively converted, eg, under the influence of a catalyst, especially an enzyme. "Enzyme thermistors" do not exactly fit with the definition of a biosensor, because they do not consist of a transducer surrounded by an immobilized enzyme, but represent a thermistor at the end of a small enzyme reactor in a flow system, and often the expression is even used for a complete setup, including a thermostat and a reference for the compensation of unspecific heat production, eg, from friction, adsorption, desorption, and turbulence effects. The first publications on calorimetric devices for substrate determination and enzyme thermistors appeared between 1974 and 1976 [174–76], and since then the principle has not changed, even though more sensitive and stable thermistors and more sophisticated set-ups are available. Two excellent reviews including aspects of instrumentation, methods, and application have appeared in recent years [177], and this section is mainly referring to those articles. A typical performance is displayed in Figure 14-10. In a thermostat (stability $\pm 0.01\ °C$) a buffer (flow rate $0.5-5$ mL min^{-1}) is temperature equilibrated with the surroundings, then it passes though the "enzyme reactor" (column with immobilized enzyme; $0.3-1$ mL), at the end of which the thermistor is glued with a heat-conducting epoxy resin to a gold capillary. The reference (selected thermistor with identical characteristics) is either mounted to the end of a similar column without enzyme, to the entrance of the enzyme column, or to the interior wall of the thermostate. The signal is produced from adjustment of

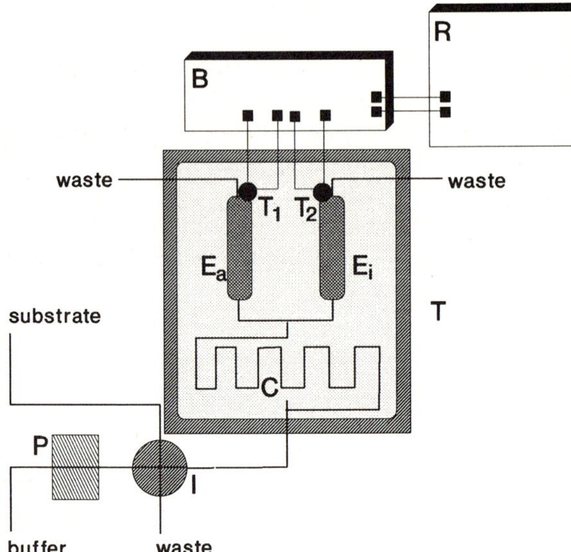

Figure 14-10.
Principle of measuring setup with two "enzyme thermistors" in compensation. P = pump; I = sample injector; T = thermostate; C = temperature equilibration coil; E_a and E_i = active and inactive enzyme columns, T_1 and T_2 = thermistors; B = Wheatstone bridge and amplification; R = recorder.

the resistances through a Wheatstone bridge, and it can be as high as 100 mV (recorder input) for a temperature change of ± 0.01 °C. Details of fabrication, sources of parts, and even complete setups have been given in [177]. The sample is added from a dosage loop (0.1–0.5 mL) or injected through a septum (20–50 µL).

As the enthalpy of enzyme-catalyzed reactions is between 30 and 100 kJ mol^{-1}, an amount of substance of 10^{-7} mol, corresponding to 0.1 mL of a 10^{-3} M solution, will lead to a maximum temperature raise of 0.2 °C, provided that there is no dilution, no heat loss, and total conversion of the substance in question by the enzyme. In practice, about 50% of the calculated maximum value would be realistic. Ideally, with a thermistor device differences of ± 0.001 °C would be measurable; however, owing to the disturbances mentioned earlier, temperature differences of ≥ 0.005 °C must be accepted as measuring signals. This means that the lower detection limit for a substrate under the above-mentioned experimental conditions, would be 10^{-5} M. As a matter of fact, in practice concentration changes in this range are at the noise level, but, measurements have been performed with linear responses beginning from this value up to 200 mM solutions, with 1–5% accuracy. The sample frequency can be up to 20 h^{-1}.

The sensitivity for certain substrates can be increased, when a second heat-producing reaction or a recycling process [178] is coupled to the enzyme reaction in question. The neutralization of a base or an acid, formed by hydrolysis, is an example, and as a matter of fact the heat yield of these reactions depends on the buffer in which they are performed. Correspondingly, the sensitivity of oxidase reactions is enhanced when the fission of H_2O_2 ($\Delta H = 100$ kJ mol^{-1}) by catalase is coupled. Such reactions are of high value to increase the signal-to-noise ratio, which is important in this measuring method, especially with biological samples, and which can, as already mentioned, have many causes and can only be overcome by compensation circuits.

Enzyme thermistors have so far been described for the determination of urea, penicillin, and other amides [179], for glucose, sucrose, cholesterol, uric acid, and lactate, and they have been used preferentially for medical analyses [180]. However, process control [181], environmental monitoring, and detection of enzyme activities in eluates from chromatographic columns [177] are further applications of these sometimes "sensible", but sensitive and universal biosensors.

14.3.3 Applications of Potentiometric Electrodes and Field-Effect Transistors

Potentiometric techniques involve the measurement of a potential under equilibrium conditions where no electrolysis and charge transport over the electrode/solution junction occur at the indicator electrode. In reality this potential is the potential difference between the indicator electrode and a reference electrode, normally a metal electrode in contact with a corresponding insoluble salt (secondary kind of electrode) which has an invariant half-cell potential. In general, the saturated calomel electrode (SCE) (Hg/Hg_2^{2+}/saturated KCl; 244 mV vs. the normal hydrogen electrode (NHE)) and the silver/silver chloride electrode (Ag/AgCl/ saturated KCl; 198 mV vs. NHE) are used as reference electrodes, and the measuring set-up is illustrated in Figure 14-11. Two different types of potentiometric measurements with respect to the potential-generating mechanism have to be distinguished, using ion-selective electrodes and noble-metal electrodes immersed in a solution of a redox couple. Ion-selective electrodes have a symmetrical arrangement with the solution placed on either side of the membrane, while noble-metal electrodes, eg, in coated-wire electrodes, have a nonsymmetrical arrangement with one side of the membrane contacted by the sample and the other by a solid material. In Figure 14-12 the potential-determining processes of these types of electrodes are shown schematically.

An ion-selective electrode consists of a semipermeable membrane in contact with a reference solution on one side and the sample solution on the other. The membrane has to be selectively permeable to either a cation or an anion, but the penetration of the related counter ion must be restricted. Thus, charge separation occurs at the interface leading to a potential difference (Donnan potential) which contains the analytically useful information. Within the membrane the diffusion of an ion is promoted by a concentration gradient, and when the mobilities of the cations and anions vary greatly, a diffusion potential is additionally developed by charge separation. The change in the membrane potential predominates, under well-defined conditions (pH, ionic force, temperature), over changes in the overall cell potential due to concentration differences in the substance in question in the analyte. Hence, the cell potential is proportional to the potential drop over the ion-selective membrane.

Ion-selective membranes attain their "permselectivity" from ion-exchange, dissolution, or complexation phenomena. Different types of "membranes" are available for the construction of ion-selective electrodes: glass and other solid state rods (crystals), liquid or polymer ion exchangers, or dissolved ionophores. Many electrodes are commercially available with selectivities for different ions, mainly H^+, alkali metal cations, heavy metal ions, and halides or pseudohalides. Also gas-sensing electrodes may be constructed from an ion-selective electrode and a gas-permeable membrane [182]. Ion selective electrodes and gas-selective electrodes

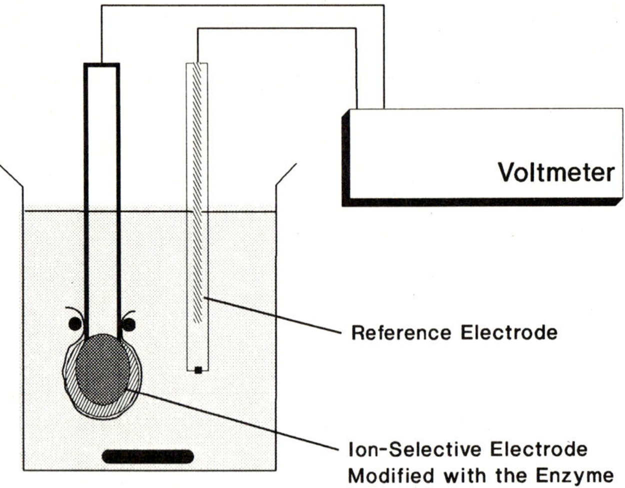

Figure 14-11. Schematical setup for measurements with potentiometric enzyme electrodes.

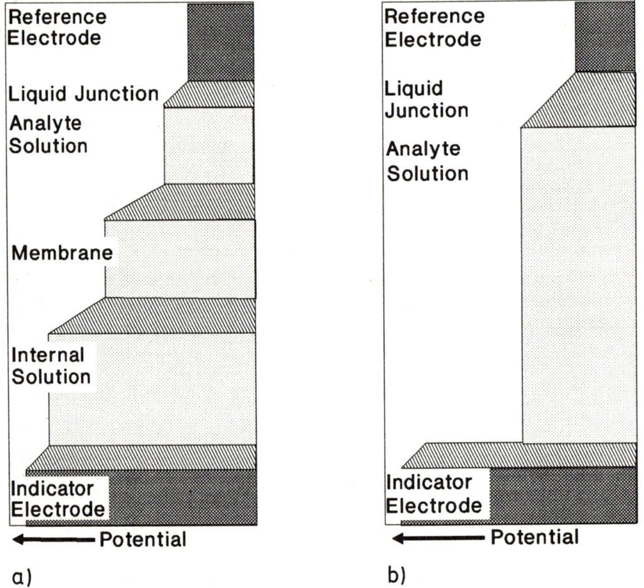

Figure 14-12. Source of potential differences being responsible for the measured overall potential with a) an ion-selective electrode and b) a redox electrode.

serve as a basis for the development of potentiometric enzyme electrodes by their combination with appropriate enzyme reactions (see Table 14-2).

By immobilizing an enzyme on or in close proximity to the ion-selective membrane, biocatalytic membrane electrodes are assembled. The consumption of a substrate and/or the formation of products by the enzymatic reaction changes the activity of an ion or a gas, for which the membrane used is selective. Thus, the selectivtiy of a potentiometric enzyme electrode is given by the combination of the selectivities of the biocatalyst and that of the ion-selective membrane. A large number of enzymes have been used, but for the connected potential generation only a few electrodes are available, including electrodes for NH_3 or NH_4^+, originating from hydrolyses of urea, creatinine, and other amides, from oxidations or oxidative deaminations of amines and amino acids, and from NH_3-lyase reactions, for CO_2 from urea hydrolysis and amino acid or other decarboxylation reactions, for H^+ from hydrolyses of esters, lactones, and amides, and from some oxidations in dehydrogenase reactions [183]. The working range of the corresponding biosensors is typically within 2–3 orders of magnitude, and a lower detection limit of 10^{-5}–10^{-4} M is common.

Potentiometric enzyme electrodes constructed from redox electrodes are typically noble metal electrodes, showing Nernstian behavior (slope of 59 mV per concentration decade) for changes in the concentration ratio of the oxidized and reduced forms of a redox couple in the analyte solution. The potential measured with these electrodes for any redox couple should be independent of the electrode material used but, in practice, preconditioning procedures and the pretreatment of the enzyme electrode have significant effects on the response characteristics. This was demonstrated for a glucose electrode, bearing immobilized glucose oxidase in a cross-linked polymer gel on different pretreated platinum surfaces [184, 185]. ESCA studies demonstrated that differences in the potentiometric response were due to differences in the surface oxidation state and to contamination by carbonaceous substances resulting from the treatment [186]. Hence, surface functions of the electrode seem to be necessary for the mechanism of potential generation.

The negative slope (-40 mV/decade) observed with such a platinum electrode suggests that the observed potential changes are due to a net reduction reaction of electrode surface groups [187]. With graphite as electrode material in an analogously prepared glucose oxidase electrode, the observed slope is positive, probably owing to oxidation of hydroquinone or aldehyde groups at the surface. Recently, potential formation at a coated-wire electrode with glucose oxidase immobilized in a matrix of graphite particles embedded in a polymer was investigated by AC impedance spectrometry [188]. The results indicated that adsorption and subsequent oxidation of H_2O_2 at the platinum surface seem to be the most important factors for potential formation. Obviously, present limitations of potentiometric glucose oxidase platinum electrodes concern the reproducibility of the surface potential and potential drift with repeated applications.

The trend to miniaturization and to internal electronic amplification of the potentiometric measuring principle led to the development of ion-sensitive field-effect transistors (ISFEs) (Figure 14-13). Theory and fabrication of FETs have been extensively reviewed [189–192]. ISFETs are derived from metal oxide semiconductor field-effect transistors (MOSFETs) by replacing the metal gate with an electrolyte junction and a reference electrode. The gate insulator (SiO_2, Si_3N_4, Ta_2O_5, Al_2O_3) is directly exposed to the analyte solution or is coated with an ion-selective membrane. Hence, the threshhold voltage of such an ISFET is a function of the local pH value at the gate insulator. The surface hydroxyl groups of the gate material

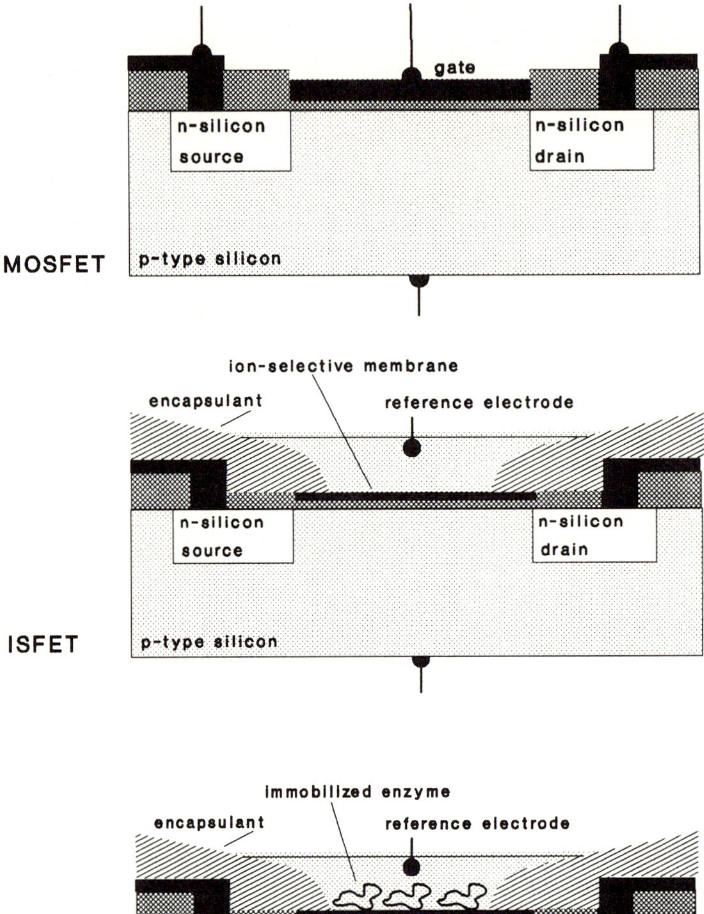

Figure 14-13. The hierarchy of field-effect transistors.

act as sites for chemical reactions due to an acid-base equilibrium. Alteration of surface charge resulting from protonation and deprotonation reactions affects the surface potential, and above the threshold potential the p-semiconducting channel between the n-semiconducting source and drain contacts is inverted, allowing the flux of a current between source and drain that is proportional to the change in the gate potential. The change in the surface potential is of the order of 25 mV/decade with SiO_2 as gate material to 55 mV/decade with Ta_2O_5. The response is determined by the kinetics of the surface reactions, which are in general very fast, hence leading to extremely short response times [193].

The disadvantages of ISFETs are that they are sensitive to temperature changes and, owing to their open structure, also to light. Hence, drift rate of the order of $0.1-1$ mV h^{-1} are typically observed [194]. In addition, for the response of a pH-sensitive FET the buffer capacity has a fundamental influence on the sensitivity, the linear range, and the concentration range. The limited commercial success of ISFET devices is mainly attributed to the fact that the connections to the chip have to be made within the liquid analyte and the concomitant encapsulation problems. Partial compensations can be achieved by differential measurements with a reference FET, which can be easily integrated on the same chip. Nevertheless, in practical applications rigid control over sample pH and buffer capacity is essential.

By analogy with the concept of ion-selective electrodes, FETs can become ion-sensitive after coating of their gate with corresponding membranes. Such a device may be used in combination with the selectivity of an enzymatic reaction catalyzed by an enzyme immobilized on the surface of the gate insulator (Figure 14-13). As example, penicillinase was immobilized within a membrane-gel on the gate of an ISFET, and the device was selective for penicillin owing to local pH changes as a consequence of the enzymatically catalyzed reaction [195]. Analogously, pH-ISFETs have been used with urease [196] and glucose oxidase [197, 198].

The factors determining the sensor response are the diffusional mass transport of all species involved, the enzyme kinetics, and the local pH value within the gel. In order to obtain a signal that is independent of potential changes in the solution, the enzyme gel must have a minimum thickness. The reproducible deposition of the enzyme membrane itself should be performed by IC technology-compatible methods, eg, spin coating followed by photolithography [199] or an ink-jet nozzle [200]. This is an important advantage over individual membrane casting techniques, thus exploiting the full potential of microfabricated sensors. A further advantage over macroscopic electrodes is the possibility of the construction of multi-sensors, either multible sensors with redundancy measurements or sensor arrays with different immobilized enzymes.

14.3.4 Optical Detectors

Color reactions and optical measurements are among the most versatile tools in biochemical analysis. The correlations between structure or energy states of organic compounds and electromagnetic radiation are well established, and qualitative and quantitative information can be obtained from measurements of adsorption, emission, reflection, flurorescence, and luminescence, the wavelength ranges of UV, visible, and IR radiation.

The availability of optical fibers as light conductors has greatly increased the possibilities for these optical measurements, in that they can be done remote from the analyzer, with high sensitivity and without the risk of interference from electrical noise. Optical fibers imply the possibility of equipment miniaturization, and they are flexible and chemically indifferent, and therefore especially suitable for applications in medicine, eg, for in vivo sensing of pH, $p(CO_2)$, and $p(O_2)$ [201, 202]. The theory of light transduction by optical fibers and the instrumentation for the construction of sensors have been treated in a Chapter 12 of this volume and are also the topic of several recent reviews [203-205]. Therefore the main aim of this section is to concentrate on the biological part of biosensors or bioanalytical methods in which fiber optics are or can be used (Figure 14-14).

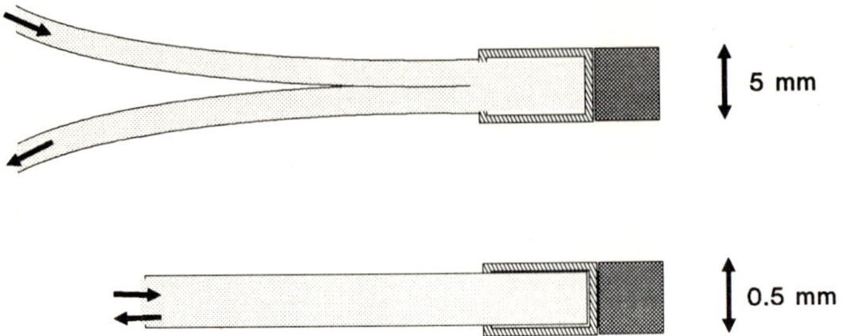

Figure 14-14. Biocatalytic optodes based on a bifurcated fiber bundle (top) or a single fiber (bottom). In the recognition part the indicator and an enzyme are immobilized. The reactive layer is permeable to the substrate but protected against light.

A chemical sensor can be obtained from a single optical fiber or from a bundle of these fibers by immobilization of an "indicator" on the tip, preferably by its binding to a polymer. Small latex unispheres, serving as reflectors, can additionally be entrapped [206]. The indicator may be a pH-indicator dye (phenol red, bromothymol blue) or a pH-modulated fluorescent dye (fluorescein, hydroxyperylenesulfonic acid or umbelliferone derivatives), or it may be a fluorescent dye whose response is quenched by O_2 (pyrene, perylene, and anthracene derivatives) [207]. For monitoring NH_3 or CO_2 [208] the pH indicator is surrounded by a selective gas-permeable membrane. The resulting probes for measuring pH values (and NH_3 or CO_2 concentrations) and O_2 concentrations are called, by analogy with other sensors, optodes or optrodes. Biocatalytic optical biosensors may be developed from hydrolases, lyases, or oxidases, as already indicated in Table 14-1 for corresponding electrochemical devices.

The best known realizations are optodes for penicillin [206, 209, 210], urea [211], and glucose [212]. A very sophisticated version of these sensors [213] uses functionalized optical fibers to which biotin is covalently bound. To this affinant, using the biotin-avidin system, the enzyme (penicillinase, esterase, urease) is connected, itself covalently bearing the indicator. The slopes of the calibration graphs for these optodes depend on the surrounding pH value and the buffer capacity; however, linear responses are obtained for 0.1–10 mM substrate concentrations.

Optodes provided with non-fluorescent esters of fluorophores have been used for the determination of external enzyme activities. The fluorophores are liberated by the enzymes and then "seen" by the optical fiber [214]. As examples of $p(O_2)$-modulated optical biosensors, a glucose probe [215] and an ethanol probe [216] can be mentioned; sensors based on glucose, alcohol, and other oxidases were reviewed by Opitz and Lübbers [217]. The advantages of these O_2-dependent optical biosensors are that, unlike corresponding amperometric sensors, they do not consume O_2 and that they are strictly diffusion limited in their response. Fiber-optical devices are also available for the determination of substrates of dehydrogenases: the NADH fluorescence produced by the immobilized enzyme is measured as a function of time [218, 219].

In addition to these catalytic sensors, optical fibers also provide the prerequisites for the construction of affinity sensors. In the classical example, a glucose sensor [220], the lectin concanavalin A, is covalently attached to the internal wall of a membrane stitched on the tip of the fiber, and it binds by affinity the fluorescein-labelled macromolecule dextran. The indicator is displaced from the lectin by glucose, which can permeate the membrane, and enters the free lumen in front of the optical fiber where it now can be "seen". It is assumed that this principle is of general value.

Another promising type of optical affinity sensor has recently been described by Trettnak and Wolfbeis [221, 222]. The fluoresence of flavoproteins (glucose oxidase, lactate monooxygenase) changes with the interaction between enzyme and substrate, and this can be used for substrate detection. As the reduced prosthetic group is reoxidized by O_2, the process is reversible, so in this case the affinant is self-regenerating.

Optical effects are also observed during the interaction between transportation or receptor proteins and their "substrates", eg, hormones. Opitz and Lübbers [217] used the quenching velocity of the protein's intrinsic fluorescence for the determination of hormones which were admitted to the optode's tip in a diffusion-limited way. Probably this principle would again be of general value when extended to the antigen-antibody reaction. Also, there seem to be good possibilities for the adaption of optical fibers for DNA probes [223]. Single-stranded DNA labeled with fluorescent dyes undergo energy transfer and hence changes in the fluorescence spectra when they are hybridized by a complementary strand. The principle of energy transfer is promising, but practical realization in the form of fiber-optic probes has not yet been achieved.

A large field for optical biosensor modules is their use in analytical flow systems, either with special optical cells within the instrument or with optical fibers for remote measurement. An "optoelectronic biosensor", proposed by Lowe and Goldfinch [224], is a flow-through cell mounted in a compact device between a light source and a photodiode, containing enzyme-indicator membranes (eg, with penicillinase and bromocresol green). The linear response of this flow sensor is between 0.5 and 5 mM substrate.

Many possibilities are also provided by flow systems using bioluminescence for substrate determinations. In corresponding devices the enzymes catalyzing the light-producing reactions are immobilized in or close to a flow-through cell within a luminometer as a part of a flow system which contains the enzymes for the conversion of substrates and the production of secondary substances for the light-generating reaction. The luciferase from fireflies reacts with ATP and luciferin to give intermediates in an excited state. The latter decays, emitting light with nearly a 90% quantum yield [225]. The luminometric ATP determination is one of the most sensitive bioanalytical processes known and it permits the determination of any substrate that can be coupled to an ATP-producing or -consuming reaction.

Of comparable general importance is the bacterial luciferase system [226–228], which opens up the opportunity to combine any $NAD(P)^+$-dependent enzyme-catalyzed reaction with a luminometric measurement. Even the chemiluminescent luminol reaction can be used for biosensing, because it can be coupled to any oxidase reaction that produces H_2O_2 [225, 229]. The logical further development of these systems towards "real" optical biosensors has recently been reported by Blum et al. [230], who immobilized the light-producing systems onto the tip of optical fibers and thus obtained fiber-optic luminescence probes.

Future possibilities for optical biosensors are extensive, as has been shown by Lundström and Gustafsson [231], who used a native fish scale, the pigment cells of which aggregate on addition of noradrenaline and then change the light transmission, as an optical bioindicator for this hormone.

14.3.5 Piezoelectric Sensors

Piezoelectricity is observed with certain anisotropic crystals, eg, quartz. When a mechanical stress is exerted on this material in a particular direction, electrical charges are induced on the surface. Conversely, an applied electric field leads to a mechanical deformation of the crystal; in particular, an AC voltage under resonance conditions will induce oscillation of the crystal. Generally a gold layer is used as electrode material to vibrate the crystal. Based on this effect, microbalances have been constructed from a quartz crystal in an oscillating circuit, and any mass change on the electrode surface will proportionally change the vibrational resonance frequency. For AT-cut quartz crystals, an empirical equation describes the correlation between mass and frequency change and is hence the base for analytical mass determinations by these devices [232]:

$$\delta F = 2.3 \cdot 10^6 \, f^2 \, \delta M \, A^{-1} \tag{14-5}$$

where δF is the change in the frequency of the crystal, f the resonant frequency, δM the mass of the adsorbed material, and A the area of the crystal. If the sensitivity is defined as the change in the resonant frequency with the change in the mass, for a 10-MHz device 0,226 Hz cm^2 ng^{-1} can be calculated. Thus, the lower detection limit (three times the noise level) would be 6.6 ng cm^{-2}, assuming a 5-Hz noise level for this crystal.

Oscillating quartz crystals are thus the ideal basis for affinity sensors. The general approach in exploiting the piezoelectric effect for a (bio)sensor is to coat a piezoelectric crystal with a material that exhibits high selectivity to the substance to be determined [233]. So far, most practical applications have involved coatings for the detection of gases or volatile compounds, eg, SO_2, CO, HCl, aromatic and aliphatic hydrocarbons, and environmental pollutants [234]. However, the selectivity of the adsorptive layers is not always satisfactory and therefore nonspecific adsorptions are the major problem to be envisaged. Especially the application of these devices to the determination of dissolved substances is very difficult, mainly owing to binding of solvent molecules.

Bioaffinant surfaces are more specific, and their application to piezoelectric crystals has been achieved by adsorption, polymer-binding, or cross-linking of antibodies and enzymes. The loadings of these affinants with their corresponding complementary substances can only be measured under strictly standardized conditions, compensating nonspecific adsorptions. Usually twin systems are conceived, having one crystal for monitoring of the analyte and the other for compensating the binding of unspecific molecules, eg, water [234].

A corresponding device for the determination of volatile organophoshorus insecticides [235] uses the binding of these substances to cholinesterase (alternating pulses of analyte gas and pure carrier). The sensor detects ppb to ppt concentrations of the insecticides; its response time is 1 min with a recovery of 10 min, and its lifetime is 21 d. A similar sensor using a coating of an antibody against parathion shows a frequency change of 400 Hz for a concentration of 36 ppb parathion, a response time of 2–3 min, a recovery time of 1–2 min, and a life time of 21 d. In this context it should be pointed out that antibody-based sensors used in the gas phase have relatively satifactory reversibility, which is not the case with the corresponding affinity sensors used in the liquid phase.

Nevertheless, the use of antibodies as immobilized affinants is the most common way to obtain piezoelectric biosensors [236]. To overcome the problems due to solvent adsorption,

Guilbault recommended drying the antibody-modified crystal before and after reaction with the dissolved antigen [236]. A further enhancement of affinity and sensitivity can be achieved by means of the streptavidin-biotin affinity system, and this seems even to open up the possibility of using piezoelectric affinity sensors in solution [237, 238].

However, all these sensors suffer from the very slow kinetics of the dissociation of the complexes. Thus renewal of the transducer surface for repeated measurements with the same device is very difficult. Dissociation of the complexes can be attained by applying highly concentrated urea solutions to the sensor, but even then not more than a few measurements are possible with the same device, as has been shown with a sensor for *Salmonella typhimurium* [236]. As a consequence, the application of immunological reactions for these and other transducer surfaces is in general limited to one-shot measurements. On the other hand, as antibodies can be obtained against nearly any substance, affinity sensors, mainly those based on piezoelectric crystals, offer so many possibilities that their further development is a promising challenge.

14.4   Binding of Biomolecules to Transducer Surfaces

As has already been pointed out in preceding sections, the selectivity, the long-term stability, and the reliability of a biosensor are dependent on the biochemical recognition elements fixed closely to the surface of the transducer in question. The immobilization procedures used must be applicable to enzymes, cofactors, microorganisms, antibodies, lectins, organelles, tissue slices, or liposomes. Adsorption, entrapment behind a semipermeable membrane or within a polymeric gel, cross-linking with bifunctional reagents, and covalent binding have been reported for the immobilization of these biochemical recognition systems (Figure 14-15). Extensive reviews on immobilization methods [239–243], the characterization of immobilized biomolecules [244] and the fixation of microorganisms [245, 246] and plant cells [247] have been published.

As large numbers of different and sometimes very special immobilization procedures have been reported, it seems almost impossible to compare the results properly. It turns out that no immobilization method is suitable for solving all the problems that arise with the immobilization of different enzymes onto different transducer surfaces. Suggestions have been made for improving the comparability of different procedures [244], and it is highly recommended that workers who have developed new methods should publish the results obtained in conformity with these proposals. In the most cases, however, the optimum method for each protein has to be evaluated empirically and adapted to the problem in question.

As the development of a "third generation" of biosensors presupposes direct immobilization of the biochemical systems onto the surface of the transducer, this section concentrates on the covalent immobilization of enzymes to functionalized solid supports, especially to electrode materials.

Any immobilization method applied must guarantee as far as possible that the activity of the biomolecule is maintained, that the accessibility of the substrate to the active sites of the bound biomolecules is not sterically hindered, and that mass transport of substrates and products through the enzyme layer is possible. It should also be ensured that the biological com-

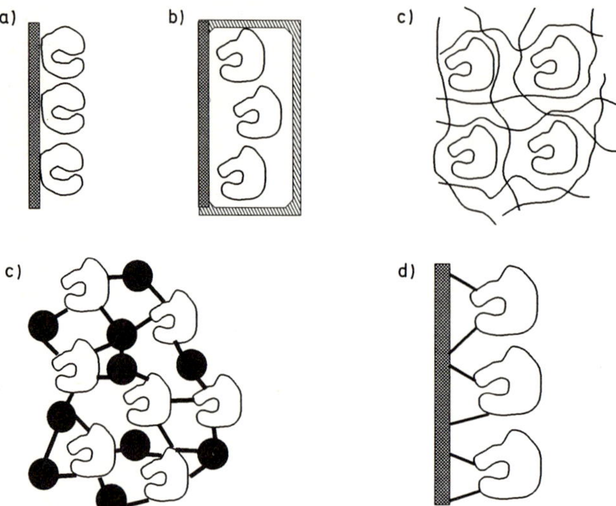

Figure 14-15. Possible methods for the immobilization of biochemical recognition elements. a) adsorption. b) physical entrapment behind a semipermeable membrane. c) entrapment into a ramified network of a polymer. d) crosslinking with an inert protein by bifunctional reagents. e) covalent binding to a solid support.

pounds retain their stability or even should be stabilized by the immobilization reaction. In the case of amperometric enzyme electrodes, transport of electrons or mediators between the solid surface and the enzyme must be possible.

Covalent binding of enzymes to solid supports requires the availability of suitable functional groups, both on the support and on the enzyme, and of non-invasive coupling reactions. As the active conformation of an enzyme is maintained by noncovalent interactions, the immobilization procedures must, in the presence of proteins, only apply conditions that do not affect these bindings.

Functional groups available in biomolecules originate from the amino acid's side-chains or from non-proteinic residues. They include ε-amino groups from lysine, carboxyl groups from aspartate and glutamate, sulfhydryl groups from cysteine, phenolic hydroxyl groups from tyrosine, the imidazole group from histidine, the indole group from tryptophan, and hydroxylic groups from serine and threonine. Taking into account the different average percentage compositions of these side-chains and the fact that they are partly buried in the interior of the macromolecule, one can assume that most substitution reactions occur with the ε-amino group of lysine, followed by the functional groups of cysteine, tyrosine, histidine, aspartate, and glutamate [248]. In the case of glycoproteins such as glucose oxidase, invertase, glucoamylase, and peroxidase, aldehyde groups can be introduced by the oxidation of the sugar residues with $NaIO_4$ or $NaClO_4$ [249], and these groups can serve as functionalities for subsequent coupling reactions. A knowledge of the three-dimensional structure and the location of functional groups within the protein can be of great value for the appropriate selection of the optimum immobilization procedure for a given protein. Often protection of the active site by the substrate is advantageous.

In the case of polymers or solid supports, intrinsic functionalities can sometimes be used as anchoring groups for the coupling with biomolecules. In other cases suitable functional groups must be introduced by substitution or activation. Several excellent reviews on the chemical modification of electrodes have covered all possibilities in this respect [250–254]. For special emphasis, modification procedures are shown here that may play a role in electrochemical biosensor development on the basis of different electrode materials (Figure 14-16).

For amperometric devices, preferably gold, platinum, graphite, and glassy carbon are used, for potentiometric electrodes in addition glass and polymeric membranes, and as gate insulators for field effect transistor SiO_2, Si_3N_4, and Ta_2O_5 are materials of choice. On gold, platinum, carbon, and silica-based materials oxygen functionalities are generally present, and their number can be increased by chemical and/or electrochemical oxidation [255], by treatment in a radiofrequency plasma [256, 257], or even by simple mechanical polishing of the surface in air. In an aqueous medium platinum and silica for example always bear hydroxyl groups which can be used for further functionalization by silanization with polyfunctional chloro- or alkoxysilanes ($YSiX_2R$) [258–260] (Figure 14-16a). The method is a very versatile means of introducing various functionalities and spacers of different lengths. Some complications can occur from possible cross-linking reactions of the reagent leading to polysiloxanes, which can form an insulating gel on the electrode surface. Monolayers can be attained when strictly dried solvents are used and the excess of the organosilane is thoroughly removed from the surface before exposing the electrode to the atmosphere. Although silane coupling techniques have been reported to yield highly stable surface functionalities, hydrolytic cleavage of the Pt-O-Si bond seems to occur. Figure 14-17 demonstrates the loss of glucose oxidase activity from a silanized platinum surface in comparison with the decay in the case of the native enzyme in solution [261].

The selective chemical or electrochemical oxidation of carbon surfaces leads to the formation of various oxygen-containing functionalities such as aliphatic and aromatic hydroxyl groups, carbonyl, quinoid groups, and carboxylic groups, the relative amount of which depends on the oxidation conditions applied. Hydroxyl groups are accessible for reactions with cyanuric chloride [262] (Figure 14-16b) and with cyanogen bromide (Figure 14-16c), and they can be activated with 1,1′-carbonyldiimidazole [263] (Figure 14-16d), or with tresyl chloride [264] (Figure 14-16e). Aldehyde groups can react with amines under the formation of the corresponding Schiff's bases (Figure 14-16f) and carboxylic groups can be activated with carbodiimides (Figures 14-16g) or can be converted to the acid chloride with thionyl chloride or PCl_5 (Figure 14-16h). Another possibility for the functionalization of electrode surfaces is the in situ polymerization of either insulating or conducting coatings onto the electrode surface. Chemical, electrochemical, photo-induced, or plasma-induced polymerization procedures have been described, among which the application of electrochemical techniques for the syntheses of non-conducting coatings yields only thin layers. On the other hand, the conducting polymer polypyrrole can be grown potentiostatically or galvanostatically on platinum, gold, glassy carbon, graphite, and semiconductor surfaces (Figure 14-18) [265, 266]. The morphology and hence the chemical and physical properties of these polymer films can be controlled by the polymerization conditions, especially the oxidation potential applied, the counter ion incorporated during the polymerization, the temperature, the reaction time, and the solvent composition. Functionalities on polypyrrole can be obtained by polymerization of functionalized monomers [267–269] or by substitution reactions of the polymer coating on the electrode surface. For example, nitration of polypyrrole and subsequent electrochemical

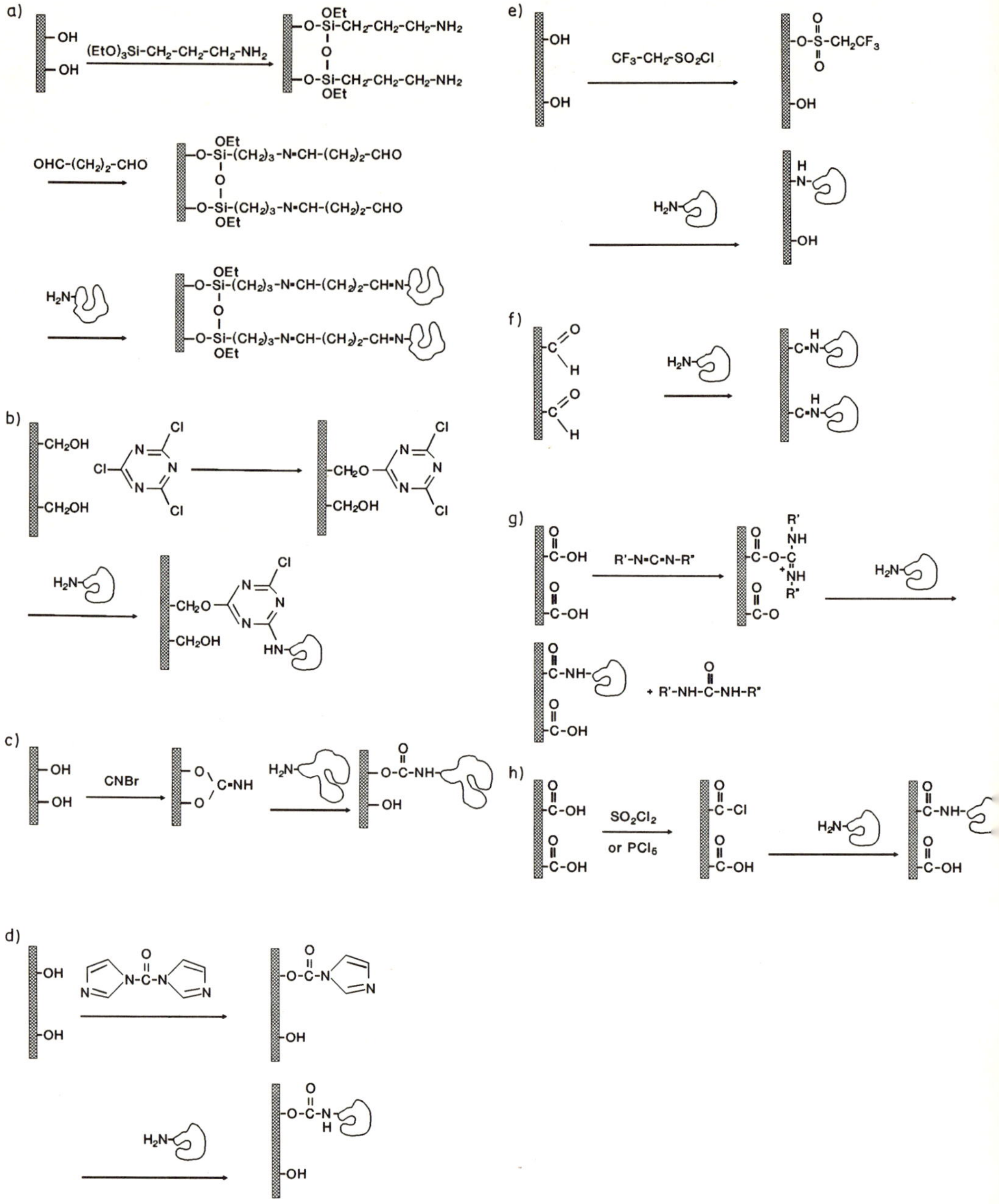

Figure 14-16. Functionalization and activation of electrode materials. a) silanization of supports bearing hydroxylic groups; b) reaction with cyanuric chloride; c) activation with cyanogen bromide; d) binding via amide bonds after activation with 1,1-carbonyldiimidazole; e) activation with good leaving groups such as tresyl chloride; f) formation of Schiff bases between aldehyde groups on the support and amino groups at the biomolecule; g) activation of carboxylic groups with carbodiimides or h) via the acid chlorides.

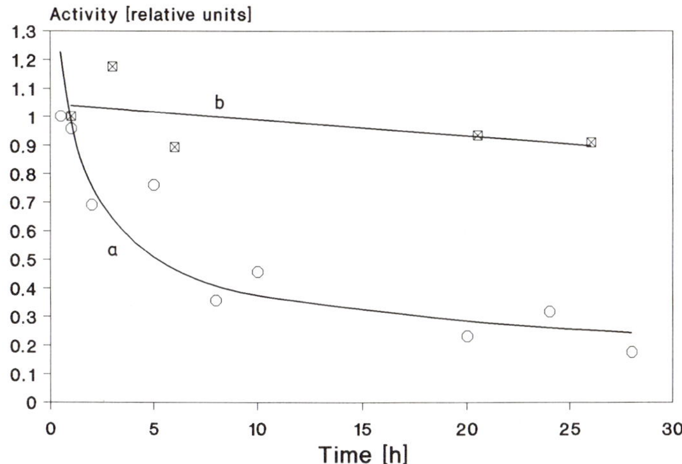

Figure 14-17. Loss of the activity of glucose oxidase a) covalently bound to a platinum surface functionalized with 3-aminopropyltriethoxysilane (APTES) and glutardialdehyde in comparison with b) the native enzyme in buffer solution.

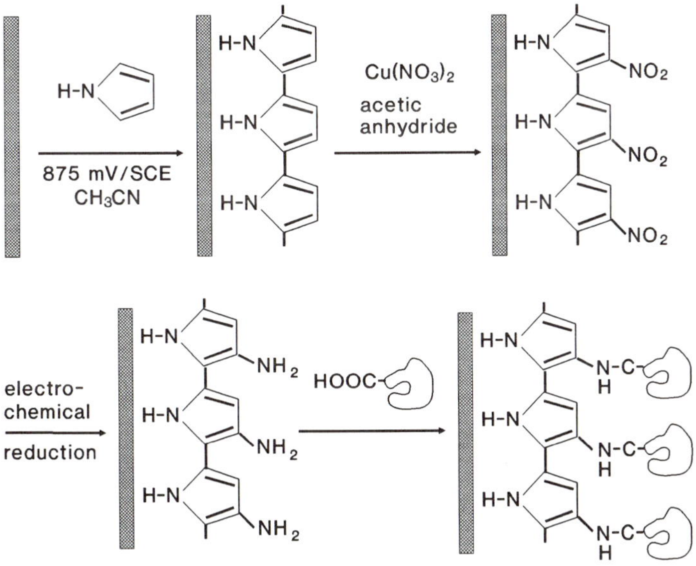

Figure 14-18. Covalent binding of enzymes to the conducting polymer polypyrrole. The reaction sequence shows the functionalization of the electrochemically deposited polymer film by nitration and subsequent electrochemical reduction.

reduction lead to amino groups on the electrode surface which have been used successfully for the covalent binding of enzymes [270]. In some cases, the functionalities thus provided at the surface of a transducer material cannot immediately be bound to those on the enzyme, either for chemical reasons or owing to the size of the enzyme and concomitant steric hindrance. In this case bifunctional spacer molecules with suitable groups can be reacted with the enzyme or the support and help to prepare the final covalent coupling of compatible functional groups. Most frequently, glutardialdehyde, diamines, and dicarboxylic acids are used. For example, carboxylic groups formed by electrochemical oxidation on graphite surfaces have been reacted with 1,6-diaminohexane, leaving free amino groups for the binding of glucose oxidase via oxidized oligosaccharide side-chains [271], or the amino groups at polypyrrole films were modified by reaction with glutardialdehyde to bind glucose oxidase via the lysine residues under the formation of Schiff bases [271].

Two of the main problems that arise with the modification of electrode surfaces with enzymes, redox-active compounds, mediators, or polymers are the chemical and physical characterization of the coatings and the reproducibility of the procedures. Although electrochemical techniques (cyclic voltammetry, differential-pulse voltammetry, chronoamperometry, rotating electrode techniques, impedance spectrometry) give preliminary informations about the nature of the electrode modifications attained, about possible electrode processes, and about effective diffusion coefficients of the analyte, quantitative data must be obtained from surface-sensitive analytical techniques such as X-ray photoelectron spectroscopy (XPS), Auger electron spectroscopy (AES), secondary ion mass spectrometry (SIMS), fast-atom bombardment mass spectrometry (FAB-MS), or Fourier transform infrared (FT-IR) spectrometry [251, 272]. Unfortunately, most of these techniques need an ultra-high vacuum, so that biological molecules can be destroyed.

In conclusion, immobilization of enzymes to solid supports has to take into consideration both the functionalities available at the biocatalyst and the surface. Optimum coupling conditions are different for most enzymes, so unfortunately general procedures cannot be given.

14.5 Coupling of Enzyme Reactions and Mass Transfer in Immobilized Layers

Biosensors using immunological active proteins or intact biological chemoreceptors have not yet reached the stage of routine application. However, these recognition elements offer unique sensing properties, and experimental research directed at their exploitation is being supported by theoretical considerations [273–275]. Among the recognition elements used in biosensors, enzymes clearly dominate. This is reflected by the number of measurable analytes, the analytical characteristics, and the extent of practical application. In addition to many empirical approaches, a large number of publications reinforcing the development of sensors with theoretical treatments have been published. This has led to general criteria for optimizing sensor design on the basis of the analysis by coupling enzyme reactions with mass-transport processes. The fundamental aspects of the kinetics of enzyme-catalyzed reactions have been outlined in Section 14.2. In the following section the specific features of enzyme-catalyzed reactions in heterogeneous systems and their influence on the sensor performance will be discussed.

14.5.1 Immobilization Effects in Biosensors

The initial rate of substrate conversion by an enzymatic reaction taking place in homogeneous solution increases linearly with increasing enzyme concentration. The reaction rate is influenced by substrate diffusion only at extremely large degrees of conversion. When the enzyme is immobilized, the measured reaction rate does not only depend on the substrate concentration and the kinetic constants K_M and v_{max}, and on diffusion, but also on so-called immobilization effects. These effects are due to the following alterations of the enzyme by immobilization [276].

1. Conformational changes of the enzyme caused by immobilization usually decrease the affinity to the substrate (increase in K_M). Further, partial inactivation of all or the complete inactivation of some of the enzyme molecules may occur (decrease in v_{max}). These two cases of a conformation-induced decrease of v_{max} may be distinguished by measuring the activity of the resolubilized enzyme.

2. Ionic, hydrophobic, or other interactions between the enzyme and the matrix (microenvironment effects) may also result in changed K_M and v_{max} values. These essentially reversible effects are mostly caused by variations in the dissociation equilibria of charged groups in the active center.

3. A nonuniform distribution of substrate and/or product between the enzyme matrix and the surrounding solution affects the measured kinetic constants.

4. In biosensors the biocatalyst and the signal transducer are in close contact with each other. The enzyme reaction takes place in a layer separated from the bulk solution. The substrate molecules reach the membrane system of the biosensor by convection and diffusion from the solution. The rate of this external transport process essentially depends on the degree of mixing. In the multilayer system in front of the sensor the substrate and product molecules are transferred by diffusion. Slow mass transfer to and within the enzyme matrix leads to differences between the concentrations of the reaction partners in the bulk solution and in the matrix.

Diffusion, partition, and enzyme reactions influence the sensor characteristics in a complex manner. The effect of enzyme immobilization on the reaction rate is described by the following terminology. *Apparent* or *effective* kinetics are observed when internal or external diffusion affects the overall rate. *Inherent* kinetics prevail when only partitioning (and not mass transfer) effects are present. *Intrinsic* kinetics describe the enzyme-catalyzed reaction when no partitioning effects or diffusion limitation are present.

However, steric or conformational constraints may cause differences in the intrinsic kinetics of the native enzyme in homogeneous solution.

The theory of the coupling of enzyme-catalyzed reactions with transport processes has been investigated for the following limiting cases [277]:

1. External diffusion limitation by mass transfer through layers in front of the enzyme membrane, eg, a semipermeable membrane or the boundary layer at the solution/biosensor membrane interface.

2. Internal limitation by diffusion within the enzyme layer or by the enzyme reaction.

A quantitative measure of the significance of solution boundary layer effects is the mass-transfer Biot number:

$$Bi = P_S/P_m = \frac{Dd}{\delta D_{eff}}, \tag{14-6}$$

where P_S denotes the permeability of the solution layer of thickness δ, P_m the membrane permeability for the diffusive transport of the solute within the membrane of thickness d, D the molecular diffusion coefficient, and D_{eff} the effective diffusion coefficient, both in solution. For membranes in contact with liquid phases the Biot number depends on the flow conditions and membrane properties. For a membrane-covered, rotating disk electrode, where the solution permeability can be varied over a wide range by changing the rotation rate, ω, the influence of P_S and P_m on the limiting current, i_D, is reflected by the following relation:

$$i_D = K_1 \left(\frac{1}{1 + P_S/P_m} \right) \omega^{1/2}, \tag{14-7}$$

where K_1 is a constant. The membrane permeability can be determined by extrapolating to infinite rotation rate ($\delta \rightarrow 0$) where mass transport is totally membrane-limited by using the typical reciprocal Levich plot:

$$\frac{1}{i_D} = K_2 \, \omega^{-1/2} + \frac{1}{P_m}, \tag{14-8}$$

K_2 being a constant.

If the membrane can be approximated as a homogeneous phase containing a linear distribution of solute, P_m is defined as

$$P_m = \frac{\alpha D_m}{d} = \frac{D_{eff}}{d}, \tag{14-9}$$

where D_m is the diffusion coefficient in the membrane and α the partition coefficient which represents the equilibrium ratio of solute concentration in the membrane to that in the solution. P_m can be determined by volumetric methods allowing the measurement of the diffusion coefficient within the membrane from the membrane diffusion resistance, $1/P_m$. Using regenerated cellulose of the Cuprophane type, the coefficients listed in Table 14-5 were determined for some analytically relevant substances.

In the operation of most biosensors the hydrodynamic conditions are adjusted in a way that mass transfer from the solution to the membrane system is fast compared with the internal mass transfer. Variations of the diffusion resistance of the semipermeable membrane can be used to optimize the sensor performance. A semipermeable membrane with a molecular cut-off of 1000–10000 daltons and a thickness of 10–20 μm only slightly influences the response time and sensitivity. In contrast, thicker membranes, eg, of polyurethane or charged material, significantly increase the measuring time but may also lead to an extension of the linear measuring range.

Table 14-5. Diffusion and partition coefficients for solutes in membranes [278].

Solute	$D/10^{-6}$ cm^2 s^{-1}	$D_{\rm m}/10^{-6}$ cm^2 s^{-1}	α
Oxygen	23.3	3.9–4.7	0.6
Glucose		1.5[a]	
Hydroquinone	11,6	2.2	0.69
Hexacyanoferrate(III)	6.3	0.5–0.7	0.88

(a) from [279].

For biosensors based on transducers which do not consume the cosubstrate or product (eg, potentiometric electrodes or optoelectronic detectors), the following relationship between the product concentration at the transducer surface, $P^{\rm d}$, and the substrate concentration in the measuring solution, S^0, has been derived [280]:

$$p^{\rm d} = S^0 \frac{D_{\rm S}}{D_{\rm p}} (1 - \text{sech} \sqrt{f_{\rm E}}) \quad \text{for } S^0 \ll K_{\rm M} . \tag{14-10}$$

with

$$f_{\rm E} = \frac{v_{\rm max} d^2}{K_{\rm M} D_{\rm S}} .$$

Provided that external diffusion is not limiting, $P^{\rm d}$ depends linearly on S^0 and the ratio of the substrate and product diffusion coefficients ($D_{\rm S}$ and $D_{\rm P}$), and nonlinearly on a square-root expression, the so-called Thiele modulus (the square of which is the enzyme loading factor, $f_{\rm E}$).

The latter parameter expresses the ratio of the rate of the enzymatic reaction, $v_{\rm max}/K_{\rm M}$, to that of diffusion, $D_{\rm S}/d^2$. It indicates whether the process in an enzyme layer is determined by enzyme kinetics or by substrate diffusion. At $f_{\rm E} < 25$ the process is *kinetically controlled*. In this case the substrate concentration does not become zero in any part of the enzyme layer, ie, the enzyme sensor signal is mainly a function of the "active" enzyme concentration. Therefore, effectors (activators, inhibiting factors, including H^+ and OH^-), the enzyme loading, that is, the amount of enzyme in front of the transducer, and the time-dependent enzyme inactivation directly affect the measuring signal. At $f_{\rm E} > 25$ *internal diffusion control* is reached. Any substrate molecule diffusing into the enzyme layer is converted therein; only part of the enzyme is acting catalytically.

Diffusion controlled sensors exhibit the following characteristics:

1. The sensitivity remains constant as long as an enzyme reserve is present.
2. The sensitivity does not depend on inhibitors and pH variations.
3. The temperature has only a minor influence since the activation energy of diffusion is lower than that of the enzyme reaction.

At high substrate concentrations ($S^0 \gg K_{\rm M}$) the enzyme reaction rate attains a limiting value, $v_{\rm max}$. Therefore, the enzyme sensor signal reaches a concentration-independent value corresponding to the product concentration at the transducer surface. Analogous relationships have been established for amperometric enzyme electrodes, where either the reaction product or a cosubstrate is converted at the electrode [277].

From the analysis of the coupling of enzyme reactions and mass transfer, the following conclusions may be drawn for the design of biosensors:

1. The substrate concentration at which deviations from the analytically usable linear measuring range occurs depends on the extent of diffusion limitation. According to the Michaelis-Menten equation, with kinetic control a linear dependence may only be expected for substrate concentration below K_M. With diffusion control the decrease in substrate concentration in the enzyme layer caused by slow substrate diffusion results in an extended linear range. It should be noted, however, that for two-substrate reactions deviations from linearity may also be produced by cosubstrate consumption.

2. At low substrate concentrations the sensitivity of kinetically controlled sensors increases linearly with increase in v_{max}. Consequently, the application of several identical enzyme layers over one another enhances the measuring signal. When the amount of enzyme becomes sufficiently high to provide complete substrate conversion, the transient to diffusion control is attained. Under these conditions a decrease in the diffusion resistance by decreasing the layer thickness results in an increased sensitivity. Nevertheless, a membrane-covered enzyme electrode is only 10–50% as sensitive as a bare electrode for an analogous electrode-active substance.

3. Owing to the excess of enzyme in the membrane, a diffusion-limited enzyme sensor has a higher functional stability than a kinetically controlled sensor. With the former, 2000–10 000 measurements per enzyme membrane can be performed whereas kinetically controlled sensors typically permit only 200–500 measurements.

4. With internal diffusion control the time to reach a steady-state current is >1.5 times the characteristic membrane diffusion time:

$$t_{\text{steady state}} > 1.5 \; d^2/D \; . \tag{14-11}$$

According to this equation, for an amperometric oxygen-sensing electrode covered by a membrane of 100 μm thickness the steady-state current will be attained 38 s after changing the oxygen concentration ($D = 3.9 \cdot 10^{-6}$ cm^2/s). Half of the steady-state value is reached at

$$t_{1/2} = 0.1 \; d^2/D \; , \tag{14-12}$$

ie, after a substantially shorter exposure of the sensor to the analyte. Since at about $t_{1/2}$ the current-time curve passes through its inflection point, derivative measurements with peak evaluation require only a fraction of the time needed for steady-state signal evaluation.

To summarize, it may be concluded that optimum sensitivity and response time can be achieved by applying a high enzyme activity in thin membranes.

14.5.2   Characterization of Immobilized Enzymes in Biosensors

14.5.2.1   Recovery of Enzyme Activity

To establish the effectiveness of different immobilization methods, it is useful to determine the proportion of enzyme remaining active after immobilization. Since the determination of the activity of the intact enzyme membrane only gives an apparent value, the membrane should be resolubilized before measurement.

The remaining activity of oxidase membranes can be determined by measuring the initial rate of H_2O_2 formation or oxygen consumption amperometrically using a defined portion of a resolubilized membrane. Gelatin membranes may be easily resolubilized by gently shaking in buffer at 37–40 °C. Thus, 70–90% of the initial activities have been found after resolubilization of gelatin membranes embodying glucose oxidase [281], uricase [282], and lactate monoxygenase [283]. Only 22% of the initial activity was found after entrapment of glucose oxidase (GOD) in photopolymerized polyacrylamide and resolubilization. With GOD cross-linked to silk together with bovine serum albumin by glutaraldehyde, only 3% of the enzyme was found to retain its activity. The high residual activity in gelatin indicates the protective effect of the "native" environment created by the matrix. Even the rather unstable cytochrome P-450 system of liver microsomes maintains 60% of its activity when entrapped in gelatin [284]. In contrast, the activity drop of GOD in polyacrylamide indicates inactivation of the enzyme by radical-forming reagents involved in the photopolymerization. The low residual activity determined after glutaraldehyde cross-linking is probably caused by incomplete resolubilization. Malpiece et al. [285] obtained a remaining GOD activity of 75% in preparations immobilized by the same method but without silk as carrier.

14.5.2.2   Effectiveness Factor

The initial rate of product formation or substrate consumption by the intact membrane or the complete sensor reflects the enzyme activity acting in the measuring process. Comparison with the residual activity permits the excess of enzyme to be estimated. With a GOD electrode the initial rate of H_2O_2 accumulation has been determined in a double measuring cell containing air-saturated buffer [286]. After glucose injection an enzyme-free electrode polarized to +600 mV indicates the rate of H_2O_2 accumulation; the other electrode is covered with a GOD membrane but is not polarized so as to permit the total H_2O_2 formed to diffuse into the measuring cell.

With a gelatin membrane entrapped between two dialysis membranes and containing 46 U/cm^2 of enzyme, the H_2O_2 formation corresponds to only 110 mU/cm^2, ie, less than 1% of the initial enzyme activity (Figure 14-19). This indicates a large excess of enzyme in the membrane. Consequently, the membrane is diffusion controlled. The low apparent activity may be attributed mainly to the diffusion resistance of the dialysis membrane for glucose. On the other hand, the measured activity of the membrane containing 46 U/cm^2 is already about 70% of the activity used for immobilization. This value approaches that expected for pure kinetic control of the process [287]. The apparent activity of urease immobilized in a cellulose triacetate membrane was found to be 66% of the initially applied enzyme activity [288].

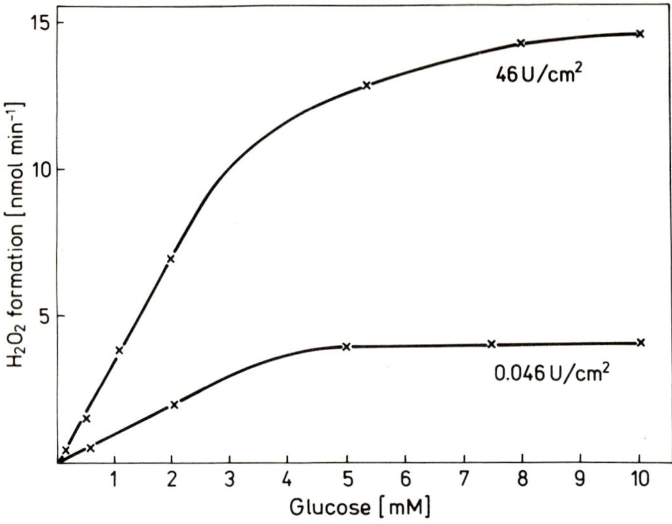

Figure 14-19. Dependence of the hydrogen peroxide accumulation rate on glucose concentration as determined in a measuring cell containing a GOD electrode and an enzyme-free electrode. A membrane area of 0.13 mm^2 was exposed to the measuring solution. The gelatin-immobilized enzyme (46 U/cm^2, ie, 6 U per electrode, or 46 mU/cm^2, ie, 6 mU per electrode) was sandwiched between two dialysis membranes. Reproduced from [281] with permission from Academic Press.

Table 14-6. Apparent enzyme activities and K_M values of adsorbed layers and enzyme membranes.

Enzyme	Immobilization	Apparent enzyme activity in mU cm^{-2}	Apparent K_M value soluble immobilized in mM		References
GOD	Gelatin entrapment	110	3.8	7.5	[281]
GOD	Collagen, covalent	60–80		3.0	[289]
GOD	Cellulose acetate, covalent	340			[290]
GOD, BQ-modified	Cellulose acetate	>1000			[291]
GOD	PVA entrapment	160–700			[292]
GOD	Spectral carbon, adsorbed	150–200			[293]
GOD	Carbon, covalent	50–170		3.1–19.1	[294]
β-Galactosidase	Gelatin entrapment	1000			[295]
Urease	Cellulose triacetate, entrapment	3–30	2.3	2.4	[288]
Cholesterol oxidase	Collagen, crosslinked	3			[296]
Creatinine amidohydrolase	Cellulose acetate, covalent	1140	35	278	[297]
Creatine amidinohydrolase	Cellulose acetate, covalent	110	13.5	64.9	[297]
Sarcosine oxidase	Cellulose acetate, covalent	13	6.7	2.4	[297]

In Table 14-6 the apparent activities of enzymes entrapped in or covalently fixed to membranes are compared with those of enzymes directly adsorbed or fixed to electrode surfaces. It may be concluded that different immobilization procedures lead to approximately identical apparent activities. The advantage of the direct fixation to the transducer surface is the low diffusion resistance of the monomolelcular enzyme layer. On the other hand, enzyme membranes are more stable because of their inherent enzyme excess.

14.5.2.3 Enzyme Loading Test and Stability of the Immobilized Enzyme

The variation of the enzyme loading is a means of determining the enzyme amount minimally required for maximum sensitivity. This test further reveals the magnitude of the enzyme reserve of diffusion-controlled sensors.

Figure 14-20 shows the results of a loading test on GOD entrapped in a gelatin layer of 30 μm thickness between two dialysis membranes each 15 μm thick. The stationary currents for 0.14 mM glucose (lower part of the linear measuring range) and for 5 mM glucose (saturation) increase linearly with increasing enzyme loading from 46 mU/cm^2 to 1 U/cm^2. At higher GOD loadings a saturation value is attained. The following values have been used to calculate the enzyme loading factor:

- thickness, $d = 30$ μm;
- Michaelis constant for glucose, $K_M = 10$ mM;
- diffusion coefficient of glucose, $D_S = 1.5 \cdot 10^{-6}$ cm^2/s.

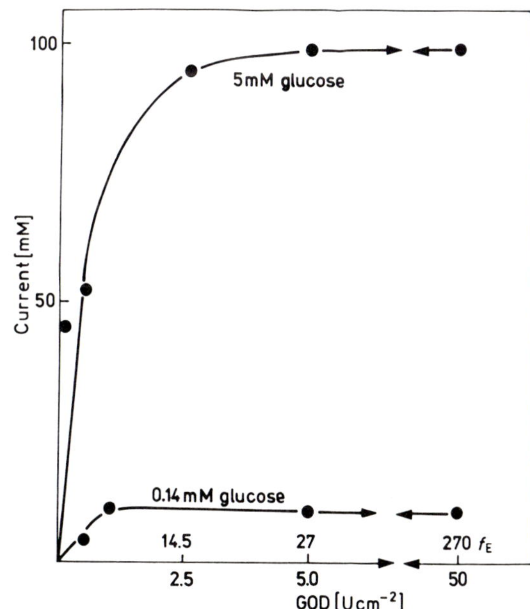

Figure 14-20.
Enzyme loading test of gelatin-entrapped GOD as performed with two different glucose concentrations at pH 7 and 25 °C. Reproduced from [281] with permission from Academic Press.

As is evident from Figure 14-20, the transition from the linear region to saturation occurs at f_E between 7 and 20. This agrees with the theoretically predicted value and indicates that above 1 U/cm^2 the function of the GOD electrode is controlled by internal diffusion. Owing to differences in the K_M values and the layer thickness, the transition from kinetic to diffusion control of different enzyme electrodes takes place at rather different enzyme activities. This is shown in Table 14-7 for gelatin-entrapped enzymes.

The enzyme loading to a major extent determines the stability of a biosensor. An enzyme reserve is built up by employing more enzyme activity in front of the electrochemical probe than is minimally required to achieve diffusion control. As long as this reserve lasts, the sensitivity will remain essentially constant. This is only significant, however, for sensors for substrate determinations. If effectors of the biocatalytic sensing reactions are to be measured, kinetic control is desired, which permits the enzyme loading to be varied only in a relatively narrow range.

The factors that are generally important to know are the operational stability (also termed the useful lifetime) and the storage stability. Comparison of literature data on these properties is sometimes difficult because the experimental conditions employed to establish the biosensor stability vary widely. It appears most practical to use the sensor intermittently for analysis and

Table 14-7. Transition from kinetic to diffusion control of some enzyme electrodes.

Enzyme	K_M in mM	Loading at transient from kinetic to diffusion control in U cm^{-2}	References
GOD	10	1	[281]
Uricase	0.017	0.17	[282]
Urease	2.4	16	[298]
Lactate monooxygenase	7.2	1	[299]
β-Galactosidase		2	[295]
Lactate dehydrogenase	0.14	0.1	[283]

Table 14-8. Operational stability of enzyme electrodes.

Enzyme	Activitiy used for immobilization in U cm$^{-2(a)}$	Operational stability number of measurements	time in d	References
GOD	50 (PUR)	1000–3000	10	[300]
	50 (gel)	3000–10000	30	[301]
Lactate oxidase	40 (PUR)	3000	14	[302]
Urease	15 (PUR)		5	[303]
	20 (PVA)		22	[303]
	12 (CTA)		28	[288]
Lactate monooxygenase	10 (gel)	600	55	[299]
Lactate dehydrogenase	50 (gel)	$\geq$ 600	$\geq$ 55	[304]
Pyruvate kinase	50 (gel)		14	[305]
Cytochrome b$_2$	50 (gel, PVA)		14	[306]

(a) Abbreviations of carrier materials: PUR = polyurethane, gel = gelatine, PVA = poly(vinyl alcohol); CTA = cellulose triacetate

to state the number of assays carried out over the time of use. An exception are sensors intended for *in situ* application, such as implantable sensors, where the time period for continuous, uninterrupted use should be given. The working stabilities of various enzyme sensors are surveyed in Table 14-8.

14.5.2.4   Concentration Dependence of the Sensor Signal

The linear measuring range of biosensors extends over 2–5 concentration decades. The lower detection limit of simple amperometric enzyme electrodes is about 100 nM, whereas potentiometric sensors may only be applied down to 100 µM. This shows that the sensitivity is affected not only by the enzyme reaction but also by the transducer.

The increase in the measuring value of the amperometric glucose electrode with increasing substrate concentration reflects the course of a Michaelis-Menten curve and reaches a concentration-independent saturation corresponding to the maximum rate, v_{max}. The sensitivity of the GOD electrode depends on the enzyme loading (Figure 14-21) [281]. The substrate concentration giving rise to the half-maximum current in air-saturated solution is between 1.4 and

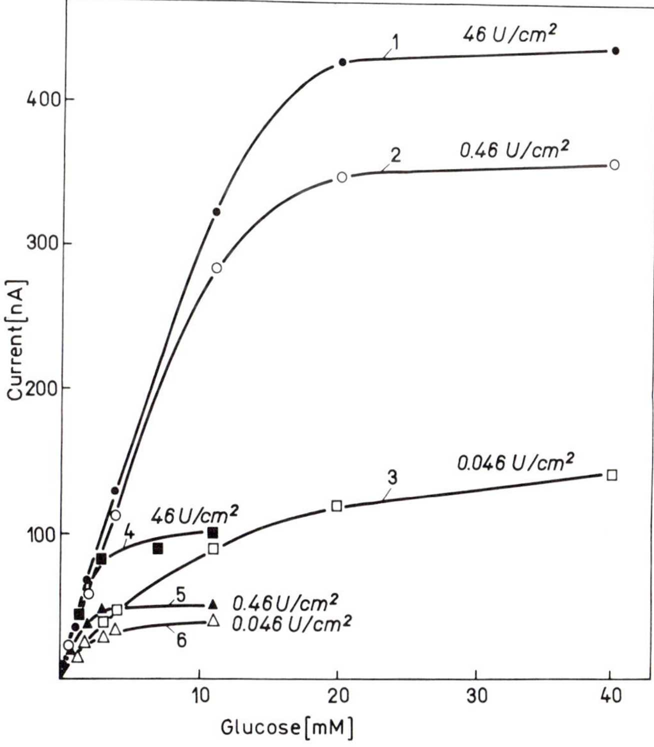

Figure 14-21. Dependence of the stationary current of a GOD electrode on glucose concentration at different enzyme loadings. Curves 1–3: oxygen-saturated solution; curves 4–6: air-saturated solution. Reproduced from [281] with permission from Academic Press.

1.8 mM glucose. The linear range extends to 2 mM glucose in the measuring cell. In this region, saturation of the measuring solution with oxygen increases the measuring signal only by 10%. At low glucose concentrations the cosubstrate concentration (ca. 200 µM at air saturation) only slightly influences the enzyme reaction. In contrast, in the saturation region above 2 mM glucose, the current rises by a factor of 4.5. At the same time, the linear range is extended by oxygen saturation.

At low enzyme loadings the plot of the reciprocal of current versus the reciprocal of glucose concentration gives a straight line and thus follows the Michaelis-Menten equation (Figure 14-22); from this curve an apparent K_M (glucose) of 7.5 mM may be calculated. The K_M of soluble GOD at air saturation has been determined to be 3.8 mM. The higher K_M of the immobilized enzyme indicates a superposition of diffusional and kinetic limitation. An increase in the apparent K_M values has also been described for other enzymes immobilized by different methods (see Table 14-6). This increase of the apparent K_M increases the linear measuring range.

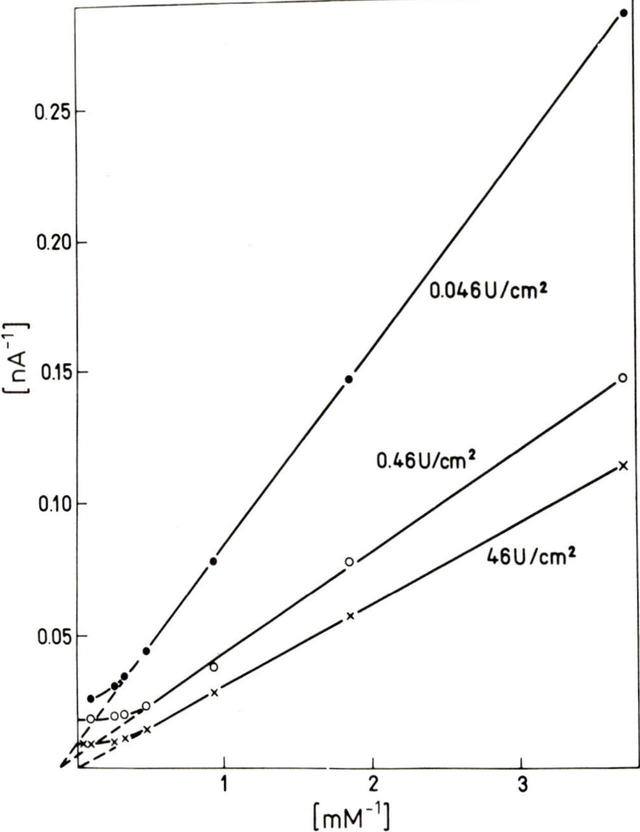

Figure 14-22. Electrochemical Lineweaver-Burk plot of GOD electrodes with different enzyme loadings, determined in air-saturated solution. Reproduced from [281] with permission from Academic Press.

14.5.2.5 Measuring Time

At present the maximum sample throughput of commercial enzyme electrode-based analyzers is about 100 per hour. The response times of GOD sensors obtained with different experimental setups are shown in Figure 14-23. An electrode composed of a hydrogen peroxide-sensing probe and GOD immobilized in a 100 µm thick polyacrylamide membrane sandwiched between two dialysis membranes reaches 95% of the steady-state current about 2 min after sample injection [307]. A comparable response time is found with a sensor using the enzyme entrapped in a gelatin layer of 20 µm thickness. However, the measuring time can be diminished to 60 s per sample by employing the kinetic measuring principle [358]. Application of an analogous glucose electrode containing the enzyme in a polyurethane membrane with a characteristic diffusion time of 24 s and an enzyme loading factor of more than 100 (internal diffusion control) results in a response time for the steady-state current of 10–15 s, both in a stirred measuring cell and in a flow injection analysis (FIA) manifold. Therefore, a sample throughput of 80 h^{-1} can be realized in the steady-state mode. Reproducible, peak-shaped signals are also obtained when the volume injected is not sufficient to reach the steady state. In this way the measuring time at only 1% carry-over is diminished to 12 s by reducing the sample volume injected to 1.5 µL [309]. Thus, a sample throughput as high as 300 h^{-1} is possible.

The FIA system has been extended to the measurement of lactate [310]. A throughput of 200 lactate samples per hour can be analyzed with good precision and negligible carry-over. Such a high sample frequency can be achieved even with double-membrane type sensors. Therefore, the application of biosensors of the second or third generation will allow a further speeding up of the measuring process. In fact, amperometric microelectrodes for glucose and enzyme field-effect transistors for urea having steady-state response times of only 3–4 s are

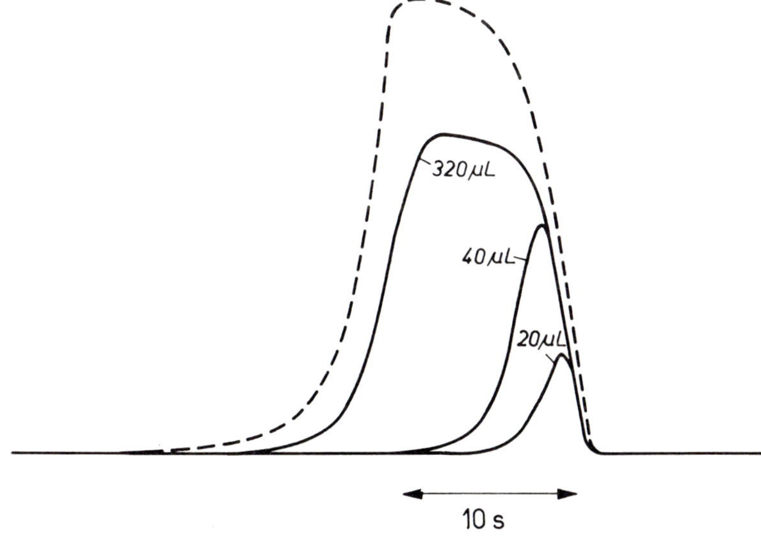

Figure 14-23. Response time of GOD electrodes applied in a stirred measuring cell (dashed line) and in a flow-injection analysis manifold (solid line) with varying injection volume.

under development [311, 312]. Combination of these biosensors with optimized FIA devices should permit the analysis of 700 samples per hour, a level recently reported for a flow-stream analyzer using dissolved enzymes [313].

14.5.2.6 pH Dependence

A large excess of enzyme in the membrane keeps the effect of *p*H variations on the measuring process small. Therefore, the *p*H profiles in the linear measuring range and with diffusion control should be substantially less sharp than those of the respective enzyme in solution [277]. The results obtained with a GOD-gelatin membrane [281] agree with this assumption (Figure 14-24). With 0.14 mM glucose the curve is almost as flat as that observed on injection of H_2O_2. On the other hand, with 10 mM glucose a pronounced maximum is found. At this saturating concentration the signal depends on the enzyme activity and therefore distinctly on *p*H. The *p*H optimum of immobilized GOD is about 0.9 *p*H more alkaline than that of the dissolved enzyme. Obviously the formation of gluconic acid within the enzyme membrane causes a local *p*H decrease, thus shifting the optimum to higher *p*H in the solution. Analogous *p*H dependences have been observed for other enzyme sensors [277].

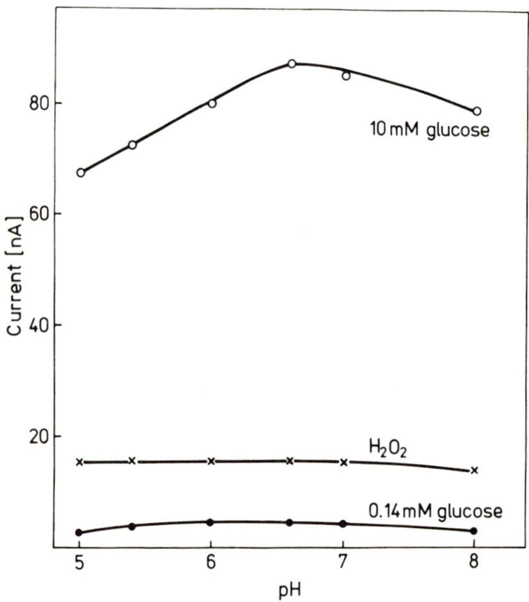

Figure 14-24.
pH dependence of the response of a GOD electrode to high and low glucose concentrations, and to hydrogen peroxide.

14.5.2.7 Temperature Dependence

The rate of enzyme reactions increases with increase in temperature up to a certain optimum, above which the effect of thermal inactivation dominates over that of the increase in the collision frequency. Enzyme stabilization by immobilization is frequently reflected by an increase in the optimum temperature for substrate conversion. If kinetic and diffusional control are superimposed, the higher activation energy results in a substantial acceleration of the enzyme reaction with increase in temperature. Hence, the slower enhancement of the diffusion rate makes mass transfer the limiting factor. Therefore, the activation energy determined at lower temperatures is ascribed to the enzyme reaction and that at higher temperatures to diffusion. In addition, the temperature profile is affected by temperature-dependent conformational changes of the enzyme and decreasing solubility of the cosubstrate.

The glucose sensor with the GOD-gelatin membrane exhibits a temperature optimum at about 40 °C [281]. Below the optimum the Arrhenius plot (Figure 14-25) gives parallel straight lines for different glucose concentrations and enzyme loadings. Probably the difference between the activation energy of H_2O_2 diffusion, 33.5 kJ/mol, and that of GOD-catalyzed glucose oxidation, 25.5 kJ/mol, is too small to give rise to two separate linear regions. This is why purely diffusion-controlled GOD electrodes are not significantly different from kinetically controlled electrodes with regard to activation energy.

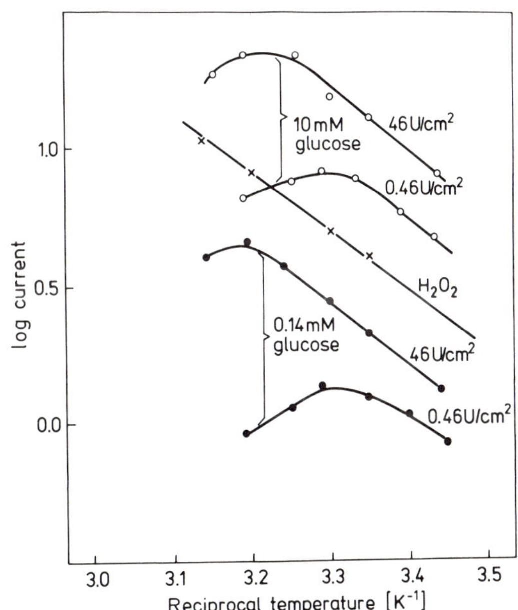

Figure 14-25.
Arrhenius plot of the temperature dependence of the response of a GOD electrode with different enzyme loadings to different glucose concentrations and to hydrogen peroxide.

14.6 Amplification and Filtering of Chemical Signals

14.6.1 Types of Coupled Enzyme Reactions Used in Biosensors

Since not all enzyme-catalyzed reactions involve transducer-active compounds such as H^+, oxygen, or hydrogen peroxide, only a limited number of substances can be determined by using monoenzyme sensors. Owing to their high susceptibility to nonspecific effects, transducers indicating general reaction effects, eg, thermistors or piezoelectric detectors, are also not always practicable. In such cases coupled enzyme reactions for analyte conversion provide a favorable alternative (Figure 14-26). In *enzyme sequences*, for example, the primary product of the analyte conversion is further converted enzymatically with the formation of a measurable secondary product or another reaction effect. On this basis, whole families of sensors have been developed, which combine glucose-, lactate-, or alcohol-producing primary enzyme reactions with their conversion by the respective oxidases. This type of sequential coupling resembles that occurring in metabolism, where energy-rich substrates are degraded in a stepwise manner (such as in glycolysis, photosynthesis, or the citrate cycle), and enzyme cascades are responsible for signal amplification in many receptor systems. In the latter, the chemical modification of enzymes leads directly to a cascade-like increase in the reaction rate (see 14.2.). So far, this powerful principle has not been used in biosensors; however, the so-called *apoenzyme electrodes*, in which the recombination of an apoenzyme with its prosthetic group results in a dramatic increase of the reaction rate, may be regarded as an analog.

Another type of sequential coupling is provided by *cycling reactions* (Figure 14-26b). The product of the primary enzyme reaction is regenerated to the substrate of this reaction, ie, the analyte, in a second, enzyme-catalyzed reaction. These cycles are based on the dependence of the two enzymes on different cofactors; thus, the required free enthalpy exists for both reactions. The analyte molecule may be considered as a catalyst of the reaction between the two cofactors. This renders the rate of cofactor conversion and enthalpy production much higher than it is in a single enzyme reaction. Therefore, these cycling reactions lead to a substantial increase in sensitivity.

The principle of signal amplification by using the free enthalpy of energy-rich compounds forms the basis of signal processing in nervous systems. Although the signal transmission makes use of mechanisms different from that mentioned above, recognition and signal amplification are also based on enzyme reactions. On the other hand, substrate cycles, such as the glucose-6-phosphate cycle in glycolysis, are important for rapid energy supply.

Coupled enzyme reactions can also be used to filter chemical signals by eliminating disturbances of the enzyme or transducer reaction caused by constituents of the sample. Com-

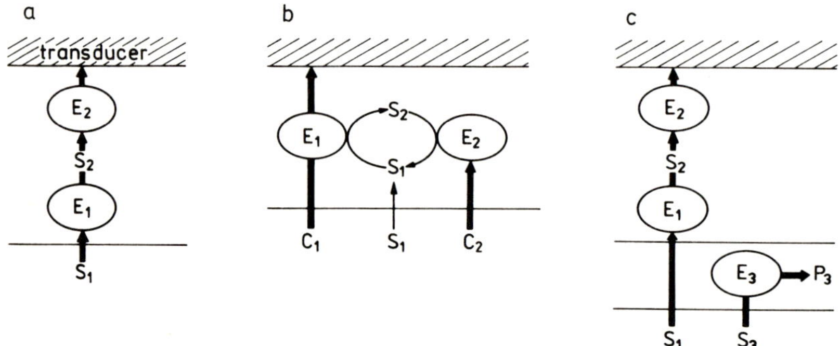

Figure 14-26. Basic principles of the coupling of enzyme reactions in biosensors. a) Sequential coupling; b) amplification by recycling; c) elimination of interfering compounds. E represent enzymes. S Substrates (with S_1 being the analyte and S_3 a potentially interfering substrate), C cosubstrates, and P_3 the (inert) product of the interferent conversion.

pounds that interfere with the signal transduction, eg, ascorbic acid with electrochemical hydrogen peroxide indication, can be transformed into inert products by reaction with an (eliminator) enzyme. Since the conversions of analyte and interferent proceed in parallel, both the eliminator and the indicator enzyme may be co-immobilized in one layer. On the other hand, constituents of the sample which are at the same time intermediate products of coupled enzyme reactions, eg, endogenous glucose in disaccharide measurement, can be eliminated before they reach the indicator enzyme layer (Figure 14-26c). For this purpose, several enzyme layers have to be used to ensure that the intermediate product formed by the indicator enzyme is not converted to an inactive substance. Therefore, the anti-interference layer has to be arranged at the solution side of the sensor.

14.6.2 Apoenzyme Sensors for the Measurement of Prosthetic Groups

Generally, the binding of a prosthetic group to the apoenzyme is characterized by high affinity and extraordinary specificity. In apoenzyme sensors, the prosthetic group to be detected is recognized and bound by the apoenzyme, thus creating active enzyme molecules which then serve to amplify the measuring signal. At substrate saturation and with an excess of apoenzyme, the reaction rate is proportional to the concentration of the holoenzyme, ie, to the amount of the prosthetic group to be analyzed.

The assay is particularly sensitive when the apoenzyme is used as a soluble reagent. FAD concentrations as low as 10^{-12} M have been measured with dissolved apo-GOD and electrochemical indication of the H_2O_2 formed in the GOD reaction [314]. Vitamins B_6 (pyridoxal phosphate, PLP) and B_1 have been assayed in a similar manner by using apo-tyrosine decarboxylase [315] and apo-pyruvate decarboxylase [316]. Under optimum conditions, PLP concentrations as low as 1 nM are detectable with an amplification of 10^5. Marker enzymes, such as alkaline phosphatase, can be determined by using an inactive derivative of a prosthetic group, such as phosphoric acid esters of PQQ, and the respective apoenzyme, eg, apo-(PQQ) glucose dehydrogenase. The reaction of the marker enzyme forms the active prosthetic group which is subsequently bound to the apoenzyme, thus leading to a cascade-like substrate conversion. This principle may also be used in DNA hybridization tests [317]. An apoenzyme electrode for Cu^{2+} has been developed by combining immobilized apo-tyrosinase with an oxygen probe [318]. The detection limit of the sensor is 50 ppm. However, the reusability of such an apoenzyme membrane appears questionable because the enzyme activity accumulates during the sensor operation whereas the measuring principle requires kinetic control of substrate conversion to gain a linear dependence of the sensitivity on the activity of the holoenzyme.

Jasaitis et al. [319] proposed a carbon electrode with covalently bound alkaline phosphatase for the determination of Zn^{2+}. Apo-phosphatase was generated by treating the electrode with EDTA. As shown by the formation of electrode-active hydroquinone from hydroxyphenyl phosphate, addition of a zinc ion-containing sample restored part of the enzymatic activity of alkaline phosphatase within 30 s. After each measurement the electrode was regenerated by treatment with EDTA. The detection limit was as low as 0.8 µM Zn^{2+}.

14.6.3   Amplification by Analyte Recycling

Analyte recycling in biosensors works analogously to the cofactor recycling known from enzymatic analysis with dissolved enzymes [320]. In a bienzyme sensor (see Figure 14-26b) the substrate to be determined is converted in the reaction of enzyme 1 to a product which is in turn the substrate of enzyme 2. The latter catalyzes the regeneration of the substrate to be determined, which thus becomes available for enzyme 1 again, and so forth. One of the coreactants is detectable directly or via an additional reaction. Assuming that enzyme 1 is present in sufficiently high concentration to assure diffusion control, an amplification is achieved by switching on enzyme 2. This can be easily accomplished by addition of its cosubstrate, C_2. In such systems the analyte acts as a catalyst, being shuttled between both enzymes in the overall reaction of both cosubstrates. In this way, significantly more cosubstrate will be converted than the amount of analyte present in the enzyme membrane. Hence the change in the parameter indicated at the transducer will greatly exceed that obtained with one-way analyte conversion. The ratio of the sensitivity in the linear measuring range of the amplified and the unamplified regime is termed the *amplification factor.* Specifically, in enzyme electrodes, where the sensitivity limit is determined by diffusion, the employment of cyclic enzyme reactions gives rise to a sensitivity enhancement by overcoming just this limit. On the other hand, the upper limit of linearity is decreased. The enzyme excess present in the membrane is included in the substrate conversion. Therefore, the amplification factor decreases with progredient enzyme inactivation during operation of the sensor. As will be seen below, the application of recycling systems to real samples is restricted by their susceptibility to both substrates of the enzymatic cycle. Therefore, in most cases the alternative substrate would have to be removed. A preferred area of application appears to be the measurement of enzyme activity, eg, in enzyme immunoassay. For example, using a marker enzyme which generates the substance to be cycled with a turnover number of 1000 s^{-1}, in a volume of 1 µL as little as 10000 enzyme molecules could be detected after a 1-min incubation.

Table 14-9 shows the enzyme sensors developed so far using biocatalytic analyte recycling for signal amplification. The first of such sensors have been studied for measurement of pyridine nucleotides, glucose, and lactate [321, 324]. The $NAD^+/NADH$ sensor involves the oxidation-reduction of the analyte by horseradish peroxidase and glucose dehydrogenase:

$$NAD^+ + \text{glucose} \rightarrow \text{gluconolactone} + NADH + H^+$$

$$NADH + O_2 + H^+ \rightarrow NAD^+ + H_2O_2. \tag{14-13}$$

The consumption of oxygen in the oxidation reaction of horseradish peroxidase is followed amperometrically. The amplification by the recycling of the pyridine nucleotide permitted as little as 1.2 µM NAD^+ to be determined, with an amplification factor of 40. In a similar system, also designed for the measurement of NADH, NADH oxidase has been used together with alcohol dehydrogenase [333]. Winquist et al. [334] have demonstrated the possibility of coupling a dehydrogenase with H_2 hydrogenase in conjunction with a hydrogen-sensitive metal oxide capacitor.

The fact that the glucose dehydrogenase reaction can be driven backward by using a more acidic buffer and an excess of NADH has been utilized to recycle gluconolactone formed in the GOD-catalyzed reaction [321]. The signal was amplified eightfold. This bienzyme sensor

Table 14-9. Substrate recycling in biosensors.

Analyte	Enzymes	Form of application	Transducer	Amplification factor	References
Glucose	GOD + glucose dehydrogenase	Membrane	O_2 electrode	10	[321]
NADH/NAD$^+$	HRP + glucose dehydrogenase	Membrane	O_2 electrode	60	[321]
Lactate/pyruvate	Cytochrome b_2 + lactate dehydrogenase	Membrane	Pt electrode (+0.25 V)	10	[321]
Lactate	Lactate oxidase + lactate dehydrogenase	Reactor	Thermistor	1000	[322]
		Membrane	O_2 electrode	4100	[323]
				250	[324]
Glutamate	Glutamate dehydrogenase + alanine aminotransferase	Membrane	Modified carbon electrode	15	[325]
			O_2 electrode	60	[325]
	glutamate oxidase + glutamate dehydrogenase	Membrane	O_2 electrode	20	[326]
ADP/ATP	Pyruvate kinase + hexokinase	Membrane with LDH + LMO*	O_2 electrode	220	[327]
Ethanol	Alcohol oxidase + alcohol dehydrogenase	Membrane	O_2 electrode		[328]
Benzoquinone/ hydroquinone	Cytochrome b_2 + laccase	Membrane	O_2 electrode	500	[329]
ATP	Pyruvate kinase + hexokinase + lactate dehydrogenase + lactate oxidase	2 reactors	Thermistor	1700	[330]
Malate/Oxaloacetate	Lactate monooxygenase + malate dehydrogenase	Membrane	O_2 electrode	3	[331]
NADH	Lactate dehydrogenase	Membrane with LMO	O_2 electrode	170	[332]

* Abbreviations: LMO = lactate monooxygenase, LDH = lactate dehydrogenase

can serve as a model for following different interactions between the coupled enzymes (Figure 14-27):

i. If no cofactor is present, the total glucose concentration is indicated only via the GOD reaction.

ii. When the reduced cofactor (NADH) is added to the glucose-containing measuring solution, GOD and glucose dehydrogenase cycle the glucose molecules and amplify the signal.

iii. When the oxidized cofactor (NAD$^+$) is added to the glucose-containing measuring solution, GOD and glucose dehydrogenase compete for the common substrate glucose

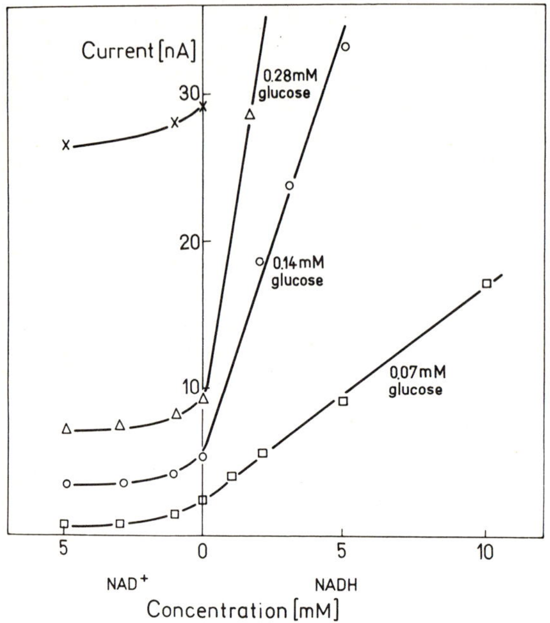

Figure 14-27.
Dependence of the current response of a GOD-glucose dehydrogenase electrode on glucose concentration at different concentrations of NAD^+ and NADH.

because in this case the forward reaction of glucose dehydrogenase is favored [335]. The more oxidized cofactor is present, the stronger glucose dehydrogenase competes with glucose oxidase for glucose and the less H_2O_2 is produced. Increasing concentrations of NAD^+ result in a decreased oxygen consumption and H_2O_2-formation and therefore a lower sensitivity and an extended linear measuring range for glucose.

In a sensor for lactate a bienzyme system composed of cytochrome b_2 for lactate oxidation to pyruvate, and lactate dehydrogenase for conversion of pyruvate back to lactate has been used [321]. Hexacyanoferrate(III) served as electron acceptor for cytochrome b_2. The reduced mediator was reoxidized at the electrode, thus giving a measuring signal depending on the analyte concentration. Attempts to determine both substrates of the recycling system have shown that, at tenfold amplification for lactate, the sensitivities for lactate and pyruvate are almost identical. The same recycling scheme has also been used in connection with Fe-EDTA as electron mediator in place of hexacyanoferrate(III) [336].

Further studies of lactate sensors [323, 324] revealed the superiority of lactate oxidase over cytochrome b_2 in giving much higher signal amplifaction. This system has been studied in more detail. Mizutani et al. [324] obtained an amplification factor of up to 250, thus obtaining a lower detection limit of 5 nM lactate. The enzymes were immobilized either by entrapment in photo-cross-linked poly(vinyl alcohol) bearing stilbazolium groups (PVA-Sa) or by chemical attachment to cellulose triacetate membranes. The PVA-Sa membranes provided substantially higher amplification, obviously because the immobilized enzyme activity yielded by entrapment was much higher than that by chemical binding. The authors mathematically modeled the sensor behavior and concluded that very high activities of both enzymes would be necessary to obtain maximum amplification. This was confirmed by investigation of the influence of enzyme loading on the amplification factor with the lactate dehydrogenase/lactate oxidase couple immobilized in gelatin [323]. According to theoretical considerations [337], the amplification factor G is given by the following equation:

$$G = k_1 \cdot k_2 \cdot L^2 / 2 (k_1 + k_2) \cdot D \qquad (14\text{-}14)$$

where k_1 and k_2 are kinetic parameters, D the diffusion coefficient of the analyte, and L the membrane thickness. By using Equation (14-14) the values given in Table 14-10 for the characteristic diffusion time, L^2/D, and the amplification factor were obtained. The mean value of the characteristic diffusion time of 90 s is reasonable in comparision with the characteristic diffusion time of 30 s for lactate in a 20 µm thick polyurethane layer [338]. The maximum amplification factor of 4100 permits lactate concentrations as low as 1 nM to be determined with reasonable precision. As employed in an enzyme thermistor, the recycling system has been used for the determination of minute amounts of pyruvate formed from phosphoenolpyruvate by alkaline phosphatase in an immunoassay procedure [339].

Table 14-10. Amplification factor and characteristic diffusion time in the lactate recycling system as a function of the enzyme loading.

V_{max} in U cm^{-2}	G	L^2/D in s
0.01	2	60
0.1	50	150
1.0	100	30
10.0	4100	123

Investigation of a recycling system for ATP and ADP using hexokinase (HK) and pyruvate kinase (PK) [327] has shown that the sensitivities for both substrates of a cycle are not necessarily the same. For ADP an amplification factor of 220 and a detection limit of 0.25 µmol/L were found; in contrast, ATP concentrations as low as 0.1 µmol/L could still be assayed. In this sensor a recycling system has been coupled with an enzyme sequence (lactate monooxygenase and lactate dehydrogenase) by co-immobilizing all the required enzymes in one membrane. Thus, it was demonstrated that enzymatic amplification sensors do not necessarily require recycling systems that produce electrode-active species. The amplification scheme for ATP/ADP can be combined with that for lactate/pyruvate to form a double amplification system (Figure 14-28), thus further decreasing the detection limit for the purine nucleotides [327, 330].

An enzyme electrode based on co-immobilized cytochrome b_2 and laccase [329] permits some mechanistic aspects of substrate recycling in enzyme electrodes to be studied in more detail. The advantage of this system is that the cosubstrate, oxygen, in addition to the analytes, hydroquinone and benzoquinone, are electrochemically active. Depending on the electrode potential, either hydroquinone is regenerated at the electrode from the benzoquinone produced, or the remaining hydroquinone that permeates through the laccase layer is indicated at the electrode. Therefore, both the effect of substrate recycling and the breakthrough of hydroquinone can be investigated. In the presence of the cytochrome b_2 substrate, lactate, benzoquinone is enzymatically reduced to hydroquinone. At the *p*H optimum of cytochrome b_2 (*p*H 6.5) and lactate saturation a maximum amplification for hydroquinone of 500 has

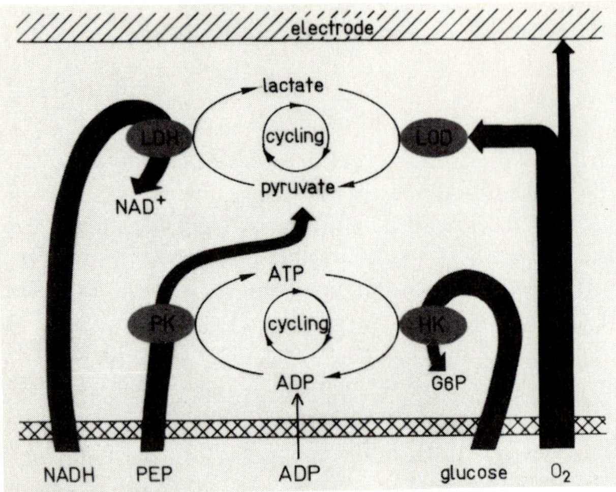

Figure 14-28. Schematic view of an enzyme electrode using a double recycling system for the determination of ATP or ADP. PK = pyruvate kinase; HK = hexokinase; LOD = lactate oxidase; LDH = lactate dehydrogenase; PEP = phosphoenolpyruvate. Reproduced from [327] with permission from Marcel Dekker, Inc.

been obtained (Figure 14-29). The linear concentration dependence of the amplified signal levels off at about 2.4 µM hydroquinone. It reaches almost the saturation current found in the unamplified system in the absence of lactate. Therefore, even under effective hydroquinone recycling, only part of the oxygen inside the bienzyme membrane is consumed.

The concentration dependence of the oxidation current at +100 mV shows threshold characteristics. In the absence of lactate, addition of hydroquinone up to 1 mM does not lead to a typical current increase. Obviously the laccase converts its substrate completely to benzoquinone, which is not detectable at this potential. Above 1.2 mM hydroquinone the current increase reflects the breakthrough of unreacted substrate. In the presence of lactate part of the benzoquinone formed in the laccase-catalyzed reaction is recycled to hydroquinone. Therefore, the threshold is found at lower hydroquinone concentrations.

The recycling of glutamate between co-immobilized glutamate oxidase and glutamate dehydrogenase has been employed in a glutamate sensor based on an oxygen electrode [326]. In the presence of NAD(P)H and ammonium ions, α-ketoglutarate formed in the oxidase reaction is converted by glutamate dehydrogenase back to glutamate, which is again oxidized. In accordance with theory, the sensitivity of the electrode to glutamate increases with increasing thickness of the enzyme membrane, the amplification factor ranging from 5 (at 30 µm thickness) to 20 (at 90 µm thickness). The linear concentration dependence levels off at about 50 µM glutamate or α-ketoglutarate, whereas the limit is 1.0 mM for the simple, unamplified glutamate oxidase sensor. The current decrease at these concentrations represents about 65 and 90% oxygen consumption for the amplified and the unamplified mode, respectively.

Alcohol oxidase has been co-immobilized with alcohol dehydrogenase in order to increase the sensitivity of alcohol determination [328]. In the presence of oxygen and NADH, ethanol is recycled between the two enzymes. A particular advantage of this system, which overcomes

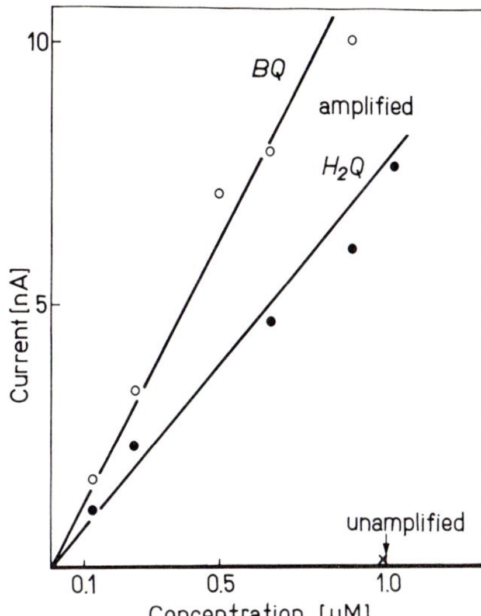

Figure 14-29.
Calibration graph for a cytochrome b$_2$-laccase electrode for benzoquinone (BQ) and hydroquinone (H$_2$Q) in the presence, and for H$_2$Q in the absence ($\times$) of lactate.

the problem of the low substrate specificity of alcohol oxidase, is that the recycling is restricted to ethanol, because methanol is converted only by the oxidase but not by the dehydrogenase. Conversely, isopropanol is oxidized by the dehydrogenase but not by the oxidase. Thus, combination of the two enzymes serves to improve the selectivity of the sensor for ethanol.

For the recycling of malate, malate dehydrogenase has been co-immobilized with lactate monooxygenase and combined with an O$_2$ sensor [331]. Lactate monooxygenase is capable of catalyzing the oxidation of malate to oxaloacetate which, in the presence of NADH, is converted back to malate by malate dehydrogenase. By this substrate recycling the oxygen consumption in the enzyme membrane is augmented, leading to an increase in the sensitivity of the electrode for malate. Since the activity of lactate monoxygenase towards malate is low [340], the amplification factor is only 3.

In addition to the enzyme pairs mentioned, more complex cycling systems have been designed by coupling a second recycling step mediated through a chemical or an electrochemical reaction. Combination of the reaction catalyzed by alanine aminotransferase and the backward reaction of glutamate dehydrogenase leads to the generation of a large amount of NADH from a small amount of L-glutamate. The NADH produced in the dehydrogenase-catalyzed reaction has been reoxidized by two different amperometric detection systems, both incorporating mediators [325]. On the one hand, NADH reacts with the N,N-dimethyl-7-amino-1,2-benzophenoxazinium (Meldola Blue, MB$^+$) to produce the reduced form of the mediator, which is immediately reoxidized at a graphite electrode. The mediator has been immobilized on the surface of the electrode by adsorption.

In another indication system, NADH is reoxidized by the N-methylphenazinium ion (NMP$^+$) inside the membrane and the reduced mediator is reoxidized by molecular oxygen. The decrease in oxygen concentration is measured by a Clark-type oxygen electrode. The amplification reaction is started by addition of L-alanine and for both sensors it results in a greatly enhanced response. The amplification factor for the MB$^+$-modified sensor is 15 within the linear range, resulting in a lowering of the detection limit to 0.5 µM. The response of the oxygen sensor containing the NMP^+ mediator is amplified by a factor of more than 60 in the linear range; the lower detection limit is 0.1 µM for glutamate and the response is linear up to 50 µM. In the oxygen sensor the NADH produced in the enzyme membrane is reoxidized by the mediator and the recycling can take place in the whole membrane volume whereas the reoxidation occurs only at the graphite surface when the MB^+-modified sensor is employed. Therefore, the NAD^+ concentration will be higher in the membrane with the former than with the latter electrode. The NAD^+ recycling results in a shift to the formation of α-ketoglutarate in spite of the unfavorable equilibrium constant. The glutamate dehydrogenase-alanine aminotransferase pair could be also connected via the pyruvate formed to an amplification system composed of the lactate oxidase-lactate dehydrogenase cycle.

The principle of enzymatic amplification can be drastically simplified by conducting the two partial reactions of the cycle with *only one* enzyme. Using this approach, a lactate dehydrogenase sensor for NADH determination has been devised [332]. The enzyme is immobilized in a gelatin membrane and coupled to an oxygen probe, where it catalyzes the oxidation of NADH by pyruvate:

$$NADH + H^+ + pyruvate \rightarrow NAD^+ + lactate, \tag{14-15}$$

and, in the presence of *glyoxylate,* the reduction of the NAD^+ formed [341]:

$$NAD^+ + glyoxylate \rightarrow NADH + H^+ + oxalate. \tag{14-16}$$

During this coupled reaction (Figure 14-30), in the membrane large amounts of lactate and oxalate are formed which can be indicated by the coupled reaction of lactate monooxygenase or oxalate oxidase. By using the former, the recycling of NADH in the "monoenzyme cycle" has been shown to give an amplification factor of 170. Similar systems may be envisaged that use other dehydrogenases or other pairs of substrates [342].

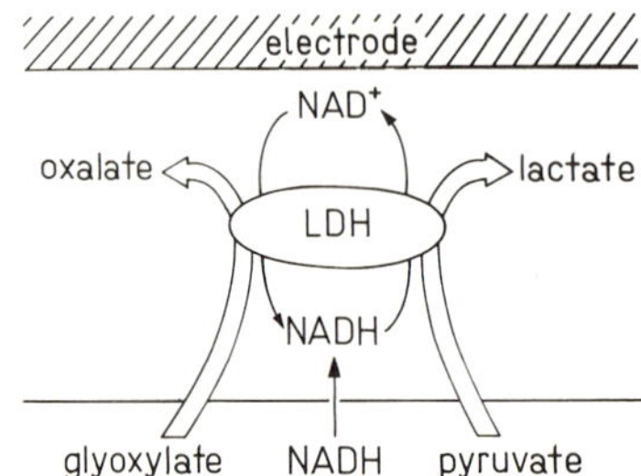

Figure 14-30.
Scheme of NADH recycling in a lactate dehydrogenase (LDH) electrode.

14.6.4   Enzymatic Elimination of Interfering Substances

In biosensors, interactions between the immobilized biomolecule and the sample, which is often a highly complex matrix, may cause undesired binding events or measuring effects. Particularly in biosensors that use coupled enzyme reactions the substrates of each reaction will interfere; therefore, increasing complexity of biosensors results in a decreased selectivity. Interferences can also occur on the level of the transducer reaction.

The effect of such interferences on the measuring signal can be eliminated by performing difference measurements with a reference transducer [343]. For the rejection of interferents, employment of permselective membranes has been devised [344]. An elegant alternative is the application of *enzymatic anti-interference systems* containing enzymes that in front of the sensor catalyze the conversion of the interfering compounds to inert products. Such systems have been developed in conjunction with analytical enzyme reactors in addition to enzyme electrodes. In assay systems based on immobilized creatinine iminohydrolase reactors with electrochemical NH_3 detection for the determination of creatinine in serum, the superposition of the signal caused by endogenous ammonia can be avoided by mixing the sample with α-ketoglutarate and pumping it through reactors containing immobilized glutamate dehydrogenase [345]. These reactors are capable of removing endogenous NH_3 according to the reaction

$$NH_3 + H^+ + NADH + \alpha\text{-ketoglutarate} \rightarrow \text{L-glutamate} + NAD^+ . \qquad (14\text{-}17)$$

Such anti-interference reactors have also been developed using GOD for the removal of endogenous glucose in samples containing different saccharides [346, 347]. In biospecific electrodes the eliminator enzymes are directly integrated in the sensor in a membrane-immobilized form and separated from the indicator enzyme layer by a semipermeable membrane. The eliminator-enzyme membrane has to be diffusion-controlled so as to assure complete conversion of the interferent penetrating into the membrane. In this manner the flux of the analyte to the indicator electrode is "filtered" (see Figure 14-26c). If there is sufficient immobilized enzyme activity, the anti-interference layer is able to prevent the interferent from reaching the indicator enzyme layer or the electrode surface. Eliminator and indicator enzymes may use the same co-substrate. The elimination capacities of some of these anti-interference membranes are shown in Table 14-11.

The first enzymatic anti-interference layer was developed to permit the electrochemical determination of catecholamines in brain tissue at a graphite electrode [351]. A layer of ascor-

Table 14-11. Filtering of interferents in biosensors.

Interferent	Eliminator enzymes	Elimination up to in mM	Indicator enzymes	Analyte	References
Glucose	GOD + catalase	2	Invertase + GOD	Sucrose	[348]
			glucoamylase + GOD	α-amylase	[349]
	hexokinase	2	Glucoamylase + GOD	Maltose	[229]
Lactate	Lactate monooxygenase	0.7	Lactate dehydrogenase + cytochrome b₂	Pyruvate	[350]
Ascorbic acid	Ascorbic acid oxidase			Catecholamines	[351]
	laccase	20	GOD	Glucose	[352]
Ammonia	Glutamate dehydrogenase	0.2	Creatinine deiminase	Creatinine	[353]

bate oxidase was attached to the electrode to provide oxidation of ascorbic acid before it could reach the electrode surface. The catecholamines could easily diffuse through the membrane to the electrode.

Since many enzyme electrodes are based on glucose measurement by GOD after sequential or competitive analyte conversion, endogenous glucose is a prominent interfering compound. To eliminate endogenous glucose an anti-interference layer containing co-immobilized GOD and catalase can be used [348, 349]. β-D-Glucose and oxygen are converted by GOD to the electrode-inactive gluconolactone and the electrode-active H_2O_2; the latter is cleaved in the catalytic reaction to non-interfering H_2O and oxygen. With this anti-interference membrane, glucose interference was completely eliminated up to a final concentration of 2 mM, for example in samples containing both glucose and sucrose, such as sugar-beet juice and instant cocoa [348]. Here, the sucrose is measured by using a membrane with co-immobilized invertase and glucose oxidase situated next to the electrode behind the glucose oxidase/catalase layer. The two membranes are separated by a dialysis membrane. Starch and α-amylase in fermentation samples are also directly measurable in the presence of minor amounts of glucose [349]. In this case, the indicator membrane consists of co-immobilized glucoamylase and glucose oxidase. A GOD-catalase system has also been used for glucose elimination in a fructose sensor involving co-immobilized glucose isomerase and GOD [353].

If the sample contains large amounts of glucose, the oxygen consumption in the anti-interference layer can result in a lack of oxygen in the indicator enzyme membrane, which would tend to reduce the measuring range of the sensor in a competitive way. In order to avoid this disadvantage, an alternative glucose anti-interference layer has been recommended, containing hexokinase [329]. It requires only ATP as cosubstrate and is impermeable to glucose up to 2 mmol/L. The membrane has been employed in connection with a glucoamylase-GOD membrane for maltose assay in the presence of glucose. Another advantage is the possibility of using both oxygen-indicating and hydrogen peroxide-indicating transducers.

Glucose measurements in urine and fermentation samples by means of GOD electrodes based on hydrogen peroxide detection usually suffer from interferences by anodically oxidizable compounds. These can be oxidized in the measuring solution by reaction with hexacyanoferrate(III). However, the hexacyanoferrate(II) ion formed is also oxidizable at an electrode potential of $+600$ mV. In order to prevent the hexacyanoferrate(II) from reaching the electrode, laccase has been co-immobilized with GOD in the sensor membrane [352]. Thus, the mediator is reoxidized in a laccase-catalyzed reaction with consumption of oxygen. The system is capable of shielding the electrode from ascorbic acid at concentrations up to 2 mmol/L.

Enzymatic anti-interference layers containing oxidases can also be used to eliminate oxygen or prevent its diffusion into the electrode-near space [354]. This permits the polarographic determination of organic compounds, eg, NAD^+, pyruvate, or methylviologen, by their cathodic reduction without tedious oxygen removal by nitrogen bubbling or other methods.

The examples presented show that the high chemical selectivity of biocatalysts is an important analytical tool not only in analyte recognition but also for the elimination of interfering substances.

14.7 Applications of Biosensors

14.7.1 Present State and General Trends

Between 12 and 15 billion US$ per year are spent for analytical purposes worldwide. In this sum the analytical usage of enzymes in clinical chemistry, food and cosmetic industry, and biotechnology for the routine measurement of about 80 substances, mainly low-molecular mass metabolites but also effectors, inhibitors, and the activity of enzymes themselves, is included. A wide range of immunoassays for low-molecular mass haptens, macro-molecules, and microorganisms have been made available in recent years through the enormous progress in immunological research, especially by the preparation of monoclonal antibodies. About 1 billion immunoassays are sold per year.

The development of immobilization methods has provided an impetus to the routine use of enzymes and antibodies in analytical chemistry. The main advantages of these immobilized reagents are their reusability, simplicity, and safety of handling, and a significant simplification of analyzers. Whereas in traditional enzymatic analysis spectrophotometric methods dominate, test strips and biospecific electrodes are at the leading edge in the analytical application of immobilized enzymes. This situation may be expected to last until the mid 1990's. A further breakthrough in biosensors can be expected in areas where high economic benefits are expected. One of the most promising fields of biosensor application is biotechnology, but probably sensor technologies will expand enormously into other areas. According to a recent prognosis [355], in 1990 the biosensor market in Western Europe will reach 440 milliom US$.

14.7.2 Application in Medicine

14.7.2.1 Clinical Diagnostics

Most clinical laboratory analyses aim at metabolites in body fluids in the micro- and millimolar concentration range. A better understanding of various diseases requires the assay of steroids, drugs and their metabolites, hormones, and protein factors present in the range 10^{-11}–10^{-9} M. At present the concentrations of these substances can only be determined by means of immunoassays. Stat determination and continuous in vivo monitoring are particularly important in intensive care medicine, surgery, and life-threatening situations.

Test strips, which are available for the determination of about ten low-molecular mass substances (metabolites, drugs, and electrolytes) and eight enzymes [356], can be considered as precursors of optoelectronic biosensors. Efficient optoelectronic sensors based on immobilized dyes have been devised for the determination of glucose, urea, penicillin, and human serum albumin [357]. Other approaches use immobilized luciferase or horseradish peroxidase to assay ATP or NADH or, when coupled with oxidases, to measure uric acid or cholesterol. These principles have not yet been generally accepted for use in routine analysis. Thermistor devices involving immobilized enzymes or antibodies for a number of clinically relevant substances have also been described. Thermometric enzyme linked immunosorbent assays are being routinely employed for monitoring the production of monoclonal antibodies.

With regard to practical application, enzyme electrodes are by far in the forefront of the biosensor field. At present, 15–20 analyzers based on enzyme electrodes are commercially available worldwide (Table 14-12) for the measurement of glucose, galactose, uric acid, choline, ethanol, lysine, lactate, pesticides, sucrose, and lactose, and for the activity of α-amylase. Application of such analyzers usually permits the enzyme demand per sample to be reduced to less than 1 µg. Typically, the devices are suitable for the measurement of only one analyte. One of the few exceptions is the ICA-LG 400 Lipid Analyzer from Toyo Jozo (Japan), which is capable of measuring a whole group of analytes, namely cholesterol, triglycerides, and phospholipids. Serum samples have to be preincubated with the respective hydrolases, ie, cholesterol esterase, lipoprotein lipase, and phospholipase D. The measurement is carried out by using enzyme electrodes involving cholesterol oxidase, glycerokinase, glycerophosphate oxidase, and choline oxidase, all immobilized in front of an oxygen probe. Using a sample volume of 30 µL a measuring frequency of 400 per hour is obtained. Although the prospects for this method appear exciting, the analyzer has not yet reached the market.

Table 14-12. Enzyme electrode-based analyzers. *CV:* imprecision

Company, country	Model	Analyte	Linear range in mM	Sample frequency in h^{-1}	Serial *CV* in %	Stability
Yellow Springs	23 A	Glucose	1–45	40	<2	300 samples
Instrument Co., USA	23 L	Lactate	0–15	40		
	27	ethanol	0–60	20	<2	
		lactose galactose sucrose }	0–55	20	2	
Zentrum für Wissen-	Glukometer	Glucose	0.5–50	60–90	1.5	>1000 samples
schaftlichen Geräte-	GKM	uric acid	0.1–1.2	40	2	10 d
bau, FRG		Glucose	0–27	80–90	1.7	>500 samples
Fuji Electric, Japan	GLUCO 20	α-amylase		30	4–5	
	UA-300 A	uric acid		50–60	3	
Daiichi, Japan	AutoSTAT GA-1120	Glucose	1–40	60–120	3	
Radelkis, Hungary	OP-GL-7110S	Glucose	1.7–20	40	5–10	240 d
Sigma, USSR	EXAN	Glucose	2–30	30	>3	
La Roche, Switzerland	LA 640	Lactate	0.5–12	20–30	>5	40 d
Omron Tateisi, Japan	HER-100	Lactate	0–8.3		>5	>10 d
Seres, France	Enzymat	Glucose	0.3–22	60		
		choline	1.0–29	60		
		L-lysine	0.1–2	60		
		D-lactate	0.5–20	60		
Tacussel, France	Gluco-processeur	Glucose	0.05–5	90	<2	>2000 samples
Prüfgeräte-Werk	ADM 300	Glucose	1–100	80	<2	>2000 samples
Medingen, FRG	ECA 20	Glucose	0.6–60	120–130	<1.5	10 d
(Eppendorf, ERG)	(ESAT 6660)	lactate	1–30	120	<2	14 d
		uric acid	0.1–1.2	80	<2	10 d

Glucose

About 4% of the population of industrialized countries has diabetes, and rapid, selective, and reliable methods for the determination of blood glucose are of utmost importance for diabetes screening and treatment. Worldwide a potential market for 500 million glucose sensors is anticipated [355].

Glucose analyzers based on bioelectrochemical sensors are being offered commercially in the USA, Japan, France, the USSR and the FRG (see Table 14-12). The analyzers developed by Yellow Springs Instrument Co. (YSI) (USA), Fuji Electric (Japan), Daiichi (Japan), and Sigma (USSR) measure the true glucose concentrations in plasma and serum samples. However, the use of untreated whole blood as sample material would be a considerable advantage over that of plasma or prediluted samples. With direct injection of whole blood into the Fuji instrument the measured values are too low by 13% [358]. This tendency is confirmed by the correlation equation obtained with the Daiichi AutoSTAT instrument as applied to whole blood samples:

$$y = (0.793\,x + 0.47)\ \text{mM}. \tag{14-18}$$

To correct for these systematic deviations, for use of the YSI analyzer a table is required that takes into account the packed cell volume [359].

Similarily to other analyzers, the Glukometer (ZWG, Berlin, FRG) is well suited for the analysis of single samples and small sample series because the sample solutions are manually injected into the background buffer by using pipettes. Before being injected, the blood is diluted tenfold [360], since without dilution the glucose content of erythrocytes is not completely accounted for. The average serial imprecision is 1.7% and the day-to-day imprecision about 3%. For hospital samples excellent agreement with a reference method is obtained. Based on the experiences gained during 5 years of routine use of the Glukometer, the Prüfgeräte-Werk Medingen (Dresden, FRG) developed the mircocomputer-based ECA 20 Enzyme-Chemical Analyzer shown in Figure 14-31. This instrument is suitable for glucose deter-

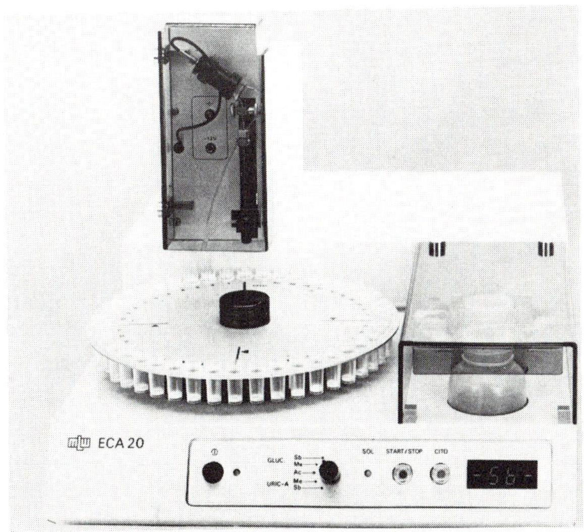

Figure 14-31.
The ECA 20 Enzyme-Chemical Analyzer (Prüfgeräte-Werk Medingen. FRG), also named ESAT 6660 (Eppendorf, FRG).

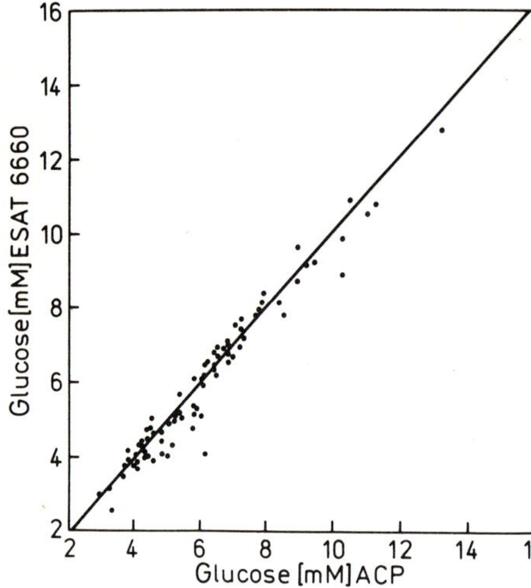

Figure 14-32.
Comparison of blood glucose concentration values obtained with the Eppendorf ACP method and with the ESAT 6660.

mination in the concentration range 0.6–60 mM with a day-to-day imprecision below 3% and an excellent correlation with the highly specific Eppendorf ACP method (Figure 14-32):

$$y = [(1.003 \pm 0.006)\,x - (0.015 \pm 0.045)]\,\text{mM};$$

$$r = 0.996\ (n = 196). \tag{14-19}$$

One hundred and twenty samples can be processed per hour; a stationary value is obtained within 60 s. Only 5–20 µL of blood are required for a double determination. The glucose oxidase membrane used is stable for at least 2000 measurements. A modified variant of this instrument, the ESAT 6660, is marketed by Eppendorf (FRG).

Hydrogen peroxide detection in enzyme electrodes for urine glucose assay is subject to severe interferences by reducing substances, especially as the normal glucose concentration in urine is only 0.2 mM. Such interferences can be eliminated by using a cellulose nitrate-modified GOD membrane in conjunction with a hexacyanoferrate(III)-containing background buffer [361] which is able to oxidize disturbing substances to electrochemically inert products, thus leading to good agreement of the glucose values with those found by using the hexokinase method.

Recently, the first commercial second-generation glucose sensor was introduced by Genetics International (UK) [362]. The sensor is based on a ferrocene-modified GOD electrode strip (Figure 14-33). For glucose determination a drop of blood is applied to the strip, which is then introduced into a pen-sized readout instrument. The imprecision in the normal concentration range is 3.9%. Lower precision has been found in the hypoglycemic range. The following correlation with an unspecified method was obtained:

$$y = (1.04x + 0.9)\,\text{mM};\ r = 0.985. \tag{14-20}$$

The simple handling makes the sensor well suited for use in the doctor's office and for patient home diagnostics.

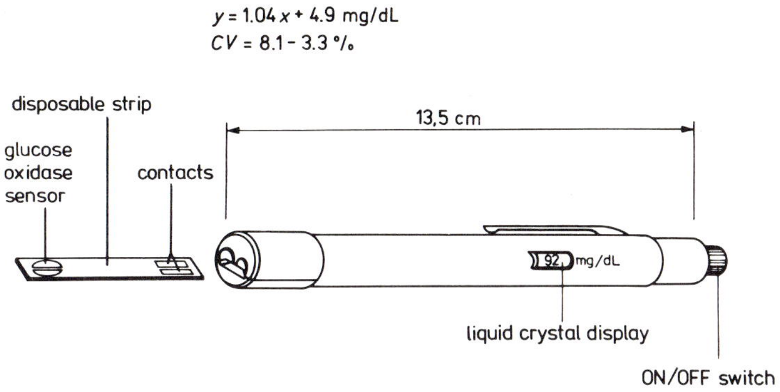

Figure 14-33. Pen-sized glucose analyzer developed at Cranfield Institute of Technology (UK) for Genetics International.

Urea

The concentration of urea in blood (blood urea nitrogen, BUN) is an important parameter in clinical chemistry for assessing kidney failure. Since urease enzyme sensors rely mainly on indication of the pH change during urea hydrolysis, problems in their application to body fluids are connected with the susceptibility of the sensors to disturbances by sample pH and buffer capacity.

A potentiometric urea electrode has been employed in an enzyme difference analyzer for urea determination in serum [363]. The difference between the potential changes of a urease-covered and a bare pH glass electrode is evaluated. Using a fixed-time regime, a measuring frequency of 20–25 h^{-1} and a *linear* range for 1 : 120-diluted samples of 1–20 mM is obtained. For serum measurement the sensor system is calibrated with urea dissolved in 0.0185 M Tris-HCl buffer (pH 7.0) having a buffer capacity of 8 mM. In this manner, disturbances by individually different sample pH values and variations in the buffer capacity are kept within the noise of the measuring system. The imprecision for 20 successive determinations in serum with a urea concentration of 6.5 mM ist 2.1%. The urease sensor has a useful lifetime of 28 d when stored at room temperature between the measurements.

Petersson [364] developed a urea analyzer for undiluted blood samples by using a urease-covered ammonium ion-selective electrode in a FIA system. Forty samples per hour can be analyzed in a measuring range up to 40 mM and with a serial imprecision of 1%. The sensor is stable for 25 d. The correlation coefficient with a routinely used method is 0.99.

An amperometric urea sensor based on the pH dependence of the anodic oxidation of hydrazine [365] can be utilized in the Glukometer GKM 02 for hemodialysis monitoring. For urea concentration in dialyzate the following correlation with the Berthelot method was obtained:

$$y = (0.9912x + 0.125) \text{ mM}; \; r = 0.997 \; (n = 67). \tag{14-21}$$

A very successful sensor for urea measurement in blood has been developed for Hitachi (Japan) [366]. Two ammonium-sensitive electrodes containing nonactin in a PVC membrane are integrated in a FIA manifold in a differential circuit. One of the electrodes is covered with a polyester membrane bearing immobilized urease. The lifetime of this membrane is 2 months. Using a sample volume of only 10 µL 60 samples per hour can be analyzed.

Lactate

The determination of lactate is becoming increasingly popular in the diagnosis of shock and myocardial infarction and in neonatology and sports medicine. Therefore, great efforts are being made to develop sensor-based lactate analyzers which may readily be used at the bedside. The first enzyme electrode-based lactate analyzer was developed in 1976 by La Roche (Switzerland) (see Table 14-8). It contains cytochrome b_2 in a small reaction chamber in front of a platinum electrode polarized at $+0.25$–0.40 V. The lactate values measured with the analyzer correlate fairly well with those obtained with the spectrophotometric reference method using deproteinized blood:

$$y = (1.007x + 0.024)\ \text{mM};\ r = 0.9813\ (n = 53). \tag{14-22}$$

The analyzer has also been employed to measure lactate in muscle biopsy specimens [367]. Other lactate analyzers use lactate oxidase (LOD). In the YSI 23 L instrument (USA) LOD is immobilized between a cellulose acetate membrane and a polycarbonate membrane [368], the latter serving to exclude high-molecular mass interferents. Lactate measurement in whole blood pipetted immediately after withdrawal into the phosphate buffer of the analyzer yields the following correlation with values obtained with deproteinized blood [369]:

$$y = (0.95x - 0.17)\ \text{mM};\ r = 0.994\ (n = 179). \tag{14-23}$$

Surprisingly, although the sensor measures only the lactate content of the plasma, whereas the results of the reference method reflect the concentration of lactate in both plasma and erythrocytes, the agreement between the two methods is good. The analyzer has also been used successfully for lactate determination in spinal fluid [370].

In the HER-100 (Omron Tateisi, Japan) an asymmetric cellulose acetate membrane is used, bearing LOD covalently bound by γ-aminopropyltriethoxysilane and cross-linked with glutaraldehyde [371]. The membrane is highly selective for hydrogen peroxide. The analyzer is suitable for lactate assay in human serum. Polyurethane-immobilized LOD is used for whole blood lactate determination in both the Glukometer and in the ECA 20 (ESAT 6660) [372]. Dilution of the samples with a hypotonic buffer provides for both complete inhibition of glycolysis and immediate hemolysis. As shown by the correlation equations, the method appears to be fairly reliable:

$$\text{GKM: } y = (1.042x - 0.023)\ \text{mM};\ r = 0.995\ (n = 70). \tag{14-24a}$$

$$\text{ECA (ESAT): } y = (0.994x - 0.076)\ \text{mM};\ r = 0.992\ (n = 244). \tag{14-24b}$$

The lactate values are available within 30 s after withdrawal of blood from a patient.

The measurement of the lactate/pyruvate ratio in plasma is possible by using a lactate dehydrogenase-lactate monooxygenase sequence electrode [373]. The sensor is equally sensitive to lactate and pyruvate (Figure 14-34), because of the high enzyme loading and the

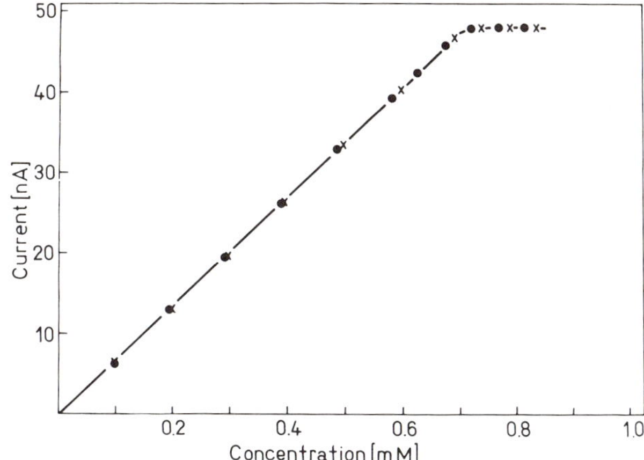

Figure 14-34. Calibration graph of a lactate dehydrogenase/lactate monooxygenase electrode for (●) lactate and (×) pyruvate. (See also Figure 14-36 for sensor configuration.) Reproduced from [423] with permission from Springer-Verlag.

equality of the diffusion coefficients of lactate and pyruvate. Determination of both substrates in a sample requires about 3 min.

Uric Acid

Since uric acid is a risk factor for gout and other diseases, diagnosis of hyperuricemia is increasingly important. The Glukometer has been equipped with a uricase membrane and employed for uric acid assay in serum [374]. Satisfactory agreement with the uricase-catalase reference method was obtained; the deviation of the mean value was as low as $+2.4\ \mu M$. The reagent costs of the method amount to only one tenth of those required for the manual photometric method.

The Fuji Electric (Japan) UA-300 analyzer uses a uricase membrane fixed to a hydrogen peroxide-selective layer [375]. As little as 20 μL of blood serum are required, and a sample throughput of 50–60 per hour with an imprecision of 3% is achieved. The correlation to the uricase-catalase method is reflected by the following equation:

$$y = (1.1x + 0.41)\ \text{mM};\ r = 0.97. \tag{14-25}$$

The Exan (USSR) enzyme electrode-based analyzer [376] is also suitable for the measurement of uric acid.

Determination of Enzyme Activities

The measurement of enzyme activities plays a key role in clinical chemistry because increased enzyme activities in body fluids often indicate tissue and cell damage. Enzyme activity determination is usually carried out by measuring the initial rate of the enzyme reaction of interest in the presence of a saturating substrate concentration. With sensors two distinct procedures are used:

1. The product is indicated after a defined reaction period outside the measuring cell.
2. The enzyme-catalyzed reaction is allowed to proceed in the measuring cell, the reaction rate being indicated by electronic differentiation of the current-time curve.

α-Amylase. α-Amylase catalyzes the stepwise hydrolysis of starch and oligosaccharides to maltose. Since assays for α-amylase use heterogeneous substrates, the results are often ambiguous.

The Fuji Electric (Japan) α-amylase analyzer [375] is based on a GOD electrode (see Table 14-12). The sensor first measures the endogenous glucose concentration of the sample and, after addition of maltopentose and α-glucosidase (maltase), the rate of glucose liberation. The formation of low-molecular-mass products of α-amylase-catalyzed starch hydrolysis can be assayed by using a glucoamylase-GOD electrode [377]. The hydrolysis products can easily diffuse into the bienzyme membrane where they are successively degraded to glucose by glucoamylase. The continuously increasing concentration of the reaction products is reflected by the steadily increasing current after the response to endogenous glucose (Figure 14-35). As only the β-anomer of glucose is formed, the sensitivity of the method is higher than that with α-glucosidase. Removing the endogenous glucose during the incubation period with GOD significantly simplifies the analysis. It correlates with the iodine-starch method as follows:

$$y = (1.077x - 0.998) \text{ U/L}; \; r = 0.947 \; (n = 21). \tag{14-26}$$

Transaminases. The reactions of alanine aminotransferase (ALAT) and aspartate aminotransferase (ASAT) can be monitored by sensing their products, pyruvate, oxaloacetate, and glutamate, with enzyme electrodes. A bienzyme electrode composed of oxaloacetate decarboxylase and pyruvate oxidase adsorbed at a PVC membrane and coupled to a hydrogen peroxide-indicating electrode has been shown to be applicable to the sequential determination of both transaminases [378]. The time required for one sequential measurement is 4 min. The correlation coefficient between the sensor and the optical method is 0.99.

Employment of glutamate oxidase permits the sequential assay of both transaminases without the need for co-immobilizing a second enzyme. Using a glutamate oxidase sensor, preincubation of the sample for 30 min has been found to be required [379]. Optimization of this type of sensor permitted the preincubation time to be lowered to 10 min [380].

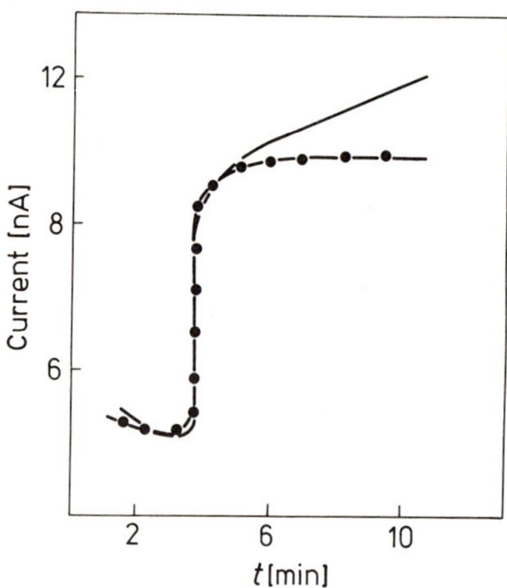

Figure 14-35.
Response of a glucoamylase – GOD electrode to glucose standard solution (11.11 mM, −) and to control serum containing 11.13 mM glucose and 0.3 U/mL of α-amylase (●). Reproduced from [377] with permission from Marcel Dekker, Inc.

Lactate Dehydrogenase. Lactate analyzers such as the YSI 23 L and the ECA 20 are applicable to LDH assay after incubation of the serum sample with NADH and pyruvate. In such methods the endogenous lactate has to be measured beforehand. This has been studied using lactate oxidase [381] as well as lactate monoxygenase [373] electrodes. With the latter, a linear measuring range up to 1200 U/L has been obtained. The imprecision for 20 analyses of a serum sample containing 252 U/L was 1.2%. The correlation was

$$y = (1.11x - 17.4) \text{ U/L}; \ r = 0.999 \ (n = 30). \tag{14-27}$$

The procedure permits 15–20 combined measurements of lactate and LDH per hour to be carried out.

Pyruvate Kinase. Pyruvate kinase activity in hemolyzed erythrocytes has been determined by using a lactate dehydrogenase-lactate monooxygenase sequence electrode [382]. The enzymes were immobilized in gelatin and attached to an oxygen probe (Figure 14-36). Since the sample material contains only negligible amounts of lactate and pyruvate, the pyruvate kinase activity can be directly derived from the current change on addition of the pyruvate kinase substrates, phosphoenolpyruvate and ADP, and the LDH cofactor, NADH. The measuring time is about 4 min. The relative standard deviation for 6.6 units of pyruvate kinase per gram of hemoglobin is 3.1%. The results agree well with those of the spectrophotometric method. Since the sensor signal is linear over the normal range of 2.1–6.9 U/g hemoglobin, and only decreased activities are of clinical relevance, the sensor is suitable for measuring any clinically possible value.

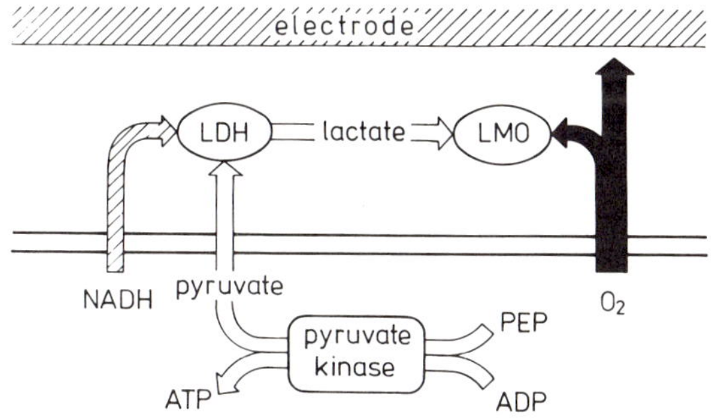

Figure 14-36.
Schematic representation of a lactate dehydrogenase (LDH) – lactate monooxygenase (LMO) sensor for the determination of pyruvate kinase activity. PEP = phosphoenolpyruvate.

Cholinesterase. Cholinesterase can be assayed by determining the choline liberated in the enzymatic reaction by using immobilized choline oxidase. Further, the direct electrochemical registration of thiocholine iodide, the product of the cholinesterase-catalyzed hydrolysis of butyrylthiocholine iodide, has been used [384]. The Glukometer has been adapted to this reaction system by polarizing the platinum electrode to 470 mV versus an AgI electrode in 0.1 M potassium iodide solution. The formation of thiocholine iodide causes an increase in the oxidation current. After a transient phase of about 20 s the reaction rate becomes constant. In the kinetic mode a constant measuring value proportional to the reaction rate is obtained. A

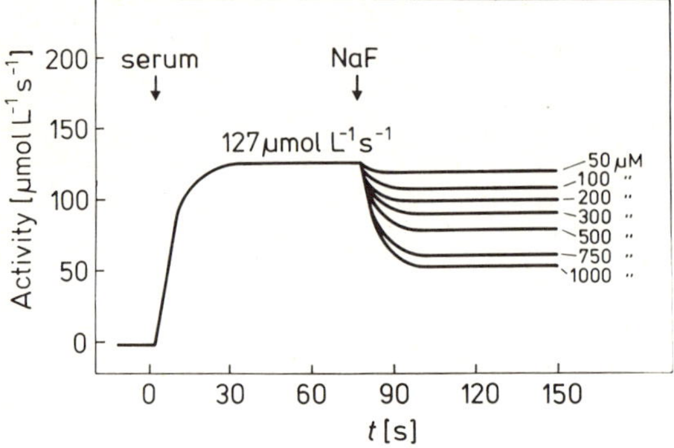

Figure 14-37.
Enzymatic-electrochemical determination of sodium fluoride as a model inhibitor using a cholinesterase-containing blood serum sample. Reproduced from [384] with permission from Springer-Verlag.

good correlation with the standard reference method has been found for both serum cholinesterase and the isozyme in erythrocytes:

$$y = (1.0108x - 4.1) \text{ U/L}; \quad r = 0.994 \ (n = 27). \tag{14-28}$$

For the determination of inhibitors of cholinesterase, a serum sample of known enzyme activity is used. The reaction rate decreases on addition of an inhibitor, and the remaining activity is indicated (Figure 14-37). This reaction system is utilized commercially in the Thorn EMI (England) and Midwest Research Instruments (USA) bioalarm instruments both incorporating immobilized cholinesterase.

14.7.2.2   On-Line Monitoring

Blood Glucose

For the monitoring of diabetes during stress situations such as surgery, traumata, or myocardial infarction, glucose-controlled insulin infusion systems are highly desirable. Pathophysiological mechanisms tending to increase the insulin demand in an unforeseeable manner can lead to life-threatening states of the organism. Therefore, euglycemia should be adjusted during and after stress situations.

For about 15 years, research has been directed at the development of implantable glucose sensors for continuous monitoring of the blood glucose level. An artificial beta-cell for perioperative glucose control, the Biostator, has been commercialized by Life Science Instruments (USA) [385]. The equipment consists of an on-line glucose analyzer based on a H_2O_2-detecting GOD electrode which is integrated in a computer-controlled feed-back system and can be used for up to 2 d. An alternative approach has been devised for perioperative monitoring of diabetes [360, 386]. This monitoring system, named Glucon, is based on a modified Glukometer glucose analyzer coupled with a computer for dialog-orientated control, and infusion pumps for insulin and glucose. The infusion rates, insulin action factor, and time course of the glucose concentration are recorded and thus readily available to the physician.

Up to now, the in vivo application of glucose sensors has been hampered by immunological reactions of the organism against the implanted material. Therefore, the biocompatibility of the material to be used has to be carefully studied. Thus, it was shown that the coverage of needle-type glucose sensors with silanized membranes increases the biocompatibility of the probe [387]. As judged by scanning electron microscopy, deposition of protein on the membrane was less drastic in tissue than in the bloodstream [388].

Another problem with the development of implantable sensors is the need to calibrate the sensor ex vivo. This requires a high sensor stability since the sensor has to be calibrated before implantation. The longest lifetime so far reported for enzymes immobilized in an implanted sensor was between 6 and 10 d [389]. Kessler et al. [390] developed a glucose sensor with an extremely low oxygen consumption and a stability of 3 months, which appears favorable for implantation. Another presumably implantable sensor is based on ferrocene as electron acceptor for GOD [391, 392], which eliminates the need for oxygen. The sensor exhibits a linear measuring range of 1–30 mM. However, experiments with the sensor subcutaneously implanted in animals revealed a rapid drop in sensitivity [393] (see 14.3.1.2). The linear range of a glucose electrode has been expanded up to 40 mM by covering the enzyme membrane with a perforated hydrophobic polyethylene membrane [394]. When applied subcutaneously at a pO_2 of 2–5 kPa, the sensor was stable for several hours.

Needle-type glucose sensors may be advantageously introduced into interstitial fluid [389, 394, 395]. As shown in animal experiments, in this manner a measuring range of 3.3–27.7 mM can be obtained [395]. Combination of a cellulose diacetate membrane containing GOD with a hydrophilic alginate-polylysine-alginate membrane provides a response time of 5 min and a stability of 7 d [389]. The increased lifetime of the sandwich membrane is due to better biocompatibility.

Recently, a glucose sensor system consisting of an oxygen electrode and a glucose oxidase electrode has been implanted into the *vena cava* of a dog [396]. The sensors operated in the potentiostatic mode and were connected to an implantable telemetry system. Both the enzyme stability and the power consumption allowed operation of the system for 3 months.

Urea Determination in the Artificial Kidney

The continuous measurement of urea in blood during dialysis is an important precondition for monitoring of the treatment of patients with chronic kidney failure. An enzyme electrode for the semi-continuous assay of urea in dialyzate has been designed and integrated in an ar-

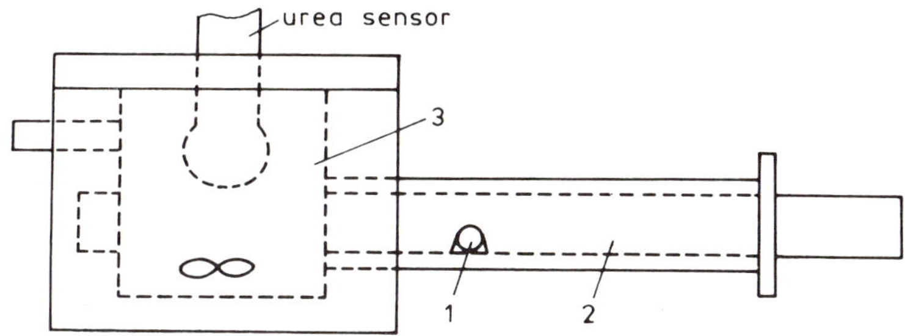

Figure 14-38. Measuring cell of the urea module with dialyzate sampler. The sample is transferred into the hole (1) of the movable piston (2) from the dialyzate cycle into the measuring cell (3).

tificial kidney [360]. The electrode is inserted in a flow-through cell (Figure 14-38) hydraulically connected with the hemodialyzer. The urea module allows 20 measurements per hour to be carried out, thus providing sufficiently exact monitoring of the time course of the urea concentration. The enzyme electrode is stable for more than 1 week when used in the hemodialyzer for about 4 h per day. The imprecision obtained for the measurement of urea in dialyzate is about 3% and the correlation coefficient with the reference method is 0.99. Practical use of this unit permits the direct control of the efficiency of hemodialysis treatment, which can thus be optimally adapted to the patient's needs.

Determiniation of Lactate and Pyruvate

Three enzyme electrodes, for the determination of lactate, pyruvate, and glucose, have been introduced into an artificial pancreas [397, 398]. Patient blood was dialyzed and pumped through the measuring cells. For the glucose and lactate sensors, oxygen probes were used, whereas the low pyruvate concentrations in blood (40–120 µM) required the use of a hydrogen peroxide electrode for assembling the pyruvate sensor. The set-up enables the three substrate concentration to be monitored during treatment of diabetics. The values obtained within a treatment period of 16 h agreed satisfactorily with those measured by reference methods.

14.7.3   Food Analysis, Process Control, and Environmental Monitoring

Quality assessment of food and fodder products requires analyses for protein, carbohydrates, and fat. The enzyme electrode-based analyzers originally developed for clinical chemistry have found only limited application in food analysis, because they are only suitable for the determination of one parameter, mostly glucose or a disaccharide. The increasing concern about food quality requires new types of biosensors allowing residual and hygiene control and on-line measurement of age and freshness.

A peculiarity of food analysis is the presence of large amounts of potentially interfering compounds in many foodstuffs. For example, assay of glucose and sucrose in instant drinks with the Glukometer [399] revealed considerable interferences from the large amounts of ascorbic acid and vanillin present in the samples. Therefore, such samples have to be analyzed by means of oxygen rather than hydrogen peroxide probes. Further, it has been proposed to saturate food and fermentation samples by bubbling with air directly in the reaction vessel [400]. Concomitantly with air saturation, hydrolysis of sucrose, glucosinolate, or starch may be performed, the concentrations of which are to be measured. This method permits up to 60 samples per hour to be analyzed automatically with good precision.

Enzyme sensors involving hydrogen peroxide-sensing electrodes can be readily employed for the determination of sucrose in sugar-beet juice [399] and of lactose in milk [401]. The ECA 20 Enzyme-Chemical Analyzer as furnished with a β-galactosidase-GOD electrode, is able to determine the lactose content of up to 1000 milk samples per hour with an imprecision below 2%. In another device, the Glucoprocesseur (Tacussel, France), interferences from reducing substances are compensated for by using a difference measurement technique. The analyzer has been applied to the assay of glucose, lactate, and oxalate in foodstuffs [402]. The alcohol analyzer from YSI (USA) is applicable to the analysis of drinks [403, 404] and human blood [405]. However, the enzyme membrane used has a poor working stability.

In biotechnological processes, high yields can be achieved by process control involving on-line determination of a multitude of parameters. In fermentation processes, eg, of antibiotics, a defined time course of the concentrations of nutrients such as carbohydrates, amino acids, phosphates, and ammonium, and also hormones has to be followed strictly. A knowledge of the product concentration permits a direct evaluation of the state of the bioprocess. For most of the low-molecular mass substances of interest, such as amino acids, sugars, phosphates, penicillin, and gluconic acid, enzyme electrodes have been described. In contrast, major difficulties have to be dealt with in the sensor-based determination of high-molecular mass compounds such as proteins and antibodies. Finally, the application of biosensors of any kind in fermenters is connected with significant problems:

 i) direct sterilization of biosensors is impossible;
 ii) discrete measurements are necessary for calibration of the sensor;
 iii) in most cases the analyte concentration exceeds the linear range of the sensor;
 iv) various interfering substances have to be expected;
 v) the sensor stability is affected by mechanical and thermal stress.

The dynamic range of *in situ* analysis in fermenters may be restricted by oxygen limitation in the culture broth. To overcome this restriction, an oxygen-stabilized glucose electrode has been developed by placing a galvanic, oxygen-generating electrode directly in the enzyme membrane [406, 407]. A common O_2 probe serves as a reference electrode. The O_2 consumption caused by glucose oxidation is compensated for by electrolytic formation of O_2 from water. Therefore, the sensor is applicable to solutions with varying oxygen content. Even in anaerobic media the required cosubstrate may be formed electrolytically. The sensor has been successfully employed for continuous *in situ* glucose measurement during *Candida utilis* fermentation [407]. Agreement of the measured glucose values with those determined by a reference method was satisfactory.

In an "externally buffered" enzyme electrode (Figure 14-39), substrate-free buffer is continuously pumped between the dialysis membrane and the enzyme layer [408], ie, the sample is diluted before it reaches the enzyme. The buffer flow-rate may be used to adjust the measuring range and sensitivity. The configuration of the sensor permits it to be sterilized. While the

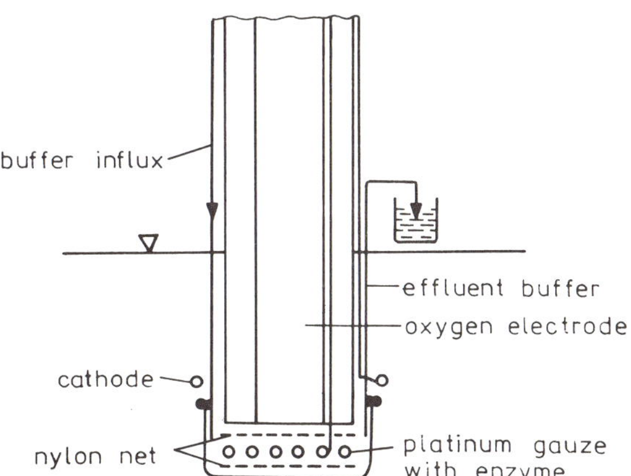

Figure 14-39.
Scheme of externally buffered
enzyme electrode.

membrane is protected by continuously flowing buffer, the other parts of the sensor can be treated with a solution of ethanol and H_2SO_4.

The measuring range of a glucose sensor has been expanded by coupling of GOD or glucose dehydrogenase to a ferrocene-modified electrode [409]. The lifetime of this sensor can be significantly enhanced by immobilizing GOD covalently to alkylamine groups at the electrode surface [410]. This type of sensor was introduced into a fermenter via a sterilizable housing separated from the fermentation broth by a polycarbonate membrane. The usable, but nonlinear, measuring range of the sensor reaches up to 100 mM; the lifetime is 14 d. A sensor system involving an alcohol oxidase electrode and an enzyme-free oxygen probe has been used for continuous ethanol assay in alcohol fermentation [411]. The measuring range of this configuration is narrow so that only the initial phase of ethanol formation can be monitored.

These examples indicate the *in situ* applicability of enzyme electrodes; however, numerous problems have still to be solved. At present, coupling of enzyme sensors for fermentation control in a bypass arrangement appears to be more favorable [412]. Following this concept, an invertase thermistor incorporating a sterilizable filter unit has been developed [413] for the monitoring of alcoholic fermentation by immobilized yeast cells. Another thermistor has been successfully used for on-line glucose measurement under real cultivation conditions of *Cephalosporium acremonium* [414]. Similar calorimetric devices are suitable for other fermentation processes and in environmental analysis.

An enzyme electrode has been employed in a bypass to control the concentration of glucose in various bioprocesses [415]. The samples are periodically withdrawn, deaerated with nitrogen, and diluted, without seperation of biomass. Glucose is analyzed by discrete measurement with the use of the artificial electron acceptor benzoquinone. The method requires correction of the measurement by means of an enzyme-free sensor.

Process control in human and animal cell cultures is very important because the required nutrients, eg, fetal calf serum, are extremly expensive [416]. For optimization of cell culture, the HER-100 lactate analyzer (Omron Tateisi, Japan) has been used together with a glucose electrode for lactate and glucose analysis in the growth medium of human melanoma cells [417]. Investigation of the process over 7 d showed that, as a result of glycolysis, with increasing cell number the concentration of glucose decreased and that of lactate increased.

The possibility of withdrawing representative samples without affecting the sterility of the bioreactor is an essential prerequisite for on-line monitoring. Appropriate filtration equipment is available from Braun Melsungen (FRG) and Control Equipment (USA). Researchers at Massachusetts Institute of Technology (USA) designed a mechanical sampling system that automatically leads the sample through a sterilized chamber before dilution. The equipment has been combined with the Enzymat (Seres, France) in order to monitor the production of monoclonal antibodies against fibronectin by hybridoma cells [418]. Every 30 min the concentrations of glucose, lactate, and glutamine are measured in parallel. The enzymes are coimmobilized in glutaraldehyde-cross-linked gelatin membranes in front of oxygen electrodes. The sensor for glutamine determination consists of coimmobilized glutaminase from *E. coli* and glutamate oxidase from *Streptomyces* sp. The measured data permit a relationship between the substrate concentrations, ATP flux, and cell growth to be derived.

The PM-1000 On Line Biotec Analyzer for automatic process control in industrial fermentation and biotechnology research (Nippon General Trading, Japan) combines the enzyme electrodes contained in the M 100, AS-200, and AD-300 laboratory analyzers (Toyo Jozo, Japan) with a unit for sterile filtration and a computer. By exchanging the enzyme electrode, the concentrations of glucose, ethanol, L-lactate, glycerol, sucrose, lactose, pyruvate, ascorbic acid, or L-amino acids can be monitored. The sensors use the respective oxidases and, in addition, glycerokinase for glycerol determination and β-galactosidase for lactose determination. For sucrose assay, invertase is combined with pyranose oxidase to avoid disturbances by endogenous glucose.

The general approach to overcoming problems with in situ applications of biosensors is therefore sampling with sterile devices and *on-line* analysis in automated external flow injection analysis (FIA) setups [419–421] (Figure 14-40) often combined with sample pretreatment. Enfors and Cleland [471] distinguished in this context between the possibilities of directly pumping the fermentation broth to the sensor or pumping an external buffer circuit extracting the analyte by means of a dialysis coil, inserted into the fermenter (Figure 14-41). They also derived mathematical expressions for the correlations between the analyte concentrations in fermentation broths and dialysis liquids. FIA can be used with enzyme electrodes

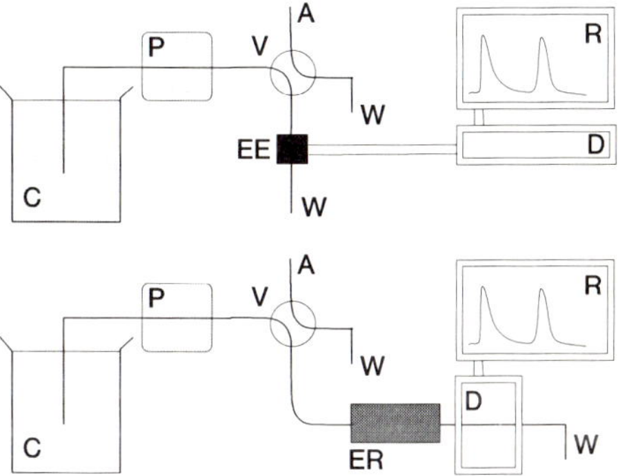

Figure 14-40. Simple flow-injection device for use with enzyme electrodes (top) or integrated immobilized enzyme reactors (bottom). The carrier C (buffer) is pumped through a valve V, where a given amount ("pulse") of analyte A can be injected into the flow stream. The concentration pulse of the substrate is "seen" by the enzyme electrode EE, or converted to a product in the enzyme reactor ER and indicated as a signal in the recorder R. W = waste.

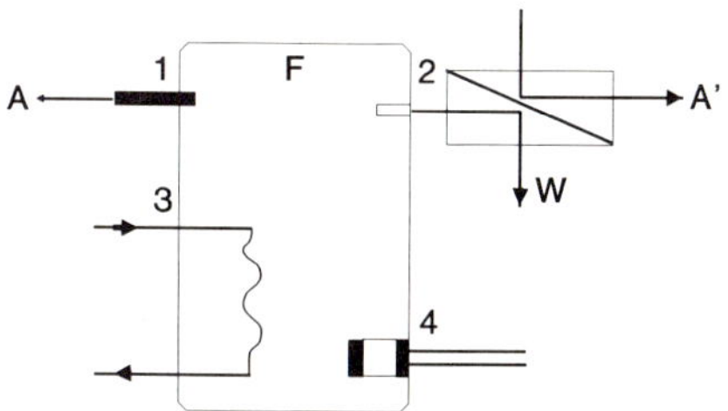

Figure 14-41. Sampling from a fermenter for on-line analysis (after [366]). 1. Direct removal of fermentation broth (analyte A); 2. indirect sampling by ultrafiltration, dialysis, electrodialysis, pervaporation, providing an analyte A' of proportional concentration, normally diluted; 3. indirect sampling by extraction of fermentation broth by external buffer; 4. in situ measurement by means of an enzyme electrode or using a sterile housing with inserted electrode. F = fermenter, W = waste.

or systems with small enzyme reactors apart from the transducer. FIA has the advantage that the enzyme is not permanently in contact with the analyte or reagent, and that the "pulsing" of one of them provides signals that can easily be amplified and permit the compensation of zero line drifts (for general principles of FIA, see [422]). Ethanol in beer has been determined by means of an alcohol oxidase electrode integrated in a FIA system [423], while the use of dehydrogenases in connection with electrocatalytic or fluorimetric NADH oxidation required the separation of enzyme and detector [424]. This principle has been further developed, especially in automated multi-substrate analyzers, and has been applied to the determination of ethanol and acetaldehyde in wine [425], and of different sugars in fruits [426], alcoholic beverages [427], and milk products [428]. Recent examples demonstrate the combination of sampling systems — even overcoming the problem of membrane clogging by bacteria and macromolecules by automated filter renewal — with versatile analyzers based on enzyme electrodes [429] or other analytical systems [430]. Not only clogging of the sampling system but also clogging of the electrode membrane and interferences from low-molecular mass substances with the enzyme require the development of "problem-orientated" devices for the sample pretreatment [431]. This can contribute to the reliability and lifetime of biosensors in the analyzer. Recent examples are the separation of amino acids from a fermentation broth prior to analysis [432] and the separation of volatile compounds (eg, ethanol) by distillation [433] or pervaporation [434].

Environmental protection requires a large arsenal of analytical methods to assess the quality of soil, water, and air. The most prominent analytes in this area are organic waste, heavy metals, and toxic gases. Indication of the wastewater constituents assimilable by microbes, ie, a parameter similar to the biological oxygen demand (BOD), can be obtained with microbial sensors. Conventional BOD determination requires 5 d and is therefore unsuitable for process control. Hence, sensors for rapid BOD estimation have been developed by using immobilized cells of *Bacillus subtilis* and *Trichosporon cutaneum* [435, 436]. They measure the acceleration of respiration resulting from nutrient supply. The sensors are calibrated in a solution containing equimolar concentrations of glucose and glutamic acid (Figure 14-42), where the signal depends linearly on concentrations between 4 and 100 mg L^{-1}. The short measuring time makes the sensors well suited to monitoring wastewater treatment. Similar microbial sensors are being used routinely for analyses of effluent water in Japan [437]. However, a limitation to this approach is that the organic wastewater components are converted with different reaction velocities; macromolecules, such as starch and proteins, are not indicated at all, but this drawback might be overcome by enzymatic sample pretreatment.

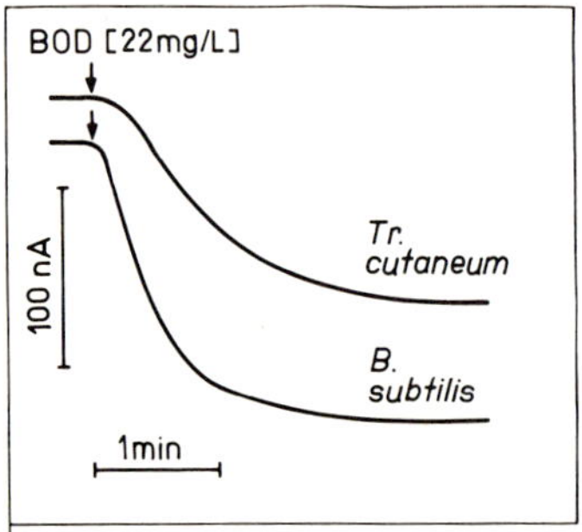

Figure 14-42.
Response curves of *Bacillus subtilis*- and *Trichosporon cutaneum*-based sensors to glucose-glutamic acid standard solution containing 22 mg/L biochemical oxygen demand (BOD). Reproduced from [420] with permission from Springer-Verlag.

14.8 Perspectives

Up to now the practical application of biosensors has been almost entirely limited to oxidase-based amperometric monoenzyme electrodes and pH-shifting by hydrolases. However, the internal coupling of different biocatalytic reactions in biosensors will lead to a greatly extended applicability and a substantial improvement of the analytical performance characteristics. Exciting results might be achieved by applying the concepts of chemical and genetic modification of enzymes. Further, the site-to-site directed fixation of artificially coupled enzymes could improve the speed and practicability of coupled substrate conversions.

Another direction for expanding the field of biosensors is the adoption of complex biological recognition elements. Using organelles, intact cells, or tissue slices, "group effects" such as toxicity, mutagenicity, or content of nutrients become accessible to sensor technology. Further, the excellent specificity of antibodies, which can be directed against a wide arsenal of substances, opens up new horizons. In this respect, immunosensing using low-affinity monoclonal antibodies combined with direct indication of the complex formation appears promising for continuous monitoring of analytes at low concentration levels. Since electrochemical processes are involved in the neuronal signals of the chemical senses, it seems possible to apply directly chemoreception for recognition, processing, and transduction of signals in biosensors. Chemical signals related to olfaction or taste might become accessible to quantification by such receptrodes.

Electrochemical biosensors may be expected to maintain their leading position up to the end of the century. In this respect, the availability of transducers, eg, ion-selective field-effect transistors prepared by mass-production technology, will result in widespread application. In addition to "one-shot" use, multifunctional sensing in minute volumes will be realized. In addition to electrodes, optical, thermometric, and piezoelectric transducers are likely to become exploited in the next generation of sensor. Inexpensive equipment to be used in all areas where material has to be detected and quantified will be produced by integrating the fixation of the biocomponent with the micromechanical fabrication of the analyzers.

14.9 References

[1] Turner, A. P. F., in: *Biosensors — Fundamentals and Applications,* Turner, A. P. F., Karube, I., Wilson, G. S. (eds.); Oxford: Oxford University Press, 1987, p. V.

[2] Kricka, L. J., "Molecular and Ionic Recognition by Biological Systems", in: *Chemical Sensors,* Edmonds, T. E. (ed.); Glasgow: Blackie and Sons, 1988, pp. 3–14.

[3] Stryer, L., "Die Sehkaskade", *Spektrum der Wissenschaft* **9** (1987) 86.

[4] Schnapf, J. L., Baylor, D. A., "Die Reaktion von Photorezeptoren auf Licht", *Spektrum der Wissenschaft* **6** (1987) 116.

[5] Wilden, U., Hall, S. W., Kühn, H., "Phosphodiesterase activation by photoexcited rhodopsin is phosphorylated and binds the intrinsic 48-kDa protein of rod outer segments", *Proc. Natl. Acad. Sci. USA* **83** (1986) 1174–1178.

[6] Pace, U., Hamski, E., Salomon, Y., Lancet, D., "Odorant-sensitive adenylate cyclase may mediate olfactory reception", *Nature* **316** (1985) 255–258.

[7] "Sensorische Transduktionsprozesse", in: *Biophysik,* Hoppe, W., Lehmann, W., Markl, H., Ziegler, H. (eds.); Berlin: Springer, 1977, pp. 391–414.

[8] "Die Grundprinzipien der Signalweiterleitung", in: *Membranrezeptoren und ihre Effektorsysteme,* Repke, H., Liebman, C. (eds.); Weinheim: VCH, 1987, pp. 175–221.

[9] Evans, P., „Receptor and Ion Channels", *J. Exp. Biol.* **124** (1986) 1–4.

[10] Berridge, M. J., "Die Signalübertragung in der Zelle", *Spektrum der Wissenschaft* **12** (1985) 136.

[11] Catterall, W. A., "Molecular Properties of Voltage-Sensitive Sodium Channels", *Ann. Rev. Biochem.* **55** (1986) 953–985.

[12] "Neurotoxine als Sonden für Ionenkanäle", in: *Einführung in die Neurochemie,* Hucho, F., (ed.); Weinheim: VCH, 1987, pp. 122–124.

[13] Biggio, G., Spano, P. F., Toffano, G., Gessa, G. L., "Voltage-Sensitive Ion Channels: Modulation by Neurotransmitters and Drugs", in: *Symposia in Neuroscience,* Fidia Research Series: Berlin: Springer, 1988.

[14] Carafoli, E., Penniston, J. T., "Das Calcium Signal", *Spektrum der Wissenschaft* **1** (1986) 76.

[15] Reuter, H., "Modulation of Ion Channels by Phosphorylation and Second Messengers", *NIPS* **2** (1987) 168–171.

[16] Tedesco, J. L., Krull, U. J., Thompson, M., "Molecular Receptors and Their Potential for Artificial Transduction", *Biosensors* **4** (1989) 135–167.

[17] Ligler, F. S., Fare, T. L., Seib, K. D., Smuda, J. W., Singh, A., Ahl, P., Ayers, M. E., Dalziel, A., Yager, P., "Fabrication of Key Components of a Receptor-Based Biosensor", *Med. Instrumen.* **22** (1988) 247–256.

[18] Buch, R. M., Rechnitz, G. A., "Intact Chemoreceptor-Based Biosensors: Responses and Analytical Limits", *Biosensors* **4** (1989) 215–230.

[19] Adler, J., "How Mobile Bacteria Are Attracted and Repelled by Chemicals. An Approach to Neurobiology", *Biol. Chem. Hoppe-Seyler* **368** (1987) 163–173.

[20] Riedel, K., Renneberg, R., Wollenberger, U., Kaiser, G., Scheller, F., "Microbial Sensors. Fundamentals and Applications for Process Control", *J. Chem. Tech. Biotechnol.* **44** (1989) 85–106.

[21] Marcus, R. A., Sutin, N., "Electron Transfer in Chemistry and Biology", *Biochim. Biophys. Acta* **811** (1985) 265–322.

[22] Schmidt, H.-L., Günther, H., "Structure and Electrochemistry of Oxidoreductases", *Phil. Trans. R. Soc. Lond.* **B 316** (1987) 73–84.

[23] Adman, E. T., "A Comparison of the Structures of Electron Transfer Proteins", *Biochim. Biophys. Acta* **549** (1979) 107–144.

[24] Koppenol, W. H., Margoliash, E., The Asymmetric Distribution of Charges on the Surface of Horse Cytochrome C. Functional Implications, *J. Biol. Chem.* **257** (1982) 4426–4437.

[25] Concar, D. W., Hill, H. A. O., Moore, G. R., Whitford, D., Williams, R. J. P., "The Modulation of Cytochrome C Electron Self-Exchange by Site-Specific Chemical Modification and Anion Binding", *FEBS Lett.* **206** (1986) 15–19.

[26] Capaldi, R. A., Malatesta, F., Darley-Usmar, V. M., "Structure of Cytochrome C Oxidase", *Biochim. Biophys. Acta* **726** (1983) 135.

[27] Carpeillere-Blandin, C., "Transient Kinetics of the One-Electron Transfer Reaction between Reduced Flavocytochrome B_2 and Oxidized Cytochrome C. Evidence for the Existence of a Protein Complex in the Reaction", *Eur. J. Biochem.* **128** (1982) 533–542.

[28] Millett, F., De Jong, C., Paulson, L., Capaldi, R. A., "Identification of Specific Carboxylate Groups on Cytochrome C Oxidase That Are Involved in Binding Cytochrome C", *Biochemistry (Wash.)* **22** (1983) 546–552.

[29] Gutmann, F., "Some Aspects of Charge Transfer in Biological Systems", in: *Modern Bioelectrochemistry* Vol. 6, Gutmann, F., Keyzer, H., (eds.); New York: Plenum Press, 1986.

[30] Koppenol, W. H., Margoliash, E., "The Asymmetric Distribution of Charges on the Surface of Horse Cytochrome C. Functional Implications", *J. Biol. Chem.* **257** (1982) 4426–4437.

[31] In: *Bioelektrochemische Membraneelektroden,* Schindler, J. G., Schindler, M. M. (eds.); Berlin: W. de Gruyter, 1983, pp. 40–116.

[32] Lehn, J. M., "Surpamolekulare Chemie — Moleküle, Übermoleküle und molekulare Funktionseinheiten", *Angew. Chem.* **100** (1988) 91–116.

[33] In: *Electrochemical Sensors in Immunological Analysis.* Ngo, T. T. (ed.); New York: Plenum Press, 1987.

[34] Cammann, K., personal communication.

[35] Lerner, R. A., Framonto, A., "Katalytische Antikörper", *Spektrum der Wissenschaft* 5 (1988) 78–87.

[36] Cook, G. M. W., "Surface carbohydrates. Molecules in Search of a Function", *J. Cell. Sci. Suppl.* 4 (1986) 45–70.

[37] Los, H., Sharon, N., "Lectins as Molecules and as Tools", *Ann. Rev. Biochem.* 55 (1986) 35–67.

[38] Quiocho, F. A., "Carbohydrate-binding Proteins: Tertiary Structures and Protein-Sugar Interactions", *Ann. Rev. Biochem.* 55 (1986) 287–315.

[39] Downs, M. E. A., Kobayashi, S., Karube I., "New DNA Technology and the DNA Biosensor", *Analyt. Lett.* 20 (1987) 1897–1927.

[40] Lehn, J.-M., "Supramolecular Chemistry: Receptors, Catalysts and Carriers", *Science* 227 (1985) 849–856.

[41] Moody, G. J., Thomas, J. D. R., "Organic Sensor Materials in Entangled and Polymer-Bound Matrices for Ion-Selective Electrodes", in: *Chemical Sensors* Edmonds, T. E. (ed.); Glasgow: Blackie and Sons, 1988, pp. 75–116.

[42] Schmidtchen, F. P., "Molekulare Wirte für Anionen", *Nachr. Chem. Tech. Lab.* 36 (1988) 9–17.

[43] Franke, J., Vögtle, F., "Complexation of Organic Molecules in Water", in: *Biomimetic and Bioorganic Chemistry II,* Vögtle, F., Weber, E. (eds.); Berlin: Springer, 1986, pp. 133–170.

[44] Hamilton, A. D., Pant, N., Muehldorf, A., "Artificial Receptors for Biologically Active Molecules", *Pure & Appl. Chem.* 60 (1988) 533–538.

[45] Koga, K., Sasaki, S., "Functionalization of Crown Ethers. An Approach to the Enzyme Model for Peptide Synthesis", *Pure & Appl. Chem.* 60 (1988) 539–543.

[46] Thompson, M., Dorn, W. H., "Selective Chemical Transduction Based on Chemoreceptive Control of Membrane Ion Permeability", in: *Chemical Sensors,* Edmonds, T. E. (ed.); Glasgow: Blackie and Sons, 1988, pp. 168–190.

[47] Egger, M., Eggl, P., Sackmann, E., "Preparation and Investigation of Asymmetric Bilayers on Solid Supports", in: *Biosensors. International Workshop 1987,* GBF-Monographs, Vol. 10, Schmid, R. D., Guilbault, G. G., Karube, I., Schmidt, H.-L., Wingard, L. B. (eds); Weinheim: VCH, 1987, pp. 337–338.

[48] Wong, H. E., *Langmuir-Blodgett Film Technology in the Development of Selective Chemical Sensors,* Ph. D. Thesis, University of Toronto, 1987.

[49] Sackmann, E., *Ber. Bunsenges. Phys. Chem.* 89 (1985) 1198–1208.

[50] Egger, M., Heyn, S. P., Gaub, H. E., "2d Pattern Formation of Lipid-Anchored Fab'-Fragments", Biophys. J. 57 (1990) 669–673.

[51] Sudholter, E. J. R., "A Key Step in the Development of Long-Lifetime Membrane Modified ISFETs", in: *Abstract EG-Workshop; The Interface between Biology and Sensors, Cranfield, 1988.*

[52] Kallury, R. K. M. R., Krull, U. J., Thompson, M., "Syntheses of Phospholipids Suitable for Covalent Binding to Surfaces", *J. Org. Chem.* 52 (1987) 5478–5480.

[53] Schmidt, H.-L., Kittsteiner-Eberle, R., "Biosensoren", *Naturwissenschaften* 73 (1986) 314–321.

[54] Hall E. A. H., "Recent Progress in Biosensor Development", *Int. J. Biochem.* 20 (1988) 357–362.

[55] Frew, J. E., Hill, H. A. O., "Electrochemical Biosensors", *Anal. Chem.* 59 (1987) 933A–944A.

[56] Pickup, J. C., Lancet (1985) 817–820. Biosensors: A Clinical Perspective.

[57] Turner, A. P. F., "Biosensors: Principles and Potentials", in: *Chemical Aspects of Food Enzymes,* Andrews, A. T. (ed.); 1987, pp. 259–270.

[58] Eddowes, M. J., Hill, H. A. O., "Binding as a Prerequisite for Rapid Electron Transfer Reactions of Metalloproteins", *Am. Chem. Soc. Adv. Chem. Ser.* 201 (1982) 173–197.

[59] Eddowes, M. J., Hill, H. A. O., Uosaki, K., "The Electrochemistry of Cytochrome C. Investigation of the Mechanism of the 4,4'-Bipyridyl Surface Modified Gold Electrode", *Bioelectrochem. Bioenerg.* 7 (1980) 527–537.

[60] Bard, A. J., Faulkner, L. R., in: *Electrochemical Methods. Fundamental and Applications,* New York: Wiley, 1980.

[61] Lyklema, J., "Proteins at Solid-Liquid Interfaces a Colloid-Chemical Review", *Colloids Surf.* 10 (1984) 33–42.

[62] Eddowes, M. J., Hill, H. A. O., "Investigation of Electron-Transfer Reactions of Proteins by Electrochemical Methods", *Biosci. Rep.* 1 (1981) 521–532.

[63] Eddowes, M. J., Hill, H. A. O., "Factors Influencing the Electron-Transfer Rates of Redox Proteins", *Faraday Disc. Chem. Soc.* **74** (1982) 331–341.

[64] Taniguchi, I., Toyosawa, K., Tamaguchi, M., Yasukouchi, K., "Voltammetric Response of Horse Heart Cytochrome C at a Gold Electrode in the Presence of Sulphur-Bridged Bipyridines", *J. Electroanal. Chem. Interfac. Electrochem.* **140** (1982) 187–193.

[65] DiGleria, K., Hill, H. A. O., Lowe, V. J., Page, D. J., "Direct Electrochemistry of Horse-Heart Cytochrome C at Amino Acid-Modified Gold Electrodes", *J. Electroanal. Chem.* **213** (1986) 333–338.

[66] Taniguchi, I., Iseki, M., Toyosawa, K., Yamaguchi, H., Yasokouchi, K., "Purines as New Promoters for the Voltammetric Response of Horse Heart Cytochrome C at Gold Electrodes", *J. Electroanal. Chem.* **164** (1984) 385–391.

[67] Hill, H. A. O., Page, D. J., Walton, N. J., "Intra-Molecular Hydrogen Bonding in Surface-Modified Gold Electrodes and the Effect of Specific Anions on the Electrochemistry of Cytochrome C", *J. Electroanal. Chem.* **208** (1986) 395–400.

[68] Haldjian, J., Bianco, P., Pilard, R., "Modification of the Gold Electrode for the Electrochemical Study of Cytochrome C", *Electrochim. Acta* **28** (1983) 1823–1828.

[69] Allen, P. M., Hill, H. A. O., Walton, N. J., "Surface Modifiers for the Promotion of Direct Electrochemistry of Cytochrome C", *J. Electroanal. Chem.* **178** (1984) 69–86.

[70] Hill, H. A. O., Whitford, D., "Direct Electrochemistry of Native and 4-chloro-3,5-dinitrophenyl (CDNP)-Substituted Cytochrome C at Surface-Modified Gold and Pyrolytic Graphite Electrodes", *J. Electroanal. Chem.* **235** (1987) 153–167.

[71] Durliat, H., Comtat, M., "Investigation of Electron Transfer Between Platinum and Large Biological Molecules by Thin-Layer Spectroelectrochemistry", *Anal. Chem.* **54** (1982) 856–861.

[72] Taniguchi, J., Iseki, M., Yamaguchi, H., Yasukouchi, K., "Surface Enhanced Raman Scattering from bis(4-pyridyl)disulfide- and 4,4′-bipyridine-modified Gold Electrodes", *J. Electroanal. Chem.* **186** (1985) 299–307.

[73] Bowden, E. F., Hawkridge, F. M., Blount, H. N., "Interfacial Electrochemistry of Cytochrome C at Tin Oxide, Indium Oxide, Gold, and Platinum Electrodes", *J. Electroanal. Chem.* **161** (1984) 355–376.

[74] Armstrong, F. A., Hill, H. A. O., Oliver, B. N., "Surface Selectivity in the Direct Electrochemistry of Redox Proteins. Contrasting Behaviour at Edge and Basal Planes of Graphite", *J. Chem. Soc. Chem. Commun.* (1984) 976–977.

[75] Armstrong, F. A., Cox, P. A., Hill, H. A. O., Lowe, V. J., "Metal Ions and Complexes as Modulators of Protein-Interfacial Electron Transport at Graphite Electrodes", *J. Electroanal. Chem.* **217** (1987) 331–366.

[76] Bowden, E. F., Hawkridge, F. M., Chlebowski, J. F., Bancroft, E. E., Thorpe, C., Blount, H. N., "Cyclic Voltammetry and Derivative Cyclic Voltabsorptometry of Purified Horse Heart Cytochrome C at Tin-Doped Indium Oxide Optically Transparent Electrodes", *J. Am. Chem. Soc.* **104** (1982) 7641–7644.

[77] Harmer, M. A., Hill, H. A. O., "The Direct Electrochemistry of Horse Heart Cytochrome C, Ferredoxin and Rubredoxin at Ruthenium Dioxide and Iridium Dioxide Electrodes", *J. Electroanal. Chem.* **170** (1984) 369–375.

[78] Reed, D. E., Hawkridge, F. M., "Direct Electron Transfer Reactions of Cytochrome C at Silver Electrodes", *Anal. Chem.* **59** (1987) 2334–2339.

[79] Armstrong, F. A., Hill, A. O., Oliver, B. N., Whitford, D., "Direct Electrochemistry of the 'Blue' Copper Protein Plastocyanin", *Biochem. Soc. Tans.* **14** (1986) 44–45.

[80] Hill, H. A. O., Page, D. J., Walton, N. J., Whitford, D., "Direct Electrochemistry, at Modified Gold Electrodes, of Redox Proteins Having Negatively-Charged Binding Domains: Spinach Plastocyanin and a Multi-Substituted Carboxydinitrophenyl Derivative of Horse Heart Cytochrome C", *J. Electroanal. Chem.* **187** (1985) 315–324.

[81] Armstrong, F. A., Hill, H. A. O., Oliver, B. N., Walton, N. J., "Direct Electrochemistry of Redox Proteins at Pyrolytic Graphite Electrodes", *J. Am. Chem. Soc.* **106** (1984) 921–923.

[82] Armstrong, F. A., Hill, H. A. O., Walton, N. J., "Direct Electrochemical Reduction of Ferredoxin Promoted by Mg^{2+}", *FEBS Lett.* **145** (1982) 241–244.

[83] Turner, A. P. F., Ramsay, G., Higgins, I. J., "Applications of Electron Transfer Between Biological Systems and Electrodes", *Biochem. Soc. Trans.* **11** (1983) 445–450.

[84] Hill, H. A. O., Walton, N. J., Higgins, I. J., "Electrochemical Reduction of Dioxygen Using a Terminal Oxidase", *FEBS Lett.* **126** (1981) 282–284.

[85] Hill, H. A. O., Walton, N. J., Whitford, D., Coleman, J. O. D., "The Measurement of Changes in pH Associated with Electrochemically Driven Respiration in Rat Liver Mitochondria", *J. Inorg. Biochem.* **23** (1985) 303–309.

[86] Cass, A. E. G., Davis, G., Hill, H. A. O., Nancarrow, D. J., "The Reaction of Flavocytochrome B2 with Cytochrome C and Ferricinium Carboxylate. Comparative Kinetics by Cyclic Voltammetry and Chronoamperometry", *Biochim. Biophys. Acta* **828** (1985) 51–57.

[87] Hilll, H. A. O., Oliver, B. N., Page, D. J., Hopper, D. J., "The Enzyme-Catalysed Electrochemical Conversion of p-Cresol into p-Hydroxybenzaldehyde", *J. Chem. Soc. Chem. Commun.* (1985) 1469–1471.

[88] Fultz, M. L., Durst, R. A., "Mediator Compounds for the Electrochemical Study of Biological Redox Systems", *Anal. Chim. Acta* **140** (1982) 1–18.

[89] Szentrimay, R., Yeh, P., Kuwana, T., "Evaluation of Mediator-Titrants for the Indirect Coulometric Titration of Biocomponents", in: *Electrochemical Studies of Biological Systems*, Vol. 38, Sawyer, D. (ed.); Washington: Am Chem. Soc. Symp., 1977, pp. 143–169.

[90] Prince, R. C., Linkletter, S. J. G., Dutton, P. L., "The Thermodynamic Properties of Some Commonly Used Oxidation-Reduction Mediators, Inhibitors and Dyes, as Determined by Polarography", *Biochim. Biophys. Acta* **635** (1981) 132–148.

[91] Davis, G., "Electrochemical Techniques for the Development of Amperometric Biosensors", *Biosensors* **1** (1985) 161–178.

[92] Nicholson, R. S., "Theory and Application of Cyclic Voltammetry for Measurement of Electrode Reaction Kinetics", *Anal. Chem.* **37** (1965) 1351–1355.

[93] Andrieux, C. P., Dumas-Bouchiat, J. M., Saveant, J. M., "Catalysis of Electrochemical Reactions at Redox Polymer Electrodes. Kinetic Model for Stationary Voltammetric Techniques", *J. Electroanal. Chem.* **131** (1982) 1–35.

[94] Kamin, R. A., Wilson, G. S., "Rotating Ring-Disk Enzyme Electrode for Biocatalysis Kinetic Studies and Characterization fo the Immobilized Enzyme Layer", *Anal Chem.* **52** (1980) 1198–1205.

[95] Clark, L. C., Lyons, C., "Electrode Systems for Continuous Monitoring in Cardiovascular Surgery", *Ann. N. Y. Acad. Sci.* **102** (1962) 29–45.

[96] Updike, S. J., Hicks, G. P., "Reagentless Substrate Analysis with Immobilized Enzymes", *Science* **158** (1967) 270–272.

[97] Ianiello, R. M., Yacynych, A. M., "Immobilized Enzyme Chemically Modified Electrode as an Amperometric Sensor", *Anal. Chem.* **53** (1981) 2090–2095.

[98] Cleland, N., Enfors, S.-O., "Control of Glucose-Fed Batch Cultivations of E. coli by Means of an Oxygen Stabilized Enzyme Electrode", *Eur. J. Appl. Microbiol. Biotechnol.* **18** (1983) 141–147.

[99] Bourdillon, C., Bourgeois, J. B., Thomas, D., "Covalent Linkage of Glucose Oxidase on Modified Glassy Carbon Electrodes. Kinetic Phenomena", *J. Am. Chem. Soc.* **102** (1980) 4231–4235.

[100] Romette, J. L., Froment, B., Thomas, D., "Glucose Oxidase Electrode. Measurements of Glucose in Samples Exhibiting High Variability in Oxygen Content", *Clin. Chim. Acta* **95** (1979) 249–253.

[101] Gough, D. A., Lucisano, J. Y., Tse, P. H. S., "Two-Dimensional Enzyme Electrode Sensor for Glucose", *Anal. Chem.* **57** (1985) 2351–2357.

[102] Tse, P. H. S., Gough, D. A., "Transient Response of an Enzyme Electrode Sensor for Glucose", *Anal. Chem.* **59** (1987) 2339–2344.

[103] Wollenberger, U., Scheller, F., Pfeiffer, D., Bogdanovskaya, V. A., Tarasevich, M. R., Hanke, G., "Laccase/Glucose Oxidase Electrode for Determination of Glucose", *Anal. Chim. Acta* **187** (1986) 39–45.

[104] Gorton, L., "A Carbon Electrode Sputtered with Palladium and Gold for the Amperometric Detection of Hydrogen Peroxide", *Anal. Chim. Acta* **178** (1985) 247–253.

[105] Jönsson, G., Gorton, L., "An Amperometric Glucose Electrode Based on Adsorbed Glucose Oxidase on Palladium/Gold Modified Graphite", *Anal. Lett.* **20** (1987) 839–855.

[106] Bennetto, H. P., DeKeyzer, D. R., Delaney, G. M., Koshy, A., Mason, J. R., Razack, L. A., Stirling, J. L., Thurston, C. F., "An Amperometric Biosensor for Laboratory Determination of Glucose", *Internat. Anal.* **8** (1987) 22–27.

[107] Cardosi, M. F., Turner, A. P. F., "Glucose Sensors for The Management of Diabetes Mellitus", *The Diabetes Annual/3*, Alberti, K. G. M. M., Krall, L. P. (eds); Amsterdam: Elsevier Science Publishers, 1987, pp. 560–577.

[108] Sternberg, R., Barrau, M.-B., Gangiotti, L., Thévenot, D. R., „Study and Development of Multilayer Needle-Type Enzyme-Based Glucose Microsensors", *Biosensors* **4** (1988) 27–40.

[109] Churchouse, S. J., Battersby, C. M., Mullen, W. H., Vadgama, P. M., "Needle Enzyme Electrodes for Biolocigal Studies", *Biosensors* **2** (1986) 325–342.

[110] Gernet, S., Koudelka, M., DeRooij, N. F., "Fabrication and Characterization of a Planar Electrochemical Cell and its Application as a Glucose Sensor", *Sensors & Actuators* **18** (1989) 59–70.

[111] Koudelka, M., Gernet, S., DeRooij, N. F., „Planar Amperometric Enzyme-Based Glucose Microelectrode", *Sens. Actuators* **18** (1989) 157–165.

[112] Mascini, M., Mazzei, F., "Amperometric Sensor for Pyruvate with Immobilized Pyruvate Oxidase", *Anal. Chim. Acta* **192** (1987) 9–16.

[113] Yao, T., Wasa, T., "Flow-Injection System for Simultaneous Assay of Free and Total Cholesterol in Blood Serum by Use of Immobilized Enzymes", *Anal. Chim. Acta* **207** (1988) 319–323.

[114] Pilloton, R., Nwosu, T. N., Mascini, M., "Amperometric Determination of Lactic Acid. Applications on Milk Samples", *Anal. Lett.* **21** (1988) 727–740.

[115] Aleksandrovskii, Y. A., Bezhikina, L. V., Rodionoy, Y. U., "Comparative Study of Reactions Catalysed by Glucose Oxidase in the Presence of Different Electron Acceptors", *Biokhimiya* **46** (1981) 708–716.

[116] Schlapfer, P., Mundt, W., Racine, P., "Electrochemical Measurement of Glucose using Various Electron Acceptors", *Clin. Chim. Acta* **57** (1974) 283.

[117] Turner, A. P. F., Hendry, S. P., Cardosi, M. F., "Tetrathiafulvalene: A New Mediator for Amperometric Biosensors", *World Biotech. Report* **1** (1987) 125–137.

[118] Kulys, J. J., Cenas, N. K., "Oxidation of Glucose Oxidase from *Penicillium Vitale* by One- and Two-Electron Acceptors", *Biochim. Biophys. Acta* **744** (1983) 57–63.

[119] Schelter-Graf, A., Huck, H., Schmidt, H.-L., "Rasche und genaue Bestimmung von Ethanol mittels einer Oxidase-Elektrode in einem Strömungs-Injektions-System", *Z. Lebensm. Unters. Forsch.* **177** (1983) 356–358.

[120] Ikeda, T., Hamada, H., Senda, M., "Electrocatalytic Oxidation of Glucose at a Glucose Oxidase-Immobilized Benzoquinone-Mixed Carbon Paste Electrode", *Agric. Biol. Chem.* **50** (1986) 883–890.

[121] Senda, M., Ikeda, T., Hiasa, H., Miki, K., "Amperometric Biosensors Based on a Biocatalyst Electrode with Entrapped Mediator. *Anal. Sci.* **2** (1986) 501–506.

[122] Cass, A. E. G., Davis, G., Green, M. J., Hill, H. A. O., "Ferricinium Ion as an Electron Acceptor for Oxido-Reductases", *J. Electroanal. Chem.* **190** (1985) 117–127.

[123] Turner, A. P. F., "Amperometric Biosensors Based on Mediator Modified Electrodes", *Methods in Enzymology (Immobilized Enzymes an Cells)* Vol. 137, Colowick, S. P., Kaplan, N. O., Mosbach, K., (eds.); San Diego: Academic Press, 1988, pp. 90–103.

[124] Stankovich, M. T., Schopfer, L. M., Massey, V., "Determination of Glucose Oxidase Oxidation-Reduction Potentials and the Oxygen Reactivity of Fully Reduced Semiquinoid Forms", *J. Biol. Chem.* **253** (1978) 4971–4979.

[125] Cass, A. G., Davis, G., Francis, G. D., Hill, H. A., Aston, W. J., Higgins, I. J., Plotkin, E. V., Scott, L. D. L., Turner, A. P. F., "Ferrocene-mediated Enzyme Electrode for Amperometric Determination of Glucose", *Anal. Chem.* **56** (1984) 667–671.

[126] Lange, M. A., Chambers, J. Q., "Amperometric Determination of Glucose With a Ferrocene-Mediated Glucose Oxidase/Polyacrylamide Gel Electrode", *Anal. Chim. Acta* **175** (1985) 89–97.

[127] Schuhmann, W., Wohlschläger, H., Lammert, R., Schmidt, H.-L., Löffler, U., Wiemhöfer, H.-D., Göpel, W., "Leaching of Dimethylferrocene, a Redox Mediator in Amperometric Enzyme Electrodes", *Sens. Actuators* **B1** (1990) 571–575.

[128] Brooks, S. L., Ashby, R. E., Turner, A. P. F., Calder, M. R., Clarke, D. J., "Development of an On-Line Glucose Sensor for Fermentation Monitoring", *Biosensors* **3** (1987/88) 45–56.

[129] Barbaric, S., Kozulic, B., Leustek, I., Pavlovic, B., Cesi, V., Mildner, P., "Crosslinking of Glycoenzymes Via Their Carbohydrate Chains", *3rd Eur. Congr. Biotechnol.* **1** (1984) 307–312.

[130] Schuhmann, W., Huber, J., Schmidt, H.-L., unpublished.

[131] Yeary, R. A., "Chronic Toxicity of Dicyclopentadienyliron (Ferrocene) in Dogs", *Toxicol. Appl. Pharmacol.* **15** (1969) 666–667.

[132] Pickup, J. C., Shaw, G. W., Claremont, D. J., "Implantable Glucose Sensors: Choosing the Appropriate Sensing Strategy", *Biosensors* **3** (1987/88) 335–346.

[133] Pickup, J. C., Shaw, G. W., Claremont, D. J., "Potentially-Implantable, Amperometric Glucose Sensors with Mediated Electron Transfer: Improving the Operating Stability", *Biosensors* **4** (1989) 109–119.

[134] Matthews, D. R., Holman, R. R., Bown, E., Steemson, J., Watson, A., Hughes, S., Scott, D., "Pen-Sized Digital 30-second Blood Glucose Meter", *Lancet* (1987) 778–779.

[135] Dicks, J. M., Aston, W. J., Davis, G., Turner, A. P. F., "Mediated Amperometric Biosensors for D-Galactose, Glycolate and L-Amino Acids Based on a Ferrocene-Modified Carbon Paste Electrode", *Anal. Chim. Acta* **182** (1986) 103–112.

[136] Frew, J. E., Green, M. J., Hill, H. A. O., "New Electrochemical Techniques Applied to Medicine and Biology", *J. Chem. Tech. Biotechnol.* **36** (1986) 357–363.

[137] McNeil, C. J., Spoors, J. A., Cocco, D., Cooper, J. M., Bannister, J. V., "Thermostable Reduced Nicotinamide Adenine Dinucleotide Oxidase: Application to Amperometric Enzyme Assay", *Anal. Chem.* **61** (1989) 25–29.

[138] Green, M. J., Hill, H. A. O., "Amperometric Enzyme Electrodes", *J. Chem. Soc. Faraday Trans.* **82** (1986) 1237–1243.

[139] Green, M. J., Davis, G., Hill, H. A. O., "Creatine Kinase Assay Using an Enzyme Electrode", *J. Biomed. Eng.* **6** (1984) 176–177.

[140] D'Costa, E. J. D., Higgins, I. J., Turner, A. P. F., "Quinoprotein Glucose Dehydrogenase and Its Application in an Amperometric Glucose Sensor", *Biosensors* **2** (1986) 71–87.

[141] Mullen, W. H., Churchhouse, S. J., Vadgama, P. M., "Enzyme Electrode for Glucose Based on the Quinoprotein Glucose Dehydrogenase", *Analyst* **110** (1985) 925–928.

[142] Jaegfeldt, H., "Adsorption and Electrochemical Oxidation Behaviour of NADH at a Clean Platinum Electrode", *J. Electroanal. Chem.* **110** (1980) 295–302.

[143] Samec, Z., Elving, P. J., "Anodic Oxidation of Dihydronicotinamide Adenine Dinucleotide at Solid Electrodes; Mediation by Surface Species", *J. Electroanal. Chem.* **144** (1983) 217–234.

[144] Elving, P. J., Bresnahan, W. T., Moiroux, J., Samec, Z., "NAD/NADH as a Model Redox System: Mechanism, Mediation, Modification by the Environment", *Bioelectrochem. Bioeng.* **9** (1982) 365–378.

[145] Cenas, N. K., Rozgaite, J., Pocius, A., Kulys, J. J., "Electrocatalytic Oxidation of NADH and Ascorbic Acid on Electrochemically Pretreated Glassy Carbon Electrodes", *J. Electroanal. Chem.* **154** (1983) 121–128.

[146] Tse, D. C.-S., Kuwana, T., "Electrocatalysis of Dihydronicotinamide Adenosine Diphopshate with Quinones and Modified Quinone Electrodes", *Anal. Chem.* **50** (1978) 1315–1318.

[147] Jaegfeldt, H., Kuwana, T., Johansson, G., "Electrochemical Stability of Catechols with a Pyrene Side Chain Strongly Adsorbed on Graphite Electrodes for Catalytic Oxidation of Dihydronicotinamide Adenine Dinucleotide", *J. Am. Chem. Soc.* **105** (1983) 1805–1814.

[148] Malinauskas, A., Kulys, J. J., "Alcohol, Lactate and Glutamate Sensors Based on Oxidoreductases with Regeneration of Nicotinamide Adenine Dinucleotide", *Anal. Chim. Acta* **98** (1978) 31–37.

[149] Torstensson, A., Gorton, L., "Catalytic Oxidation of NADH by Surface-Modified Graphite Electrodes", *J. Electroanal. Chem.* **130** (1981) 199–207.

[150] Gorton, L., Torstensson, A., Jaegfeldt, H., Johansson, G., "Electrocatalytic Oxidation of Reduced Nicotinamide Coenzymes by Graphite Electrodes Modified with an Adsorbed Phenoxazinium Salt, Meldola Blue", *J. Electroanal. Chem.* **161** (1984) 103–120.

[151] Huck, H., "Katalytische Oxidation von NADH an Graphitelektroden durch adsorbierte Phenoxazinderivate", *Fresenius Z. Anal. Chem.* **313** (1982) 548–552.

[152] Huck, H., Schelter-Graf, A., Danzer, J., Kirch, P., Schmidt, H.-L., "Bioelectrochemical Detection Systems for Substrates of Dehydrogenases", *Analyst* **109** (1984) 147–150.

[153] Gorton, L., Johansson, G., Torstensson, A., "A Kinetic Study of the Reaction between Dihydronicotinamide Adenine Dinucleotide (NADH) and an Electrode Modified by Adsorption of 1,2-Benzophenoxazine-7-One", *J. Electroanal. Chem.* **196** (1985) 81–92.

[154] Kulys, J. J., "Enzyme Electrodes Based on Organic Metals", *Biosensors* **2** (1986) 3–13.

[155] Albery, W. J., Bartlett, P. N., Cass, A. E. G., "Amperometric Enzyme Electrodes", *Phil. Trans. Royal Soc. Lond.* **B 316** (1987) 107–119.

[156] Carlson, B. W., Miller, L. L., "Oxidation of NADH by Ferrocinium Salts. Rate-Limiting One-Electron Transfer", *J. Am. Chem. Soc.* **105** (1983) 7354–7453.

[157] Carlson, B. W., Miller, L. L., Neta, P., Grodkowski, J., "Oxidation of NADH Involving Rate-Limiting One-Electron Transfer", *J. Am. Chem. Soc.* **106** (1984) 7233–7239.

[158] Powell, M. F., Wu, J. C., Bruice, T. C., "Ferricyanide Oxidation of Dihydropyridines and Analogues", *J. Am. Chem. Soc.* **106** (1984) 3850–3856.

[159] Yon Hin, B. F. Y., Lowe, C. R., "Catalytic Oxidation of Reduced Nicotinamide Adenine Dinucleotide at Hexacyanoferrate-Modified Nickel Electrodes", *Anal. Chem.* **59** (1987) 2111–2115.

[160] Degani, Y., Heller, A., "Direct Electrical Communication between Chemically Modified Enzymes and Metal Electrodes. 1. Electron Transfer from Glucose Oxidase to Metal Electrodes via Electron Relays, Bound Covalently to the Enzyme", *J. Phys. Chem.* **91** (1987) 1285–1289.

[161] Degani, Y., Heller, A., "Direct Electrical Communication between Chemically Modified Enzymes and Metal Electrodes. 2. Methods for Bonding Electron-Transfer Relays to Glucose Oxidase and D-Amino-Acid Oxidase", *J. Am. Chem. Soc.* **110** (1988) 2615–2620.

[162] Bartlett, P. N., Whitaker, R. G., Green, M. J., Frew, J., "Covalent Binding of Electron Relays to Glucose Oxidase", *J. Chem. Soc. Chem. Commun.* (1987) 1603–1604.

[163] Foulds, N. C., Lowe, C. R., "Enzyme Entrapment in Electrically Conducting Polymers. Immobilization of Glucose Oxidase in Polypyrrole and Its Application in Amperometric Glucose Sensors", *J. Chem. Soc. Faraday Trans.* **82** (1986) 1259–1264.

[164] Umana, M., Waller, J., "Protein-Modified Electrodes. The Glucose Oxidase/Polypyrrole System", *Anal. Chem.* **58** (1986) 2979–2983.

[165] Bartlett, P. N., Whitaker, R. G., "Electrochemical Immobilization of Enzymes. Part 2. Glucose Oxidase Immobilized in Poly-N-Methylpyrrole", *J. Electroanal. Chem.* **224** (1987) 37–48.

[166] Yon Hin, B. F. Y., Lowe, C. R., Sethi, R. S., "Multi-Analyte Microelectronic Biosensors", *Sens. Actuators* **B1** (1990) 550–554.

[167] Foulds, N. C., Lowe, C. R., "Immobilization of Glucose Oxidase in Ferrocene-Modified Pyrrole Polymers", *Anal. Chem.* **60** (1988) 2473–2478.

[168] Schuhmann, W., Lammert, R., Uhe, B., Schmidt, H.-L., "Polypyrrole, a New Possibility for Covalent Binding of Oxidoreductases to Electrode Surfaces as a Base for Stable Biosensors", *Sens. Actuators* **B1.** (1990) 537–541.

[169] Albery, W. J., Bartlett, P. N., Craston, D. H., "Amperometric Enzyme Electrodes. Part 2. Conducting Salts as Electrode Materials for the Oxidation of Glucose Oxidase", *J. Electroanal. Chem.* **194** (1985) 223–235.

[170] Mauk, A. G., Scott, R. A., Gray, H. B., "Distances of Electron Transfer to and from Metalloprotein Redox Sites in Reactions with Inorganic complexes", *J. Am. Chem. Soc.* **102** (1980) 4360–4363.

[171] Kulys, J. J., Samalius, A. S., "Kinetics of Biocatalytic Current Generation", *Bioelectrochem. Bioenerg.* **10** (1983) 385–398.

[172] Kulys, J. J., Samalius, A. S., Svirmickas, G. J. S., "Electron Exchange between the Enzyme Active Center and Organic Metals", *FEBS Letters* **114** (1980) 7–10.

[173] Kulys, J. J., Svirmickas, G.-J. S., "Reagentless Lactate Sensor Based on Cytochrom B_2", *Anal. Chim. Acta* **117** (1980) 115–120.

[174] Mosbach, K., Danielsson, B., *Biochim. Biophys. Acta* **364** (1974) 140.

[175] Schmidt, H.-L., Krisam, G., Grenner, G., "Microcalorimetric Methods for Substrate Determination in Flow Systems with Immobilized Enzymes", *Biochim. Biophys. Acta* **429** (1976) 283–290.

[176] Krisam, G., Schmidt, H.-L., "Development and Properties of Caloric Systems for Substrate Determinations with Immobilized Enzymes", in: *Applications of Calorimetry in Life Sciences*, Lamprecht, I., Schaarschmidt, B. (eds.); Berlin: Walter de Gruyter, 1977, pp. 39–47.

[177] Danielsson, B., Mosbach, K., „Enzyme Thermistors", in: *Methods in Enzymology* Vol. 137, Colowick, S. P., Kaplan, N. O., Mosbach, K. (eds.); San Diego: Academic Press, 1988, pp. 181–197.

[178] Scheller, F. W., Renneberg, R., Schubert, F., "Coupled Enzyme Reactions in Enzyme Electrodes Using Sequence, Amplification, Competition, and Antiinterference Principles", in: *Methods in Enzymology* Vol. 137, Colowick, S. P., Kaplan, N. O., Mosbach, K. (eds); San Diego: Academic Press, 1988, pp. 29–44.

[179] Decristoforo, G., "Flow-Injection Analysis for Automated Determination of β-Lactams Using Immobilized Enzyme Reactors with Thermistors or Ultraviolet Spectrophotometric Detection", in: *Methods in Enzymology* Vol. 137, Colowick, S. P., Kaplan, N. O., Mosbach, K. (eds.); San Diego: Academic Press, 1988, pp. 197–217.

[180] Satoh, I., "Biomedical Applications of the Enzyme Thermistor in Lipid Determinations", in: *Methods in Enzymology* Vol. 137, Colowick, S. P., Kaplan, N. O., Mosbach, K. (eds.); San Diego: Academic Press, 1988, pp. 217–225.

[181] Mandenius, C. F., Danielsson, B., "Enzyme Thermistors for Process Monitoring and Control", in: *Methods in Enzymology* Vol. 137, Colowick, S. P., Kaplan, N. O., Mosbach, K. (eds.); San Diego: Academic Press, 1988, pp. 307–318.

[182] Lunte, C. E., Heineman, W. R., "Electrochemical Techniques in Bioanalysis", *Topics Current Chem.* **143** (1988) 3–48.

[183] Kuan, S. S., Guilbault, G. G., in: *Biosensors – Fundamentals and Applications,* Turner, A. P. F., Karube, I., Wilson, G. S. (eds); Oxford: Oxford University Press, 1987, pp. 135–152.

[184] Castner, J. F., Wingard Jr., L. B., "Alterations in Potentiometric Response of Glucose Oxidase Platinum Electrodes Resulting from Electrochemical or Thermal Pretreatments of a Metal Surface", *Anal. Chem.* **56** (1984) 2891–2896.

[185] Wingard Jr., L. B., Castner, J. F., Yao, S. J., Wolfson Jr., S. K., Drash, A. L., Liu, C. C., "Immobilized Glucose Oxidase in the Potentiometric Detection of Glucose", *Appl. Biochem. Biotechnol.* **9** (1984) 95–104.

[186] Proctor, A., Castner, J. F., Wingard Jr., L. B., Hercules, D. M., "Electron Spectroscopic (ESCA) Studies of Platinum Surfaces Used for Enzyme Electrodes", *Anal. Chem.* **57** (1985) 1644–1649.

[187] Wingard Jr., L. B., Castner, J., "Potentiometric Biosensors Based on Redox Electrodes", in: *Biosensors – Fundamentals and Applications,* Turner, A. P. F., Karube, I., Wilson, G. S. (eds.); Oxford: Oxford University Press, 1987, pp. 153–162.

[188] Xie, S.-L., Schmidt, H.-L(unpublished).

[189] Janata, J., Huber, R. J., "Ion-Sensitive Field Effect Transistors", *Ion-Selective Electrode Rev.* **1** (1979) 31–79.

[190] Janata, J., "Chemically Sensitive Field Effect Transistors", in: *Solid State Chemical Sensors,* Janata, J., Huber, R. J. (eds.); Orlando: Academic Press, 1985, pp. 65–118.

[191] Sibbald, A., "Recent Advances in Field-Effect Chemical Microsensors", *J. Molecular Electronics* **2** (1986) 51–83.

[192] Van der Schoot, B. H., Bergveld, P., "ISFET Based Enzyme Sensors", *Biosensors* **3** (1987/88) 161–186.

[193] Bergveld, P., "The Operation of an ISFET as an Electronic Device", *Sens. Actuators* **1** (1981) 17–29.

[194] Arnoux, C., Buser, R., Decroux, M., Van den Vlekkert, H. H., De Rooij, N. F. "Analysis of the Structure and Drift of Al_2O_3 Layers Used as a pH-Sensitive Material For pH-ISFETs", *Proc. 4th International Conference on Solid-State Sensors and Actuators (Transducers '87), Tokyo, Japan, June 2-5, 1987,* pp. 751–754.

[195] Caras, S., Janata, J., "Field Effect Transistor Sensitive to Penicillin", *Anal. Chem.* **52** (1980) 1935–1937.

[196] Miyahara, Y., Moriizumi, T., Ichimura, K., "Integrated Enzyme FETs For Simultaneous Detection of Urea and Glucose", *Sens. Actuators* **7** (1985) 1–10.

[197] Caras, S. D., Petelenz, D., Janata, J., "pH-based Enzyme Potentiometric Sensors. Part 2. Glucose-Sensitive Field Effect Transistor", *J. Anal. Chem.* **57** (1985) 1920–1923.

[198] Hanazato, Y., Inatomi, K.-I., Nakako, M., Shiono, S., Maeda, M., "Glucose-Sensitive Field Effect Transistor with a Membrane Containing Co-Immobilized Gluconolactonase and Glucose Oxidase", *Anal. Chim. Acta* **212** (1988) 49–59.

[199] Shiono, S., Hanazato, Y., Nakako, M., "Urea and Glucose Sensors Based on Ion Sensitive Field Effect Transistor with Photolithographically Patterned Enzyme Membrane", *Anal. Sci.* **2** (1986) 517–521.

[200] Kimura, J., Kawana, Y., Kuriyama, T., "An Immobilized Enzyme Membrane Fabrication Method Using an Ink Jet Nozzle", *Biosensors* **4** (1988) 41–52.

[201] Martin, M. J., Wickramasinghe, Y. A. B. D., Newson, T. P., Crowe, J. A., "Fibre-Optics and Optical Sensors in Medicine", *Med. & Biol. Eng. & Comput.* **25** (1987) 597–604.

[202] Wolfbeis, O. S., "Fibre-Optic Sensors in Biomedical Sciences", *Pure & Appl. Chem.* **59** (1987) 663–672.

[203] Harmer, A. L., Narayanaswamy, R., "Spectroscopic and Fibre-Optic Transducers", in: *Chemical Sensors,* Edmonds, T. E. (ed.); Glasgow: Blackie, 1988, pp. 275–294.

[204] Seitz, W. R., "Optical Sensors Based on Immobilized Reagents", in: *Biosensors. Fundamentals and Applications,* Turner, A. P. F., Karube, I., Wilson, G. S. (eds.); Oxford: Oxford University Press, 1987, pp. 599–616.

[205] Seitz, W. R., "Chemical Sensors Based on Immobilized Indicators and Fiber Optics", *CRC Critical Reviews in Analytical Chemistry* **19** (1988) 135–173.

[206] Polster, J., Höbel, W., Pappberger, A., Schmidt, H.-L., "Fundamentals of Enzyme Substrate Determinations by Fiber Optics Spectroscopy", SPIE Vol. 1172 "Chemical, Biochemical and Environmental Sensors" (1989) pp. 273–286.

[207] Wolfbeis, O. S., Weis, L. J., Leiner, M. J. P., Ziegler, W. E., "Fiber-Optic Fluorosensor for Oxygen and Carbon Dioxide", *Anal. Chem.* **60** (1988) 2028–2030.

[208] Munkholm, C., Walt, D. R., Milanovich, F. P., "A Fiber-Optic Sensor for CO_2 Measurement", *Talanta* **35** (1988) 109–112.

[209] Kulp, T. J., Camins, I., Angel, S. M., "Enzyme-Based Fiber Optic Sensors", *SPIE* **906** (1988) 134–138.

[210] Yerian, T. D., Christian, G. D., Ruzicka, J., "Flow Injection Analysis as a Diagnostic Tool for Development and Testing of a Penicillin Sensor", *J. Anal. Chem.* **60** (1988) 1250–1256.

[211] Petersson, B. A., Andersen, H. B., Hansen, E. H., "Determination of Urea in Undiluted Blood Samples by Flow Injection Analysis using Optosensing", *Anal. Lett.* **20** (1987) 1977–1994.

[212] Trettnak, W., Leiner, M. J. P., Wolfbeis, O. S., "Fibre-Optic Glucose Sensor with a pH Optrode as the Transducer", *Biosensors* **4** (1988) 15–26.

[213] Luo, S., Walt, D. R., "Avidin-Biotin Coupling as a General Method for Preparing Enzyme-Based Fiber-Optic Sensors", *Anal. Chem.* **61** (1989) 1069–1072.

[214] Wolfbeis, O. S., "Fiber-Optic Probe for Kinetic Determination of Enzyme Activities", *Anal. Chem.* **58** (1986) 2874–2876.

[215] Trettnak, W., Leiner, M. J. P., Wolfbeis, O. S., "Optical Sensors. Part 34. Fibre Optic Glucose Biosensor with an Oxygen Optrode as the Transducer", Analyst **113** (1988) 1519–1523.

[216] Wolfbeis, O. S., Posch, H. E., "Optical Sensors. Part 20. A Fibre Optic Ethanol Biosensor", *Fresenius Z. Anal. Chem.* **332** (1988) 255–257.

[217] Opitz, N., Lübbers, D. W., "Electrochromic Dyes, Enzyme Reactions and Hormone-Protein Interactions in Fluorescence Optic Sensor (Optode) Technology", *Talanta* **35** (1988) 123–128.

[218] Walters, B. S., Nielson, T. J. Arnold, M. A., "Fiber-Optic Biosensors for Ethanol, Based on an Internal Enzyme Concept", *Talanta* **35** (1988) 151–156.

[219] Narayanaswamy, R., Sevilla F., "An Optical Fibre Probe for the Determination of Glucose Based on Immobilized Glucose Dehydrogenase", *Anal. Lett.* **21** (1988) 1165–1175.

[220] Schultz, J. S., Mansouri, S., "Optical Fiber Affinity Sensors", in: *Methods in Enzymology* Vol. 137, Colowick, S. P., Kaplan, N. O., Mosbach, K. (eds.); San Diego: Academic Press, 1988, pp. 349–366.

[221] Trettnak, W., Wolfbeis, O. S., "Fully Reversible Fibre-Optic Glucose Biosensor Based on the Intrinsic Fluorescence of Glucose Oxidase", *Anal. Chim. Acta* **221** (1989) 195–203.

[222] Trettnak, W., Wolfbeis, O. S., "A Fully Reversible Fiber Optic Lactate Biosensor Based on the Intrinsic Fluorescence of Lactate Monooxygenase", *Fresenius Z. Anal. Chem.* **334** (1989) 427–430.

[223] Downs, M. E. A., Kobayashi, S., Karube, I., "New DNA Technology and the DNA Biosensor", *Anal. Lett.* **20** (1987) 1897.

[224] Lowe, C. R., Goldfinch, M. J., "Solid-Phase Optoelectronic Biosensors", in: *Methods in Enzymology* Vol. 137, Colowick, S. P., Kaplan, N. O., Mosbach, K. (eds); San Diego: Academic Press, 1988, pp. 338–348.

[225] Kurkijärvi, K., Turunen, P., Heinonen, T., Kolhinen, O., Raunio, R., Lundin, A., Lövgren, T., "Flow-Injection Analysis with Immobilized Chemiluminescent and Bioluminescent Columns", in: *Methods in Enzymology* Vol 137, Colowick, S. P., Kaplan, N. O., Mosbach, K. (eds); San Diego: Academic Press, 1988, pp. 171–181.

[226] Idahl, L.-A., Sandström, P.-E., Sehlin, J., "Measurements of Serum Glucose Using the Luciferin/Luciferase System and a Liquid Scintillation Spectrometer", *Anal. Biochem.* **155** (1986) 177–181.

[227] Roda, A., Girotti, S., Ghini, S., Carrea, G., "Continuous-Flow Assays with Nylon Tube-Immobilized Bioluminescent Enzymes", in: *Methods in Enzymology* Vol. 137, Colowick, S. P., Kaplan, N. O., Mosbach, K. (eds); San Diego: Academic Press, 1988, pp. 161–171.

[228] Girotti, S., Roda, A., Ghini, S., Piacentini, A. L., Carrea, G., Bovara, R., "A Sensitive Continuous-Flow Bioluminescent System for Determining Ethanol in Serum and Saliva", *Anal. Chim. Acta* **183** (1986) pp. 187–196.

[229] Blum, L. J., Plaza, J. M., Coulet, P. R., "Chemiluminescent Analyte Microdetection Based on the Luminol-H_2O_2 Reaction Using Peroxidase Immobilized on New Synthetic Membranes", *Anal. Lett.* **20** (1987) 317–326.

[230] Blum, L. J., Gautier, S. M., Coulet, P. R., "Alternate Determination of ATP and NADH with a Single Fiber-Optic Sensor", *Sens. Actuators* **B1** (1990) 580–584.

[231] Lundström, I., Gustafsson, A., "Fish Scales as Biosensors", *Sens. Actuators* **B1** (1990) 533–536.

[232] Sauerbrey, G. Z., "Use of a Quartz Vibrator for Weighting Thin Films on a Microbalance", *Z. Physik* **155** (1959) 206–210.

[233] Bastiaans, G. J., "Piezoelectric Transducers", in: *Chemical Sensors,* Edmonds, T. E. (ed.); Glasgow: Blakie, 1988, pp. 295–319.

[234] Clarke, D. J., Blake-Coleman, B. C., Calder, M. R., "Principles and Potential of Piezo-Electric Transducers and Acoustical Techniques", in: *Biosensors – Fundamentals and Applications*, Turner, A. P. F., Karube, I., Wilson, G. S., (eds.); Oxford: Oxford University Press, 1987, pp. 551–571.

[235] Ngeh-Ngwainbi, J., Foley, P. H., Kuan, S. S., Guilbault, G. G., "Parathion Antibodies on Piezoelectric Crystals", *J. Am. Chem. Soc.* **108** (1986) 5444–5450.

[236] Guilbault, G. G., "Immobilization Methods for Piezoelectric Biosensors", *Bio/Technology* **7** (1989) 349–351.

[237] Wilchek, M., Bayer, E. A., "The Avidin-Biotin Complex in Bioanalytical Applications", *Anal. Biochem.* **171** (1988) 1–32.

[238] Berg, P., Näbauer, A., Müller, E., Woias, P., "Biosensors Based on Piezoelectric Crystals", *Sens. Actuators* **B1** (1990), 508–509.

[239] Mosbach, K. (ed.), "Immobilized Enzymes", in: *Methods in Enzymology* Vol. 44, Colowick, S. P., Kaplan, N. O., (series eds.); San Diego: Academic Press, 1976.

[240] Trevan, M. D., in: *Immobilized Enzymes. An Introduction and Applications in Biotechnology*, Chichester: John Wiley & Sons, 1980.

[241] Chibata, I. (ed.), in: *Immobilized Enzymes. Research and Development;* New York: John Wiley, 1978.

[242] Messing, R. A. (ed.), in: *Immobilized Enzymes for Industrial Reactions;* New York: Academic Press, 1975.

[243] Scouten, W. H., "A Survey of Enzyme Coupling Techniques, in: *Methods in Enzymology* Vol. 135, Mosbach, K., Colowick, S. P., Kaplan, N. O. (eds); San Diego: Academic Press, 1987, pp. 30–65.

[244] Buchholz, K., Klein, J., "Characterization of Immobilized Biocatalysts", in: *Methods in Enzymology* Vol. 135, Mosbach, K., Colowick, S. P., Kaplan, N. O. (eds); San Diego: Academic Press, 1987, pp. 3–30.

[245] Chibata, I., Wingard Jr., L. B. (eds.), "Immobilized Microbial Cells", in: *Applied Biochemistry and Bioengineering* Vol. 4, Wingard Jr., L. B., Goldstein, L., Katchalski-Katzir, E. (series eds.); New York: Academic Press, 1983.

[246] Venkatsubramanian, K. (ed.), in: *Immobilized Microbial Cells; ACS Symposium Series* Vol. 106; Washington: American Chemical Society, 1979.

[247] Hulst, A. C., Tramper, J., "Immobilized Plant Cells: A Literature Survey", *Enzyme Microb. Technol.* **11** (1989) 546–558.

[248] Srere, P. A., Uyeda, K., "Functional Groups on Enzymes Suitable for Binding to Matrices", in: *Methods in Enzymology* Vol. 44, Mosbach, K., Colowick, S. P., Kaplan, N. O. (eds); San Diego: Academic Press, 1976, pp. 11–19.

[249] Royer, G. P., "Immobilization of Glycoenzymes through Carbohydrate Chains", in: *Methods in Enzymology* Vol. 135, Mosbach, K., Colowick, S. P., Kaplan, N. O., (eds.) San Diego: Academic Press, 1987, pp. 141–146.

[250] Murray, R. W., "Chemically Modified Electrodes", *Acc. Chem. Res.* **13** (1980) 135–141.

[251] Albery, W. J., Hillmann, A. R., "Modified Electrodes", *Ann. Rep. Roy. Soc. Chem. C* **78** (1981) 377–437.

[252] Murray, R. W., "Modified Electrodes. Chemically Modified Electrodes for Electrocatalysis", *Phil. Trans. R. Soc. Lond.* **A 302** (1981) 253–265.

[253] Schreurs, J., Barendrecht, E., "Surface-Modified Electrodes", *Recl. Trav. Chim. Pays-Bas* **103** (1984) 205–219.

[254] Murray, R. W., Ewing, A. G., Durst, R. A., "Chemically Modified Electrodes, Molecular Design for Electroanalysis", *Anal. Chem.* **59** (1987) 379 A–388 A.

[255] Engstrom, R. C., Strasser, V. A., "Characterization of Electrochemically Pretreated Glassy Carbon Electrodes", *Anal. Chem.* **56** (1984) 136–141.

[256] Evans, J. F., Kuwana, T., "Introduction of Functional Groups onto Carbon Electrodes via Treatment with Radio-Frequency Plasma", *Anal. Chem.* **51** (1979) 358–365.

[257] Schreurs, J., Van den Berg, J., Wonders, A., Barendrecht, E., "Characterization of a Glassy-Carbon-Electrode Surface Pretreated with rf-Plasma", *Recl. Trav. Chim. Pays-Bas* **103** (1984) 251–259.

[258] Cameron, G. M., Marsden, J. G., "Silane Coupling Agents", *Chem. Brit.* **8** (1972) 381–385.

[259] Haller, I., "Covalently Attached Organic Monolayers on Semiconductor Surfaces", *J. Am. Chem. Soc.* **100** (1980) 8050–8055.

[260] Kallury, K. M. R., Krull, U. J., Thompson, A., "X-ray Photoelectron Spectroscopy of Silica Surfaces Treated with Polyfunctional Silanes", *Anal. Chem.* **60** (1988) 169–172.

[261] Uhe, B., Schuhmann, W., Schmidt, H.-L., unpublished.

[262] Wieck, H., Ianniello, R. M., Osborn, J. A., Yacynych, A. M., "Thermal Studies of Carbonaceous Electrode Materials Chemically Modified with Cyanuric Chloride", *Anal. Chim. Acta* **140** (1982) 19–27.

[263] Hearn, M. T. W., "1,1'-Carbonyldiimidazole-Mediated Immobilization of Enzymes and Affinity Ligands", *Methods in Enzymology* Vol. 135, Mosbach, K., Colowick, S. P., Kaplan, N. O. (eds.); San Diego: Academic Press, 1987, 102–117.

[264] Nilsson, K., Mosbach, K., "Tresyl Chloride-Activated Supports for Enzyme Immobilization", in: *Methods in Enzymology* Vol. 135, Mosbach, K., Colowick, S. P., Kaplan, N. O. (eds.) San Diego: Academic Press, 1987, pp. 65–78.

[265] Diaz, A. F., "Electrochemical Preparation and Characterization of Conducting Polymers; *Chemica Scripta* **17** (1981) 145–148.

[266] Street, G. B., Clarke, T. C., Krounbi, M., Kanazawa, K., Lee, V., Pfluger, P., Scott, J. C., Weiser, G., "Preparation and Characterization of Neutral and Oxidized Polypyrrole Films", *Mol. Cryst. Liq. Cryst.* **83** (1982) 253–264.

[267] Bryce, M. R., Chissel, A., Kathirgamanthan, P., Parker, D., Smith, N. R. M., "Soluble, Conducting Polymers from 3-Substituted Thiophenes and Pyrroles", *J. Chem. Soc. Chem. Commun.* (1987) 466–467.

[268] Haimerl, A., Merz, A., "Ferrocen-modifizierte Polypyrrolfilme durch elektrochemische Copolymerisation", *Angew. Chem.* **98** (1986) 179–180.

[269] Stefan, K.-P., Schuhmann, W., Parlar, H., Korte, F., "Synthese neuer 3-substitutierter Pyrrole", *Chem. Ber.* **122** (1989) 169–174.

[270] Schuhmann, W., Lammert, R., Uhe, B., Schmidt, H.-L., "Polypyrrole, a New Possibility For Covalent Binding of Oxidoreductases to Electrode Surfaces as a Base for Stable Biosensors", *Sens. Actuators* **B1** (1990) 537–541.

[271] Schuhmann, W., Kittsteiner-Eberle, R., *Biosensors & Bioelectronics* **6** (1990), in press. Evalution of Polypyrrole/Glucose Oxidase Elecrodes in Flow-Injection Systems for Sucrase Determination.

[272] Göpel, W., "Entwicklung chemischer Sensoren: Empirische Kunst oder systematische Forschung", *Technisches Messen* **52** (1985) 47–58, 92–105, 175–182.

[273] Eddowes, M. J., *Biosensors* **3** 1987/88) 1–16.

[274] Tedesco, J. L., Krull, U. J., Thompson, M., *Biosensors* **4** (1989) 135–168.

[275] Buch, R. M., Rechnitz, G. A., *Biosensors* **4** (1989) 215–230.

[276] Kobayashi, T., Laidler, K., *Biotechnol. Bioeng.* **16** (1974) 77–92.

[277] Carr, P. W., Bowers, L. D., *Immobilized Enzymes in Analytical and Clinical Chemistry;* New York: Wiley, 1980.

[278] Gough, D. A., Leypoldt, J. K., *AIChE J.* **26** (1980) 1013–1019.

[279] Schulmeister, T., Scheller, F., *Anal. Chim. Acta* **170** (1985) 279–285.

[280] Blaedel, W. J., Kissel, T., Boguslaski, R., *Anal. Chem.* **44** (1972) 2030-2034.
[281] Scheller, F., Renneberg, R., Schubert, F., in: *Methods in Enzymology* Vol. 137, Part D, Mosbach, K. (ed.); New York: Academic Press, 1988, pp. 29-43.
[282] Jänchen, M., Dissertation B, Universität Leipzig, 1987.
[283] Weigelt, D., Dissertation, Universität Berlin, 1988.
[284] Schubert, F., Scheller, F., in: *Methods in Enzymology* Vol. 137, Part D, Mosbach, K. (ed.); New York: Academic Press, 1988, pp. 152-160.
[285] Malpiece, Y., Sharan, M., Barbotin, J.-N., Personne, P., Thomas, D., *J. Biol. Chem.* **255** (1981) 6883-6887.
[286] Scheller, F., Pfeiffer, D., Seyer, I., Kirstein, D., Schulmeister, T., Nentwig, J., *Bioelectrochem. Bioenerg.* **11** (1983) 155-165.
[287] Scheller, F., Wollenberger, U., Schubert, F., Pfeiffer, D., Renneberg, R., Jänchen, M., Walzel, G., Weise, H., Bertermann, K., *Priklad. Biokhim. Mikrobiol.* **18** (1983) 454-470.
[288] Hamann, H., Scheller, F., Kühn, M., *Electroanalysis* (1989) 535-539.
[289] Thevenot, D. R., Sternberg, R., Coulet, P. R., Laurent, J., Gautheron, D. C., *Anal. Chem.* **51** (1979) 96-100.
[290] Tsuchida, T., Yoda, K., *Enzyme Microb. Technol.* **3** (1981) 326-330.
[291] Sternberg, R., Bindner, D., Welsin, G., Thevenot, D., *Anal. Chem.* **60** (1988) 2781-2786.
[292] Matsumoto, K., Mizuguchi, H., Ichimura, K., *Kobunshi Ronbunshu* **41** (1984) 221-224.
[293] Hintsche, R., Scheller, F., *Studia Biophys.* **119** (1987) 179-182.
[294] Razumas, V. J., Jasaitis, J. J., Kulys, J. J., *Bioelectrochem. Bioenerg.* **12** (1984) 297-306.
[295] Pfeiffer, D., Makower, A., Meiske, C., Scheller, F., *5th Bucher Symposium, Berlin, December 1988,* Poster.
[296] Bertrand, C., Coulet, P. R., Gautheron, D. C., *Anal. Lett.* **12** (1979) 1477-1488.
[297] Tsuchida, T., Yoda, K., *Clin. Chem.* **29** (1983) 51-55.
[298] Kirstein, D., Scheller, F., Olsson, B., Johansson, G., *Anal. Chim. Acta* **171** (1985) 345-350.
[299] Weigelt, D., Schubert, F., Scheller, F., *Analyst* **112** (1987) 1155-1158.
[300] Pfeiffer, D., Bertermann, K., Dissertation, Academy of Sciences of the GDR, Berlin, 1982.
[301] Scheller, F., Wollenberger, U., Schubert, F., Pfeiffer, D., Bogdanovskaya, V. A., in: *Biosensors International Workshop 1987, GBF Monographs* Vol. 10, Schmid, R. D. (ed.); Weinheim: Verlag Chemie, 1987, pp. 39-50.
[302] Schubert, F., Pfeiffer, D., Wollenberger, U., Kühnel, S., Hanke, G., Hauptmann, B., Scheller, F., in: *Biosensors: Applications in Medicine, Environmental Monitoring, and Process Control, GBF Monographs,* Schmidt, R. D., Scheller, F. (eds.); Weinheim: Verlag Chemie, 1989, pp. 11-15.
[303] Kirstein, L., Dissertation, Humboldt-Universität Berlin, 1987.
[304] Weigelt, D., Schubert, F., Scheller, F., *Fresenius Z. Anal. Chem.* **328** (1987) 259-262.
[305] Weigelt, D., Schubert, F., Scheller, F., *Anal. Lett.* **21** (1988) 225-239.
[306] Schubert, F., unpublished work, 1987.
[307] Scheller, F., Schubert, F., Pfeiffer, D., Riedel, K., Kirstein, D., Gruß, R., Weigelt, D., *Biotech. 87, London, May 1987, Proceedings* Vol. 1, Part 3; London: Online Publications, 1987, pp. 117-124.
[308] Bertermann, K., Scheller, F., Pfeiffer, D., Jänchen, M., Lutter, J., *Z. Med. Lab.-Diagn.* **22** (1981) 83-88.
[309] Olsson, B., Lundbäck, H., Johansson, G., Scheller, F., Nentwig, J., *Anal. Chem.* **58** (1986) 1046-1052.
[310] Scheller, F., Schubert, F., Olsson, B., Gorton, L., Johansson, G., *Anal. Lett.* (1986) 1691-1703.
[311] Hintsche, R., Neumann, G., Dransfeld, I., Kampfrath, G., Scheller, F., *Biosensors* **4** (1989) 327-334.
[312] Ikariyama, Y., Shimada, N., Yamauchi, S., *Anal. Lett.* **21** (1988) 953-964.
[313] Mottola, H. M., *Analyst* **112** (1987) 719-727.
[314] Ngo, T. T., Lenhoff, H. M., *Anal. Lett.* **13** (1980) 1157-1166.
[315] Hassan, S. S. M., Rechnitz, G. A., *Anal. Chem.* **53** (1981) 512-515.
[316] Seegopaul, P., Rechnitz, G. A., *Anal. Chem.* **55** (1983) 1929-1938.
[317] Higgins, I. J., Bannister, J. V., Turner, A. P. F., in: *Biosensors International Workshop 1987, GBF Monographs* Vol. 10, Schmid, R. D. (ed.); Weinheim: Verlag Chemie, 1987, pp. 23-32.
[318] Mattiasson, B., Nilsson, H., Olsson, B., *J. Appl. Biochem.* **1** (1979) 377-384.
[319] Jasaitis, J. J., Razumas, V. J., Kulys, J. J., *Anal. Chim. Acta* **152** (1983) 271-274.

[320] Lowry, O., Passonneau, J. P., Shulz, D. W., Rock, M. K., *J. Biol. Chem.* **236** (1961) 2746–2755.
[321] Schubert, F., Kirstein, D., Schröder, K. L., Scheller, F., *Anal. Chim. Acta* **169** (1985) 391–396.
[322] Scheller, F., Siegbahn, N., Danielsson, B., Mosbach, K., *Anal. Chem.* **57** (1985) 1740–1743.
[373] Wollenberger, U., Schubert, F., Scheller, F., Danielsson, B., Mosbach, K., *Stud. Biophys.* **119** (1987) 167–170.
[324] Mizutani, F., Yamanka, T., Tanabe, Y., Tsuda, K., *Anal. Chim. Acta* **177** (1985) 153–166.
[325] Schubert, F., Kirstein, D., Scheller, F., Appelqvist, R., Gorton, L., Johansson, G., *Anal. Lett.* **19** (1986) 1273–1288.
[326] Wollenberger, U., Scheller, F., Pavlova, M., Böhmer, A., Passarge, M., Müller, H.-G., *Biosensors* **4** (1989) 381–391.
[327] Wollenberger, U., Schubert, F., Scheller, F., Danielsson, B., Mosbach, K., *Anal. Lett.* **20** (1987) 657–668.
[328] Hopkins, T. R., *Int. Biotechnol. Lab.* **3** (1985) 20–25.
[329] Scheller, F., Wollenberger, U., Schubert, F., Pfeiffer, D., Bogdanovskaya, V. A., in: *Biosensors International Workshop 1987, GBF Monographs* Vol. 10, Schmid, R. D. (ed.); Weinheim: Verlag Chemie, 1987, pp. 39–50.
[330] Kirstein, D., Danielsson, B., Scheller, F., Mosbach, K., *Biosensors* **4** (1989) 231–239.
[331] Schubert, F., Wollenberger, U., Scheller, F., Müller, H.-G., in: *Biosensors and Bioinstrumentation,* Wise, D. L. (ed.); New York: Marcel Dekker, 1990, pp. 19–37.
[332] Schubert, F., Scheller, F., Krasteva, N. G., *Electroanalysis* **2** (1990) 347–351.
[333] Mizutani, F., Tsuda, K., *Nippon Kagaku Kaishi* **3** (1987) 531–534.
[334] Winquist, F., Danielsson, B., Lundström, I., Mosbach, K., in: *Methods of Enzymology* Vol. 137, Colowick, S. P., Kaplan, N. O., Mosbach, K. (eds.); San Diego: Academic Press, 1988, pp. 232–247.
[335] Renneberg, R., Scheller, F., *International Meeting on Chemical Sensors, Fukuoka, 1983, Proceedings;* Amsterdam: Elsevier, pp. 11–15.
[336] Vidziunaite, R. A., Kulys, J. J., *Trud. Akad. Nauk Lit. SSR, Ser. B* **90** (1985) 84–91.
[337] Kulys, J. J., Sorochinski, V. V., Vidziunaite, R. A., *Biosensors* **2** (1986) 135–146.
[338] Olsson, B., Lundbäck, H., Johansson, G., Scheller, F., Nentwig, J., *Anal. Chem.* **58** (1986) 1046–1052.
[339] Danielsson, B., *5th Bucher Symposium, Berlin, December 1988, Lecture.*
[340] Takemori, S., Nakanishi, M., Katagiri, M., in: *Methods in Enzymology* Vol. 41, Part B, Wood, W. A., (ed.); New York: Academic Press, 1975, pp. 329–333.
[341] Romano, M., Cerra, M., *Biochim. Biophys. Acta* **177** (1969) 421–426.
[342] Gestrelius, S., Mansson, M.-O., Mosbach, K., *Eur. J. Biochem.* **57** (1975) 529–535.
[343] Bertrand, C., Coulet, P. R., Gautheron, D. C., *Anal. Chim. Acta* **126** (1981) 23–34.
[344] Tsuchida, T., Yoda, K., *Enzyme Microb. Technol.* **3** (1981) 326–330.
[345] Mascini, M., Fortunati, S., Moscone, D., Palleschi, G., *Anal. Chim. Acta* **171** (1985) 175–182.
[346] Matsumoto, K., Kamikado, H., Matsubara, H., Osajima, Y., *Anal. Chem.* **60** (1988) 147–151.
[347] Olsson, B., Stalbom, B., Johansson, G., *Anal. Chim. Acta* **179** (1986) 203–211.
[348] Scheller, F., Renneberg, R., *Anal. Chim. Acta* **152** (1983) 265–269.
[349] Renneberg, R., Scheller, F., Riedel, K., Litschko, E., Richter, M., *Anal. Lett.* **16** (1983) 877–890.
[350] Weigelt, D., Schubert, F., Scheller, F., *Analyst* **112** (1987) 1155–1158.
[351] Nagy, G., Rice, M. E., Adams, R. N., *Life Sci.* **31** (1982) 2611–2616.
[352] Wollenberger, U., Scheller, F., Pfeiffer, D., Bogdanovskaya, V. A., Tarasevich, M. R., Hanke, G., *Anal. Chim. Acta* **187** (1986) 39–45.
[353] Kihara, K., Yasukawa, E., *Anal. Chim. Acta* **183** (1986) 75–80.
[354] Scheller, F., Schubert, F., Pfeiffer, D., Renneberg, R., in: *Enzyme Engineering 8,* Laskin, A. R., Mosbach, K., Thomas, D., Wingard, L. B. (eds.); New York: The New York Academy of Sciences, 1987, pp. 240–248.
[355] Tschannen, R., Hoeren, R., Schröder, N., *Biotechnologie* **2** (1987) 18–20.
[356] Libeer, J. C., *J. Clin. Chem. Clin. Biochem.* **23** (1985) 645–650.
[357] Lowe, C. R., Goldfinch, M. J., Lias, R. J., *Biotech 83, London 1983, Proceedings;* Northwood: Online Publications, 1983, pp. 633–640.
[358] Niwa, M., Itoh, K., Nagata, A., Osawa, H., *Tokai J. Exp. Clin. Med.* **6** (1981) 403–414.
[359] Mason, M., *Biotec 87, Düsseldorf, March 1987, Abstracts;* Heidelberg: Springer-Verlag.

[360] Scheller, F., Pfeiffer, D., Kühn, M., Hamann, H., Fahrenbruch, B., Klimes, N., Nentwig, J., Möricke, R., Kiesewetter, M., Jänchen, M., Scholz, P., Müller, E., Uffrecht, E., Brunner, J., Hanke, G., *Z. Klin. Med.* **41** (1986) 565–568.

[361] Hanke, G., Nentwig, J., Scheller, F., *Z. Med. Lab.-Diagn.* **29** (1988) 390–394.

[362] McCann, J., *Biotech 87, London, May 1987, Proceedings* Vol. 1, Part 2; London: Online Publications, 1987, pp. 41–50.

[363] Hamann, H., Dissertation, Academy of Sciences of the GDR, Berlin, 1987.

[364] Petersson, B. A., *Anal. Chim. Acta* **209** (1988) 239–248.

[365] Kirstein, D., Kirstein, L., Scheller, F., *Biosensors* **1** (1985) 117–130.

[366] Tokinaga, D., Kobayashi, T., Katori, A., Karasawa, Y., Yasuda, K., *International Meeting on Chemical Sensors, Fukuoka, 1983, Proceedings;* Amsterdam: Elsevier, pp. 626–630.

[367] Denis, C., Dormois, D., Linossier, M.-T., Geyssant, A., *J. Physiol. (Paris)* **80** (1985) 168–172.

[368] Clark, L. C., Noyes, L. K., Grooms, T. A., Moore, M. S., *Crit. Care Med.* **12** (1984) 461–464.

[369] Weil, M. H., Leavy, J. A., Rackow, E. C., Halfman, C. J., Bruno, S. J., *Clin. Chem.* **32** (1986) 2175–2177.

[370] Clark, L. C., Noyes, L. K., Grooms, T. A., Gleason, C. A., *Clin. Biochem.* **17** (1984) 288–291.

[371] Tsuchida, T., Takasugi, H., Yoda, K., Takizawa, K., Kobayashi, S., *Biotechnol. Bioeng.* **27** (1985) 837–841.

[372] Schubert, F., Pfeiffer, D., Wollenberger, U., Kühnel, S., Hanke, G., Hauptmann, B., Scheller, F., in: *Biosensors: Applications in Medicine, Environmental Monitoring, and Process Control, GBF Monographs,* Schmidt, R. D., Scheller, F. (eds); Weinheim: Verlag Chemie, 1989, 11–15.

[373] Weigelt, D., Schubert, F., Scheller, F., *Fresenius Z. Anal. Chem.* **328** (1987) 259–262.

[374] Jänchen, M., Scheller, F., Pfeiffer, D., in: *Biosensors: Applications in Medicine, Environmental Monitoring, and Process Control, GBF Monographs,* Schmidt, R. D., Scheller, F. (eds.); Weinheim: Verlag Chemie, 1989, 17–25.

[375] Osawa, H., Akiyama, S., Hamada, T., *1st Sensor Symposium, Fukuoka, 1981, Proceedings.* Fukuoka, pp. 163–168.

[376] Kulys, J. J., Laurinavicius, V. A., Pesliakiene, M. W., Cureviciene, V. A., *Anal. Chim. Acta* **148** (1983) 13–18.

[377] Pfeiffer, D., Scheller, F., Jänchen, M., Bertermann, K., Weise, H., *Anal. Lett.* **13** (1980) 1179–1200.

[378] Kihara, K., Yasukawa, E., Hayashi, M., Hirose, S., *Anal. Chem.* **56** (1984) 1876–1880.

[379] Yamauchi, H., Kusajabe, H., Midorikawa, Y., Fujishima, T., Kuninaka, A., *International Congress on Biotechnology, München, 1983, Proceedings;* Weinheim: Verlag Chemie, 1983, pp. I-705–710.

[380] Wollenberger, U., Scheller, F., Böhmer, A., Passarge, M., Müller, H.-G., *Biosensors* **4** (1989), 381–391.

[381] Mizutani, F., Sasaki, K., Shimura, Y., *Anal. Chem.* **55** (1983) 35–38.

[382] Weigelt, D., Schubert, F., Scheller, F., *Anal. Lett.* **21** (1988) 225–239.

[383] Gruß, R., Scheller, F., *Z. Med. Lab.-Diagn.* **28** (1987) 333–335.

[384] Gruß, R., Scheller, F., *Fresenius Z. Anal. Chem.* **333** (1989) 29–32.

[385] Fogt, E. J., Dodd, L. M., Jenning, E. M., Clemens, A. H., *Clin. Chem.* **24** (1978) 1366–1371.

[386] Kiesewetter, M., Möricke, R., Wernstedt, J., Lembke, K., *30th International Scientific Colloquium, Ilmenau, 1985, Lecture.*

[387] Mullen, W. H., Keedy, F. H., Churchhouse, S. J., Vadgama, P. M., *Anal. Chim. Acta* **183** (1986) 59–66.

[388] Vadgama, P. M., *2nd International Conference Bio 87 Sensing and Control, London, 1987, Proceedings.*

[389] Shichiri, M., Yamasaki, Y., Kawamori, R., Ueda, N., Sekiya, M., in: *Biosensors International Workshop 1987, GBF Monographs* Vol. 10, Schmid, R. D. (ed.); Weinheim: Verlag Chemie, 1987, pp. 95–100.

[390] Kessler, M., Höper, J., Volkholz, H. J., Sailer, D., Demling, L., *Hepato-Gastroenterol.* **31** (1984) 285–287.

[391] Cass, A. E. G., Davis, G., Francis, G. D., Hill, H. A. O., Aston, W. J., Higgins, I. J., Plotkin, E. V., Scott, L. D. L., Turner, A. P. F., *Anal. Chem.* **56** (1984) 667–671.

[392] Davis, G., Hill, H. A. O., Higgins, I. J., Turner, A. P. F., in: *Implantable Sensors for Closed-Loop Prosthetic Systems,* Ko, W. H., (ed.); Mount Kisco: Futura Publishing Co., 1985, pp. 189–195.

[393] Pickup, J. C., *2nd International Conference Bio 87 Sensing and Control, London, 1987, Proceedings.*

[394] Abel, P., Müller, A., Fischer, U., *Biomed. Biochim. Acta* **43** (1984) 577–584.

[395] Shichiri, M., Kawamori, R., Hakui, N., Asakawa, N., Yamasaki, Y., Abe, H., *Biomed. Biochim. Acta* **43** (1984) 561–568.

[396] McKean, B. D., Gough, D. A., *IEEE Trans. Biomed. Eng.* **35** (1988) 526–532.

[397] Mascini, M., in: *Biosensors International Workshop 1987, GBF Monographs* Vol. 10, Schmid, R. D. (ed.); Weinheim: Verlag Chemie, 1987, pp. 87–94.

[398] Mascini, M., Mazzei, F., Moscone, D., Calabrese, G., Massi-Benedetti, M., *Clin. Chem.* **33** (1987) 591–593.

[399] Scheller, F., Karsten, C., *Anal. Chim. Acta* **155** (1983) 29–36.

[400] Weise, H., Kreibich, G., Scheller, F., *Acta Biotechnol.* **7** (1987) 61–69.

[401] Pfeiffer, D., *5th Bucher Symposium, Berlin, December 1988, Poster.*

[402] Coulet, P., in: *Biosensors International Workshop 1987, GBF Monographs* Vol. 10, Schmid, R. D. (ed.); Weinheim: Verlag Chemie, 1987, pp. 75–80.

[403] Mason, M., *J. Assoc. Off. Anal. Chem.* **66** (1983) 981–984.

[404] Mason, M., *Am. J. Enol. Vitic.* **34** (1983) 173–175.

[405] Greenway, C. V., Pushka, K., Sitar, D. S., Rogers, L. K., *Can. J. Physiol. Pharmacol.* **66** (1988) 307–311.

[406] Enfors, S.-O., *Enzyme Microb. Technol.* **3** (1981) 29–32.

[407] Enfors, S.-O., *Appl. Biochem. Biotechnol.* **7** (1982) 113–119.

[408] Cleland, V., Enfors, S.-O., *Anal. Chem.* **56** (1984) 1880–1884.

[409] Turner, A. P. F., *Biotech. Europe, London, 1985, Proceedings*; Pinner: Online Publications, 1985, pp. 181–192.

[410] Brooks, S., Ashby, R., Turner, A. P. F., Calder, M., Clarke, D. J., *Biosensors* **3** (1987/88) 45–56.

[411] Verduyn, C., Zomerdijk, T., van Dijken, J., Scheffers, W., *Appl. Microbiol. Biotechnol.* **19** (1984) 181–184.

[412] Klopper, W. J., Angelino, S. A. G. F., Knol, W., Minekus, M., *21st European Brewery Convention Congress, Madrid, 1987, Proceedings,* Oxford: Oxford Information Printing, 1987, pp. 87–97.

[413] Mandenius, C. F., Danielsson, B., Mattiasson, B., *Biotechnol. Lett.* **3** (1981) 629–634.

[414] Wehnert, G., Sauerbrei, A., Bayer, T., Scheper, T., Schügerl, K., *Anal. Chim. Acta* **200** (1987) 73–78.

[415] Geppert, G., Asperger, L., *Bioelectrochem. Bioenerg.* **17** (1987) 399–407.

[416] Merten, O. W., *Animal Cell Biotechnol.* **3** (1988) 75–140.

[417] Tsuchida, T., Takasugi, H., Yoda, K., Takizawa, K., Kobayashi, S., *Biotechnol. Bioeng.* **27** (1985) 837–841.

[418] Romette, J. L., in: *Biosensors. International Workshop 1987, GBF Monographs* Vol. 10, Schmid, R. D., Guilbault, G. G., Karube, I., Schmidt, H.-L., Wingard, L. B. (eds.); Weinheim: VCH, 1987, pp. 81–86.

[419] Schmidt, H.-L., Saschewag, I., "Possibilities of Process Control with Biosensors", in: *Biochemical Engineering − A Challenge for Interdisciplinary Cooperation,* Chmiel, H., Hammes, W. P., Bailey, J. E. (eds.); Stuttgart: Gustav Fischer Verlag, 1987, pp. 132–138.

[420] Clarke, D. J., Calder, M. R., Carr, R. J. G., Blake-Coleman, B. C., Moody, S. C., Collinge, T. A. "The Development and Application of Biosensing Devices for Bioreactor Monitoring and Control", *Biosensors* **1** (1985) 213–320.

[421] Enfors, S.-O., Cleland, N., "Enzyme Sensors for Fermentation Monitoring: Sample Handling", in: *Methods in Enzymology* Vol. 137, Colowick, S. P., Kaplan, N. O., Mosbach, K. (eds.); San Diego: Academic Press, 1988, pp. 298–307.

[422] Ruzicka, J., Hansen, E. H., "Flow Injection Analysis", in: *Chemical Analysis: A Series of Monographs on Analytical Chemistry and its Applications* Vol. 62, 2nd Ed. Winefordner, J. D., Kolthoff, I. M., series (eds.); New York: John Wiley, 1988.

[423] Schelter-Graf, A., Huck, H., Schmidt, H.-L., "Rasche und genaue Bestimmung von Ethanol mittels einer Oxidase-Elektrode in einem Strömungs-Injektions-System", *Z. Lebensm. Unters. Forsch.* **177** (1983) 356–358.

[424] Schelter-Graf, A., Schmidt, H.-L., Huck, H., "Determination of the Substrates of Dehydrogenases in Biological Material in Flow-Injection Systems with Electrocatalytic NADH Oxidation", *Anal. Chim. Acta* **163** (1984) 299–303.

[425] Lazaro, F., Luque de Castro, M. D., Valcarcel, M., "Individual and Simultaneous Determination of Ethanol and Acetaldehyde in Wines by Flow Injection Analysis and Immobilized Enzymes", *Anal. Chem.* **59** (1987) 1859–1863.

[426] Matsumoto, K., Kamikado, H., Matsubara, H., Osajima, Y., "Simultaneous Determination of Glucose, Fructose, and Sucrose in Mixtures by Amperometric Flow Injection Analysis with Immobilized Enzyme Reactors", *Anal. Chem.* **60** (1988) 147–151.

[427] Ruz, J., Luque de Castro, M. D., Valcarcel, M., "Determination of Glucose in Alcoholic Beverages by Flow Injection with Two Internally Coupled Injection Valves and an Enzyme Reactor", *Anal. Chim. Acta* **211** (1988) 281–285.

[428] Linares, P., Luque de Castro, M. D., Valcarcel, M., "Sequential Determination of Glucose and Fructose in Foods by Flow-Injection Analysis with Immobilized Enzymes", *Anal. Chim. Acta* **202** (1987) 199–205.

[429] Geppert, G., Asperger, L., "Automatic On-Line Measurement of Substrates in Fermentation Liquids with Enzyme Electrodes", *Studia Biophys.* **119** (1987) 159–162.

[430] Ghoul, M., Ronat, E., Engasser, J.-M., "An Automatic and Sterilizable Sampler for Laboratory Fermentors: Application to the On-Line Control of Glucose Concentration", *Biotechnol. Bioeng.* **28** (1986) 119–121.

[431] Ogbomo, I., Prinzing, U., Schmidt, H.-L., "Prerequisites for the On-Line Control of Microbial Processes by Flow-Injection Analysis", *J. Biotechnol.* **14** (1990) 63–70.

[432] Schmidt, H.-L., Horita, Y., Kittsteiner-Eberle, R., Ogbomo, I., "Flow Systems For the Simultaneous Concentration Control of Different Substrates During Fermentation", in: *Biosensors – International Workshop, GBF-Monographs* Vol. 10, Schmid, R. D., Guilbault, G. G., Karube, J., Schmidt, H.-L., Wingard, L. B. (eds.); Weinheim: VCH Verlagsgesellschaft, 1987, pp. 113–118.

[433] Birch, S. W., Turner, A. P. F., Ashby, R. E., "An Inexpensive On-Line Alcohol Sensor for Fermentation Monitoring and Control", *Proc. Biochem.* (1987) 37–42.

[434] Prinzing, U., Ogbomo, I., Lehn, C., Schmidt, H.-L., "Fermentation Control with Biosensors in Flow-Injection Systems. Problems and Progress", *Sens. Actuators* (1989) **B1** (1990) 542–545.

[435] Riedel, K., Renneberg, R., Liebs, P., Kaiser, G., *Stud. Biophys.* **119** (1987) 163–166.

[436] Riedel, K., Renneberg, R., Kühn, M., Scheller, F., *Appl. Microbiol. Biotechnol.* **28** (1988) 316–318.

[437] Karube, I., *Sci. Technol. Jpn.* **july/sept.** (1986) 32–35.

15 Chemical Sensors and Instrumentation in Analytical Chemistry

Sections 1 and 2: WOLFGANG GÖPEL, Universität Tübingen, FRG
Section 3: MICHAEL OEHME, Norwegian Institute for Air Research, Lillestrøm, Norway

Contents

15.1   Introduction

Chapters 15–26 deal with specific applications of chemical and biochemical sensors in selected typical application fields. In many of these fields, analytical chemistry has been and still continues to be the most important approach for problem solving. As a result, the current market for sensors is orders of magnitudes lower than the market for high-cost instrumentation in analytical chemistry.

By definition, analytical chemistry is the science of signal production and interpretation necessary for the qualitative and quantitative characterization of materials. According to the general concept formulated by Malissa [1], in which an analytical signal is seen to be the result of an interaction between sample and reagent (generally, any source of energy), signal production can be described by the simple relationship $M_i + R \rightarrow x_i$, where M_i is the sample, R is the reagent, and x_i is the signal.

A resulting complex of signals is called analytical information. Signals may be processed by mathematical methods, eg, transformations or derivations, to yield data smoothing, feature selection, and other results (see also Chapter 6). The choice of mathematical methods used to process data depends on the conditions of the analytical measurement. To translate analytical information as defined above into chemical information (qualitative and quantitative data), a decoding procedure must be applied [2] (see Figure 15-1). The position of analytical knowledge in the overall scheme of analysis is shown in Figure 15-2 [2].

For an overview, Table 15-1 characterizes the different branches of analytical chemistry with their typical problems and methods. Table 15-2 gives a brief summary of the most important techniques used in analytical chemistry, and Table 15-3 lists typical analytical instruments in selected different application fields.

A definition of chemical sensors has already been given in Chapter 1. A clear distinction between sensors and instrumentation in analytical chemistry cannot be made, however, for several reasons:

− Sensors are often only one part in the complete measuring setup including filtering, data preprocessing, and data evaluation (see Chapter 1).

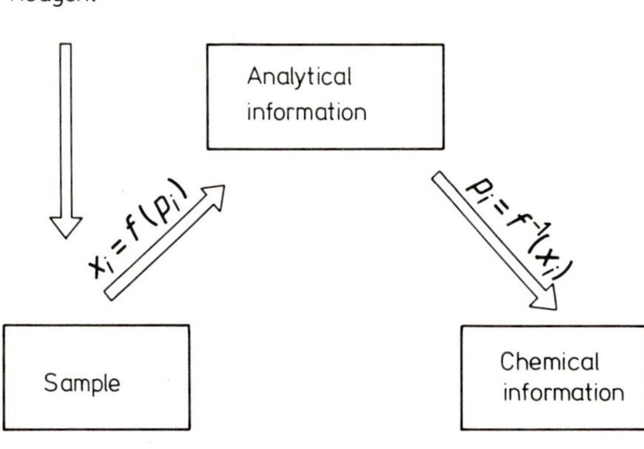

Figure 15-1.
Once obtained from the sample and reagent, analytical information must be translated into chemical information (qualitative and quantitative data). For simplification, amounts of species are characterized by p_i and measuring signals by x_i.

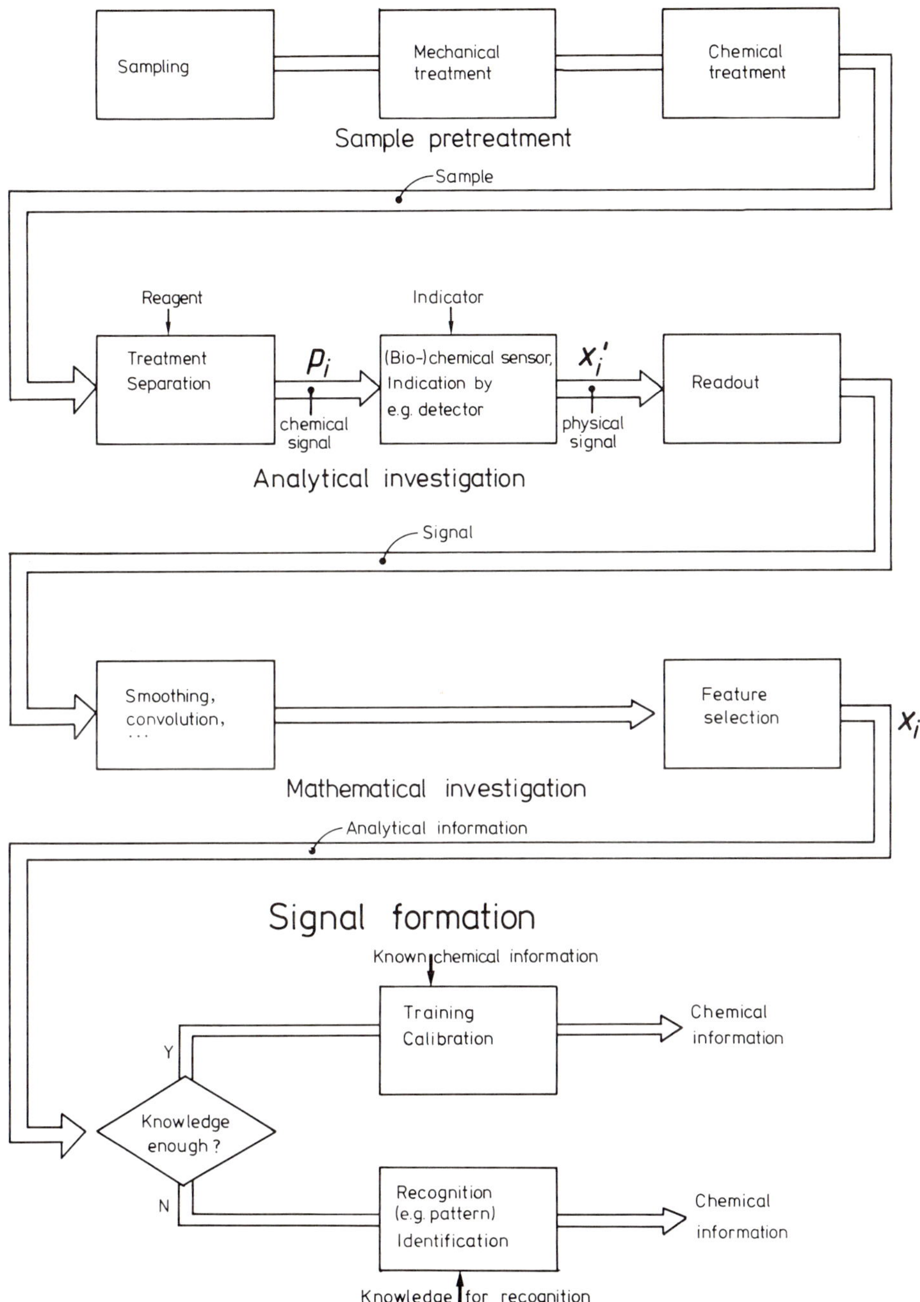

Figure 15-2. The schematic diagram of the analytical process of signal formation and signal interpretation.

Table 15-1. Specialization in analytical chemistry (typical examples).

Typical methods	Typical problems
Separation methods	Environmental analysis
Mass spectrometry	Food analysis
Atomic spectroscopy	Biochemical analysis
Molecular spectroscopy	Pharmaceutical analysis
Structure elucidation	Clinical analysis
Chemometrics	Material science
Trace analysis	Automated analysis
Microanalysis	Toxicology
Surface analysis	Forensic analysis
Electroanalysis	Geochemical analysis
Thermal analysis	Extraterrestrial analysis

Table 15-2. Abbreviations and acronyms of techniques and important terms used in analytical chemistry.

AAS	Atomic absorption spectrometry
AES	Atomic emission spectrometry; Auger electron spectroscopy
AFS	Atomic fluorescence spectrometry
ARIES	Angular resolved ion and electron spectroscopy
ASIA	Atomizer, source, ICPs in AFS
ATR	Attenuated total reflection
CADI	Computer-assisted dispersive infrared
CARS	Coherent anti-Stokes Raman spectroscopy
CCC	Counter-current chromatography
CD	Circular dichroism
CI	Chemical ionization
CIR	Cylindrical internal reflection
COMAS	Concentration-modulated absorption spectrometry
CP-MAS	Cross-polarization-magic angle spinning
CSN-ICP	Conductive solids nebulizer inductively coupled plasma
CSP	Chiral stationary phase
DAC	Diamond anvil cell
DCCC	Droplet counter-current chromatography
DCI	Direct current ionization
DCP	Direct current plasma
DMS	Dynamic mass spectrometer
DRIFTS	Diffuse refelctance infrared Fourier transform spectroscopy
DTA	Differential thermal analysis
EDS	Energy dispersive spectroscopy
EDXRF	Electron diffraction x-ray fluorescence
EELS	Electron energy loss spectroscopy
EIEIO	Easily ionized element interface observation
ELISA	Enzyme-linked immunosorbent assay
ENDOR	Electron nuclear double beam resonance
EPR	Electron paramagnetic resonance
ES	Electron spectroscopy
ESCA	Electron spectroscopy for chemical analysis

Table 15-2. continued.

ESD	Electron-stimulated desorption
ESIE	Electron-stimulated ion emission
ESR	Electron spin resonance
ETA	Electrothermal atomization
EXAFS	Extended x-ray absorption fine structure
FAB-MS	Fast atom bombardment mass spectrometry
FD	Field desorption
FDIR	Fast dispersive infrared
FIR	Far-infrared spectroscopy
FTIR	Fourier transform infrared
FTIR	Frustrated total internal reflection
FTMS	Fourier transform mass spectrometry
FTR	Frustrated total reflection
GC	Gas chromatography
GC-IR	Gas chromatography-infrared
GC-MS	Gas chromatography-mass spectrometry
GC-NMR	Gas chromatography-nuclear magnetic resonance
GDMS	Glow discharge mass spectrometry
GFAAS	Graphite furnace atomic absorption spectroscopy
GPC	Gel permeation chromatography
GPMAS	Gas phase molecular absorption spectrometry
HDC	Hydrodynamic chromatography
HIXSE	Heavy ion-induced x-ray satellite emission
HPLC	High-performance liquid chromatography
HPTLC	High-performance thin-layer chromatography
HREELS	High-resolution electron energy loss spectroscopy
HRGC	High-resolution gas chromatography
IC	Ion chromatography
ICAP	Inductively coupled argon plasma
ICP	Inductively coupled plasma
ICP-AES	Inductively coupled plasma atomic emission spectroscopy
ICP-ES	Inductively coupled plasma emission spectrometry
ICP-MS	Inductively coupled plasma mass spectrometry
ICRS	Ion cyclotron resonance spectroscopy
INEPT	Insensitive nuclei enhanced by polarization transfer
IR	Infrared spectroscopy
IRE	Internal reflection element
IRS	Internal reflection spectroscopy
ISE	Ion-selective electrode
ISS	Ion-scattering spectroscopy
LAMMA	Laser microprobe mass analysis
LASER	Light amplification by stimulated electromagnetic radiation
LC	Liquid chromatography
LC-IR	Liquid chromatography-infrared
LC-MS	Liquid chromatography-mass spectrometry
LEED	Low-energy electron diffraction
LIDAR	Light detection and ranging
LIMA	Laser ionization mass analysis

Table 15-2. continued.

LIMS	Laser ionization mass spectrometry
LIMS	Laboratory information management system
LRS	Laser Raman spectrometry
MBAS	Molecular beam atom scattering
MCD	Magnetic circular dichroism
MIKE	Mass-analyzed ion kinetic energy
MIP	Microwave-induced plasma
MIR	Multiple internal reflection
MIR	Mid-infrared spectroscopy
MRI	Magnetic resonance imaging
MS	Mass spectrometry
NCI	Nitrogen chemical ionization
NIR	Near-infrared spectroscopy
NIRA	Near-infrared reflectance analysis
NMR	Nuclear magnetic resonance
NQR	Nuclear quadrupole resonance
OAS	Optoacoustic spectroscopy
OES	Optical emission spectrometry
ORD	Optical rotatory dispersion
PAS	Photoacoustic spectroscopy
PASCA	Positron annihilation spectroscopy for chemical analysis
PDMS	Plasma desorption mass spectrometry
PEM	Photoelastic modulator
PES	Photoelectron spectroscopy
PIGME	Particle-induced gamma ion emission
PIXE	Particle-induced x-ray emission
PNMR	Proton nuclear magnetic resonance
RBS	Rutherford back-scattering
RGA	Residual gas analysis
ROA	Raman optical activity
RRS	Resonance Raman spectrometry
RTP	Room-temperature phosphorescence
RTPL	Room-temperature phosphorescence in liquids
SAM	Scanning Auger microscopy
SAX	Selected areas x-ray photoelectron spectroscopy
SEM	Scanning electron microscope
SERS	Surface-enhanced Raman spectrometry
SFC	Supercritical fluid chromatography
SIM	Scanning ion microscopy
SIMS	Secondary ion mass spectrometry
SNR	Signal-to-noise ratio
SSMS	Spark-source mass spectrometry
STAT	Slotted-tube atom trap
TIMS	Thermal ionization mass spectrometry
TLC	Thin-layer chromatography
TOF	Time-of-flight mass spectrometry
TPD	Temperature-programmed desorption
UPS	Ultraviolet photoelectron spectroscopy

Table 15-2. continued.

UV-VIS	Ultraviolet-visible spectroscopy
VUV	Vacuum ultraviolet
WDS	Wavelength-dispersive spectroscopy
XANES	X-ray absorption near edge structure
XPS	X-ray photoelectron spectroscopy
XRD	X-ray diffraction
XRF	X-ray fluorescence

Table 15-3. Selected application fields and analytical methods and instruments (for explanation of abbreviations, see Table 15-2).

Air pollution	GC, MS, IC, gas and particle monitoring instruments, most instruments for water analysis
Coatings	GPC, HPLC, TLC, AAS, PES, IR, NMR, surface spectroscopies, UV-VIS, XRF, thermal analysis
Ferrous analysis	AAS, AES, XPS, ES, LIMS, LAMMA, MS, XRF, GC, GPC, HPLC, TLC, IR, NMR, UV-VIS, thermal analysis
Food	TLC, GC, HPLC, GC-MS, SFC, SEC, NMR, ELISA, AAS, ICP-AES, XRF, PIXE
Geological and inorganic materials	AAS, ICP-AES, ICP-MS, DCP, MS, XRF, EDX, PIXE, SEM, SIMS, EXAFS, XANES, nuclear activation methods, chromatography and electroanalytical methods
Pesticides	GC, HPLC, TLC, GC-FTIR, HPLC-MS, GC-MS
Pharmaceuticals and related drugs	Chemometrics/robotics, chromatographic and spectroscopic methods
Synthetic polymers	GC, pyrolysis techniques, LC, MS, NMR, IR, Raman, thermal analysis, surface spectroscopies
Rubbers	NMR, IR, thermal analysis, GPC, HPLC, etc.
Water analysis	IC, GC, TLC, HPLC, MS, GC-MS, HPLC-MS, NCI-MS, photometry and spectrophotometry, enzymatic and electroanalytical methods

- Sensors are often used as the detection element of more complex analytical chemistry instruments. Typical examples are chromatographic sensors, which will be discussed separately in Section 15.3 because of their specific importance.
- The trend towards miniaturization of instruments in analytical chemistry will also lead to devices which may be called either instruments or sensors. Examples are miniaturized optical spectrometers with simplified light sources, monochromators, and detectors (see eg, Chapters 17 and 18).
- On the other hand, electrochemical sensors are widely used as "instruments" in analytical chemistry (see Table 15-4).

Only in some cases is a distinction between sensors and instruments obvious; eg, an NMR spectrometer is usually not considered to be a sensor, and a field effect transistor is usually not considered to be an analytical chemistry instrument. For the sake of completeness and because of the overlap, we decided to include some sections on analytical instrumentation in this volume, which is otherwise exclusively devoted to sensors.

Table 15-4. Worldwide analytical instrument sales (US$ millions per year)*.

Molecular analysis	1989	1985	Elemental analysis	1989	1985
High-performance liquid chromatographs	782	513	Scanning electron microscopes	205	142
Gas chromatographs	544	395	Atomic absorption spectrometers	183	148
Mass spectrometers	416	271	Plasma emission spectrometers	156	99
Liquid chromatographs	262	223	pH/ISE (ion sensitive electrodes) systems	147	115
UV-visible spectrophotometers	242	202	Electron spectrometers	139	94
Fourier transform infrared spectrometers	200	98	Energy-dispersive x-ray spectrometers	137	94
Nuclear magnetic resonance spectrometers	199	133	Transmission electron microscopes	129	104
Ion chromatographs	136	50	Arc/spark emission spectrometers	101	80
Thermal analyzers	126	86	Wavelength-dispersive x-ray spectrometers	97	88
Fluorescence spectrometers	75	48	Organic elemental analysers	87	65
Titrators	67	50	X-ray diffraction instruments	71	66
Dispersive infrared spectrometers	59	64	Electron microprobes	34	26
Near-infrared spectrometers	60	36			
Scintillation counters	55	45			
Densitometers	49	39			
Automated flow analyzers	32	24			
Voltammeters	24	20			
Raman spectrometers	14	9			
Spectropolarimeters/ORD/CD	14	10			

* Overall totals: 1989 = $ 4812; 1985 = $ 3437.

15.2   Analytical Instruments: Markets and Sales

The worldwide sales of analytical instruments are at nearly twice the level as of 5 years ago. To characterize the different shares from different parts in the world, one must consider the following: the USA supplies 40–45% of the world market, Europe 20–30%, and Japan more than 20% [3]. Of the top worldwide instrumentation manufacturers with sales of over US $ 100 million per year, five are American, five are European and four are Japanese. The instrumentation industry is becoming increasingly global.

Certain forces driving the instrumentation market can be detected in an altogether prosperous chemical industry: materials development in general, production of microelectronics, metallurgy, environmental protection, chemical technology, biomedical research, biotechnology, and the pharmaceutical and clinical sectors all have tremendous potential as consumers of instruments for their R&D efforts. Analytical instrument sales are made chiefly to three branches: industry, government, and academia.

Table 15-4 gives an overview of instruments sales world wide for molecular and elemental analysis.

Evidently chromatography [4] is by far the most imporant technique and it is the fastest growing area. Chromatrographic methods are used extensively in industrial laboratories which purchase about 70% of the devices made for separation, purification, and analysis. One of the most frequent demands in all forms of chromatography is biocompatibility. Biocompatible instruments are designed to have chemically inert, corrosion-resistant surfaces in contact with biological samples.

The largest sales of individual instrument types are in high-performance liquid chromatography (HPLC) and high-resolution gas chromatography (HRGC). The higher cost but more efficient HPLC technology has caused sales of more standard liquid chromatographs (LC) to decrease during the last years. It is biotechnology in new drug development efforts that is expected to affect HPLC sales in the near future. Expectations are that large scale-ups of fermentation and genetic engineering will lead to increased use of HPLC instruments. In addition, the developing market for on-line monitoring process control and quality control can be predicted to increase significantly during the coming years. Some of the newest innovations in GC technology are the production of more instruments with high-efficiency, high-resolution capillaries and supercritical fluid capability. Supercritical fluid chromatography (SFC) in particular has attracted a great deal of attention since its introduction around 1985. It operates using a supercritical fluid as the mobile phase and bridges the gap between GC and HPLC. The use of these mobile phases allows for higher diffusion rates and lower viscocities than liquids and a greater solvating power than gases. Another area showing tremendous growth is ion chromatography (IC). Its popularity has been enhanced through extending its applicability from inorganic systems to amino acids and other biochemical systems by the introduction of biocompatible instruments.

Mass spectrometer (MS) [5] sales have always been high, especially since MS is the principal detector in a number of "hyphenated" techniques such as GC$-$MS, MS$-$MS, and LC$-$MS. The GC$-$MS combination accounts for about 60% of MS sales since it is used widely in drug and environmental testing. Innovations in interface technology such as inductively coupled plasma (ICP)$-$MS, FC$-$MS and thermospray or particle beam interfaces for LC$-$MS have both advanced the technology and expanded the interest in applications. Recent introductions of lower cost MS instruments with automated sampling and computerized data analysis have added to the attraction of the technique for first-time users.

Spectroscopy [6] in general accounts for half of all instrument sales and is the largest overall category of instruments. It can be broken down evenly into

– optical methods and
– electromagnetic or non-optical methods.

These categories include many individual high-cost systems such as nuclear magnetic resonance (NMR) spectrometers, X-ray equipment, and electron microscopy and spectroscopy setups. Sales of spectroscopic instruments that are growing include Fourier transform infrared (FTIR), Raman NMR, plasma emission and energy-dispersive X-ray spectrometers.

FTIR spectrometers [7] in particular have been popular for the past few years. The number of suppliers of FTIRs worldwide is large, and with close competition the cost per instrument has recently dropped significantly. As a result, dispersive IR instruments have become of less importance because of the competition with the faster, more sensitive, and computerized FT methods.

Plasma emission spectrometers have shown a rapid growth. This holds also for NMR spectrometer sales because of new applications in biomedical research and more sophisticated experimental methods using increased computing power. Similarly, *Raman spectroscopy,* traditionally used in academic research, is gaining acceptance in industrial R&D and quality control applications. Materials research and surface analysis in a variety of industries keeps the sales of electron microscopic, electron spectroscopic, ion spectroscopic, and X-ray instruments growing. Details of the various techniques on surface and interface characterization which are also important in R&D of chemical sensors themselves, can be found in Chapter 3, Section 3.4.2.

The future of analytical instrumentation in general is characterized by new innovations and consequent developments in existing technologies. Among these are the introduction of FT Raman, IR dichroism, IR microscopy and NMR imaging spectrometers. Hyphenated and automated types of apparatus are also appearing on the market more frequently. New analytical techniques such as capillary electrophoresis, gel capillary electrophoresis, scanning tunnelling microscopy and related scanning techniques for imaging with atomic resolution, and atomic force microscopy in particular for the imaging of biological systems, as well as a variety of other techniques for surface and materials analysis are already or may soon be appearing as commercialized instruments (see Chapter 3, Section 3.4.2). If the chemical industry continues to do well in the next few years, so will the sales of analytical instrumentation and sensors, too.

15.3   Specific Features of Sensors and Instrumentation in Chromatography

15.3.1   Principles of Chromatographic Separation

Chromatographic techniques are used to separate complex mixtures of compounds into individual substances or compound groups with similar chemical or physical properties. Separation is effected by interaction between a stationary phase (normally liquid or solid) and the sample; this results in different retention characteristics for different substances. A mobile phase that does not normally interact with the stationary phase is used as a carrier. Figure 15-3 shows the schematic setup of a chromatographic separation system.

The compounds eluted from a chromatographic separation column show ideally a Gaussian concentration distribution due to peak broadening mechanisms in the stationary and mobile phase. A short summary of the important terms describing chromatographic separation is given in Figure 15-4 and in the corresponding literature [8-11].

Detectors which are able to detect the compounds eluted from the chromatographic system have to fulfill different requirements depending on the chromatographic technique employed. They can be summarized as follows:

1. The chromatographic separation should not be changed by detector characteristics such as dead volume or response time.

2. The detector should measure a property of the eluted compound or the mobile phase or both.

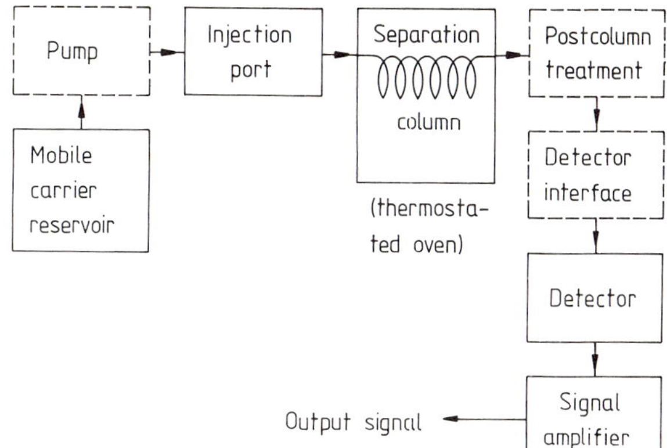

Figure 15-3.
Schematic diagram of a chromatographic separation system used in liquid, gas or supercritical fluid chromatography. The dashed boxes represent elements which are only used for some of the techniques.

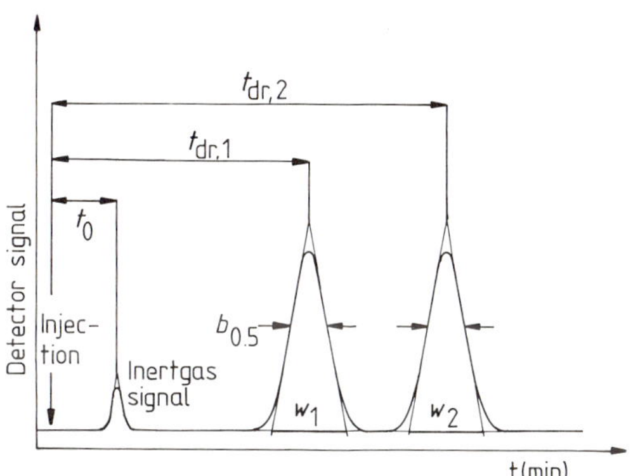

t_0: dead time of the chromatographic system

t_{dr}: retention time of signal 1,2

Capacity ratio of a compound:

$$K' = \frac{t_{dr} - t_0}{t_0}$$

Resolution of two compounds:

$$R = \frac{2\,(t_{dr,2} - t_{dr,1})}{w_1 + w_2}$$

Figure 15-4.
Important terms describing the separation in a chromatographic system. The separation efficiency of a column can be expressed by the theoretical plate number,

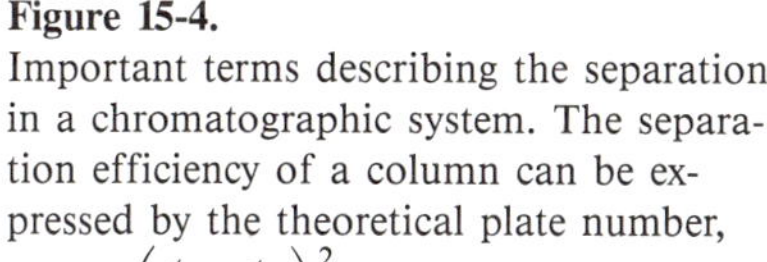

$$N_{th} = \left(\frac{t_{dr} - t_0}{b_{0.5}}\right)^2.$$

3. The detector should be compatible with the physical separation conditions such as pressure, temperature, flow rate, and reactivity of the eluent.

A number of other criteria have to be fulfilled, depending on the application range of the detector. For example, a simple detector function which indicates if and when a sample fraction elutes from a precolumn requires a non-destructive detector. However, a well-defined relationship between compound concentration and detector response is often not necessary.

A variety of chromatographic detectors have been described in the literature. Because of limited space, only the most common commercially available detector principles are described in detail here.

15.3.1.1 Chromatographic Techniques and Their Detector Requirements

Gas Chromatography (GC)

In GC, normally thermally stable but volatile compounds are separated. Both uncoated and coated solid adsorbents as well as thin films of liquids or polymers of high molecular weight are used as the stationary phase. Inert gases (He, Ar, N_2) and H_2 are used as the mobile phase, which is normally inert compared with the separated compounds. Many GC detectors are therefore based on thermal decomposition or ionization processes which convert the eluted substances into ions. Detector systems which measure a physical property of the column eluent (mobile phase and/or compound) and the eluted substance alone are also frequently used [12, 13].

Liquid Chromatography (LC)

Liquid chromatography usually employs solid porous supports with hydrophilic (silica, alumina) or lipophilic properties (silica modified by chemically bonded aliphatic chains) as stationary phases. Materials with size-exclusion properties can also be used for the separation of high molecular weight compounds. Depending on the surface characteristics of the support, both aqueous mixtures of organic solvents and mixtures of organic solvents with different polarity can be used as mobile phases. These solvents, having a known refractive index, absorbtivity in the UV region and/or thermal conductivity, pass together with the eluted compound (solute) to the detector. Many solvents contain reactive functional groups. The range of usable detectors in LC is therefore often limited to those which measure a physical property of the eluent or the solute. Only in cases where the solute is more reactive than the mobile phase can other detection principles such as amperometric or electrochemical detection be used. Detection principles which work only in the vapor phase are normally not usable, since the evaporated mobile phase produces large volumes of vapor which have to be separated from the solute.

Molecular structures which are present in both the solvent used as the mobile phase and the eluted compound often give a low but measurable detector response. The resulting background signal, which is often higher than that for GC detectors, limits the lower detection limit. This is one of the reasons why there is still a lack of highly sensitive universal LC detectors [14, 15].

Thin-Layer Chromatography (TLC)

This two-dimensional separation technique uses the same stationary phase materials and mobile phases as liquid chromatography. Separation can be carried out in either one or both directions on the support-coated plate or as circular chromatography starting in the middle of the plate. After completion of the chromatographic process the solvent is evaporated. The separated compounds are then still adsorbed on the solid stationary phase, which complicates the detection. Detectors which are able to scan over the whole plate surface and which are based on reflectance measurements in the visible or UV range are therefore used. In some instances the separated compounds are evaporated and can then be detected by a GC detector. Solvent extraction of each single separated compound would allow the use of other detectors for quantitative measurement, but this technique is normally too time consuming and is not sensitive enough because of the dilution factor involved [16, 17].

Ion and Ion-Exchange Chromatography (IC)

Organic and inorganic ions soluble in water or buffer solutions can be separated by ion-exchange materials with high or low ion-exchange capacity. Classical ion-exchange chromatography is carried out on high-capacity ion-exchange resins which require a high ion strength in the eluent (normally an aqueous buffer solution) to give reasonably short retentions. This severely limits the use of conductivity detectors because of the small conductivity difference between the eluent alone and the solute.

The introduction of low-capacity ion-exchange materials (silica- or polymer-based) provided the possibility of eluting ions with either very low concentration buffers (eg, $1 \cdot 10^{-3}$ mol Na_2CO_3 for the separation of anions) or weakly dissociated organic acids. In the latter the background is low enough to allow a direct determination of ppb to ppm amounts by direct conductivity measurement. The conductivity of inorganic buffers of low ionic strength can be reduced further by conversion, eg, Na_2CO_3 to very weakly dissociated H_2CO_3 using a cation-exchange system following anion separation [18, 19].

The use of eluents containing weakly dissociated organic compounds with strong UV-absorbing groups allows a UV-detector to be used. Inorganic anions and cations do not normally show UV absorption and the elution of such ions causes a decrease in UV absorption.

Supercritical Fluid Chromatography (SFC)

Supercritical fluids can also be used as mobile phases in chromatography [20, 21]. Stationary phases used in both GC and LC can be employed. The sample is normally injected into a mobile phase which is in the subcritical liquid state. Subsequently it is converted into a supercritical fluid by raising the temperature above the critical point.

Supercritical phases have a viscosity similar to that of gases while the density can be varied over a wide range from gas- to liquid-like properties. Both inorganic gases such as CO_2 and organic compounds such as pentane can be used as mobile phases. CO_2 has the advantage of becoming supercritical at low temperature (31 °C) and moderate pressure (7.3 MPa) and is therefore often used. Both LC and GC detectors can be employed. LC detectors tolerate such high pressures. For GC detectors such as the flame ionization detector the column eluent pressure has to be reduced to atmospheric pressure by mounting a restrictor between the column outlet and the detector.

15.3.2   General Aspects of Chromatographic Detectors

Chromatographic detectors can be divided into subgroups according to different criteria. Figure 15-5 summarizes the detector classes according to general detector characteristics.

In addition, chromatographic detectors can be classified according to their working principles, eg, light emission/absorption or ionization mechanisms.

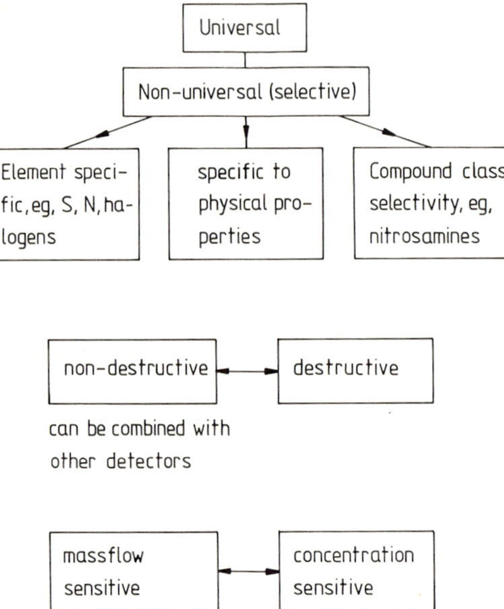

Figure 15-5.
Classification of chromatographic detectors based on general detector characteristics.

Universal detectors which can detect any kind of compound find a quantitative use only in GC. Examples are the thermal conductivity detector and the gas density balance (see Section 15.3.4.2). Most GC detectors are mass flow sensitive, which means that the output signal is directly proportional to the mass flow of a compound through the detector. The signal from such detectors is normally obtained from a direct effect of the detection system on the eluted compound, eg, ionization or transfer to an excited state.

Nearly all LC detector principles produce an output signal which is related to the concentration of the solute in the eluent and are therefore concentration sensitive.

Non-destructive detectors usually measure a physical property of the eluent. The detected compounds are neither decomposed nor is the structure changed. Ionization processes are non-destructive as long as no dissociative reaction takes place, such as cleave-off of functional groups. Non-destructive detectors can be combined with other detectors in multi-detection systems.

As mentioned before, an ideal chromatographic detector should not influence the peak shape or chromatographic resolution; detector dead volume and response time are critical parameters [22].

15.3.2.1 Detector Response Time

The response time is defined as the time period which is necessary to reach 90–95% of the change in signal level after a sudden concentration change. As a general rule, only detector time constants are acceptable which are one tenth or less of the chromatographic signal width measured at 4σ (standard deviations) of the Gaussian peak shape. As Figure 15-6 shows, both the signal area and height are distorted by longer response times.

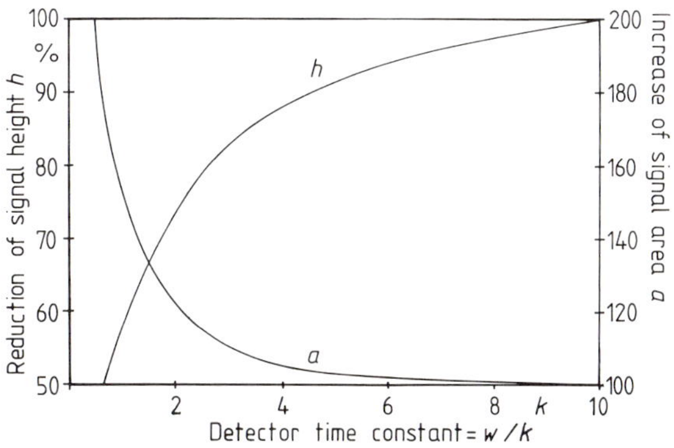

Figure 15-6.
Distortion of the height h and area a of a chromatographic signal due to a ratio $k > 0.1$ between the detector time constant c (in s) and the chromatographic peak width w (in s).

15.3.2.2 Detector Dead Volume

Signal distortion caused by the dead volume in the gas sampling lines and the detector itself can influence the chromatographic resolution obtained. Dead volumes can act as an exponential dilution chamber leading to signal tailing or can cause signal broadening because of transport dispersion in the connecting tubes. The influence of the dead volume is directly proportional to the flow rate through the system. As a general rule, a compound residence time in the detector of one twentieth of the chromatographic signal width at half-height does not influence the peak shape. The residence time R is defined as the time a signal will need to pass the detector volume V at a given flow rate w:

$$R = V/w$$

15.3.2.3 Dynamic and Linear Range

For quantitative analysis the relationship between the amount of eluted compound and the detector signal has to be well defined. For most detection principles a linear response is obtained. However, the range where a strongly linear relationship is valid is often much smaller than the overall measuring range of the detector, also called the dynamic range. It is limited by the lower and upper (detector saturation) detection limit. Most chromatographic detectors have a dynamic range over several orders of magnitude while the usable linear range which can be used after proper calibration only extends over 1–3 decades.

15.3.2.4 Detection Limit

The lower detection limit of a detector is defined as the minimum amount of compound detectable at a given signal-to-noise ratio. Ideally the detection limit should be determined independently of the chromatographic separation system using standard dilution devices. Detection limits are often expressed in g/s (mass flow-sensitive detectors) or g/mL (concentration-sensitive detectors) to obtain a value which is independent of the measuring conditions (flow rate, etc.).

The lower detection limit is influenced by the noise frequency and amplitude distribution compared with the signal band width and height. The noise level is usually measured over a given time period which is a multiple of the signal width. The noise can be expressed as peak-to-peak (p-t-p) or root-mean-square (rms) values. The latter gives about 70–80% lower noise levels. Figure 15-7 summarizes the detection limits of some selected chromatographic detectors.

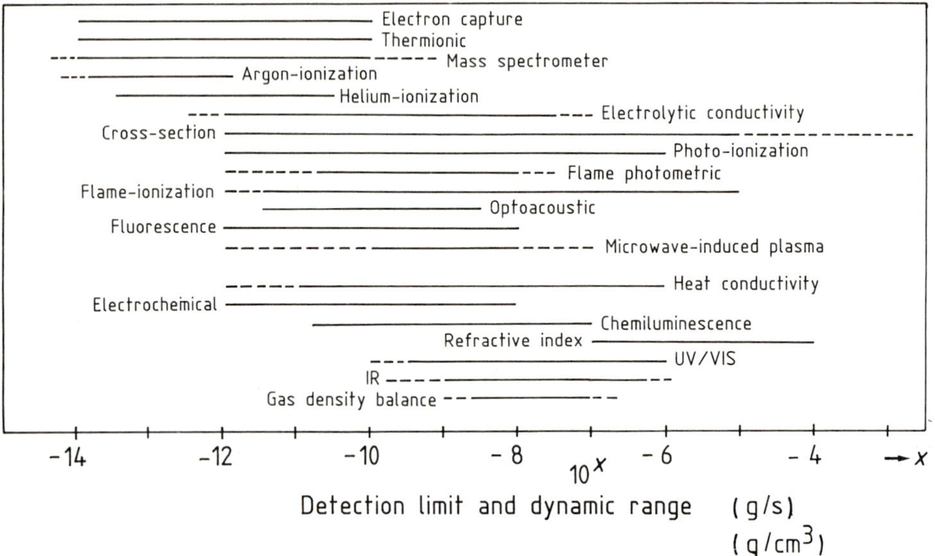

Figure 15-7. Detection limits and dynamic range of some chromatographic detectors.

15.3.3 Detector Selectivity for Selected Compounds

Selective detectors have considerably increased response factors for a single substance, certain compound classes, certain elements or selected physical properties which are closely related to specific compound substructures. The selectivity S to a given compound i compared to another compound j can be expressed as follows:

$$S_{ij} = \frac{\text{measuring signal } i \cdot \text{number of mol } j}{\text{measuring signal } j \cdot \text{number of mol } i} \cdot$$

This relationship is only valid within the linear response range of both substances.

For a practical selective chromatographic detector, a minimum selectivity of a factor of 100 is required. Selectivity is not a constant parameter of a detector but is highly dependent on factors such as detector temperature, construction, flow rate, sample background, and test compounds. In addition, detector selectivity can be influenced by co-elution of compounds from the chromatographic column. In many cases a discriminated compound may cause signal enhancement or quenching owing to interferences with the detector mechanism. Examples of this are fluorescence quenching in LC by aromatic contaminants in the sample or signal quenching of the flame photometric detector in the sulfur-selective mode by hydrocarbons.

Additional selectivity can also be achieved by coupling detectors with different selectivities for compound substructures. An example is the combination of an electron-capture detector, which is selective for compounds with a certain electron affinity, and a thermionic detector, which is selective for N-containing substances; the combination gives a high selectivity for aromatic nitro groups.

Use of selective derivatization reactions prior to the separation allows the introduction of functional groups which can be detected with high selectivity. In some cases derivatization can be carried out on-line before (pre-column) or after the separation column (post-column) [23]. Use of fluorescent reagents which react selectively with, eg, aliphatic amines is an example of pre-column derivatization.

Post-column derivatization requires a reagent addition and a reaction mechanism which does not affect the chromatographic separation by additional band broadening. Post-column addition of light-absorbing strongly complexing agents allows the detection of heavy metal cations which were separated as weak non-light-absorbing citrate complexes on an ion chromatographic column [24].

Another example of how to increase compound selectivity is post-column irradiation of explosives in an LC column eluent leading to the formation of highly fluorescent compounds.

The labeling of compounds with stable or radioactive isotopes in some cases allows the detection of trace amounts with very high selectivity.

15.3.4 Working Principle and Application Range of Selected Chromatographic Detectors

The detectors are divided into four classes based on the general detection principle or detector selectivity. A short description of the working principle and the application range is given for each detector. Detector properties such as detection limits, selectivity, and dynamic range are summarized in tables.

15.3.4.1 *Detectors Based on Light Emission/Absorption Mechanisms*

UV Absorption Detector

Working principle. The eluent from the separation column passes into a flow cell of volume 1–50 μL which is placed in the light path of a UV light source (mono- or polychromatic). Normally a reference cell (filled with air or solvent or coupled to a second column) is used in parallel to compensate for variations in light intensity, temperature, and flow rate. Two silicon photodiodes register the transmitted light from both cells as a differential signal. In the case of polychromatic light sources a UV absorption spectrum can be obtained after the flow cell by a holographic grating combined with a photodiode array with a resolution of 5 nm or better. In this way both complete UV spectra can be recorded and simultaneous detection at different wavelengths is possible [14, 25].

Practical aspects. The detector is mainly used in LC and TLC and occasionally in IC (see Section 15.3.1.1) but is not sufficiently sensitive for GC.

It detects all compounds which absorb light at 180–350 nm due to their structures containing π-electrons and/or unshared electrons.

It can also be used to detect non-UV-absorbing species in an eluent with high absorbance (inverse detection).

In TLC, reflectance of the UV light from the plate is usually measured by a spectrophotometer. The separated compound spots are detected by scanning over the whole plate with a given slit width which controls the resolution in space.

Fluorescence Detector

Working principle. Monochromatic light of a given wavelength in the UV range traverses a flow cell coupled to the separation column. Molecules in the eluent which can undergo fluorescence absorb the UV light and are promoted to an excited state. The return to the ground state is accompanied by a loss of energy in the form of light emission at a longer wavelength than the absorbed radiation. The fluorescent light is usually detected at 90 ° to the exciting light beam, which minimizes the background signal caused by light scattering from the exciting light source [14].

Practical aspects. The detector is frequently used in LC and TLC to detect compounds which show fluorescence or can be made fluorescent by derivatization. It is one of the most sensitive LC detectors.

A non-linear detector response can be caused by direct transfer to the ground state (quenching) due to intermolecular collisions with co-eluting compounds or by phosphorescence.

It is also suitable for GC detection of aromatic hydrocarbons which show very intense fluorescence.

Chemiluminescence Detector

Working principle. The energy released from a chemical reaction can excite the reaction product to a higher electronic state. The loss of energy on the return to the ground state can result in emission of light (chemiluminescence). This effect is used, eg, to detect halogen-containing compounds by adding indium vapor to a hot flame which is fed with the eluent from a GC column. The indium halides so formed emit a strong chemiluminescence spectrum. Nitrogen-containing compounds are detectable with high selectivity after pyrolysis to NO, CO_2 and H_2O (which has to be removed using a drying cartridge). NO can undergo a chemiluminescence reaction with O_2 forming NO_2 in the excited state [13].

Chemiluminescence from S- or P-containing compounds is obtained by combustion in a hydrogen-rich flame. Both the excited HPO$^-$ radicals formed and the S_2 dimers can emit a chemiluminescence spectrum which allows the selective detection of S (eg, at 394 nm) and P (526 nm) using suitable filters and a photomultiplier tube for signal amplification [26]. This flame photometric detector shows a non-linear realtionship between concentration and output signal in the sulfur mode which in theory should be quadratic (owing to the dimer formation). However, in practive, exponential coefficients of between 1 and 2 are found. Electronic linearization of the output signal is therefore necessary.

The detector is extremely sensitive to temperature and flow changes as well as the H_2/O_2 ratio in the fuel-rich flame. Deactivation of the excited species by collision with organic radicals can cause signal quenching when high hydrocarbon concentrations co-elute.

Practical aspects. Chemiluminescence detectors are only applicable as GC detectors because of signal quenching by high hydrocarbon levels such as solvents used as mobile phases in LC. They are primarily used for selective sulfur detection.

Quenching problems can be reduced by using a two-flame construction where the first flame pyrolyzes the compound and the second excites the decomposition products.

Infrared Analyzer

Working principle. The classical infrared analyzer is based on absorption of infrared radiation of wavelength 2–20 μm which traverses a flow cell of length 10–40 cm and diameter 1–3 mm (volume about 10–500 μL). BaF_2 is often used as the cell window material; this allows the passage of radiation of wavelength up to 13 μm. Depending on the wavelength, substructures such as carbonyl or hydroxyl groups can be selectively detected. The infrared detector is used mainly in GC and but is not compatible with LC because strongly IR-absorbing solvents (eg, water or methanol) are usually employed, which additionally damage the cell windows.

The use of a light pipe increases the absorption path length considerably because of wall reflection. This, used together with a Fourier transform interferometer, markedly increases the sensitivity of the analyzer. The beam from the infrared source is split using a fixed and an adjustable mirror and then recombined. The adjustable mirror allows one to vary the signal wavelength λ with a given frequency ω. After traversing the light pipe, which is connected to the separation column outlet, radiation is detected by a photoconductive cell. The split light beam is recorded as a function of the different path lengths producing so-called Michelson interferograms. These Fourier-transformed interferograms contain all the information relating to the infrared spectrum of the eluted compound which can easily be reconstructed [12, 22].

Practical aspects. The whole wavelength range can be scanned within about 1 s by the Fourier transform technique.

The sensitivity as a GC detector is reasonably good. Complete IR spectra can be detected with 50–100 ng of compound. Detection of strongly absorbing functional groups such as carbonyls is possible in the sub-nanogram range.

The detector is only slightly pressure and temperature sensitive.

As an LC detector it is only useful for compounds with very strong IR-absorbing groups and when normal-phase LC with aliphatic solvents is used.

Typical Data. Typical data for light-emitting/-absorbing detectors are given in Table 15-5.

Table 15-5. Typical data for light-emitting/-absorbing detectors.

Detection principle	Minimum detectable amount	Linear dynamic range	Selectivity characteristics
UV-VIS absorption: Molar absorption	0.1 – 1 ng	10^4 – 10^5	Low/moderate
$\varepsilon = 10$	1 µg		
$\varepsilon = 100\,000$	0.1 ng		
Fluorescence	1 pg	10^3 – 10^5	High
Chemiluminescence	0.1 – 1 ng	10^3	High
Infrared absorption	10 – 50 ng	10^4	Low/high

15.3.4.2 Detectors Measuring Physical Properties of the Eluent

Thermal Conductivity Detector

Working principle. The detection principle is based on the difference in the thermal conductivities of a pure gas and a gas mixture. Its use is therefore exclusively restricted to GC or possibly SFC. The flow cell of the detector contains a heating element the resistance of which is temperature dependent and which acts both as a heat source and temperature-measuring device. The elution of a compound with thermal conductivity different from that of the carrier gas leads to a change in the temperature of the filament. This results in a change in the electrical resistance and subsequently in a voltage drop which is proportional to the amount of compound traversing the cell. The lower detection limit decreases with increased temperature difference between the cell wall and filament as well as the thermal conductivity difference between a compound and the carrier gas. The detector is very sensitive to flow and temperature changes and requires high temperature stability of the cell block. In addition, bridge circuits with one or two reference and one or two measuring cells are used to compensate for such fluctuations. The signal from the separation column is then measured differentially in relation to the carrier gas flow through the reference gas cell [13, 28].

Practical aspects. It is only applicable as a GC detector.

Reactive compounds may cause oxidation of the filament. Use of hydrogen can lead to hydrogenation reactions of the eluted compounds.

The detector has universal detection properties and is therefore eminently suitable for the analysis of permanent gases.

Changes of the cell properties by slow surface oxidation of the filament can lead to an increased background signal owing to asymmetries in the bridge circuit.

Use of single cells which are pneumatically switched between reference and eluent gas flow (about 10 Hz sampling rate) improves the signal-to-noise ratio and avoids problems due to mismatches between reference and measuring cells.

Gas Density Balance

Working principle. The gas density balance uses the same measuring principle as the heat conductivity detector. However, as can be seen from Figure 15-8 two filaments are placed in a reference gas flow which is split into an upper and a lower flow path. The eluent from the chromatographic column is added to both flows downstream of the filaments. Elution of a compound leads to a higher density of the eluent compared with the pure carrier gas. Because of the influence of gravity, most of the eluent is mixed with the reference gas flow at the lower flow path leading to a decreased reference flow rate and a higher temperature of the filament in the lower flow path. When the eluent is lighter than the carrier gas, an increased reference gas flow passes the upper filament and leads to a lower temperature. The resulting resistance changes are proportional to the density difference and the molecular weight of the eluted compound.

Practical aspects. The gas density balance can only be used as a GC detector.

Corrosive or reactive substances do not come into contact with the filaments, which makes the detector ideally suitable for the detection of such substances.

Two measurements with two reference gases of different density allow in principle the determination of the molecular weight of the eluted compound.

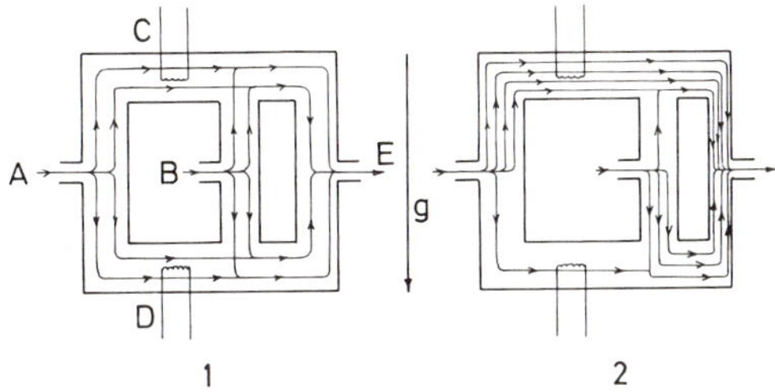

Figure 15-8. Working principle of the gas density balance. 1 No eluted compound in the carrier gas; 2 flow changes due to elution of a sample compound. A = Reference gas inlet; B = inlet for eluent from the separation column; C, D = filaments for heat conductivity measurements; E = exit; g = influence of gravity.

Electrochemical or Amperometric Detector

Working principle. The eluent from the chromatographic column passes over a flow cell equipped with a reference electrode (eg, Ag/AgCl) and a working electrode made from glassy carbon, gold, or silver. Depending on the applied electrode potential, the eluted compounds can undergo either reduction or oxidation reactions on the electrode surface. Normally a potential is chosen where the electrode reaction becomes diffusion controlled. The measured diffusion current is then directly proportional to the compound concentration in the eluent [14, 15].

The conductivity of the eluent (often water-methanol or water-acetonitrile mixtures) is usually increased by adding quaternary ammonium or other inorganic lipophilic salts. Constructions are frequently based on the wall-jet principle where the eluent is sprayed on the surface of the electrode, resulting in very low detector cell volumes [18, 25].

Practical aspects. This very sensitive and selective detector can be used for both LC- and IC applications. In LC compounds are usually detected which can undergo oxidation reactions; these include phenols, aromatic amines, and sulfur compounds.

- Anions such as CN^-, S^{2-}, Br^- and NO_2^- can be detected in IC.
- The detector is also occasionally used in GC. However, in order to make electrochemical detection possible, the eluted compounds have normally to be pyrolyzed to simple functional groups or anions which are adsorbed in aqueous solutions.
- Electrode poisoning by the sample matrix and unwanted surface reactions may cause serious problems.

Electrolytic Conductivity Detector

Working principle. The electrolytic conductivity of the eluent is measured in a flow cell. The metallic inlet and outlet tubes act as electrodes across which an alternating current voltage at a frequency of 1–10 kHz is applied. Constructions with cell volumes from 0.1 to about 10 µL are achievable. Weakly dissociated aqueous electrolytes such as phthalic acid and nicotinic acid are frequently used in IC. They allow ppm levels to be detected after electronic compensation of the background conductivity.

Alternatively, strong electrolytes such as KOH or Na_2CO_3 can be used which are converted either to very weakly dissociated H_2CO_3 or to water, respectively, after the separation column using a second ion-exchange column or a micromembrane ion exchanger. The background conductivity is then greatly reduced, allowing the direct detection of a few ppb to ppm in 0.1–5 mL of sample.

Practical aspects. Most of the common anions and alkali/alkaline earth metal cations as well as NH_4^+ can be detected directly. The same is possible for C_1 and C_2 acids and other ionic organic compounds.

The use of strong electrolytes with high conductivity allows the separation of ions with low conductivity or weakly dissociated compounds. They are detected as negative signals of lower conductivity [18, 24].

Other Detectors Measuring Physical Properties of the Eluent

Detectors which measure a physical property of the eluent such as the refractive index, rate of radioactive decay or the speed of sound waves can be used with chromatographic systems. However, in most instances the detection limits are relatively high, which reduces the application range to preparative and semi-preparative separations.

In LC the refractive index of the column eluent can be measured relative to that of the pure eluent solvent using a double-beam Fresnel refractometer. The light beam passing the measuring cell is deflected, corresponding to the diffraction in the eluent. This changes the light intensity, which is detected by a photo-cell. This detector is the only universal LC detector, since all eluted compounds influence the refractive index of an eluent. However, only separations using isocratic eluent compositions can be carried out [9, 14].

Radioactively labelled compounds (^{14}C or ^3H) can be detected in LC by adding a scintillation liquid to the eluent from the LC column. The γ- or β-radiation emitted during the radioactive decay excites the electrons of the scintillator to higher energy levels. On returning to the ground state the excess energy is lost as photons which are detected using a photomultiplier. Normally two photomultipliers are used which are placed at either end of the flow cell. Only light pulses detected by both systems are registered, thus suppressing the background noise. Owing to the high price of scintillation liquids, the eluent is usually collected and re-extracted. Radioactive compounds can also be detected after GC separation. However, the eluent has to be pyrolyzed to CO_2 and ^3H_2O (subsequently reduced to ^3H) before the radioactive radiation is determined in a gas flow cell equipped with a Geiger counter. The main drawback of this measuring technique is the large cell volume (a few milliliters) and long residence time (> 30 s) of the compound required to obtain a sufficiently high counting yield. This may have a serious influence on the GC resolution.

The ultrasonic detector can only be used for GC separations. The eluent is excited to oscillations of 4–6 MHz in a small cell of 10–40 μL volume using an oscillator crystal. The phase displacement between the oscillator and the receiver (a phase meter) is influenced by the composition of the eluent and proportional to the molar fraction (carrier gas and eluted compound). The response factor is influenced by the differences of the specific heat ratio, c_p/c_v, between carrier gas and eluent. The main drawback of this detector is its high sensitivity to pressure and temperature changes so that a temperature stability of 0.001 °C is required.

Typical Data

Typical data for some detectors based on measuring physical properties of the eluent are given in Table 15-6.

Table 15-6. Typical data for some detectors based on measuring physical properties of the eluent.

Detection principle	Minimum detectable amount	Linear dynamic range	Selectivity characteristics
Thermal conductivity	50 pg/cm^3	$10^4 - 10^5$	None
Gas density balance	10 ng/cm^3	10^2	None
Electrochemical	1 – 100 pg	$10^2 - 10^4$	Medium/high
Electrolytic conductivity	10 ng/cm^3	10^4	None
Radioactivity	0.1 ng	$10^4 - 10^5$	High
Refractive index	0.1 – 1 μg	$10^2 - 10^3$	None
Ultrasonic waves	10^{-14} mol	10^6	None

15.3.4.3 Ionization Detectors

Electron-Capture Detector

Working principle. The detector cell (Figure 15-9) contains a radioactive foil which emits primary electrons with, eg, 67 keV energy (^{63}Ni). These are slowed down by collisions with a reagent gas (N_2 or Ar + 5% CH_4) to the thermal energy level (about 0.01 eV). Compounds with a positive electron affinity can form stable negative ions by electron capture. This leads to a reduction of the electron population which is proportional to the amount of substance present in the detector. The electron density in the detector is measured by a pulsed voltage (eg, 100 μs pulse intervals, 1 μs pulse width and 40 V pulse amplitude) applied to a collector

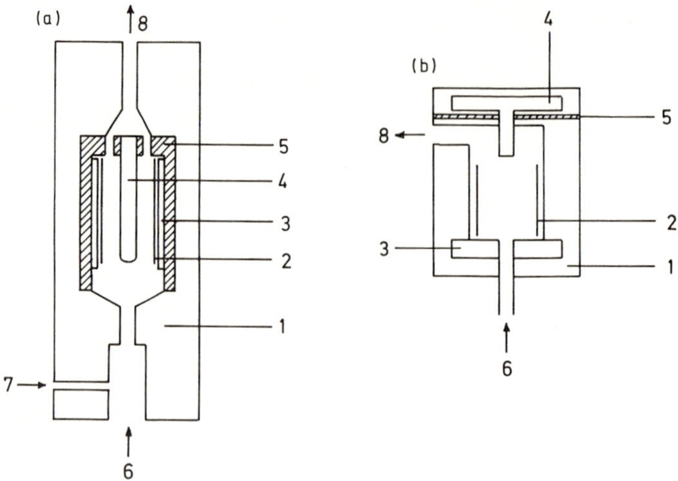

Figure 15-9. Construction principles used for electron-capture detectors. (a) Concentric form; (b) pin-cup construction. 1 = Heated detector block; 2 = radioactive foil; 3 = cathode; 4 = anode; 5 = insulation; 6 = eluent from the separation column; 7 = reagent gas; 8 = outlet.

electrode. Owing to their higher mobility only electrons and not negatively charged ions are collected. Positive ions are discharged on the cell walls. Another measuring technique varies the pulse frequency so that a constant amount of electrons is always collected (constant-current method). At low electron densities (elution of an electron-capturing compound) a higher frequency has to be applied than at high electron concentrations. The pulse frequency is therefore proportional to the amount of compound eluted. To avoid direct ionization of the compound by the β-radiation (up to 0.4 mm from the foil), detector cells of relatively large volume (>200 µL) have to be used, which makes the addition of a make-up gas necessary to reduce the residence time in the detector cell. The reagent gas usually also acts as the make-up gas [30, 31].

Practical aspects. Very low levels of electron-capturing impurities in the carrier gas (H_2O, O_2, organic compounds) can substantially reduce the amount of available thermally excited electrons in the detector. This reduces the sensitivity drastically. A very clean chromatographic separation system is therefore required.

The detector is mainly used in GC. Sometimes it is also used to detect non-polar electron-capturing compounds in LC. However, only solvents with very low electron affinities can be used. Further, the eluent has to be transferred to the vapor phase before it enters the detector.

Ar/He Ionization Detector

Working principle. The detector arrangement is very similar to that of the electron-capture detector. However, the distance between the radioactive foil and the collector anode is less than 0.5 mm. He or Ar which passes the detector is transferred to an excited metastable state by the β-radiation. Eluted compounds with an ionization potential below the excitation potential of He (20.6 eV) or Ar (11.7 eV) can be ionized, forming positive ions and electrons. When He is used an almost universal detector response is obtained. The increase in the electron concentration is determined by applying a potential of 500–1000 V/cm to the collector electrode. This GC detector is extremely sensitive but requires an extraordinarily clean separation system. Another drawback is its limited dynamic range (less than three orders of magnitude) [13].

Flame Ionization Detector

Working principle. The compounds to be detected have to be present as gases in a carrier gas containing no carbon or only oxidized carbon such as CO_2. H_2 as fuel gas is added to the eluent and the substances are burnt in a diffusion flame. Air is added to the outer core of the flame, which contains several combustion zones. The inner zone initiates thermal pyrolysis of the compounds forming fragments and $CH^\bullet$ radicals. These react with oxygen to form formaldehyde ions and free electrons in the outer reaction core, which is about 0.1 mm thick. In the surface layer of the flame CHO^+ undergoes a further reaction with water, resulting in charged H_2O clusters and CO/CO_2. Free electrons are captured by O_2 forming O_2^-. The total yield of this reaction is very low (1 ppm of all $CH^\bullet$ radicals are ionized) but very reproducible. The ions formed are collected by an electric field of about 300 V between the collector and counter electrode (eg, the flame jet) [28].

Practical aspects. All compounds which contain oxidizable carbon can be detected. A linear range over seven decades and a detection limit of a few pg/s of carbon make this mass flow-sensitive detector very suitable for GC. It has also found increased application in SFC when CO_2 is used as the mobile phase. Decompression from the supercritical state to atmospheric pressure is performed by a restriction capillary at the outlet of the separation column.

Occasionally the detector is also used in LC. However, then the eluent has to be evaporated by transfer to a heated belt or wire. Relatively nonvolatile compounds (b. p. $\approx$ 100–150 °C above the solvent boiling point) remain on the wire and are decomposed to CO_2, which is reduced to CH_4 by H_2 and a nickel catalyst. The CH_4 evolved is then detected.

Mixtures of relatively nonvolatile compounds can also be separated on silica rods. After evaporation of the solvent the separation zones are gradually heated and the released compounds are transferred to the flame by an inert gas.

Alkali Metal Salt Flame Ionization Detector

Working principle. An alkali metal salt ring (eg, RbCl) is placed above the flame of a flame ionization detector. The formation of $CH^\bullet$ radicals is then inhibited while N- and P-containing compounds can form heteroatomic radicals which react with the alkali metal atoms in the flame. The resulting increase in flame temperature leads to additional ionization of the remaining alkali metal atoms. The ions formed are discharged on a collector electrode giving a signal which is proportional to the P or N concentration in the flame. The distance between the flame and the collector is adjusted to optimize the detector response for P and/or N as well as the P/C and N/C selectivity ratio.

A modification of the alkali metal salt flame ionization detector is the thermionic detector. Instead of a flame, a low-temperature H_2 plasma is used. The alkali metal salt is embedded in a ceramic matrix which is heated to about 600–800 °C. P or N containing radicals and ions such as $^\bullet C \equiv N$ or $[O = P - O]^{-\bullet}$ are formed by surface reaction on the salt reservoir. They react further with alkali metal atoms and the resulting ions are detected [13, 32].

Practical aspects. Both detectors have found widespread application in GC for the detection of picogram amounts of N- and P-containing pesticides.

The surface reactions are inhibited for hours by halogen-containing compounds or solvents which should not be present at high concentrations in the sample.

High column bleeding or P-containing contaminants (detergents) can also poison the salt source, especially of the thermionic detector.

Photoionization Detector

Working principle. High-energy photons are formed in a low-pressure discharge lamp (10–100 Pa) and passed through an optical window into the detector flow cell. Depending on the window material and the noble gas used in the discharge lamp, photon energies between 9.2 and 11.7 eV are obtained. The eluted compound is promoted to an excited state by photon absorption and ionized. The ions formed are collected on an electrode. At 11.7 eV nearly all organic compounds except CH_4 are ionized while photons of 9.5 eV only allow aromatic, unsaturated and heteroatomic structures to be detected [33].

Practical aspects. The detector is especially suitable for portable gas chromatographs since no additional gases are needed for its operation.

A lower detection limit of a few pg/s and a linear range of six decades make the direct detection of trace levels in ambient air possible.

Detection of ppb levels of inorganic gases such as NH_3 and PH_3 is also possible.

Quenching effects by high levels of water or O_2 can lead to a certain reduction of the detector sensitivity.

Typical Data

Typical data for detectors based on ionization mechanisms are given in Table 15-7.

Table 15-7. Typical data for detectors based on ionization mechanisms.

Detection principle	Minimum detectable amount	Linear dynamic range	Selectivity characteristics
Electron-capture ionization	1 – 10 pg/s	10^4	Medium/high
Ar/He ionization	0.1 pg/s	10^3	None
Flame ionization	5 pg/s	10^6	None
Alkali metal salt FID	1 pg/s	10^4	High
Photoionization	0.1 pg/s	10^6	Low/medium

15.3.4.4   Element-Specific Detectors

Most of the element-specific detection techniques, such as atomic absorption spectrometry (AAS), inductively coupled plasma (ICP) spectrometry and microwave-induced plasma detection, can be coupled with chromatographic methods. The most critical points are the interface coupling and the removal of liquid eluents (if used). Coupling of GC and SFC with these techniques is normally relatively easy as long as the dead volume of the element-specific instrumentation does not influence the separation. Removal of the solvents used in LC is normally difficult and allows either only discontinuous operation or the analysis of compounds of lower volatility than the solvent [34-36].

15.4 References

[1] Malissa, H., *Fresenius Z. Anal. Chem.* **271** (1974) 97.

[2] Pungor, E., Kellner, R., *Anal. Chem.* **60** (1988) 623 A.

[3] Thayer, A. M., *C&EN,* Nov. 1988, p. 17.

[4] see, eg, Henschen, A., Hupe, K., Lottspeich, F., Voelter, W., *High Performance Liquid Chromatography in Biochemistry;* Weinheim: VCH-Verlagsgesellschaft, 1985; Ishii, D., *Introduction to Micro-Scale High-Performance Liquid Chromatography;* Weinheim: VCH-Verlagsgesellschaft, 1988; Sherma, J., Fried, B., *Handbook of Thin-Layer Chromatography;* New York: Marcel Dekker, 1991.

[5] see, eg, Dickworth, H., Barber, R. C., Venkatasubramanian, V. S., *Mass Spectroscopy* (2nd ed.); Cambridge: Cambridge University Press, 1986; Busch, K. L., Glish, G. L., McLuckey, S. A., *Mass Spectrometry/Mass Spectrometry; Techniques and Application of Tandem Mass Spectrometry;* Weinheim: VCH-Verlagsgesellschaft, 1989.

[6] see, eg, Willard, H. H., Merrit, L. L., Dean, J. A., Settle, F. A., *Instrumental Methods of Analysis* (6th ed.); Belmont: Wadsworth Publishing, 1981; Ernst, R. R., Bodenhausen G., Wokaun, A., *Principles of Nuclear Magnetic Resonances in One and Two Dimensions;* Oxford: Clarendon Press, 1987.

[7] Griffiths, P. R., de Haseth, J. A., *Fourier Transform Infrared Spectrometry;* New York: Wiley, 1986.

[8] Lee, M. L., Yang, F., Bartle, K. D., *Open Tubular Column Gas Chromatography;* New York: Wiley, 1984.

[9] Engelhardt, H., *High Performance Liquid Chromatography:* Berlin: Springer, 1979.

[10] Kaiser, R. E., *Planar Chromatography;* Heidelberg: Hüthig, 1986.

[11] Grob, R. L., *Modern Practice of Gas Chromatography;* New York: Wiley, 1977.

[12] Dressler, M., *Selective Gas Chromatographic Detectors;* Amsterdam: Elsevier, 1986.

[13] Oehme, M., *Gas-chromatographische Detektoren;* Heidelberg: Hüthig, 1982.

[14] Scott, R. P. W., *Liquid Chromatographic Detectors;* Amsterdam: Elsevier, 1986.

[15] Paris, N. A., *Instrumental Liquid Chromatography;* Amsterdam: Elsevier, 1983.

[16] Zlatkis, A., Kaiser, R. E., *High Performance Thin-Layer Chromatography;* Amsterdam: Elsevier, 1977.

[17] Geiss, F., *Die Parameter der Dünnschicht-Chromatographie;* Wiesbaden: Vieweg, 1972.

[18] Gjerde, D. T., Fritz, J. S., *Ion Chromatography;* Heidelberg: Hüthig, 1987.

[19] Weiss, J., *Fresenius, Z. Anal. Chem.* **327** (1987) 451-455.

[20] Schoenmaker, P. J., Verhoeven, F. C. C. J. G., *Trends Anal. Chem.* **6** (1987) 10-17.

[21] White, C. M., Houck, R. K., *J. High Resolut. Chromatogr. Chromatogr. Commun.* **9** (1986) 4-17.

[22] Ševčík, J., Lips, J. E., *Chromatographia* **12** (1979) 693-703.

[23] Brinkman, U. A. Th., *Chromatographia* **24** (1987) 190-200.

[24] Weiss, J., *Handbook of Ion Chromatography;* Sunnyvale: Dionex, 1986.

[25] Poppe, H., in: *Journal of Chromatography Library, Vol. 13, Instrumentation for High Performance Liquid Chromatography,* Huber, J. F. K. (ed.); Amsterdam: Elsevier, 1978, pp. 113-129.

[26] Farwell, S. O., Gage, D. R., Kagel, R. A., *J. Chromatogr. Sci.* **19** (1981) 358-372.

[27] Hausdorff, H. H., *J. Chromatogr.* **134** (1977) 131-146.

[28] Ševčík, J., *Journal of Chromatography Library, Vol. 4, Detectors in Gas Chromatography,* Amsterdam: Elsevier, 1976, p. 36.

[29] Nerheim, A. G., *J. Chromatogr. Sci.* **15** (1977) 465–468.
[30] Poole, C. F., *J. High Resolut. Chromatogr. Chromatogr. Commun.* **5** (1982) 454–471.
[31] Brinkman, U. A. Th., *LC-GC* **5** (1988) 476–484.
[32] Hall, R. C., *CRC Crit. Rev. Anal. Chem.* **8** (1978) 323–380.
[33] Driscoll, J. N., *J. Chromatogr. Sci.* **23** (1985) 488–492.
[34] Jiang, S., De Jonghe, W., Adams, F., *Anal. Chim. Acta* **136** (1982) 183–190.
[35] Chong, N. S., Houk, R. S., *Appl. Spectrosc.* **41** (1987) 66–74.
[36] Estes, S. A., Uden, P. C., Barnes, R. M., *Anal. Chem.* **53** (1981) 1829–1831.

16 Calibration of Gas Sensors

KLAUS KALTENMAIER, Dr.-Ing. Häfele Umweltverfahrenstechnik Karlsruhe, FRG

Contents

16.1   Introduction

Gas sensor systems hardly differ from other kinds of analyzers. However, most users often pay far too little attention to the calibration procedure. If the analyzer is badly adjusted, or too infrequently, the measurement accuracy will be limited. Proper calibration often suffices to provide essential improvements in the performance of the measuring system setup.

16.2   Routine Calibration

On routine calibration, at first zero gas is released and the output signal is set at zero [1, 2]. Calibration gas is then admitted. Next, the analyzer is adjusted to the concentration value of the calibration gas (compare the squares in Figure 16-1). The origin of the signal/concentration diagram is found with zero gas. Ideally, zero gas should not contain the component to be measured, but it should be identical in all other respects. This is a requirement that, for the most part, cannot be fulfilled.

In practice this demand cannot even be partially fulfilled because, in actual analytical tasks, various components usually vary simultaneously. In the case of less significant variations in concentration, it may be of help to use a zero gas that contains the components supplemented with the mean value of the concentration.

Most measurement tasks concern the analysis of contaminants in the air. The cross-sensitivity towards oxygen is minimal. Cross-sensitivities towards other substances in the air can also be disregarded because the concentrations of these substances, except water vapor, are insignificant. For that reason pure nitrogen or pure inert gases are commonly used as zero gas,

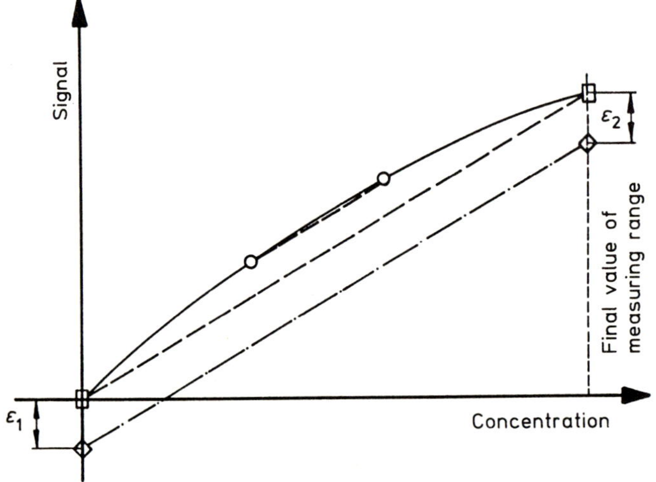

Figure 16-1. Measurement signal/concentration course on routine calibration.

even though this is only justifiable when cross sensitivities can be disregarded and nitrogen behaves like the zero gas in the method of analysis used. In this case, unlike the presentation in Figure 16-1, the lower left rhombus and the lower left square become eqivalent because the accompanying components exert no measuring effect.

Following the adjustment of the zero point, a calibration gas, consisting of the zero gas and the measuring components, will be admitted. The calibration gas should conform to the gas to be analyzed, and the concentration should amount to between 70 and 100% of the upper limit of the measuring range. Thereafter, the measuring signal can be adjusted according to the concentration value of the calibration gas employed. The measuring components are often diluted in nitrogen and not in simulated analysis gas. This is justified when ε_1 in Figure 16-1 is sufficiently small or when this numerical value is determined separately and will be corrected by calculation.

At the time of the latter measurement it will be imputed that the measurement signal may be interpolated between zero and the calibration concentration. As long as the physical bearing between measurable quantity and concentration is not linear (in the case of electrochemical cells, for example, the measuring signal is proportional to the logarithm of the concentration), the course of the signal will already be linearized by the electronics of the measuring instrument.

16.3　Calibration Elements

Although some devices need not be recalibrated unless the sensor is in continual operation, calibration normally needs to be repeated. Information concerning the proper intervals can be found in the handbooks of each device. The intervals are usually governed by the operating conditions.

Looking closer at the task of calibration, several sub-tasks become obvious:

- choosing the appropriate gases (calibration gas, as well as zero gas),
- providing or generating the desired gas mixtures,
- exposing the sensor to the calibration gas,
- receiving the sensory signal, or signals, in the case of sensor arrays,
- adjusting the value (e.g. correcting with a trimming potentiometer), or
- interpreting the calibration results (eg, determining characteristic curve parameters or establishing calibration tables or cross-sensitivity matrices).

Routine calibration is confined to two concentration points at each measuring range limit. Calibration gases with concentrations that include the expected values can also be employed. This can increase the precision in those cases where devices do not exhibit good linearity (compare the spheres in Figure 16-1).

Measuring systems such as sensor arrays of nonselective sensors or sensors with nonlinear cross-sensitivities require different feedback calibration strategies and the choice of different calibration gases. This need not be discussed here, because detailed calibration instructions come with the measuring instrument.

During the exposure to calibration gases, care must be taken that identical conditions are present during calibration and during analysis. Some analyzers are sensitive to changes in pressure and temperature at the measuring point (remember that in many measuring systems the density or partial pressure rather than the concentration, represent the measurable variable).

During calibration, it is necessary to wait for the appearance of stable measuring signals. Changes in the concentration (for example, a change from zero gas to calibration gas) lead to delayed changes in the measuring signal. The test gas first must purge the tubes, including the entire clearance volume, before it reaches the sensor, which in turn responds with delay to changes in concentration. In trace concentrations the decisive delay is due to adsorption. Many species are adsorbed on solid surfaces. The component does not reach the sensor until all surfaces arranged on the entry side are coated to saturation. The adsorption of trace concentrations ($<$ 1 ppm) of easily adsorbable components, such as ammonia, may lead to delays of as long as several hours.

Many measuring instruments operate with suction pumps; in the calibration procedure, however, the calibration gas from a cylinder streams through the instrument at an increased flow rate. Should the conduits at the unit outlet be somewhat too small in dimension, the elevated flow rate will lead to a considerable pressure burden on the sensor.

In the case of noncomplex systems, points 4 and 5 only call for the reading of measurements and adjustment of the trimming potentiometer. However, the workload may increase considerably. The appropriate procedure requires detailed knowledge of the specific system's sensors and/or sensor arrays, as well as cross-sensitivities. This is especially true in the case of the above-mentioned sensor arrays with nonselective sensors. Appropriate software tools are available for the processing of the calibration results.

16.4   Supply of Synthetic Gas Mixtures in Gas Cylinders

The method of supplying the gas mixture, a vital point in calibration, is usually left entirely up to the user — who finds little or no support in the instruction manuals. That is why the following will deal mainly with these problems.

Which gas mixtures (and how many different ones) are necessary is the determining factor when deciding how the gas can be supplied most economically. If, for example, only a small number of different gases is required (as is the case for the calibration of one device), premixed gases with defined concentrations are most likely the ideal solution. They can be bought as test gases from most gas suppliers.

As soon as a larger number of different gas mixtures is necessary (either because characteristic curves of analyzers are to be taken at several concentration points, or because a wide variety of analyzers with different gas compositions in differing concentration ranges are to be calibrated), a large number of gas mixture cylinders must be kept on stock. As gas mixtures are expensive, and a large number of gas cylinders must be stored and handled, it is worthwhile to use multi-functional calibration gas generators.

Another problem, depending on the type of gas, is that the concentration inside the cylinder only remains stable for a few months. As this concerns the adsorption phenomena on and in the container material, the fewer molecules there are in the cylinder, that is, the lower the concentration, the more problematic this effect will become. For reasons of precision, new cylinders must therefore be ordered, in spite of the fact that only an insignificant portion of the gas was actually used.

16.5 Calibration Gas Generators for Permanent-Gas Mixtures

From pure gases, dynamic calibration gas generators can produce a defined gas mixture with a defined flow rate (see Figure 16-2). With the exception of those calibration gas generators that operate on the basis of permeable membranes, pure gases or binary gas mixtures are dosed and mixed in the desired proportion. The number of required gas cylinders is reduced to the number of required components [3].

All of the calibration gas generators for laboratory use work dynamically, ie, according to the dosing principle. The desired gases are dosed with definite flow rates and then mixed inside the apparatus. With the technical equipment available in laboratories, standard volumetric flow rates ($\dot{V}_s$) or molar flow rates (F) can be proportioned much more precisely than concentrations or absolute quantities (gas suppliers, however, often manufacture gas mixtures by a weighing-in process). Calibration gas generators are offered on the basis of three different methods of gas dosage (to be described in detail later).

Characteristic of these devices is that the total gas flow is defined according to the mixing task; only in special cases can this be adapted to the needs of the individual measuring instrument. If a self-priming analyzing device with an integrated test gas pump is to be calibrated, the flow rate at the calibration gas generator is set slightly higher than the amount, necessary for the analyzing instrument. The test gas pump withdraws the needed gas flow and the surplus escapes.

The mole fraction of a component i, y_i, that arises at the outlet of the test gas generator, can be found with the following equation:

$$y_i = F_i/F_{tot} = \dot{V}_{s,i}/\dot{V}_{s,tot} \tag{16-1}$$

In the example displayed in Figure 16-2 the total flow rate yields: $\dot{V}_{s,tot} = (\dot{V}_{s,1} + \dot{V}_{s,2})$, or $F_{tot} = (F_1 + F_2)$. The molar flow rate F and the standard volumetric flow rate $\dot{V}_s$ are equivalent quantities that can be converted by using the standard molar volume V_0 as conversion factor.

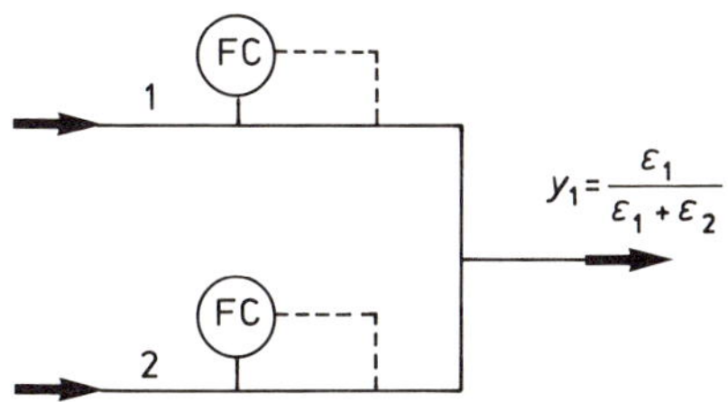

Figure 16-2.
Principles of dynamic (dosing) test gas generators.

$$F = \dot{V}_s/V_0 = \dot{V}_s/(R \cdot T_0/P_0) = \dot{V}_s/(22.4 \text{ L/mol}) \tag{16-2}$$

with R = 8.314 J/(K $\cdot$ mol), T_0 = 273.15 K and P_0 = 101 300 Pa.

The smallest possible standard volumetric flow rate is in the range of 0.1 to 1 standard cubic centimeter per minute (SCCM). In the following example, a total standard volumetric flow rate (that in close approximation is equivalent to the carrier gas standard volumetric flow rate $\dot{V}_{s,\text{carrier}}$) of about 1000 SCCM results in a minimal possible concentration of 10^{-4} or a dilution ratio of 1 : 10 000.

$$y_{i,\min} \equiv \dot{V}_{s,i,\min}/\dot{V}_{s,\text{tot}} = 0.1 \text{ SCCM}/1000 \text{ SCCM} = 1 : 10\,000 \tag{16-3}$$

The dilution ratio can be raised if the flow rate of the carrier gas is raised or an already pre-rarefied gas is used on the inlet side.

$$y_{i,\text{out}} = y_{i,\text{in}} \cdot \dot{V}_{s,\text{pr,in}}/\dot{V}_{s,\text{tot,out}} \tag{16-4}$$

with pr standing for i-containing pre-rarefied gas.

Calibration gas generators based on the permeation principle work somewhat differently. The components to be added are supplied in permeable plastic tubes as 2-phase mixtures (gaseous/liquid). The flow rate of the components to be added is determined by their diffusion through the membrane.

16.6   Calibration Gas Generators with Volumetric Techniques

The most commonly used method to proportion gas mixtures (in the percent range) works with proportioning pumps that convey flow rates sufficiently precisely [4]. In the case of trace concentrations, microvolumes are filled with gas, which is then liberated to the carrier gas at set intervals [5,6]. In both methods the pulsating technique is characteristic. Because volumetric flow rates (volumetric under operating conditions) are dosed, the molar flow rate (F in mol/min) depends on the ambient pressure. The resultig errors are insignificant because the pumps are operated in such a way that all of the proportioned gases are exposed to identical pressure, so that only the total flow rate but not the concentration is influenced.

Beats in the dosed flow rates lead to fluctuations in the concentration of the calibration gas. Principally, one cannot assume that an analyzer will show the same results in the case of oscillating concentrations (calibration gas) and in the case of stable concentrations (test gas), even if the mean time value of the concentration is the same in both cases.

The concentration pulses can be smoothened out in remixing devices. To do this, a vessel with a residence characteristic of a perfectly mixed flow vessel and a medium residence time of five to ten times the dosing interval is necessary. In the case of a total standard volumetric flow rate of 2 SLM and a pulsing interval of one minute, one needs a device that can hold approximately 10–20 L, which conforms to a delay in the system response of 5–10 min.

In dosing pumps, the gas is in contact with the pump oil; when conveying hydrocarbons, for example, these may interact with the parafinic oil and errors will result.

The volumetric gas mixing method is the only technique that offers the advantage that the concentration, or the mole fraction, remains largely unaltered when the gas changes.

Generating calibration gas mixtures with a pump is inexpensive and a universally employable solution in a situation that demands only a small number of components, but it can become quite costly if there is an increase in the number of components to be proportioned simultaneously. Pumps allow dilutions of up to a factor of 100; this limit is a result of the minimal conveying flow rate of the dosing pumps, approximately 10 cubic centimeters per minute (CCM). With other volumetric techniques, with a dosing volume of approximately 40 μL, significantly lower concentrations can be achieved.

16.7 Calibration Gas Generators with Permeation Techniques

Some calibration gas generators dose the measuring components by permeation [7, 8]. At the permeation sources the component to be proportioned is supplied in the form of a two-phase mixture (liquid/gaseous) in a small tube, and therefore these apparatuses need only one gas cylinder for the carrier gas.

Permeation sources consist of tubes made of a polymer such as polytetrafluorethylene (PTFE). This tube is hermetically sealed at both ends and is filled with the component to be proportioned. In other designs the tube is gas-tight and the stopper is permeable.

Gas molecules diffuse through a membrane when the partial pressure gradient acts as driving thermodynamic force. The flow rate is determined by the diffusion coefficient D_{membrane} and the concentration gradient in the membrane $\Delta C_i/l$:

$$F_{\text{diff}} \sim D_{\text{membrane}} \cdot \Delta C_i/l \tag{16-5}$$

The diffusion coefficient D_{membrane} depends on the gas and the type of membrane polymer. The molar weight M_i enters into the equation as a substance-specific value. The interaction between the gas and the solids influences the diffusion coefficient essentially and must be defined experimentally.

The concentration difference $\Delta C_i = \Delta P_i/RT$ results from the pressure difference $\Delta P = P_{\text{inside}} - P_{\text{outside}}$. This is defined by the internal pressure, which is equivalent to the vapor pressure at the present temperature.

A two-phase mixture must be present inside the tube. If the tubes are merely filled with gas, the internal pressure will drop along with the loss of gas, resulting in a decrease in the diffusion rate with increased use.

The permeation rate increases with increasing temperature in the permeation chamber, making it possible to attain a higher level of concentration. The temperature dependency of the diffusion rate, F_{diff}, is given by the temperature function of the vapor pressure inside, $P_V \sim \exp\{-\Delta H_V/RT\}$, which depends on the heat of vaporization (ΔH_V). The T-dependency of the diffusion coefficient can be described with an analog equation, $D_{\text{membrane}} \sim \exp\{-E_d/RT\}$.

For this reason reproducible permeation rates require the chamber with the permeation tube, as well as the entering gas, to be perfectly temperature-controlled.

The sources of permeation require a long feeding phase before the permeation rate reaches a steady state. With permeation techniques, exact gas mixtures can be generated if the entire system, depending on the application, has operated under stationary conditions for several hours beforehand, and the permeating quantity is redetermined by measuring the loss of weight. The precision of the concentration will then only be determined by the error in the dosage of the carrier gas. If the reweighing of the permeation sources is omitted, one must rely on the certified permeation rates, which are burdened with an error rate of approximately 5%.

Of all of the proportioning methods described here, the lowest dosage rates can be achieved with the permeation technique. In many permeation elements, the rates are in the range of less than a few 10^{-3} SCCM. Because permeation sources can only contain small quantities of the component to be added, such a calibration gas generator is especially suited for the preparation of gas mixtures in which the constituents are in the range of ppm, ppb or sub-ppb.

A disadvantage is that a temperature-controlled permeation chamber must be installed for each component if the concentrations of the individual components are to be varied independently.

Calibration gas generators on the basis of permeation (Figure 16-3) are often used for calibration of mobile immission analyzer units, when mobility is desired and precision is less important.

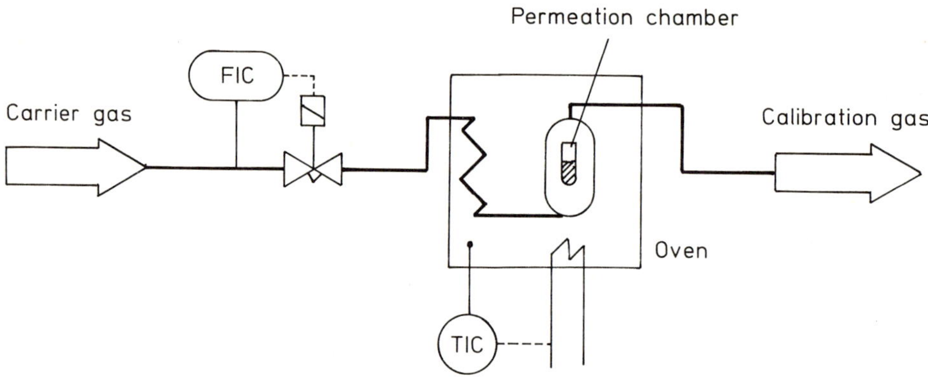

Figure 16-3. Calibration gas generator according to the permeation principle.

16.8   Calibration Gas Generators with Super-Critical Nozzles

One of the two other remaining alternatives is a device that attains the defined flow rate with the help of super-critical nozzles [9, 10, 11, 12]. The flow rate through a nozzle, F_{orifice}, can be calculated according to the following equation:

$$F_{\text{orifice}} = \alpha \cdot A \cdot P_{\text{in}} \sqrt{\frac{1}{M_i \cdot R \cdot T_{\text{in}}} \cdot \frac{2K}{K-1} \left[\left(\frac{P_{\text{in}}}{P_{\text{out}}}\right)^{2/K} - \left(\frac{P_{\text{in}}}{P_{\text{out}}}\right)^{(K+1)/K} \right]} \quad (16\text{-}6)$$

where A is the crossing section area of the orifice, α is the contraction factor, and P_{in}, P_{out} are the pressures in front and behind the orifice. The specific heat ratio $K = C_{\text{P}}/C_{\text{V}}$ and the molar weight M_i enter into the equation as substance-specific variables.

If the pressure difference at a nozzle is increased (in an imaginary experiment this will take place by decreasing the outlet pressure), the flow rate would also increase. If the pressure ratio, defined by the inlet and the outlet pressure, increases to a level above the critical pressure ratio of $P_{\text{in}}/P_{\text{out}} = [2/(K-1)]^{K/(K-1)}$, the flow rate will not increase further. At the orifice the gas reaches the speed of sound and cannot be accelerated further by lowering the outlet pressure. The volumetric flow rate is no longer a function of the back pressure. The equation for the molar flow rate still contains the inlet pressure because the density of the gas at the orifice enters into the calculation for the molar flow rate.

$$F_{\text{orifice}} = \alpha \cdot A \cdot P_{\text{in}} \cdot \sqrt{\frac{1}{M_i \cdot R \cdot T_{\text{in}}} \cdot \left[2K \cdot \left(\frac{1}{K+1}\right)^{(K+1)/(K-1)} \right]} \qquad (16\text{-}7)$$

The super-critical nozzle has an advantage over the subcritical nozzle in that only the inlet pressure P_{in} but not the back pressure, depending on the operational conditions, enters into the flow rate as a variable, thereby making further regulation of the back pressure unnecessary. The reproducibility of the proportioned flow rate is mainly defined by the pressure in front of the orifice. For this regulation extremely good long-term stability is required.

A disadvantage of the nozzle method is that the nozzles are not variable in their cross-sectional area, that is, the concentrations and total flow rate are fixed and cannot be varied continuously. The built-in nozzles are chosen and fixed at the time of conception of the system. To enable the flow rate to vary, precise, reproducibly steerable pressure regulators for the inlet pressure would be necessary, which is not economically feasible. Orifices can be manufactured with various cross-sectional areas, but there is a technical lower limit, so that the minimum standard volumetric flow rate achievable lies at approximately 1 SCCM.

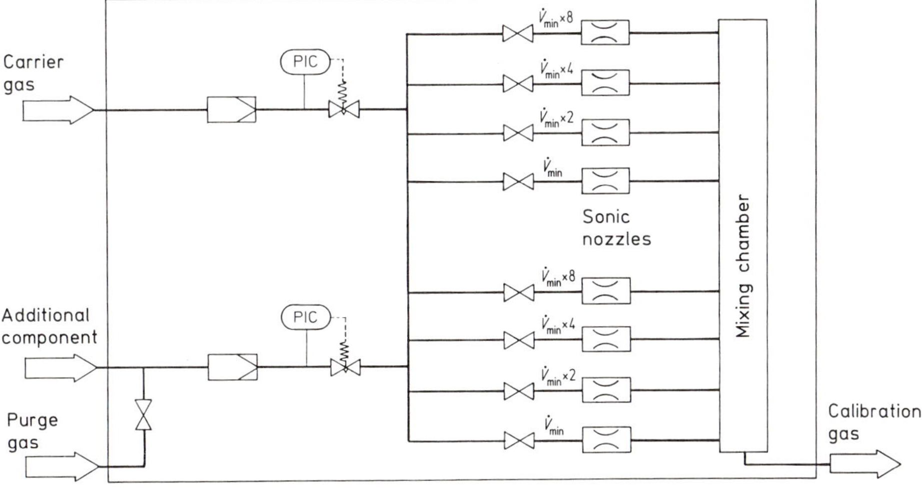

Figure 16-4. Gas diluter with super-critical nozzles.

Various devices became known under the name of gas diluter in which gas, containing the calibration component, and the carrier gas flow through a square bottleneck under super-critical conditions (Figure 16-4). By turning a knob, the cross-sectional area of one orifice can be enlarged to the same degree that the second one is made smaller via a mechanical device, enabling a variation in the proportion. Using a second knob, both orifices are simultaneously altered at the same ratio, thereby varying the total flow rate. These systems with variable orifices work at the significantly reduced rate of precision of about 2% absolute [13].

16.9   Calibration Gas Generators with Mass Flow Controllers (MFC)

In the case of the latter alternative, the set volumetric flow rate is attained via Mass Flow Controllers (MFC). Using this method, complete chains of analyzers can be automatically calibrated with ease. The MFCs receive a signal between 0 and 5 V and transform this into the corresponding flow rate. A sensor then measures the flow rate and a controller aligns the actual value with the set point via a metering valve, thereby controlling the dosing process.

Mass Flow Controllers divide the gas flow. A small part of the flow, usually a maximum of 5 SCCM, flows past the actual sensor element; the rest flows through the bypass (Figure 16-5). The MFC is adjusted to the desired flow rate with a ratio equal to $\dot{V}_{s,bypass}/\dot{V}_{s,sensor}$. This ratio stays constant for each device. The sensor flow $\dot{V}_{s,sensor}$ can be measured with a precision of 1% and a reproducibility of 0.2% referring to the maximum dosable flow rate with this MFC. Depending on the required precision, this results in a usable dynamic range of 10–100%, referring to the maximal sensor flow rate.

In the smallest flow range of the MFCs, the gas flows completely through the sensor element, which correlates to a usable dosage flow rate of 0.1–5 SCCM. Some manufacturers offer devices with which this range is decreased by a factor of 5.

The sensory element operates according to the principle of a flow calorimeter and consists of a stainless steel tube, onto which a heating source with a defined heat production is attached [14]. Precision resistance thermometers are located both upstream and downstream. If

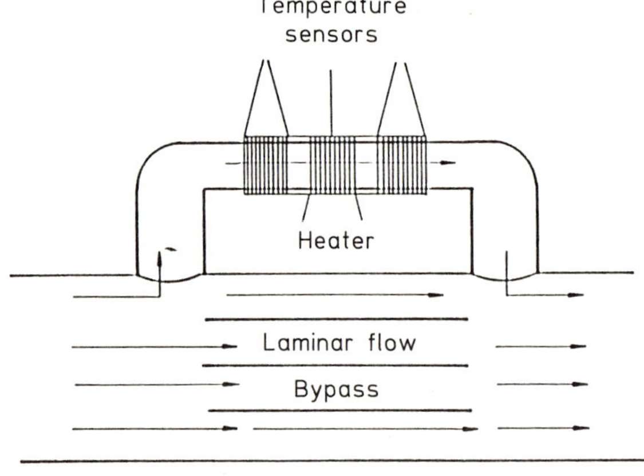

Figure 16-5.
Operating principle of the MFC sensor.

there is no gas flow, both thermometers will show the same temperature. As the gas begins to flow, a temperature difference becomes obvious: the thermometer located downstream shows a higher temperature. Without heat losses, the thermal power Q that is generated by the electric heating process, and therefore identical with the electric power $Q_{electric}$, must be the same as the quantity of heat absorbed by the gas stream per time, Q. The following simple equation results:

$$Q_{electric} = Q = F \cdot C_P \cdot \Delta T \tag{16-8}$$

Resolving after F, the molar flow rate F results directly from the temperature difference ΔT. The substance-specific heat capacity C_P also enters into the calculation of molar flow F .

The temperature and pressure dependency of the flow sensor is determined by the functional dependency of the heat capacity $C_P = f(T, P)$. The influence of changes in pressure is insignificant. Temperature changes between 18 and 40 °C can account for an error of up to 2.5 %.

If gas mixtures with concentrations varied over several orders of magnitude are needed, dosing the gas must be carried out via several dosing tracks with different flow rates. As is the case with many other measuring and control instruments, a percent rate of error, ε, is specified when using the MFC technique. As the value refers not to the actual but to the maximum flow rate of the MFC ($\dot{V}_{s,MFC-max}$), this is not a relative error but an absolute flow error ($\Delta \dot{V}_{s,MFC} = \varepsilon_{a,MFC} \cdot \dot{V}_{s,MFC-max}$) for the specified device. The relative error, ε_r, referring to the actual flow rate $\dot{V}_{s,act}$ ($\varepsilon_r = \Delta \dot{V}_{s,MFC} / \dot{V}_{s,act}$) is often more important to the user. This error, ε_r, is at its lowest value at the maximum flow rate and increases with sinking gas flow until the actual flow rate $\dot{V}_{s,act}$ is smaller than the error in the flow rate $\Delta \dot{V}_{s,MFC}$.

Usually the user can accept only a maximum relative error $\varepsilon_{r,max}$. Thus the minimum volumetric flow rate for this MFC in this application $\dot{V}_{s,MFC-min}$ can be calculated as follows:

$$\dot{V}_{s,MFC-min} = \Delta \dot{V}_{s,MFC} / \varepsilon_{r,max} \quad \text{that is} \quad \dot{V}_{s,MFC-min} = \dot{V}_{s,MFC-max} \cdot (\varepsilon_{a,MFC} / \varepsilon_{r,max}) \tag{16-9}$$

The dynamic dilution range therefore results from the following equation:
$$D_{MFC} = \dot{V}_{s,MFC-max} / \dot{V}_{s,MFC-min}.$$
If the user requires higher dilution dynamics, several dosing devices with different maximum flow rates ($\dot{V}_{s,MFC-max}$) can be arranged in parallel. With a falling gas flow rate, an MFC is used until the same flow rate can be achieved by the MFC with the next smaller flow range, $\dot{V}_{s,MFC-max}$. The same goal can be achieved if several prediluted test gases at different concentrations are provided.

16.10 Calibration Gases with Vapors

Other strategies must be implemented if the proportioning component is in a liquid state at room temperature. If very low concentrations are desired, then test gases can also be used. However, the concentration will be limited because the partial pressure of this component must lie beneath the saturated vapor pressure in the gas cylinder. Using water as an example, this means:

The maximum possible concentration of water vapor in the test gas cylinder is approximately 30 vol.-ppm. If the partial pressure in the cylinder, $P_i = y_i \cdot P_{tot}$, is above the saturated vapor pressure at the existing temperature, $P_V = f(T)$, water condenses and collects at the bottom of the cylinder while the gas becomes dehumidified. Thus the concentration of water vapor in the vapor phase inside the cylinder sinks. The above-mentioned concentration maximum can be calculated, at the lowest expected ambient temperature of $0\,°C$, from a saturated vapor pressure of 6 mbar and a total pressure of 200 bar inside the gas cylinder. The concentration increases in conjunction with sinking pressure in the cylinder and rising ambient temperature.

If higher concentrations are required in order to simulate the water vapor concentration of a hot flue gas (the dew point for water vapor can be well above the ambient temperature), other methods must be used. Using premixed gases with liquid components is difficult and expensive at high concentration levels.

Of all the proportioning principles already known from the production of calibration gas from permanent gas, only the permeation method can be used because the diffusion process is not bound to the gaseous state and even for the dosage of permanent gases the permeation tubes contain two-phase mixtures. This method, as well as the mixed gas cylinders, are limited to the trace domain because the tubes only contain insignificant amounts.

However, if highly flexible calibration gas mixtures consisting of liquid components with short response times need to be generated, vaporizers or saturators are suitable solutions.

16.11 Calibration Gas Generators According to the Saturation Technique

A gas stream is injected at the bottom of a vessel filled with liquid and bubbles through the liquid phase [15, 16]. At the same time, the liquid evaporates into the gas bubbles until the exact equilibrium saturated vapor pressure, P_V, (with sufficient duration of contact time) is reached.

$$P_V \sim \exp\{- \Delta H_V / RT\} \tag{16-10}$$

For a single compound this vapor pressure is only a function of the temperature. For a large number of substances, this correlation can be obtained from technical or scientific handbooks as approximation functions (eg, Antoine equation) or in the form of pairs of variates.

The mole fraction at the outlet of the saturator y'_V results from the vapor pressure P_V and the total pressure P_{tot} according to the following equation:

$$y'_V = P_V / P_{tot} \tag{16-11}$$

As shown in Equation (16-11) the mole fraction of the evaporated component at the outlet of the saturator does not depend on the flow rate of the gas, $y'_v \neq f(F_{in})$. If a specific concentration of one component is required, carrier gas dosing is not necessary, as long as the gas flux is maintained.

The flow rate of the proportioning component at the outlet of the saturator $F_{V,out}$ is calculated using the flow rate of the carrier gas component ($F_{carrier,in}$), and the mole fraction of the proportioning component y'_v:

$$F_{V,out} = F_{carrier,in} \cdot y'_v / (1 - y'_v) \tag{16-12}$$

and for the total flow rate $F_{tot,out}$:

$$F_{tot,out} = F_{carrier,in} / (1 - y'_v) \tag{16-13}$$

As soon as the gas leaves the temperature-controlled room, the pipes are heated to temperatures above those of the recondenser in order to prevent later condensation.

The achievable concentration range at the outlet of the saturators is limited by two values. The closer the temperature comes to the boiling point, the more small temperature deviations result in big errors in the vapor pressure. The practical upper limit of the application lies at concentrations of 50%.

On the other hand, partial pressures that are lower than the vapor pressure at the melting point cannot be attained. If the recondenser temperatures lie below -10 or $-30\,°C$, more expensive cryostats become necessary at falling temperatures.

At the melting point, the vapor pressures of most substances are in the range of several mbar, in favorable cases even below that. Assuming a hypothetical vapor pressure of 1 mbar and a total pressure of 1 bar, the lowest mole fraction possible (1000 ppm) is the result. In the case of substances such as benzene, for example, with a vapor pressure of 40 mbar at the melting point, the lowest achievable mole fraction is much higher.

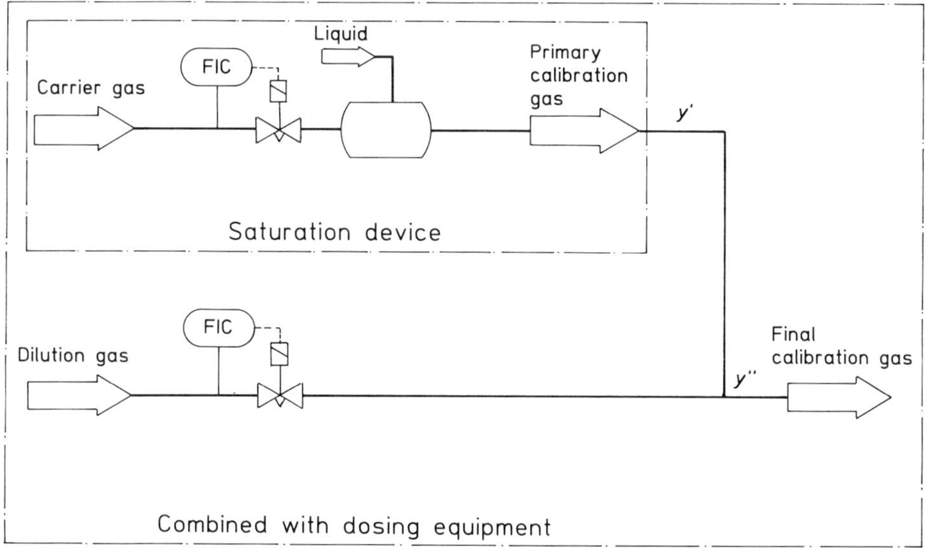

Figure 16-6. Calibration gas generator with saturator.

Lower concentrations can be reached if the saturation principle is combined with the dosing technique. In a double-track technique a small carrier gas flux (eg, 1 SCCM) is fed through the saturator and is subsequently mixed with a larger diluting flux, for example 1 SLM, (see Figure 16-6). In this way mole fractions in concentrations between 1 and 1000 ppm can be attained. After this secondary dilution in this type of system, the mole fraction of the proportioning component y''_V is:

$$y''_V = [y'_V/(1 - y'_V)]/(1 + F_{dil.,in}/F_{carrier,in}) \tag{16-14}$$

For substances with high vapor pressure at the melting point, solid matter saturators may be implemented. In order to achieve this, solid matter is brought into a temperature-controlled room as particles and gas is passed through, whereby the sublimation pressure appears in the gas phase. To reach equilibrium requires no further effort.

Practical handling of the system becomes all the more difficult the further the melting point of the substance is below room temperature. First the substance must be cooled to solidification. Subsequently the solid material must be reduced to smaller particles and filled into the temperature-controlled solid-material saturator. During this procedure the temperature of the particles should not rise above the melting point; otherwise the particles will cluster to form a compact material.

A gas stream can only be enriched with one component per saturator. Should two liquids be filled into a saturator, the ratio between both components will be different in the liquid phase than in the gas phase (except for the azeotrope point). While the saturator is in operation, the liquid phase becomes exhausted with the more volatile component and the concentration ratio shifts in time. The use of more saturators, one after another, also brings no advantages because the liquid of the second saturator could wash out the substance that was added by the first saturator. Therefore saturators have to be arranged in parrallel, and each must be equipped with separate gas-dosage tracks.

In a saturator, gas should principally be combined only with those liquids in which this gas is hardly soluble (e.g., not propane with hexane/heptane, not carbon dioxide with water). In the case of soluted gas this may lead to a drop in vapor pressure (Raoul's law) in relation to the tabeled values for the pure liquid.

As the feed gas is absorbed until saturation in the liquid, this additionally leads to a delay in the system. If the inlet gas itself is a gas mixture containing absorbable compounds in trace concentrations, it will take a long time and a great amount of gas until this compound is again found in the outlet gas.

In order to alter the proportioning components, the saturator must be emptied, cleaned, and refilled. Should rapid alterations in concentration be required, parallel proportioning tracks must be provided in order to be able to switch quickly enough from one component to another.

A system in which more than one component can be proportioned uses a steady state flowthrough saturator, conceived by Weitkamp and Dauns [17,18]. With this apparatus the liquid in the saturator is continuously renewed. Merely 1–10% of the added liquid evaporates into the gas phase, the rest flows out unused. Even if the liquid is depleted at the more volatile component, the errors will not be too significant because liquid is continually added at the desired concentration.

The disadvantage of this type of saturation construction is that a large amount of liquid leaves the saturator unused. This liquid can be analyzed in a laboratory, reset to the initial concentration, and then reused as saturator-inserted material. If such a laboratory is not available the liquid must be disposed of. Such steady-state flowthrough saturators can only be implemented if vapor pressure curves are available for multicomponent mixtures, either as a table or function, or if these values can be measured by the user.

Correct operation of the saturator cannot be guaranteed if the residence time of the gas bubbles in the liquid is not long enough to reach the saturation equilibrium within the limit of error. If not released, aerosol droplets make their way into the gas phase and later evaporate. This leads to a concentration increase.

Saturators consisting of an evaporator and a recondenser have been used successfully for such applications. In the evaporator the gas bubbles through the liquid and is loaded with the compound above the desired concentration value. Therefore the temperature in the evaporator is kept at a value of $5-10\,°C$ above the saturation temperature, corresponding to the desired concentration. Maintaining this temperature at approximately this level will suffice.

Subsequently, the gas flows through a reflux cooler located on top, which is temperature-controlled to the correct saturation temperature. The excess liquid condenses and flows back into the evaporator below. The heat exchange area is sufficient to cool the gas down to the desired temperature and to remove the condensation heat. Aerosols at the outlet of the recondenser need not be feared because the cooling process takes place relatively slowly, that is, at low temperature gradients. This results in large droplets or condensation on the wall, which can easily be separated.

As Equation (16-11) shows, errors in the total pressure inside the saturator cause substantial errors in the concentration. Because the vapor pressure itself shows only little dependency on the total pressure, this additional error can be neglected.

The installations using the calibration gas define the total pressure (P_{tot}) inside the saturator via its experimental setup. If the gas can freely be released into the atmosphere, the system pressure is the air pressure. Assuming atmospheric air pressure fluctuations of ± 15 mbar, this leads to an error rate of $\pm 1.5\%$. Remember that the actual pressure and not the standard sea level pressure of 1.1013 must be used for the total pressure. Should the altitude be 100 meters, the mean value of the atmospheric pressure is only approximately 1000 mbar.

In some applications the gas flows through relatively thin capillaries after leaving the saturator. With increasing gas velocity, the pressure in the saturator rises significantly and may reach several hundred mbar above the atmospheric pressure. If at this point a value of 1.013 bar is inserted into the equation, instead of the correct pressure in the saturator, (Equation 16-11), an error rate of more than 10% will result.

16.12 Calibration Gas Generators Using the Vaporization Technique

The dosing pump/vaporizer technique is essentially easier to handle and in many cases suitable for multicomponent gas mixtures. Here the liquid with a definite flow rate is metered into a system, analog to the proportioning system for the generation of calibration gases (see Figure 16-7). The substance is vaporized in the system and simultaneously mixed with the pro-

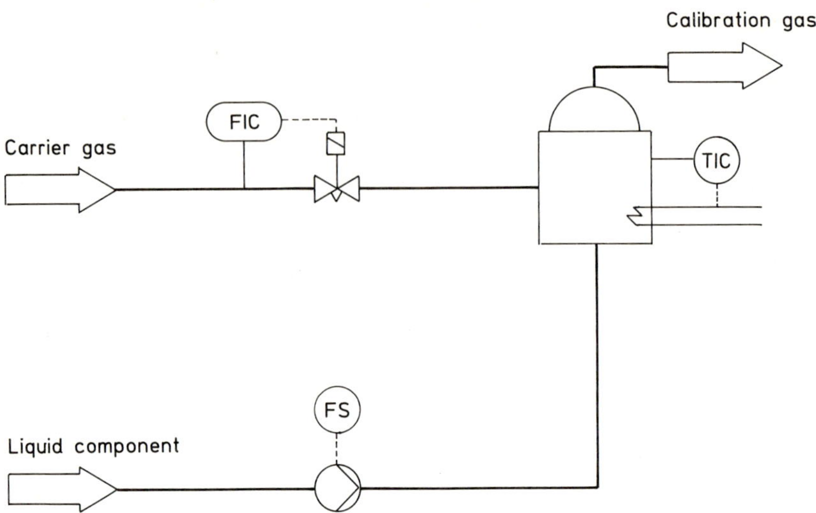

Figure 16-7. Calibration gas generator with vaporizer.

portioning gases. The resulting mole fraction is analog to Equation (16-1); the volumetric flow rate of the liquid dosed component must first be recalculated into standard gaseous conditions. The volumetric flow rate (liquid) must be multiplied by the ratio of the liquid density ($\sigma_{i,\text{liquid}}$) to the hypothetical standard gas density $\sigma_{\text{s},i,\text{gas}}$ (defined to $P_0 = 1.013$ bar and $T_0 = 0\,°C$) .

$$\dot{V}_{\text{s},i,\text{gas}} = \dot{V}_{i,\text{liquid}} \cdot \sigma_{i,\text{liquid}} / \sigma_{\text{s},i,\text{gas}} \tag{16-15}$$

Reciprocating-, double-piston, wobbling-piston, or diaphragm pumps are usually implemented — pump types operating intermittently. In the case of diaphragm pumps, the pulsations are caused by the opening and closing of the valves. Long-stroke double-piston pumps also work only quasi-continually, because the pistons are moved by stepping-motors, and especially at the smallest output rates each individual step is discernible.

A quasi-continual output can be assumed, when the concentration pulses in the calibration gas can no longer be ascertained. To achieve this, the time lag between two pulses must be short in relation to the residence time of the carrier gas in the vaporizer.

The liquid is conveyed into the vaporizer, where it is both vaporized and mixed with the carrier gas stream. Because vaporizing and mixing take place in one unit, the transfer from the liquid phase to the gas phase is supported by the inert gas. The task of the apparatus now is to continually vaporize the quasi-continually entering liquid stream. Because liquids tend to beat during vaporization, a poor vaporizer may generate pulsating concentrations at the outlet, even though the liquid was added continually.

If the liquid transport is not continuous, the vaporizer can smoothen the concentration by remixing. That is why the vaporizer should show the residence time behavior of an perfectly mixed flow vessel in relation to the carrier gas, and the medium residence time of the gas is about five to ten times longer than the conveying pulse interval.

The chosen vaporizer volume should not be too large because, together with the increasing vessel content, the system will become increasingly sluggish and it will take several hours until dosing alterations can be observed.

The lowest possible concentration range is a critical point in dosing pump/vaporizer solutions. The presently available pumps, in the lowest range, transport approximately $10 \cdot 10^{-3}$ — $100 \cdot 10^{-3}$ CCM, which, after vaporization, correlates to a gas standard volumetric flow rate of 10–100 SCCM. The lowest mole fraction range is 1 % at a total standard volumetric flow rate of 1 SLM (a typical value for gas analysis equipment).

Vaporizers can be implemented relatively easily for the proportioning of multiple components, without having to supply each component with its own complete set of proportioning tracks with pump and vaporizer. Vaporization can be executed in a single vaporizer, provided that the boiling temperatures are not too far apart ($\Delta T < 30\,°\text{C}$). However, it is not the boiling temperature at atmospheric pressure, but rather at the desired partial pressure, which is decisive.

The liquid components can all be dosed together as one mixture, using one pump, if they develop a single homogeneous liquid phase, ie, they can mix with each other perfectly. Only then a defined flow rate for each component will result. In multicomponent mixtures the substances must be combined to form homogeneous mixtures with comparable boiling points. One dosing pump is required for each mixture. One must check whether or not a single vaporizer is sufficient for the task.

The proportioning of premixed liquids leads to longer phases of output loss when the concentration is altered because the vaporizer must be emptied and refilled each time. Therefore the volumes between the liquid tank and the pump need to be considered when estimating the response time of the system. This volume first must be displaced before the new mixture can reach the vaporizer. The concept of a system in which all substances are emitted via one dosing track is suitable for experiments in which concentrations need not be changed in rapid succession.

The use of premixed liquids offers an important advantage. As explained above, the technique with dosing pump and vaporizer is limited by the minimum conveying flow rate of the dosing pump. If, for example, only traces of benzene are to be proportioned and other aromatic hydrocarbons are desired at higher concentrations, benzene may be proportioned in liquid form together with other hydrocarbons, thereby achieving lower benzene concentrations.

16.13   Cascading Dilution Steps

If concentrations in the ppb range are required, this can only be achieved, in one step, with permeation techniques, because only they allow flow rate doses of a few 10^{-3} SCCM. Otherwise the concentration of a generated precursory calibration gas can be reduced by one to three orders of magnitude via a second dilution step.

This may be achieved with MFCs, as well as with the help of super-critical nozzles; however, both alternatives require a preliminary pressure of 2–3 bar on the side of the precursory calibration gas.

Because saturators and vaporizers for laboratory use are usually made of glass equipment, higher pressures are unpopular. That is why the commercially available jet probe has become popular (see Figure 16-8). A carrier gas stream flows through an injector pump and creates an underpresure of about 0.4 bar (absolute) on the sucking side. The pressure difference of 0.6 bar (or a pressure ratio of 1 : 0.4) in relation to the precursory calibration gas is sufficient to guarantee the supercritical flow and to suck up a defined part of the flow (in the range of 1 SCCM, for instance).

In the case of a carrier gas standard volumetric flow rate of about 5 SLM, a maximal dilution ratio of 1 : 5000 can be achieved. The surplus precursory calibration gas is emitted as exhaust.

When using supercritical nozzles the preliminary pressure of the precursory calibration gas enters into the proportioned amount and thereby into the dilution. As long as the precursory calibration gas is not released into the atmosphere, atmospheric pressure fluctuations lead to an error rate of $\pm 1.5\%$. Each dilution step with cascading dilutions increases the total error of the proportion at system.

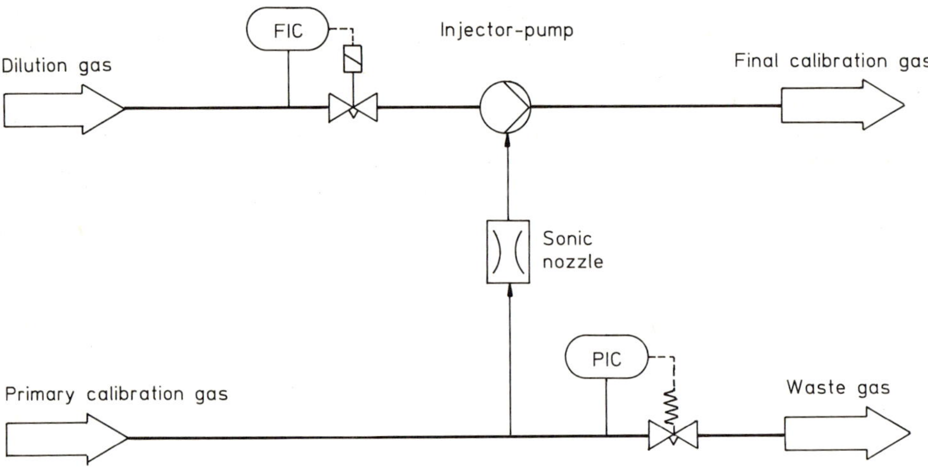

Figure 16-8. Gas dilution step with the help of a jet pump.

16.14   Zero Gas Generators

To set analyzers to zero, that is, to adjust the signal value to zero, nitrogen is usually used. In environmental analysis, in which contaminants in the air are to be determined, synthetic air is employed. When larger amounts of zero air are needed, zero air can be generated with a zero gas generator. The task of this apparatus is to condition the ambient air so that the contaminants that are to be measured later (such as CO, CH_4, NO, NO_2, SO_2, and NH_3) are removed.

Immission analysis operates in the same concentration ranges as those prevailing in the air in the immediate environment. For SO_2 and NO these concentration values are in the range of 0.1 ppb, but they can vary greatly, depending on the industry in the area, the season, and the weather conditions.

In the case of zero gas generators, the above-named contaminants are usually removed by adsorption because this separation process is the easiest to handle (for instance in comparison to washing processes). Gas is sucked in through a dust filter and is compressed to between 7 and 8 bar. If possible each adsorption step should be settled in this region, because the grades of separation rise in conjunction with the partial pressure. Water vapor is removed from the air initially, because water vapor is the most prevalent accompanying component (y_{H_2O} approximately 3% absolute).

In a pressure swing adsorption process, water vapor is adsorbed by a zeolithic adsorbent and, after saturation of the adsorbent, it is switched over to a second adsorber, filled with fresh adsorbent. Meanwhile the gas is cleaned in the second adsorber, the pressure inside the first adsorber is reduced, whereby water vapor desorbs from the adsorbent material.

In the case of zero gas generators for low flow rates and mobile employment, an adsorber which can be utilized only once is used and then it is exchanged upon saturation and disposed of, or it is regenerated externally. The other contaminants are of such minimal concentrations that theses adsorbers provide sufficient exposure time.

All components, except NO and methane, can unproblematically be removed by adsorption. If these two gases are also to be removed from the air, then special steps are necessary. An ozone oxidation step must be installed for the removal of NO. Oxygen from the air is oxidized to ozone using a UV source, and ozone oxidizes nitrogen oxide to easily adsorbable nitrogen dioxide. Surplus ozone is caused to react in a subsequent active carbon, producing CO and CO_2.

Methane and lower hydrocarbons can, for example, be converted into carbon dioxide by catalytic postcombustion by using a catalytic converter that is inserted into a portion of the tube and heated to temperatures above 300 °C.

Typical adsorbents for polar compounds (such as water vapor, ammonia, and hydrogen sulfide) are zeolithes, active carbon for water vapor and higher hydrocarbons, as well as calcium oxide for carbon dioxide.

The generating of zero air will usually not remove all components from the air (nitrogen, oxygen, and argon will remain) but merely those components that are to be analyzed later on. Thus most zero gas generators usually include only a subset of the named separation procedures.

16.15   References

[1] ISO, "Gas Analysis – Determination of Composition of Calibration Gas Mixtures – Comparison methods", *ISO/DIS 6143/1*; Geneva: ISO, 1981.
[2] ISO, "Gas Analysis – Preparation of Calibration Gas Mixtures – Dynamic Volumetric Methods – Part 1: Calibration Methods", *ISO/DIS 6145/1*; Geneva: ISO, 1986.

[3] Nelson, G., O., *"Controlled Test Atmospheres − Principles and Technics"*, Ann Arbor Science Publication, Lib. Congress Cataloque No. 73−141231, 1971.

[4] VDI, "Measurement of Gases − Calibration Gas Mixtures − Dynamic Preparation by Gas-Mixing Pumps", *VDI-Guideline 3490*, sheet 6; Berlin: Beuth, 1988.

[5] VDI, "Messen von Gasen − Prüfgase − Dynamische Herstellung durch periodische Injektion", *VDI-Guideline 3490*, sheet 7; Berlin: Beuth, 1980.

[6] ISO, "Gas Analysis − Preparation of Calibration Gas Mixtures − Dynamic Volumetric Methods − Part 3: Periodic Injections into a Flowing Gas Stream", *ISO/DIS 6145/3*: Geneva: ISO, 1986.

[7] VDI, "Messen von Gasen − Prüfgase − Herstellung durch Permeation der Beimengung in einen Grundgasstrom", *VDI-Guideline 3490*, sheet 9; Berlin: Beuth, 1980.

[8] ISO, "Gas Analysis − Preparation of Calibration Gas Mixtures − Permeation Method *ISO/DIS 6349*; Geneva: ISO, 1979.

[9] VDI, "Messen von Gasen − Prüfgase − Herstellen von Prüfgasen mit Blenden-Mischstrecken", *VDI-Guideline 3490*, sheet 16; Berlin: Beuth, 1989.

[10] ISO, "Gas Analysis − Preparation of Calibration Gas Mixtures − Dynamic Volumetric Methods − Part 6: Sonic Orifices", *ISO/DIS 6145/6*; Geneva: ISO, 1985.

[11] Klobe, M., Oppermann, F., "Ein Verfahren und ein Gerät zum kontinuierlichen und definierten Mischen von Gasen und Dosieren des entstandenen Gasgemischs", *ATM−sheet V 723-35*; München: Oldenbourg, 1971.

[12] Klobe, M., Oppermann, F., "Einrichtung zum Einstellen und Konstanthalten von Gasströmen in Gaschromatographen", *ATM−sheet V 273-29*; München: Oldenbourg, 1968.

[13] Rochat, J.-D., Zellweger, C., "Device for Controlling the Mixture and Flow of al least Two Fluids", *United States Patent, Registration No. 4,467,834*, 1984.

[14] Sullivan, J. J., Jacobs, R. P. Jr., "Flow Measurement and Control in Vacuum Systems for Microelectronics Processing" *Solid State Technology*, October 1986.

[15] VDI, "Messen von Gasen − Prüfgase − Herstellen von Prüfgasen durch Sättigungsmethoden", *VDI-Guideline 3490*, sheet 13; Berlin: Beuth, 1986.

[16] ISO, "Gas Analysis − Preparation of Calibration Gas Mixtures − Saturation Method", *ISO/DIS 6147*; Geneva: ISO, 1979.

[17] Weitkamp, J., Dauns, H., "Eine neue Methode zur Dosierung gasförmiger Mehrkomponentengemische in Strömungsapparaturen", *Chem.-Ing.-Tech.* **56**, No. 12 (1984) 929−930.

[18] Weitkamp, J., Dauns, H., "Generation of Multicomponent Gas Mixtures in Catalytic Flow Type Units: a Novel High Pressure Saturator", *Appl. Catal.* **38** (1988) 167−177.

17 Applications of Optochemical Sensors for Measuring Chemical Quantities

Otto S. Wolfbeis, Joanneum Research, Graz, Austria
Gilbert Boisdé, Commissariat à l'Energie Atomique, Saclay, France

Contents

17.1   pH and Electrolyte Sensors

The inconveniences inherent in electrodes and the recognized advantages of fiber optics, especially their immunity towards electromagnetic interference, the instantaneous transfer of light data, and the small transverse diameter of the fibers, have offered the most decisive arguments for launching into research on pH optodes. The potential of miniaturization of the optodes attracted the interest of the clinical and biomedical world in the possibilities offered for in vivo analysis.

Accordingly, most of the research on optodes has been conducted in the pH range of physiological liquids (see Section 18.2). In recent years, however, measurements have been extended to other pH ranges for industrial applications.

The techniques most frequently employed are absorption, reflectance and fluorescence, with some attempts to exploit the variation in refractive index of the sensitive polymer acting as the cladding for the fiber. The general principle of "semi-active" and "active" optodes (Chapter 12) is applied. The optical spectrum of an optode is modified by chemical action between a reagent (indicator), immobilized on the fiber or an inert support at its end, and the activity of the H^+ ions (a_{H^+}):

$$pH = -\log(a_{H^+}) = -\log(\gamma_{H^+}[H^+]) \tag{17-1}$$

where $[H^+]$ is the concentration of hydrogen ions and γ_{H^+} the activity coefficient of hydrogen ions.

The thermodynamic dissociation constant (K_a) of an indicator (HI) is given by the equilibrium

$$HI \;\rightleftharpoons\; H^+ + I^- \tag{17-2}$$

with

$$K_a = \frac{a_{H^+}[I^-]\gamma_{I^-}}{[HI]\gamma_{HI}}, \tag{17-3}$$

where $[HI]$ and $[I^-]$ are the concentrations and γ_{HI} and γ_{I^-} are the activity coefficients of the acid and conjugate base species, respectively.

The logarithmic form is the Henderson-Hasselbalch equation [1]:

$$pH = pK_a + \log([I^-]/[HI]) + \log(\gamma_{I^-}/\gamma_{HI}) . \tag{17-4}$$

For $\gamma_{H^+} \approx 1$, Sörensen's definition of pH is

$$pH = -\log[H^+] . \tag{17-5}$$

Using the concentration constant (K_a') for the equilibrium in Equation (17-2) according to the mass action law:

$$K_a' = [H^+][I^-]/[HI] \tag{17-6}$$

and in logarithmic form:

$$\text{pH} = \text{p}K'_a + \log\left([\text{I}^-]/[\text{HI}]\right) . \tag{17-7}$$

For a total concentration C of indicator given by

$$C = [\text{HI}] + [\text{I}^-] , \tag{17-8}$$

the pH function is

$$\text{pH} = \text{p}K'_a - \log\left(\frac{C}{[\text{I}^-]} - 1\right) \tag{17-9}$$

with

$$\text{p}K'_a = \text{p}K_a + \log\left(\gamma_{\text{HI}}/\gamma_{\text{I}^-}\right) . \tag{17-10}$$

Colorimetric methods with indicators have been used for many years in chemical analysis [2–5]. However, the functions described above also form the basis of many optical and potentiometric methods (electrodes). They show that the practical inaccessibility of the knowledge of ionic activity raises a problem for the analyst because of the approximation of γ_{H^+} and the neglect of the second term in Equation (17-10). This, ultimately, questions the accuracy of optical sensors [6], the future of which appears limited in certain areas [7].

This section attempts to present a broad review of this technique in the light of recent research on fiber-optic chemical sensors (FOCS). A discussion on the advantages and performance of pH optodes will be broached by considering the various applications planned and the resulting pH measurements, including titration and ionic strength. Sensors for low molecular weight electrolytes, particularly optodes for anions and cations in solution, are also considered.

17.1.1   pH Measurement

17.1.1.1   *Theoretical Aspects*

Absorption

In absorption, according to the Beer–Lambert law (see Chapter 12), and within the limits of a linear application of the absorbance (A_λ) at the wavelength λ of an indicator containing the species HI and I$^-$, the following equation applies:

$$A_\lambda = \ell\left(\sum_{i=0}^{n} (\varepsilon_i)_\lambda\, C_i\right) + B_\lambda = \ell\left\{(\varepsilon_{\text{I}^-})_\lambda\, [\text{I}^-] + (\varepsilon_{\text{HI}})_\lambda\, [\text{HI}]\right\} + B_\lambda \tag{17-11}$$

where

ℓ is the optical path length;
$(\varepsilon_i)_\lambda$ is the molar absorption coefficient of species i;
C_i is the concentration of species i;

$(\varepsilon_{I^-})_\lambda$ is the molar absorption coefficient of species I^-;

$(\varepsilon_{HI})_\lambda$ is the molar absorption coefficient of species HI;

B_λ is the interfering absorbance of the dye background and the measuring instrument; sometimes B_λ is not negligible, eg, in on-line control of slightly turbid media or in laboratories with single-beam spectrophotometers.

In solution, the zero correction helps to ensure $B_\lambda = 0$. If so, the pH is calculated from the absorbance measured (A) and the maximum absorbance of the acidic and basic forms [8,9], respectively:

$$\text{pH} = pK_a' - \log\{(A_{max} - A)/(A - A_{HI})\} \tag{17-12}$$

where

A_{max} is the absorbance at a pH high enough to ensure complete dissociation of indicators;

A is the absorbance at any given pH;

A_{HI} is the absorbance of the totally undissociated indicator.

In principle, A_{HI} is low at the wavelength at which $[I^-]$ is maximum, and $A_{HI} \approx 0$, giving approximately

$$\text{pH} = pK_a' - \log\{(A_{max} - A)/A\} . \tag{17-13}$$

Peterson et al. [10] proposed the function

$$A_I \, \ell \, \varepsilon_{I^-} \, [I^-] \qquad (\text{with } A_{HI} = 0) . \tag{17-14}$$

This function, together with Equation (17-9) and

$$D = \text{pH} - pK_a' , \tag{17-15}$$

gives the function

$$A_I = C\ell\varepsilon_I/(1 + 10^{-D}) \tag{17-16}$$

or the equation recently developed [6]:

$$A = A_{max}/(1 + 10^{-D}) \tag{17-17}$$

with

$$A_{max} = C\ell\varepsilon_I . \tag{17-18}$$

In some instances, B_λ may vary as a function of time in the case of optodes (aging of the immobilization support, for example). The term can be corrected by the use of two measurement wavelengths. Thus, Goldstein and Peterson [11] took a relative measurement between the absorption peak of the dye and the isosbestic point. By combining two Equations (17-11) of the same type for a wavelength λ_1 (absorbance A_1) and an absorption valley at λ_v (absorbance A_v), an internal spectral reference is employed:

$$A_m = (A_1 - A_v) = \ell\{(\varepsilon_1 - \varepsilon_v)_{HI}\,[HI] + (\varepsilon_1 - \varepsilon_v)_I\,\text{-}[I]\} + B_1 - B_v \qquad (17\text{-}19)$$

where $B_1 \approx B_v$ (background noise due to the support).

For an optode, it is difficult to obtain A_{max}, A_{HI} and A_I because the concentration C of the immobilized dye is unknown. This gave rise to the recent proposal of using three wavelengths, two for the measurement and one for the reference [12]. Based on measurements at wavelengths λ_1, λ_2 and λ_v, the respective absorbances A_1, A_2 and A_v can be used in equations:

$$A_p = (A_1 - A_v) \qquad (17\text{-}20)$$

and

$$A_x = (A_1 - A_2)\,. \qquad (17\text{-}21)$$

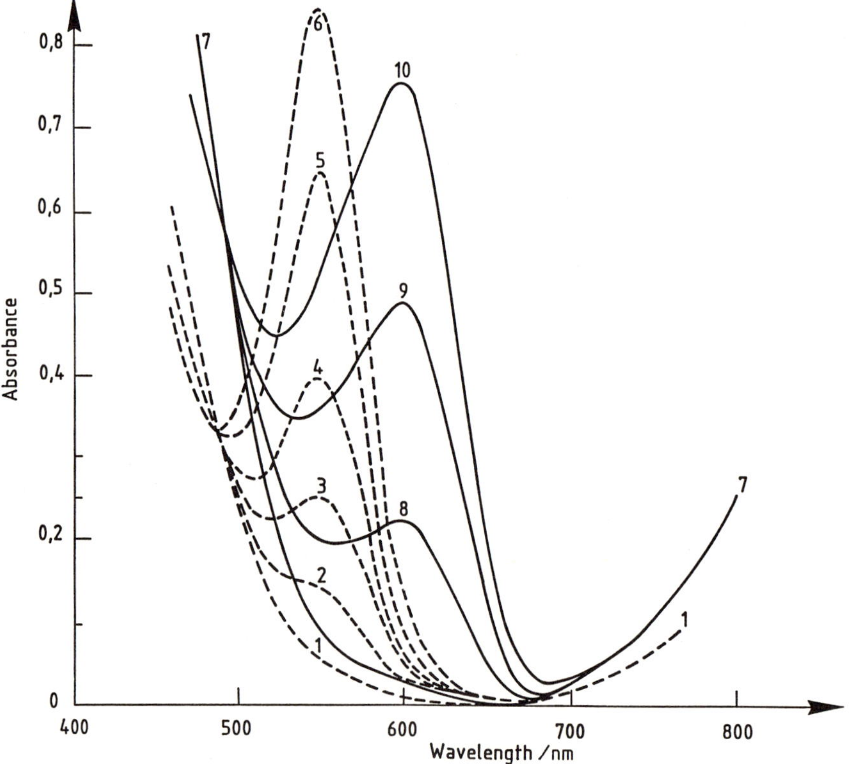

Figure 17-1. Spectral variation of pH optode with Thymol Blue (TB) immobilized on Amberlite XAD–4. pH: 1: 5.25; 2: 3.2; 3: 2.85; 4: 2.5; 5: 1.9; 6: 1.35; 7: 9.0; 8: 11.1; 9: 11.7; 10: 12.2 (from [12]).

The general pH–absorbance relationship obtained is

$$\text{pH} = pK_a' - \log\{(K_1 A_x - K_2 A_p)/(K_3 A_p - K_4 A_x)\} \tag{17-22}$$

where K_1, K_2, K_3 and K_4 are calibration coefficients depending on the choice of the wavelength.

An example of the spectral variation of a dye (Thymol Blue) immobilized on Amberlite and sensitive to two pH ranges (acidic and basic) is shown in Figure 17-1. The pH–absorbance variation is sigmoid, as shown in Figure 17-2. This function decreases or increases with pH according to the measurement range and the acidic or basic properties of the dye.

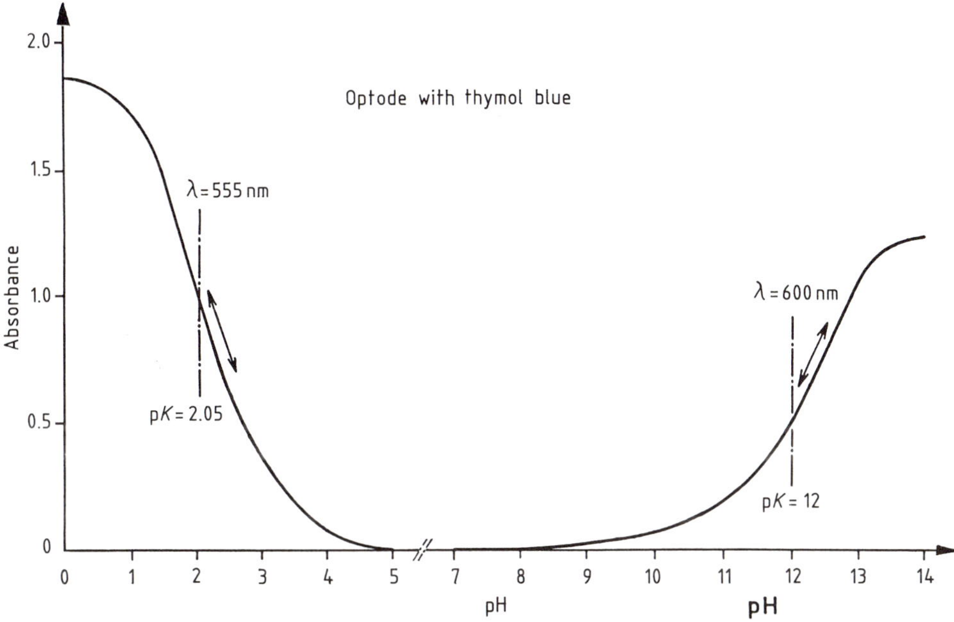

Figure 17-2. Thymol Blue optode in acidic and basic media; reference wavelength at 800 nm.

Reflectance

The response of an optode in reflectance is frequently presented in standard arbitrary units of reflected radiation (R_N):

$$R_N = (R - R_{min})/(R_{max} - R_{min}) \tag{17-23}$$

where R is the measured value, and R_{max} and R_{min} are the upper and lower limits of R. With the Kubelka-Munk function (see Chapter 12) the reflectance is not directly proportional to the concentration C:

$$(1 - R)^2/2R = KC . \tag{17-24}$$

For the popular case where an indicator is used to dye particles serving as a solid support, Narayanaswamy and Sevilla [13] allowed for four equilibrium constants:

$$[HI]_{ads} \xrightleftharpoons{K_1} [H^+]_{ads} + [I^-]_{ads}$$

$$K_2 \Big\Updownarrow \qquad\qquad K_3 \Big\Updownarrow$$

$$[HI]_{sol} \xrightleftharpoons{K_4} [H^+]_{sol} + [I^-]_{sol}$$

where $[HI]_{sol}$ is negligible owing to its high absorption on the Amberlite resin employed. They also assumed that absorption of the indicator on a non-polar surface prevents the dissociation of the solute, and compared the constants K:

$$K_1/K_4 = K_2/K_3 \tag{17-25}$$

with $K_2 < K_3$ (high absorption of the neutral form).

Fluorescence

The high sensitivity, selectivity, and low background noise in the presence of absorbing species are the main factors in favor of fluorescence.

If F_{max} is the fluorescence intensity when the dye is completely in the (fluorescent) phenolate form, F_{I^-} the intensity of the phenolate form at a given pH, and assuming absorption by one species only, the variation in pH is given by

$$pH = pK_a' - \log\left(\frac{F_{max}}{F_{I^-}} - 1\right) \tag{17-26}$$

which is identical with Equation (17-9) because

$$F_{max} = \varepsilon_1 \phi_1 [I^-] + \varepsilon_2 \phi_2 [HI] \tag{17-27}$$

and

$$F_{I^-} = \varepsilon_1 \phi [I^-] \tag{17-28}$$

where ε and ϕ are the absorption and fluorescence yields for the base (1) and acid (2) form, respectively.

To eliminate the intensity variations of the source, the influence of the loss of dye in the medium, or the quenching of the non-active form, the measurement of the ratio of the excitation intensities at two wavelengths has been suggested [14]. One of them (basic) is selective for the fluorophore and the second is influenced by the forms present (acid and base). This ratio as measured at two wavelengths λ_1 and λ_2 is thus:

$$F_{\lambda_1}/F_{\lambda_2} = \frac{(K_{I^-})_{\lambda_1} K_a'}{(K_{I^-})_{\lambda_2} K_a' + (K_{HI})_{\lambda_2} [H^+]} \cdot \tag{17-29}$$

The constants K are referenced by wavelengths (λ_1 and λ_2) and chemical species (I^- and HI).

17.1.1.2   pH Sensors

Many patents have been issued for pH optodes, both for absorption and reflectance measurements [11, 15–17] and for fluorescence [18–22]. The interaction of a pH-sensitive layer deposited as cladding on the surface of a fiber has also been proposed [23]. Figure 17-3 shows a number of arrangements developed in various laboratories.

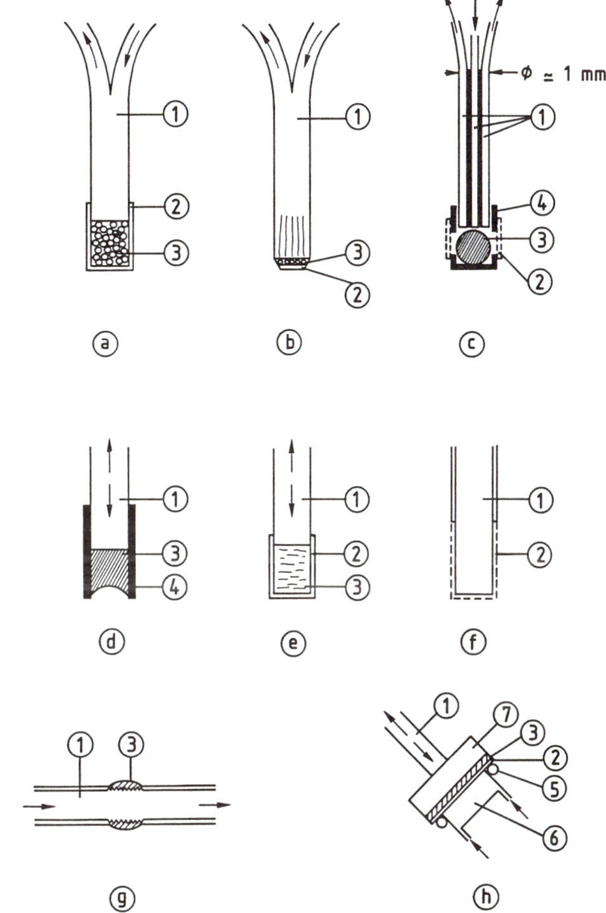

Figure 17-3.
pH optode configurations.
1: fiber; 2: membrane;
3: reagent; 4: support; 5: O-ring;
6: flow cell; 7: glass. (a) Immobilization on microspheres or gel;
(b) bundles for reflectance;
(c) single microsphere
(0,4–0,8 mm); (d) monofiber and deposit; (e) reservoir for liquids;
(f) surface treatment and immobilization; (g) cladding effect;
(h) flow cell and immobilization on membrane.

The first FOCS described for pH measurements was based on the absorption variations of Phenol Red [10]. This dye is copolymerized with a monomer of the acrylic type (5–10 µm diameter) [24]. Polystyrene microspheres (20% with diameter 1 µm) improve the diffusion of light and provide a hydrophilic structure that is advantageous for ion transfer. The ratio (R_a) of the intensities at two wavelengths (558 nm at the peak and 478 nm at the isosbestic point) is related to the concentration and pH by the equation

$$\log (R_a/K) = -C/(10^{-D} + 1) \tag{17-30}$$

(see Equations (17-15) and (17-16)).

The physiological pH range (7.0–7.4) is covered with an accuracy of ± 0.01 units [25–28]. A first version was tested in vitro and in vivo (see Chapter 18).

For a similar purpose, this type of sensor was prepared with a hypodermic needle containing the optical fiber [29]. The accuracy obtained was ± 0.013 pH units. Also, Bromocresol Green (BCG) and Bromothymol Blue (BTB) immobilized covalently on a cellulose membrane, with a variation of the signal (1.6–2 V) obtained around pK'_a (7.5 for BTB), helped to achieve standard deviations of less than three decimal places [30].

The application of pH optodes to industrial measurements in more acidic and more basic ranges has been the subject of more recent developments. Kirkbright and co-workers [16, 31, 32] investigated a broad range of dyes immobilized on an Amberlite fixed at the fiber and with a poly(tetrafluoroethylene) membrane permeable to the ions. In particular, the reflectance measured with BTB (λ = 620 nm) conforms to the requirements in the pH range 8–10.5. A shift of the pK'_a is observed when the dyes are immobilized on this resin [31]. The influence of turbidity (strong), of ionic strength (significant), and of temperature (coefficient about $0.01\,°C^{-1}$ between 20 and 30 °C, $0.02\,°C^{-1}$ between 30 and 40 °C, and $0.06\,°C^{-1}$ between 40 and 50 °C) are also considered. The response time at 99% of signal is about 5 min due to the membrane. The performance of the sensor was stable for 5 months. In the pH range 4.0–8.0, similar measurements [33] using BTB immobilized on a collodion membrane show the importance of the thickness (e) of this membrane for the optode response time (up to 30 min for e = 3 mm). While such sensors are easily fabricated, it is recognized now that whenever use is made of dyes immobilized on a hydrophobic support such as Amberlite, the respective pH sensors are prone to hysteresis and long-term drift at pHs near their pK. This contrasts the excellent reversibility and stability of pH sensors in which a hydrophilic material such as cellulose acts as the solid support, but which are more difficult to prepare.

For industrial measurements of highly acidic and highly basic pH, a miniature optode (diameter <1.5 mm) was made with silica fibers with low losses over long distances [12, 17, 34]. Thymol Blue (TB) immobilized on Amberlite is used for two pH ranges (0.8–3 and 10–13) with the same optode, as the spectral characteristics for each range are different (Figures 17–1 and 17–2). In these zones, the response time (t_r), at 99% completion, is 15–60 s for ΔpH = 0.2. For a change in pH from pH_0 to pH_1, the absorbance varies from A_0 to A_1 (with ΔA = $|A_1 - A_0|$) during the response time (t_r). The variation of a relative function $f(t_r, \Delta A) = t_r/\Delta A$ with the mean pH (pH_m) was studied for different dyes and with the same optode configuration (see Figure 17-3c) [35]. A significant increase was observed in the neutral pH range (Figure 17-4). The excellent reversibility of the sensor at highly acidic and basic pH is also decreased at neutral pH (hysteresis).

Cresol Red immobilized on an anion-exchange resin was also tested in highly acidic media (H_2SO_4 or HCl from 2 to 0.015 M) and from pH 5 to pH 9 [36]. The sensor response (λ = 570 nm) is linear from 0.025 to 0.60 M in the first interval (pH 0–1.5), and in the pH range from 6.25 to 7.20 to the nearest ± 0.02 in the second interval. The stability of the sensor (3 months) makes it possible to determine the pH of milk.

Recently, a pH sensor developed by immobilizing a direct azo dye on a porous cellulosic polymer film has been described [37]. Advantageous features of the design include a rapid equilibration time (<1.3 s) and a large dynamic range (>4 pH units with Congo Red as dye).

Many investigations have also been devoted to fluorimetry. Fluorescein is a popular fluorophore, allowing measurements between pH 4 and pH 9, for biological applications. An optode proposed by Saari and Seitz [38] uses fluoresceinamine immobilized either on porous

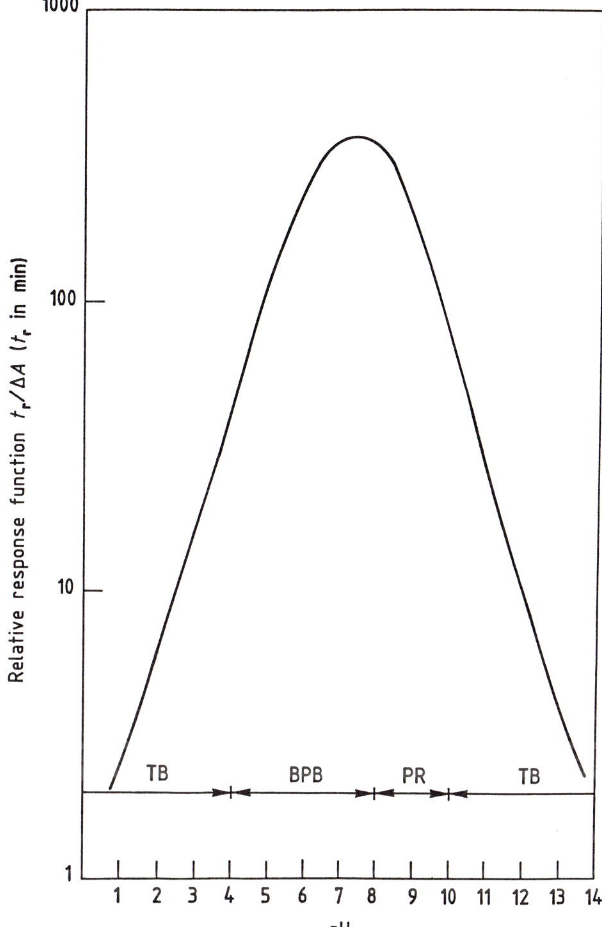

Figure 17-4.
Relative function $t_r/\Delta A$ for various pH ranges, where $pH_m = (pH_0 + pH_1)/2$, with pH_0 and pH_1 being the initial and final pH, respectively, ΔA the absorbance variation, and t_r the response time with 99% of the total signal change. TB: Thymol Blue; BPB: Bromophenol Blue; PR: Phenol Red (from [35]).

glass after oxidation of the support, or preferably on cellulose powder. However, the accuracy is limited owing to the high background noise. Above pH 8, the fluorophore is partly detached from the support. Using immobilized fluorescein at the fiber end and the remote fiber fluorimetric (RFF) technique, Hirschfeld and co-workers [39, 40] developed optodes which were said to have a resolution of 0.01–0.02 pH units and a response time shorter than 3 min. Low-loss, small-diameter (100–140 µm) silica fibers allow long-distance signal transmission (1 km). Fluorescein isothiocyanate (FITC) immobilized on an individual porous glass bead fixed to a single fiber was also studied [41]. The dynamic range, pH 3–7, of the miniature sensor can be extended by decreasing the dye loading. The response time is between 20 and 25 s. Changes in indicator fluorescence caused by nonspecific alterations are a potential problem.

Recent investigations [42] incorporate fluoresceinamine into an *N,N'*-methylenebisacryl-amide copolymer covalently attached to a glass fiber whose surface is modified by plasma deposition followed by acrylamide photopolymerization. The sensor displays a reversible response for pH ranging from 4.0 to 8.0. The response time of the sensor with the polymer modified on the fiber is less than 9 s. One disadvantage of this sensor is its lack of reproducibility. A similar optode was tested by Kawabata et al. [43] and gave an accuracy of only $\pm 0.1\%$ in the pH range 4.0–6.9. The response time of the optode (about 1 s) is increased by the response of the apparatus (30 s).

Another fluorimetric sensor, based on energy transfer, was recently developed [44]. The fluorophore (eosin) and an "absorber" (Phenol Red) are co-immobilized. An argon laser ($\lambda_e = 488$ nm) excites the eosin at a wavelength at which Phenol Red is non-absorbent. The fluorescence emitted by the fluorophore (donor) is modulated in the presence of the absorber (acceptor) either by the "inner fibers" effect (decrease in fluorescence) or by the energy transfer. These two effects generate an overlap of the fluorescence and absorption emissions. The measurement at 546 nm is taken between pH 4.5 and pH 8.6 (65% decrease in intensity). When optimized for the physiological range, an accuracy of 0.01 pH units is obtained.

4-Methylumbelliferone (4-MU) incorporated in polyacrylamide capsules permeable to protons was the earliest fluorophore investigated for intravascular fluorimetric measurements [45]. The emission spectrum displays an isosbestic wavelength at 330 nm, and the dissociated form has its peak at 365 nm. The ratio of the measured intensities is related to the pH. The response time is less than 2 minutes. This pH measurement (range 6.5–8.5) is advantageously taken with determination of the ionic strength [46] in combination with 1-hydroxypyrene-3,6,8-trisulfonic acid (HPTS) [47] (see Section 17.1.1.3).

The HPTS fluorophore has proved to be an ideal dye for neutral pH, and thus for physiological media. It has been tested [14] by calculating the ratio of the relative excitation and fluorescence intensities at 405 and 470 nm (see Section 17.1.1.1). The three sulfonated groups of the dye yield an adequate hydrophilic structure for nearly irreversible immobilization (except at high charge) on anion-exchange membranes. This sensor offers the advantage of the absence of quenching. After calibration, the standard deviation for three sera was 0.02–0.08 between pH 7.3 and pH 7.4. However, photodegradation is observed, even at daylight, thus limiting the lifetime of the sensor.

Wolfbeis and co-workers [47–49] immobilized HPTS covalently by cross-linking on cellulose. The pK'_a is lowered from 7.3 to 6.7 by this immobilization, which is a drawback with regard to biological liquids. Nevertheless, they were measured to the nearest ± 0.02 pH units (range 7.0–7.5). The response time is about 1 min for 99% of the total signal variation. However, the response rate is not symmetrical (the response time of high values toward low values is 20% greater than vice versa). Another optode, made with a coumarin-based fluorophore [48] and immobilized on porous glass, is used in the same range ($pK'_a = 7.1$). The pH is obtained to the nearest ± 0.005. The response time may be as long as 1.6 min for 90% of the signal if a cuprophane membrane is inserted. Figures 17-5a and b show the excitation spectra of HPTS- and 7-hydroxycoumarin-3-carboxylic acid (HCC)-based pH sensors.

HPTS covalently immobilized on cellulose granules and with a pK_a of 7.1 is now used as the pH indicator in the Gas Stat device (Cardiovascular Devices Inc.) for continuous measurements on the extra-corporeal circuit. The pH sensor is generally associated with pCO_2 and pO_2 measurements in a single catheter [50], which consists of three 125-μm fibers with appropriate chemistries at their ends [51].

The most recent developments are based on the simultaneous excitation of two fluorescent indicators [52]. This may make possible the realization of a triple one-fiber sensor for oxygen, carbon dioxide and pH (see Section 18.2).

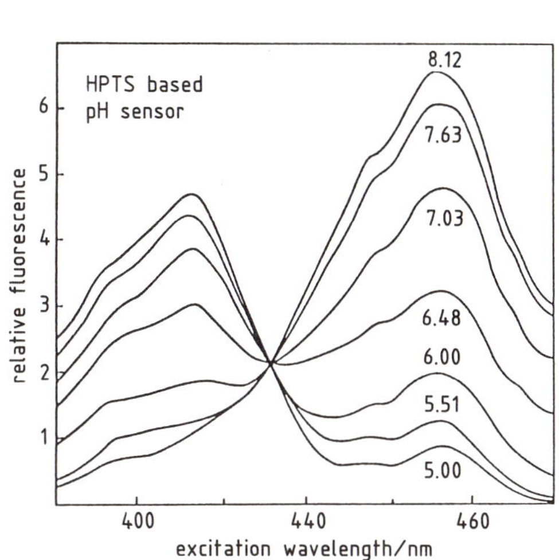

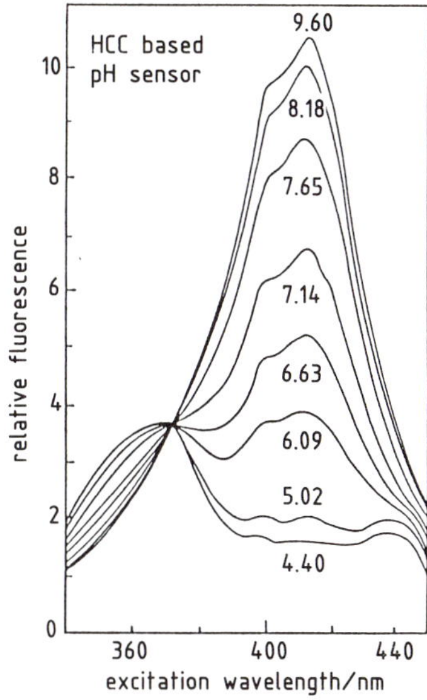

Figure 17-5. Excitation spectra of glass-immobilized HPTS and HCC at the various pH values indicated on the curves (from [48]).

A new fiber-optic pH sensor with a response time of 5 s has been described in accordance with a new principle [53]. The pH depends on the refractive index of a sensitive polymer film a polyacrylamide ion-exchange membrane, impregnated with Phenol Red as dye. The sensitive component is a directional coaxial coupler in the form of a tube (with three layers) whose cylindrical symmetry is compatible with single-mode fibers and a sensitive medium externally surrounding the guide tube. The difference in dispersion between the surroundings and the tube causes variations in wavelength at the coupling depending on the intensity fluctuations of the source. The estimated resolution is 0.02 pH units.

17.1.1.3 Application of pH Sensors

Performance and Implications for pH Measurements

Table 17-1 for absorption and reflectance techniques and Table 17-2 for fluorescence summarize the pH measurements with the main indicators used so far for biological and industrial media. Apart from the experiments of Attridge et al. [53], all the investigations are concerned with absorption (and/or reflectance) or luminescence. Both methods proved satisfactory for in vivo measurements, and the accuracies obtained are comparable (± 0.01 near the pK'_a). The lifetime of the indicator and the optode response time vary according to the authors. In absorption, this response time can differ with pH (see Figure 17-4) and decrease with Δ pH.

The advantage for absorption (and reflectance) resides in the simplicity of the apparatus and the availability of inexpensive and stable components (sources, detectors). This explains the recent developments in uncomplicated and effective instrumentation (modulable light-

Table 17-1. Typical pH measurements by absorption and reflectance using optical fibers

Indicator	Support and membrane	Fiber	λ_m (nm)	t_r (s)	pK_a	pK'_a	Range	Accuracy (pH units)	References
Phenol Red	Polaycrylamide + polystyrene–cuprophane	Plastic	558/478 or 600	40 (63%)	7.9	7.57	7.0–7.4	0.01	10, 25, 29
Bromothymol Blue	Cellulose		565		7.7	7.5	7.3–7.7	10^{-3}	30
Bromothymol Blue + other dyes	Styrene-divinylbenzene, (XAD–2)/PTFE	Plastic	593/665	300 (99%)	7.7	8.5–9.5	8.5–10		31, 32, 33
Thymol Blue, Bromophenol Blue	XAD–4 (no membrane)	Silica	555/830 664/830	<60 (95%) <600 (99%)	1.65/9.2 4.1	2.05–12 4.6	0.8–3.2 10–13 3–6	0.01 near pK'_a	12, 34
Cresol Red	XAD–2–PTFE	Plastic	570			6.7	6.2–7.2	0.02	36

Table 17-2. Typical pH measurements by fluorescence using optical fibers

Indicator	Support and membrane	Fiber	λ_e (nm)	λ_m (nm)	t_r (s)	pK_a	pK'_a	Range	Accuracy (pH units)	References
Fluoresceinamine	Porous glass–cellulose, N,N-methylenebisacryl-amide	Silica	480–488	500–600	30 <150 <9	4.4/6.7		4–8 6.8–7.8	0.02	38 39, 40, 43 42
Fluoresceine isothiocyanate	Porous glass bed	Silica	488	530	20–35	2.2/4.4/6.7		3–7	–	41
Eosin + Phenol Red	Modified surface, no membrane	Silica	488/514.5	546	4 (63%)		7.2	4.5–8.6	0.01	44
4-Methyl-umbelliferone	Polyacrylic		365		<60			6.5–8.5		45, 46
HPTS	Ion-exchange membrane glass–cellulose/ cuprophane	Glass	405/470 465	520 520 510	150 60 (99%)	7.3 7.3	6.7	6.0–8.0 6.4–7.7	0.02–0.08 0.02	14 48, 49 46
HCC	Glass–cellulose/ cuprophane	Glass	410	455	90	7.3	6.9	6.4–7.7	$5 \cdot 10^{-3}$	48, 49

emitting diodes (LED), low-noise positive-intrinsic-negative (PIN) diodes [12, 30, 54, 55] for diverse uses [12, 33, 56]. An internal reference with measurement at two wavelengths (frequency modulated) is an advantageous technique. However, absorption and reflectance suffer from a low dynamic range.

Once the effects of inner filters and quenching are controlled, luminescene appears more promising owing to its sensitivity and selectivity. Simultaneous excitation at two wavelengths is interesting, but requires a more complex installation and demultiplexing. It is preferable to have an internal reference in the system. The incorporation of the reference signal at the optode by means of two copolymerized reactants and the use of the principle of fluorescence

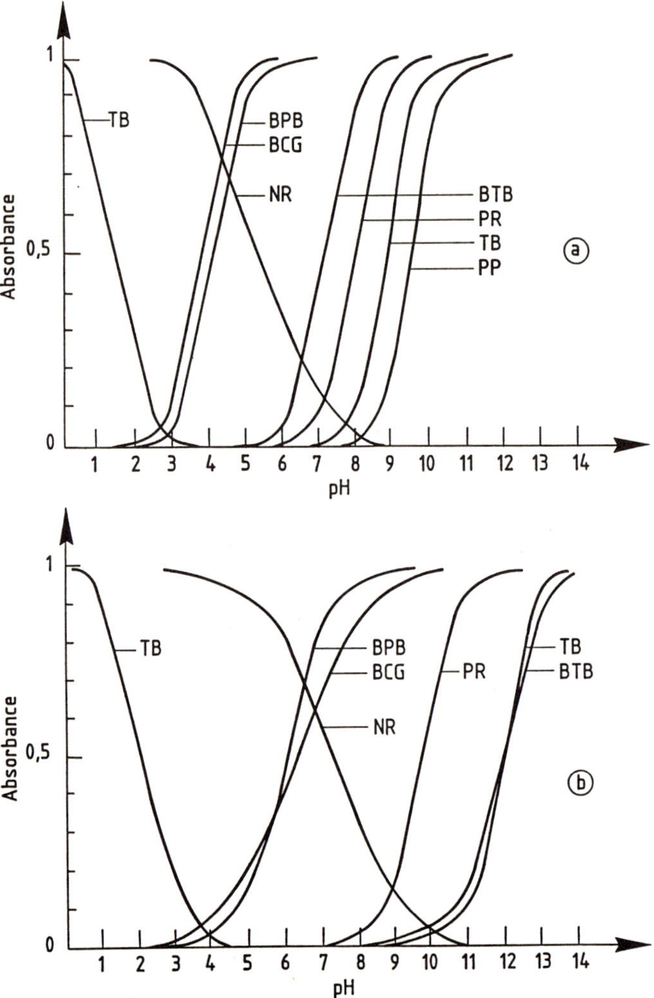

Figure 17-6. Comparison of normalized absorbance for different dyes. TB: Thymol Blue; BPB: Bromophenol Blue; BCG: Bromocresol Green; NR: Neutral Red; BTB: Bromothymol Blue; PR: Phenol Red; PP: phenolphthalein. (a) Passive optodes; (b) optode with Amberlite XAD-4 without membrane (configuration as in Figure 17-3c).

energy transfer are promising techniques. The fluorescence lifetime is a parameter that is still applied to passive optodes only. The drawbacks of luminescence stem from the greater expense of the apparatus. However, products using low-cost sources (LED) and detectors (PIN) are beginning to emerge in fluorimetry [57].

The pH range of the dyes is narrow (3 pH units). The best performance (accuracy < 0.01) is obtained in a narrow zone, generally ± 0.02 units around the pK'_a. The accuracy decreases sharply with increasing distance from the pK'_a (the relative intensity curves $f(pH)$ are nonlinear). In the outer zones (± 1.5 pK'_a units), the accuracy hardly exceeds 0.1 pH units. The use of several pK'_a values with the same indicators requires differentiated spectral responses for the two species present [12, 14, 36]. New dyes are under development [58, 59] to broaden the pH range with a single indicator for several pK'_a values and for wider spectra. Also, the pK'_a is different in solution and for a semi-active optode as presented in Figure 17-6 [35]. A summary of theoretical pK_a and pK'_a values for some optodes is given in Table 17-3.

Many problems still remain to be solved for pH FOCS. Areas still not completely under control are the techniques of immobilization and grafting on the support, permanent monitoring of the concentration of the indicators, ionophore lifetime [60], stability of maintenance of the dye on the support over a wide pH range, determination of the dissociation constant of the immobilization dye [61], optode reversibility, reproducibility of their manufacture (disposal optodes), monitoring of the influence of the medium, and the simplification of calibration procedures. Some progress has been achieved in the understanding of the influence of the ionic strength of the medium.

Table 17-3. pK'_a values for passive (without immobilization of dyes) and semi-active (active-extrinsic) optodes

Dye	Theoretical pK_a		pK'_a measured with passive optode	Semi-active optode		References
	Ref. [8]	Ref. [5], ionic strength zero		pK'_a	Immobilization support	
Thymol Blue	–	1.65	1.65	2	XAD–4	12
Bromophenol Blue	3.7	4.10	4.10	4.6–6.1	XAD–4	12
				4.1	XAD–2	31
Bromocresol Green	–	4.90		6.5	XAD–4	35
Bromocresol Purple	5.8	6.40	–	6.1	XAD–2	31
Bromocresol Blue	7.7	7.3	7.4	11.5	XAD–4	35
			7.1	8.5–9.5	XAD–2	31, 32
				7.5	Cellulose	30
HPTS			7.3 (in solution without fiber)	6.7	Cellulose	48, 49
HCC			7.3	6.9	Cellulose	48, 49
Phenol Red	–	8.0		7.6	Polyacrylamide	10
			7.9	9.6	XAD–4	35
Thymol Blue (basic)	9.6	9.20	9.5	12.1	XAD–4	12
Phenolphthalein	10.2	9.7		9.6	XAD–2	31
Thymolphthalein	11.0	–	–	9.3	XAD–2	31

Influence of Various Parameters

The ionic strength (I) of the medium whose pH is to be measured affects the pK_a' and hence the pH. The Debye–Hückel equation defines the activity coefficient γ of a Z-valent ion by [62]

$$\log \gamma_i = M Z_i^2 I^{1/2}/(1 + B d I^{1/2}) \tag{17-31}$$

whereas Lewis and Randell [see, eg, 63] define ionic strength as

$$I = 0.5 \sum_i C_i Z_i^2 \tag{17-32}$$

where C_i is the concentration and Z_i the charge of species i, M is a constant that varies with temperature, B is the dielectric constant of the solvent, and d is the ion-size parameter. Under these conditions, the variation in pK_a' with I becomes

$$pK_a' = pK_a + M (Z_{I-}^2 - Z_{HI}^2) I^{1/2}/(1 + B d I^{1/2}) . \tag{17-33}$$

This equation, which is valid for both photometry and fluorimetry, shows that the variation in pK_a' with ionic strength (I) is a major drawback for optical sensors. Methods have been developed in luminescence to overcome this drawback.

It is desirable for the chemical reaction at the fiber end not to depend on ionic strength. Offenbacher et al. [48] suggested eliminating the effect of I by creating a clearly defined, highly charged environment of amino groups, protonated at neutral pH. This result is obtained by modification of the surface conditions and silanization of the support glass, and by the choice of a singly charged indicator such as 7-hydroxycoumarin-3-carboxylic acid (HCC). In an electrolyte concentration range between 100 and 200 mM, the remaining effect is 0.003 pH units for a 20% increase in I.

A second method includes the simultaneous measurement of the pH and the ionic strength, in the presence either of two differently charged indicators [46, 64], or of two sensors, one insensitive and the second sensitive to I [48].

In the former system, the pH is obtained empirically from the pH of the two indicators 4-methylumbelliferone (4-MU) (pH_1) and HPTS (pH_2):

$$pH = pH_1 - 0.17 (pH_2 - pH_1) . \tag{17-34}$$

In fact, the influence of ionic strength on pK_a' is different for 4-MU with a single negative charge and the phenolate form of HPTS with a quadruple negative charge.

In the latter system, two sensors are used [48, 49]. One sensor consists of an HCC indicator insensitive to ionic strength (microenvironment simulating a high I), for which each molecule is surrounded by at least two or three charged ammonium groups. A second sensor is sensitive to ionic strength (decrease of 0.09 pK_a' units for a 0.05 M increase in I), for which the uncharged acetylamine groups replace the ammonium groups (Figure 17-7). The drawback with these two sensors usable at different wavelengths (λ_e = 410 nm for HCC and 465 nm for HPTS, λ = 455 nm for HCC and 520 nm for HPTS) lies in the complexity of the associated equipment.

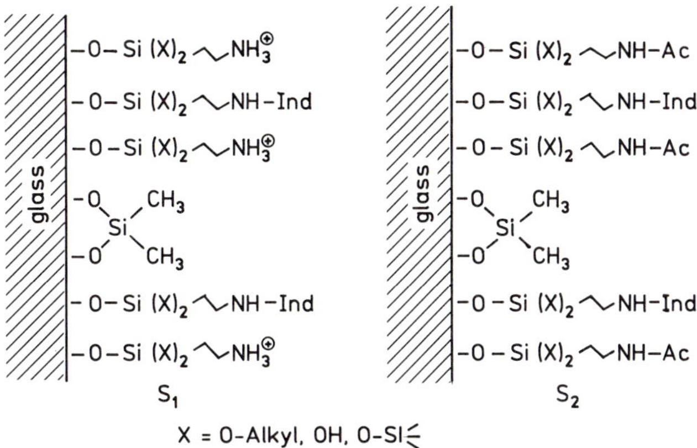

Figure 17-7. Scheme of surface chemistry of two sensors, one insensitive (left) and the other sensitive (right) to ionic strength (from [49]).

The concentration of the dye immobilized on the support induces a variation in pK'_a, because the ionic strength of the adsorbed species is different [14] This stresses the value of a direct measurement of ionic strength.

The "fluorescence energy tranfer" technique was recently developed for the determination of ionic strength [65]. This sensor uses a fluorescein-labelled dextran indicator and a poly(ethyleneimine) labelled with sulforhodamine 101 (Texas Red) confined behind a dialysis membrane. At low ionic strength, the polymers combine, causing efficient energy transfer. At high I, the polymers are dissociated and the energy transfer efficiency decreases. The fluorescence intensity ratios are measured at 520 nm (peak for fluorescein) and 620 nm (Texas Red). The response mechanism ($t = 5$ min) of this sensors is difficult to predict because the complexity of polymer-polymer interactions. The higher fluorescence intensity with I differs slightly depending on the type of ion (chloride, nitrite, fluoride, etc.) suggesting more than a simple ionic strength effect.

As the activity coefficient varies with temperature, this also applies to pK'_a and hence pH. Thus, the temperature coefficient (expressed in Δ pH/K) is -0.017 between 20 and 40 °C for the optode of Peterson et al. [10], and -0.01 to 0.06 between 20 and 50 °C as reported by Kirkbright et al. [32]. For glass electrodes, the change in the indicated pH with change in temperature is $-$pH/T, or -0.023 at pH 7 and 25 °C from the Nernst equation. According to Edmonds et al. [7], a 5 °C increase (between 25 and 30 °C) would raise the pK'_a by about 0.2. The following relationship between pK and absolute temperature (T) has been suggested for measurements in sea water between 5 and 30 °C and for salinities (S) ranging between 20 and 35‰ [66]:

$$pK'_a = a/T + b + c \log T + 4.4 \cdot 10^3 (35.2 - S) \tag{17-35}$$

where a, b and c are the thermodynamic constants (enthalpy, entropy, heat capacity).

Applications

The measurement of pH is more advanced in the biochemical area owing to the narrow pH range required. Biochemical analysis, biotechnology and industrial laboratory analysis, for which short-distance fibers (plastic or glass bundles) are adequate, are tending towards dual- or triple-purpose optodes, including temperature measurement. pH control is also essential for other measurements with which it is associated, such as glucose [67], enzymes [68], and pCO_2 (see Section 17.2). Titrimetry is another application in which optical fibers can replace electrodes. Thus acid-base and redox indicators, chelating reagents and absorption indicators are normally used in absorption and fluorescence [57]. HPTS and fluorescein, highly soluble in aqueous solution, have absorption spectra matching the emission of the blue LED, allowing the titration of strong acids and bases. The flow-injection analysis (FIA) technique associated with optical fibers is used in laboratories to optimize the lifetime of flow-sensing sensors [69], or for the determination of the pK'_a of indicators [70].

Agricultural, food (milk), environmental (sea water, geology), and industrial (process control) applications are merely at their beginning. However, there optodes are subject to interferences and many possible errors owing to the ignorance of the exact concentration of the immobilized dye, the conservation of the state of chemical bonds between the support and the dye, and other factors. Nevertheless, the miniaturization of optodes and the possibility of long-distance measurement are major assets for pH FOCS. The media for which the pH range is narrow (irrespective of the zone concerned) are naturally the first to be investigated, while awaiting better performance with broader pH ranges.

17.1.2   Electrolyte Sensors

17.1.2.1   Theoretical Aspects

With active and semi-active optodes, the chemical species (S) to be measured (analyte) reacts with a reagent phase (R_e). This reaction forms another species (SR_e) according to the reaction

$$S + R_e \underset{}{\overset{K_1}{\rightleftharpoons}} SR_e \tag{17-36}$$

with

$$K_1 [S] [R_e] = [SR_e] \tag{17-37}$$

where the constant K_1 is related to the concentrations of S, R_e and SR_e.

In a reaction of a stoichiometric type (excess of reagent), the free species (S) is completely converted into a combined species (SR_e). The resulting sensor has a limited lifetime and is generally used only once. The reagent phase is not renewed. The indicating reactions for which the reagent (or indicator) has a much lower concentration than the analyte (the concentration of which remains substantially constant) offer the advantage of producing reversible reactions.

In *direct* analysis, only one overall reaction takes place (Equation 17-37). However, this reaction is the resultant of a process of multiple interactions between the species in solution, the reagent support, and the indicator. The general case of absorption and the possibility of reaction in all the phases present are shown in Figure 17-8.

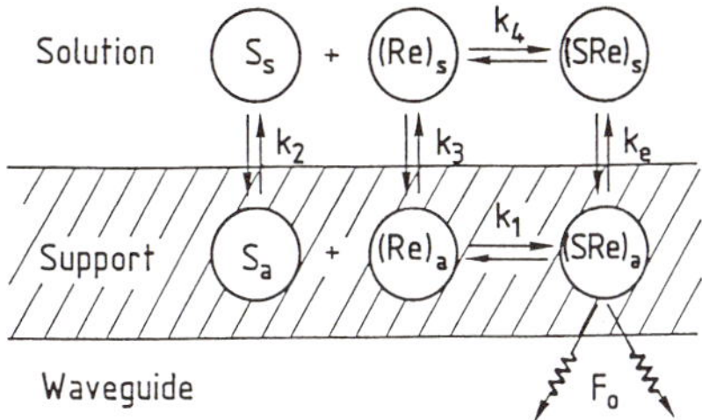

(a) Direct reaction

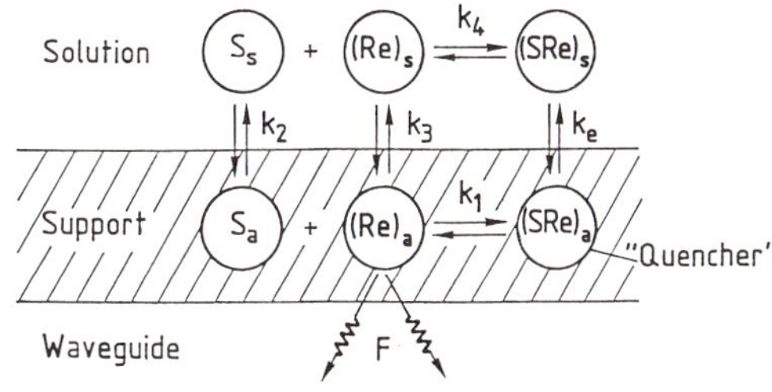

(b) Reaction with quenching

Figure 17-8.
Model of the direct analysis of multiple interactions between the species (S) and the reagent (R_e) in solution and in the support. (a) Direct reaction; (b) reaction with quenching.

The chemical equilibra corresponding to monomolecular species and in the phases s (in solution) and a (adsorbed on the support) are thus:

$$S_s + [(R_e)_s] \xrightleftharpoons{K_4} [(SR_e)_s] \ . \tag{17-38}$$

The constant K_4 corresponds to the pK' of the solutions:

$$[(R_e)_s] \xrightleftharpoons{K_3} [(R_e)_a] \tag{17-39}$$

where K_3 is higher if the reagent is strongly bonded to the support

$$[S_s] \xrightleftharpoons{K_2} [S_a] \ . \tag{17-40}$$

Equilibrium (17-40) represents the mass transfer of the analyte to the support.

$$[S_a] + [(R_e)_a] \xrightleftharpoons{K_1} [(SR_e)_a] \ . \tag{17-41}$$

Reaction (17-41) is the one actually "observed" through the fiber. $[S_s]$ is generally determined by measuring the variation in $[R_e]_a$ or $[(SR_e)_a]$, which are related to the optical value measured. The value initially known (calibration) is the total concentration of reagent (C_{R_e}), where

$$C_{R_e} = [(R_e)_s] + [(SR_e)_s] + [(R_e)_a] + [(SR_e)_a] \ . \tag{17-42}$$

Equilibria (17-38)–(17-41) and Equation (17-42) can be used to demonstrate the general equation

$$\frac{C_{R_e}}{[(R_e)_a]} = 1 + \frac{1}{K_3} + [S_s]\left(K_1 + K_2 + \frac{K^4}{K_3}\right).$$

(17-43)

For very large K_3 this equation assumes a simplified form of the type

$$\frac{C_{R_e}}{[R_e]} = (1 + [S]\,K_1)$$

(17-44)

obtained from Equilibrium (17-36) .

In the *indirect* method, the indicator includes two or more components whose interaction varies with the analyte concentration [71]. This method, called "competitive binding", can necessitate a particular structure of the optode with several compartments, as described by Schultz et al. for glucose [72]. One simplified example for monomolecular systems, for which the interface is not taken into account, is given.

For a two-component system, species (S) (analyte) and a ligand (L), the equilibrium for [S] is given by (17-36) and for the ligand by

$$[L] + [R_e] \xrightleftharpoons{K_L} [L\,R_e]$$

(17-45)

with

$$[L]\,[R_e]\,K_L = [L\,R_e]$$

(17-46)

and total concentrations C_{R_e} and C_L as

$$C_{R_e} = [R_e] + [S\,R_e] + [L\,R_e]$$

(17-47)

and

$$C_L = [L] + [L\,R_e] .$$

(17-48)

Assuming $[(L\,R_e)] = x$, the concentration of (S) is given by the equation

$$[S] = (1/K_1)\{(C_{R_e} - x)(C_L - x)\,K_L/x - 1\} .$$

(17-49)

The resolution of equations for a three-component system as described by Zhujun et al. [73] for the sodium ion is carried out on the same model. However, the determination of the constants (and hence calibration) is increasingly difficult and subject to errors with increase in the complexity of the corresponding equations.

The best application for the optode thus consists in finding specific chemical reactions with the minimum of components, preferably monomolecular. Nevertheless, the importance of quenching in fluorescence favors systems with two or three components, of which one is the fluorophore, the second the quencher, and, in certain cases, the third is a fluorescence "reverser".

17.1.2.2 Overview of Sensors

Cations

The main research projects for determining cations are on semi-active optodes. The measurement of trace species (Pu^{III}, UO_2^{2+}, rare earths) in the environment is dealt with in Chapter 18 and optodes for continuous monitoring (Cu^{II}), U and Pu valencies) in Section 17.3.

A potassium-sensitive optode uses a crown ether labeled with an azo dye and immobilized on Amberlite by encapsulating the tip by a porous PTFE membrane. The absorption is modified by chelation with potassium ions [74]. All others cation measurements use the fluorescence technique.

A potassium opto-sensor was recently described [75] for the continuous determination of electrolytes. Certain fluorescent dyes respond to an electric potential at the interface between the aqueous and lipid phases. This potential is created by the neutral ion carrier. The lipid layer is formed on a glass support by the Langmuir–Blodgett thin-film technique. This layer incorporates Rhodamine B as a dye and valinomycin as the carrier. The lipid membrane is made of arachidic acid. The fluorescence intensity decreases when this layer is exposed to potassium ions (linearity between 0.01 and 10 mM). This optode is also sensitive to sodium ions [76]. The selectivity factor of potassium in comparison with sodium ions varies from $10^{2.5}$ to 10^5, and in relation to ammonium ions by 10^2. Interferences can be compensated for by a reference optode. However, better selectivity is obtained with new lipid membrane compositions (octadecan-1-ol-valinomycin) [77]. Tetralayers (Figure 17-9) give a maximum response for K^+. The K^+/Na^+ selectivity is about 10^4 in a wide range (0.01–100 mM).

Sodium measurements use three components for the reagent phase [73]: an ionophore selective for Na^+ immobilized on silica, an anionic fluorophore (F), the ammonium salt of 8-anilino-1-naphthalenesulfonic acid and a cationic polyelectrolyte (P), copper(II)−poly-

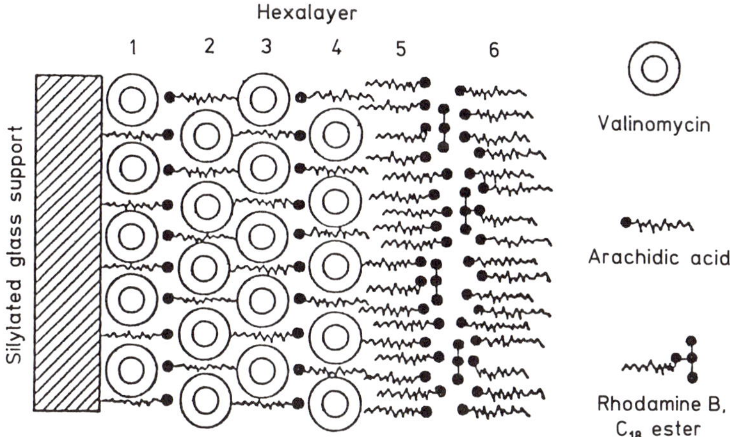

Figure 17-9. Schematic cross-section of a composite hexalayer membrane with an arachidic acid and a valinomycin tetralayer covered with an arachidic acid and Rhodamine B dye bilayer (from [77]).

(ethylenediamine). This substance quenches the fluorescence of the fluorophore (F) during the formation of a complex (PF). All the reagents are confined in a dialysis membrane permeable to sodium ions but not to the polyelectrolyte P. The addition of Na^+ causes an increase in fluorescence by decreasing the quenching effect due to PF. The formation of ion pairs is selective and reversible and varies with the Na^+ concentration in the range 20–200 mM. The response time of the optode is 1–3 min (90% of the signal).

A broadly applicable sensing scheme has been discovered by Simon's group [77b]. It makes use of neutral ion carriers, that is, substances capable of specifically recognizing ions such as K^+, NH_4^+, or Ca^{2+}. When the carrier transports such an ion into a PVC membrane, a proton is simultaneously released from a protonated dye in order to maintain electro-neutrality. The release of the proton from the dye results in a change in color which is monitored.

It is easier to complex heavier cations than cations of alkali metals. Thus the formation of a fluorescent complex helped to determine $Al^{(III)}$ [78]. Morin, which is very weakly fluorescent, can be immobilized on cellulose and then forms a highly fluorescent complex with $Al^{(III)}$. This property helped to obtain a detection limit of 10^{-6} M for a response time of 1–2 min. This response is linear between 10^{-6} and 10^{-4} M at pH 4.8. The major interferents are $Fe^{(III)}$ and $Cu^{(II)}$, which quench the fluorescence, and $Be^{(II)}$, which increases it. This enhancement was exploited to make a $Be^{(II)}$ sensor [79]. The fluorescence intensity is higher at pH 5.3. The response is linear between 10^{-6} and 10^{-5} M. The interference of $Al^{(III)}$ can be eliminated by chelation with EDTA or using the excitation spectra, which are different for $Al^{(III)}$ and $Be^{(II)}$.

8-Hydroxyquinoline sulfonate (8-HQS) immobilized on Amberlite ion-exchange resin was tested for measuring $Al^{(III)}$, $Zn^{(II)}$ and $Cd^{(II)}$ [80]. The non-fluorescence ligand begins to fluoresce after chelation. The detection limit is about $5 \cdot 10^{-7}$ M. The response time of the optode is 1.5–2.5 min. This response is a non-linear function of the ion concentration.

Many interferences occur, including those of transition metals (quenching) and of other ions sensitive to 8-HQS and EDTA. The fluorescence polarization of the ligand as a function of chelation has also been exploited for a selective sensor based quinolin-8-ol to determine $Al^{(III)}$, $Mg^{(II)}$, $Zn^{(II)}$, $Be^{(II)}$, $Ca^{(II)}$, $Cd^{(II)}$ and $Pd^{(II)}$ [81].

Optodes for determining UO_2^{2+}, whose fluorescence is enhanced in a liquid phosphate reservoir, and for $Fe^{(III)}$ (which quenches this fluorescence), have also been proposed [82] (see Chapter 18).

Another sensor, with calcein immobilized on cellulose, has been designed [79]. It is based on the observation that $Co^{(II)}$, $Cu^{(II)}$ and $Ni^{(II)}$ form nonfluorescent complexes at neutral pH (5–7), whereas calcein fluoresces strongly by itself. However, this sensor is not reversible in the pH zones analyzed. The addition of a metallic ion (non-quenching) such as $Zn^{(II)}$, which displaces the metal from the fluorescent complex, has also been proposed, and also the use of immobilized calcein as a complexometric titration indicator [83].

A recent sensor [84] uses Rhodamine GG electrostatically trapped on Nafion film. The quenching or enhancement of the fluorescence of Rhodamine is measured in the presence of various ions that are either quenchers such as $Co^{(II)}$, $Ce^{(III)}$, $Fe^{(II)}$, $Fe^{(III)}$, $Cu^{(II)}$, $Ni^{(II)}$ and NH_4^+ or "reversers" (reversal of quenching) such as H^+, Li^+, Na^+, K^+, $Ba^{(II)}$, $Ca^{(II)}$, $Mn^{(II)}$, $Zn^{(II)}$ and $Mg^{(II)}$. The detection limit is about 0,7–2.1 µM depending on the ion. The sensor has the drawback of being non-selective.

A fiber-optic sensor for calcium ions has also been proposed recently [85]. It is based on the fluorescent chelate of chlortetracycline (CTC) formed in aqueous solution at pH 7.5. The

CTC is immobilized on an anion-exchange membrane. The sensor responds to calcium ions in the range 0–400 mM. The response is reversible but other divalent cations, such as $Mg^{(II)}$, $Zn^{(II)}$ and $Sr^{(II)}$, interfere.

Anions

The chloride optode has been tested with silver fluorescinate (non-fluorescent) immobilized on collodion as a reagent. The chloride (and other halides) increases the fluorescence by the formation of AgCl [39]. In principle, this optode is an irreversible probe (see Section 18.1). A true sensor designed for continuous monitoring of halides is based on the dynamic fluorescence quenching of acridinium and quinolinium indicators immobilized on a modified glass surface [86]. This immobilization is performed preferably on a cation-exchange membrane. The emission due to the quaternized heterocyclic fluorophore is quenched by the Cl^-, Br^- and I^- ions. The sensitivity increases in the order $Cl < Br < I$. The detection limits (range 10–100 mM) are 0.15 mM for I, 0.40 mM for Br, and 10.0 mM for Cl. The response time is about 40 s (95% signal). This response is reversible but not selective (interference from sulfites and cyano derivatives, but not from sulfates, phosphates, or nitrates).

Three methods have been described for three halogens, two based on fluorescence and one on absorption. In the first [87], the fluorescence of rubrene in polystyrene is quenched by traces of iodine. This method is nonselective and the optode is also sensitive to oxygen. In another sensor, naphthoflavone in solution in a material of the silicone or PVC type serves as a sensitive layer for free halides [88]. The absorption technique uses a fiber with a liquid CS_2 core [89] to detect 10 ng of iodide using a 5 m long capillary cell with sample circulation. The fiber itself constitutes the active optode (total reflection in the liquid core). A comparison of optodes based on dynamic quenching of absorbed Rhodamine 6 G by iodide was reported [90]. Three solid supports for immobilization were used: PTFE tape, XAD resin beads and crushed XAD-4 resin. The limits of detection are 0.18–0.30 and 1.1 mM respectively. Some anions (eg, Cl^-, Br^-, CN^-) interfere at the 1-M level.

Recent investigations have been devoted to the measurement of fluorides [91]. A ternary complex colored by fluoride ion is formed from a cerium(III)–alizarin complexone binary complex (alizarin fluorine blue). The reagent is immobilized on an Amberlite polymer matrix. The reflectance is measured in a flow-cell assembly with a bifurcated optical fiber. The response is linear for 0.16–0.95 mM fluoride at pH 4.1. The response time is about 12 min. The major interferents are aluminium, iron, and phosphate.

Cyanide ion was determined [92] by the color reaction with a poly(vinyl alcohol) solution of sodium picrate deposited on the fiber surface. The measurement of the absorbance variation (due to evanescent waves) is proportional to CN^- concentration, but the probe is irreversible. A sulfate probe with non-fluorescent barium chloroanilate, which becomes fluorescent under the action of SO_4^{2-} (precipitation of sulfate), and a nitrate probe based on the quenching of the fluorescence of an aromatic amine by nitrite ions (reduction of NO_3^- by hydrazine sulfate) have also been proposed [93].

Two methods have been investigated to determine sulfides in solution (for H_2S see Section 17.2). The first uses 2,6-dichlorophenolindophenol (DIP) as a reagent immobilized on XAD-2, XAD-4, or XAD-7 resin [94, 95]. The reflectance is measured in a flow cell in the 2–20 mM range. The response time (2–25 min) varies with the sulfide concentration. The phenanthroline and dithiofluorescein complexes have also been tested. The reducing ions produce interference. The consumed reagent phase must be regenerated by oxidation. The second method uses *N,N*-dimethyl-*p*-phenylenediamine immobilized on Dowex 50-X8 resin, the final product being methylene blue [96]. The probe is based on reflectance changes in the presence of sulfide ions and acts irreversibly. The response time is not constant. The interfering ions are mainly thiosulfate, sulfites, and thiols. The concentration range measured is about 0.06–4 ppm.

Titration

FOCS are finding increasing applications in analytical laboratories for titration and the measurement of redox potential. Acid-base titration [57, 69, 70] (see Section 17.1.1.3) with glass fibers (bifurcated) is developing in fluorescence as in absorption with inexpensive solid-state components (LEDs and photodiodes) [57], [97].

Argentometric and complexometric titration methods have been investigated recently. The determination of the end-point by argentometry uses fluorescence indicators such as 6-methoxyquinoline (or acridine) that are sensitive (by quenching) to halides in the presence of silver(I) [98]. The fluorescence intensity is governed by the Stern–Volmer equation (see Section 12.3) as a function of the halogen concentration. Also, the titration of $Al^{(III)}$ at pH 4.6 (1–100 ppm range) [99] is carried out by monitoring the fluorescence of the formation of the aluminium–morin complex in the presence of the chelating agent DCTA. The titration time is about 20 min. The study of chelating reactions at the ends of optodes in the presence of DCTA or EDTA is beginning to find new applications, using either photometry under extreme conditions [100] or fluorescence [101].

A new method for the determination sulfides is based on titration by precipitation and fluorimetric detection of the end-point [102]. The indicator is a dye derived from acridine, the fluorescence of which is statically quenched by sulfide. The concentrations measured range between 1 and 10 mM by titration with silver nitrate.

17.2   Gas Sensors

The majority of methods for the continuous optical monitoring of gases can be divided into two groups. In the first, the intrinsic optical property of the gas is exploited to sense it. Typical examples include chlorine, methane, carbon monoxide, and nitrogen oxides. In the second method, an indicator is used to transduce the gas concentration into a measurable optical parameter. This approach has frequently been applied when the gas has no useful intrinsic optical property or when it is dissolved in water. Typical examples include oxygen, carbon dioxide, and sulfur dioxide.

Typical experimental arrangements for performing remote fiber absorptiometry with plain fibers are shown in Figure 17-10. In fact, the method is an extension of classical absorbance measurements to fiber technology. In addition to the flow-through cell methods shown, evanescent wave approaches have also been applied successfully. When an indicator-mediated method is applied, the fiber end or the waveguide surface is equipped with a so-called working chemistry. This, in fact, is the immobilized reagent that responds to the analyte by a color change. Such a typical working chemistry is shown in Figure 17-11. Other sensing schemes such as the interferometric determination of hydrogen (see Section 17.2.3.4) form a minor group in chemical sensing.

This section presents a representative (although not complete) compilation of existing sensing schemes for the most important gaseous analytes. Rather than discussing the existing sensors according to the methods applied for their detection and determination, this section is organized such that all methods known for a given species are listed under the respective analyte. This should facilitate the rapid search for the most appropriate method for a given species.

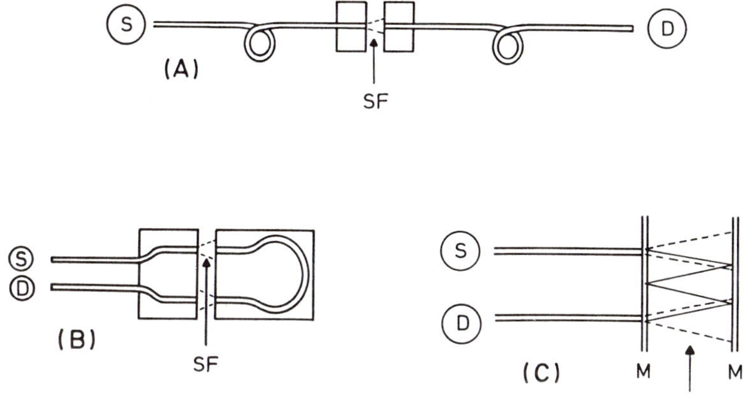

Figure 17-10. Typical kinds of flow-through cells for remote fiber absorptiometry. S: light source; SF: sample flow; D: detector; M: mirror. The path length increases on going from A to C, while the light losses increase owing to incomplete coupling into the fibers.

Figure 17-11
Fiber-optic chemical sensor with an indicator chemistry at the distal end. The overcoat provides some mechanical stability and, when dyed black, can also serve as an optical isolation to prevent ambient light from entering the fiber.

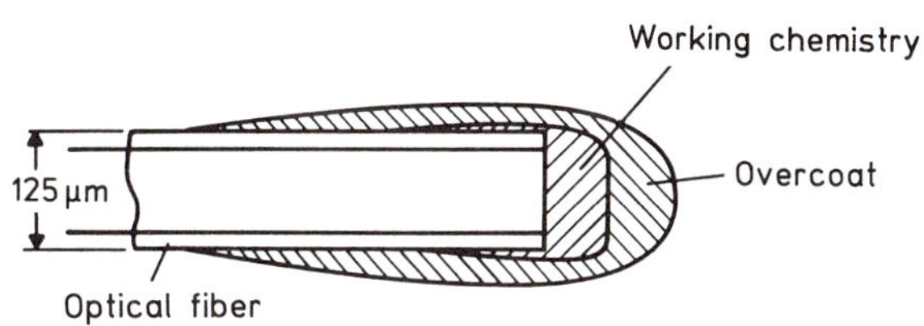

17.2.1 Oxygen Sensors

The greatest efforts in optical gas sensing have been devoted to oxygen owing to its unique importance in various fields. With a few exceptions, all optical oxygen sensors are based on dynamic quenching of luminescence. The major advantages of optical oxygen sensors over electrochemical sensors are the small size, lack of oxygen consumption, and inertness against sample flow rates, stirring, or bubbles. Also, the precision of oxygen sensors at oxygen pressures above 26 kPa is distinctly better than that of electrochemical sensors (eg, the Clark electrode).

17.2.1.1 Phosphorescence-Based Sensors

The first optical sensor used for determination of oxygen was described in the 1930s, when it was observed [103] that the room temperature phophorescence of certain dyes adsorbed on silica gel was efficiently quenched by molecular oxygen with the formation of singlet oxygen.

The effect was used to detect traces of oxygen [103, 104] produced in the photosynthetic process. Trypaflavine, benzoflavine, Euchrysine 3R, Rheonine 3A, Rhoduline Yellow, safranine, chlorophyll, and hematoporphyrin adsorbed on silica gel or aluminium oxide were the dyes that were most efficiently quenched. Unfortunately, most dyes are photolabile, and traces of water or ammonia strongly interfere. Hydrogen, nitrogen, methane, ethylene, and carbon dioxide are without influence. The method is suitable for the extremely sensitive determination of oxygen at partial pressures between 0.5 and 0.006 kPa and is characterized by a fast response time, but requires several seconds for regeneration.

Low oxygen concentrations are fequently encountered in waters heavily loaded with biodegradable waste. Zakharov and Grishaeva [105] utilized the phosphorescence quenching effect to devise an optosensor for very low oxygen levels in water. Trypaflavine (acriflavine), adsorbed on silica gel, cellulose, or ion exchangers, was immersed in the sample and the phosphorescence monitored. Although the sensor layer has not been combined with a fiber optic, it may easily be so. Obviously, the interfering effect of humidity observed by others plays no role in an underwater instrument. Quenching does usually not obey the Stern–Volmer relationship, except within a small oxygen concentration range. Using silica gel as a support, the device is able to determine oxygen in the range between 0.06 and 1.00 μg L^{-1}.

Both sensitivity and selectivity can be improved by using a hydrophobic silica gel with an average pore radius of 20–60 nm as a solid support [105]. Silica gel can be made hydrophobic by treatment with derivatization reagents such as dichlorodimethylsilane. The phosphorescence is sensitive to oxygen concentrations below 0.05 μg L^{-1}, but a large persistence was observed. A stationary value after a change in oxygen concentration was sometimes established only after 3–5 min. Stern–Volmer plots showed a distinct positive deviation from linearity at oxygen levels above 0.01 μg L^{-1}. The shape of the plot seems to depend on excitation intensity. The effect was traced back to photoabsorption of oxygen by adsorbates.

17.2.1.2 *Chemiluminescence-Based Sensors*

Freeman and Seitz [106] reported on a chemiluminescence oxygen sensor, consisting of a reagent chamber at the fiber end containing a solution of an basic amine that reacts with oxygen to yield chemiluminescence. The detection limit is estimated to be 1 mg L^{-1}. A thermoluminescence-based oxygen sensor described by Hendricks [107] is also worth mentioning. Almost all other optical sensors for oxygen described so far are based on dynamic fluorescence quenching of a suitable indicator. Unlike most other molecules, oxygen has a triplet multiplicity in its electronic ground state, resulting in more or less expressed quenching of all fluorophores.

17.2.1.3   Fluorescence-Based Sensors

Dynamic quenching of the fluorescence of polycyclic aromatic hydrocarbons has been known for a long time, but the effect was not utilized for optical sensors until 1968 when Bergman described the first oxygen fluorosensor [108], albeit without the use of fibers (see Figure 17-11). The oxygen-sensitive dye fluoranthene in a porous glass disk is excited by a UV lamp, and the intensity of the resulting emission is measured in a photocell. This intensity varies according to the oxygen tension. Vycor glass-adsorbed indicator is favored over an indicator solution in a 25-50 μm thick polyethylene or silicone film because of the lower quenching efficiency of oxygen for dissolved dyes. Thus, 63.5% of the fluorescence is quenched when fluoranthene is adsorbed on Vycor glass, whereas only 9.9% is quenched when the dye is dissolved in polyethylene.

Among the polycyclic aromatic hydrocarbons, pyrene and, less so, pyrenebutyric acid (PBA) are probably most efficiently quenched by molecular oxygen by virtue of their long fluorescence lifetimes. PBA in dimethylformamide or silicone-rubber solution has been applied to measure oxygen [109–111] in an arrangement shown in Figure 17-11. The indicator has an excited-state lifetime of more than 100 ns and is therefore subject to strong quenching by oxygen. the Nernst distribution favors the presence of PBA in the polymer phase rather than in the aqueous sample phase, but it was noted that it is slowly washed out, resulting in signal drift of the sensor.

This problem was overcome by immobilizing PBA on controlled-pore glass that had previously been sintered onto a glass slide to give a mechanically stable oxygen sensor having a very fast response of less than 20 ms for the 90% value [112]. Since straylight could not be separated quantitatively from fluorescence, its contribution (I_s) had to be taken into account by calculating the oxygen partial pressure p_{O_2} according to

$$(I_0 - I_s)/(I - I_s) - 1 = K_d p_{O_2} \, . \qquad (17\text{-}50)$$

The sensor is highly specific for oxygen. No interferences were observed with nitrogen, carbon monoxide, dinitrogen oxide, methane, carbon dioxide, and noble gases. Covering the sensitive layer with silicone-rubber resulted in a 60% reduction of the quenching constant.

Notwithstanding the stability and long decay time of pyrene and PBA, its short-wavelength excitation and emission maxima are disadvantageous with respect to suitable light sources, fiber material, and background from biological matter and fibers. It therefore became desirable to look for suitable indicators possessing more ideal spectral properties. Such an indicator was found in benzo[*ghi*]perylene, and a steam-sterilizable oxygen sensor was developed for use in bioreactors which utilizes the effect of oxygen quenching of this dye [113]. This device, which apparently is the first fiberoptic oxygen sensor, consists of a rigid light guide with an oxygen-sensing layer attached to its end, a filter, and a photodetector at the tip (Figure 17-12). The sensor layer is covered with black gas-permeable material to avoid interferences from fluorescence from the sample. The rod, when immersed in a bioreactor, gives a response to oxygen that compares favorably with respect to response time (9–65 s), drift (−0.01 to −0.09% signal loss per hour), and effects of stirring (not observed).

After screening approximately 70 dyes, Peterson et al. [114] found perylene dibutyrate (Colour Index No. 59 075) adsorbed on Amberlite XAD-4 resin beads to be a most useful probe. It has excitation and emission maxima of 468 and 514 nm, respectively, is stable, and is effi-

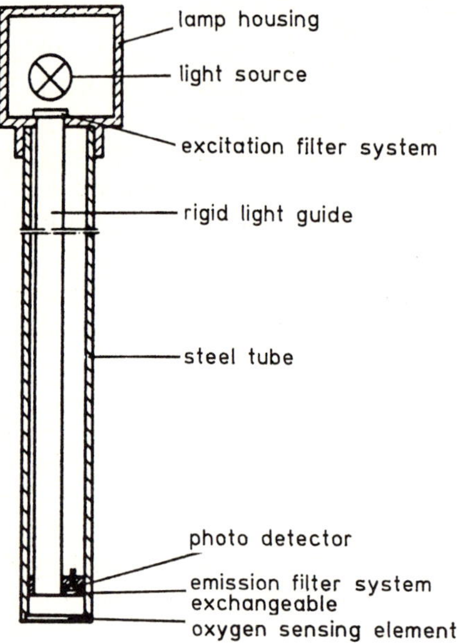

Figure 17-12.
First fiber-optic oxygen sensor [113]. The fiber is contained in a steel tube that can be flanged to a hole in a bioreactor.

ciently quenched by oxygen, thereby allowing a resolution of ± 100 Pa for oxygen partial pressures up to 20 kPa. The resulting optode, shown schematically in Figure 17-13, consists of two small fibers for guiding the exciting and fluorescing light, and fine particles bearing the fluorescent dye in a tube at the common end. To protect it from contamination, it is covered with porous polyethylene of 25 μm thickness. The response time is of the order of 1 min.

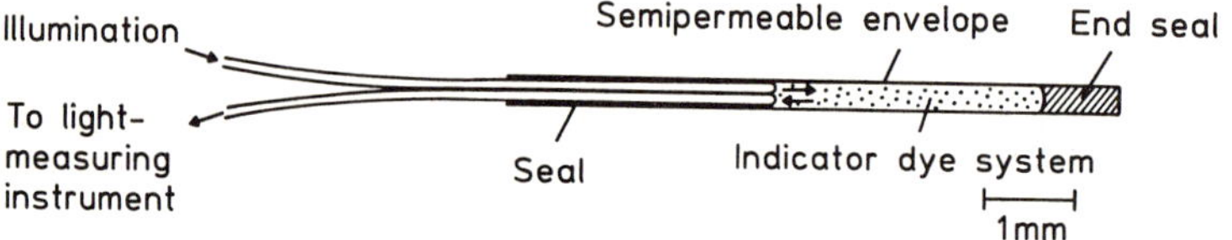

Figure 17-13. Fiber-optic invasive catheter for measuring oxygen in vivo [114].

The sensor measures the ratio of scattered blue excitation light (I_0) and green fluorescence (I). An electronic circuit processes the blue and green signals in accord with the following relationship:

$$p_{O_2} = (\text{gain}) \cdot (I_0/I - 1)^m \tag{17-51}$$

which is the Stern–Volmer equation rearranged with an exponent m added for curvature since Stern–Volmer plots are not linear. This ratio method can eliminate temperature effects and, partially, drifts resulting from photobleaching.

Whereas hydrophilic quenchers are unlikely to interfere because of an impermeable polymer envelope, the sensor is affected by hydrophobic quenchers. Thus, nitrous oxide has 2% of the effect of an equal partial pressure of oxygen. A more serious problem is caused by bromine-containing anesthetics and narcotics, which interfere strongly. The interference is cumulative, related to both concentration and time of exposure, and changes both the sensitivity and zero adjustment of the sensor. If exposure is not too severe, the effect is reversible.

Cox and Dunn [115, 116] investigated the oxygen quenching of five polycyclic aromatic hydrocarbons (PAHs), including 9,10-diphenylanthracene, decacyclene, and rubrene, in a silicone-rubber matrix. This material appears to be most ideal for oxygen-sensing layers owing to its excellent solubilizing properties for oxygen. Excitation and emission spectra in addition to fluorescence intensities as a function of fluorophore concentration in silicones of different viscosities were reported. It was found that the increase in fluorescence with increasing viscosity reaches a limiting value when the viscosity exceeds 10 P. The great influence of the polymer properties on the sensor performance was also reported.

Diphenylanthracene was considered to be a very useful indicator because it is highly fluorescent in viscous solvents, stable, displays a good Stokes shift, and presents no health hazard. Its relative fluorescence intensity in poly(dimethylsiloxane) and silicone fluids as a function of temperature and oxygen partial pressure above the membrane is plotted in Figure 17-14. Most noteworthy, the quenching process becomes more efficient with decreasing temperature, which is in contrast to the behavior in fluid solution or in the gas phase. This is possibly due to the strong increase in the oxygen solubility on going to lower temperatures.

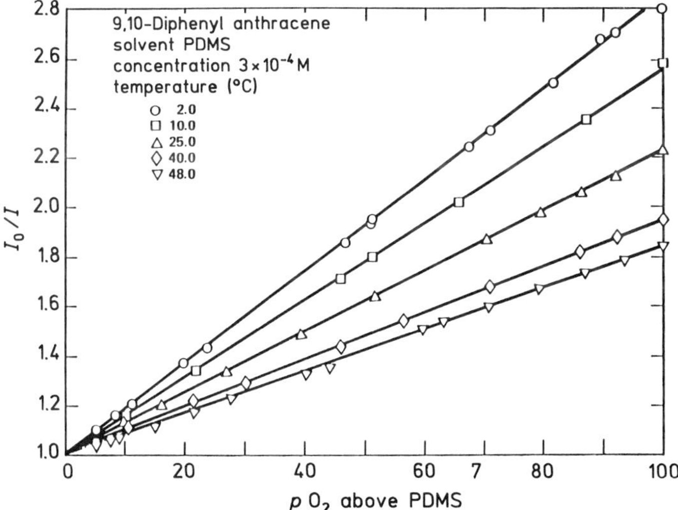

Figure 17-14. Effect of temperature on the slope of the Stern–Volmer quenching curve of diphenylanthracene in silicone-rubber solution [115].

The almost linear relationship between I_0/I and oxygen pressure indicates that, when the fluorophore is entangled or trapped in silicone, the only bimolecular deactivation process is oxygen quenching. From the temperature dependence of quenching, the solution enthalpy δH was calculated to be -3.0 kcal mol^{-1}.

Silicone-rubber is certainly a very suitable material for use in oxygen sensors because of its high solubility and permeability for this gas. All other polymers seem to have less favorable properties. However, a mixture of two different polymers does not necessarily display the properties that would be expected from the properties of the individual components. It was found, for instance, that the oxygen-quenching efficiency of PBA dissolved in mixtures of silicone and poly(vinyl chloride) (PVC) decreases almost linearly as the percentage of PVC increases [117]. If, however, the proportion of PVC exceeds 25%, the quenching efficiency becomes very high again, being almost the same at 35% as at 5% PVC. The only disadvantage of silicone-rubber is due to its excellent dissolution power for oxygen: in case of minute sample volumes, the sensor membrane causes a depletion of oxygen in the aqueous sample. In such cases, PVC is the preferred membrane material, but requires an indicator with a rather high quenching constant.

A fiber sensor for the determination of oxygen, or halothane, or both, was developed [118] on the basis of the observations that both analytes act as strong quenchers and that PTFE acts as a barrier to halothane. It consists of two sensing layers attached to the ends of two fibers. One sensor layer is covered with PTFE, giving a signal α (defined as $^a I_0 / {}^a I - 1$). The second is uncovered and gives a signal β (defined as $^b I_0 / {}^b I - 1$). Concentrations of oxygen and halothane can be calculated from α and β with the help of two simple equations.

This two-sensor technique allows the determination of oxygen in the 0–25 kPa range with a precision of ± 100–200 Pa even in the presence of halothane. Outside this range it is poorer, but still better than that of a Clark electrode. The response time is 10–15 s for 90% of the total signal change. It is useful for in vivo catheters, for continuous sensing in blood vessels, for in vitro determination of oxygen in samples containing halothane, and for breath gas analysis during inhalation narcosis.

17.2.1.4   Lifetime-Based Sensors

A fiber-optic oxygen sensor with the fluorescence decay time (rather than its intensity) as the information carrier has been described by two groups [119, 120]. In the former work, a ruthenium complex is immobilized in silicone-rubber, and quenching by oxygen is measured by either lifetime or intensity measurements. The 337-nm line of a nitrogen laser served as the excitation line, and the dye was dissolved in a silicone-rubber membrane placed in the fluorimeter. This sensing membrane is reported to be highly specific, and chlorine and sulfur dioxide were the only interferents.

In the other approach [120], a high-frequency modulated blue LED acts as a light source, and the phase shift between excitation and long-lived fluorescence (ca. 200 ns) is measured as a function of oxygen partial pressure. The detection limit is ≈ 300 Pa of oxygen. Figure 17-15 shows a schematic diagram of the experimental arrangement for performing fiber-phase fluorimetry, and Figure 17-16 compares the Stern–Volmer plots obtained by intensity and lifetime measurements.

The advantages of lifetime-based sensors are an internal referencing system which compensates for drifts resulting from bleaching and leaching, and the lack of adverse effects of light-source intensity and photodetector fluctuations, since the lifetime is independent of indicator concentration and light intensity. They are therefore likely to exhibit considerable long-term stability. Fibers with lengths up to of 20 m do not measurably distort the lifetimes when they are of the order of 1 µs or less.

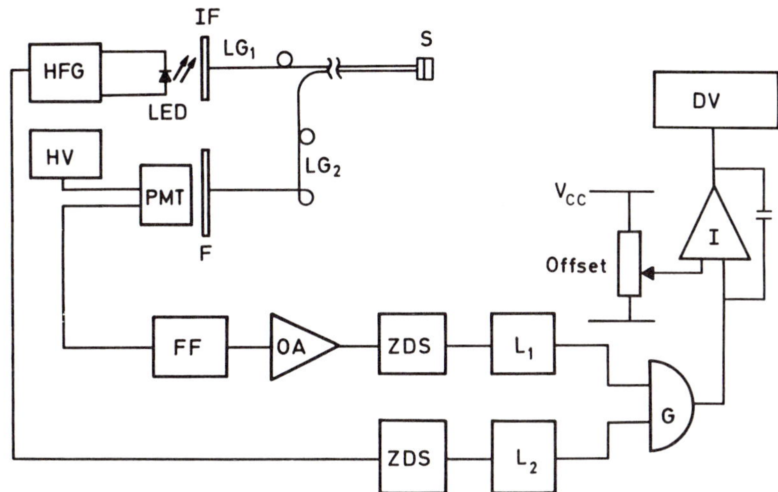

Figure 17-15. Block diagram for the oxygen sensor based on lifetime measurements [120]. HFG: high-frequency generator (455 kHz); LED: blue light-emitting diode; IF: interference filter; LG: light guide; S: sensor; F: cut-off filter; PMT: photomultiplier tube; HV: high-voltage power supply; FF: frequency filter; OA: operational amplifier; ZDS: zero detection switch; L: latch; G: AND gate; I: integrator; DV: digital voltmeter.

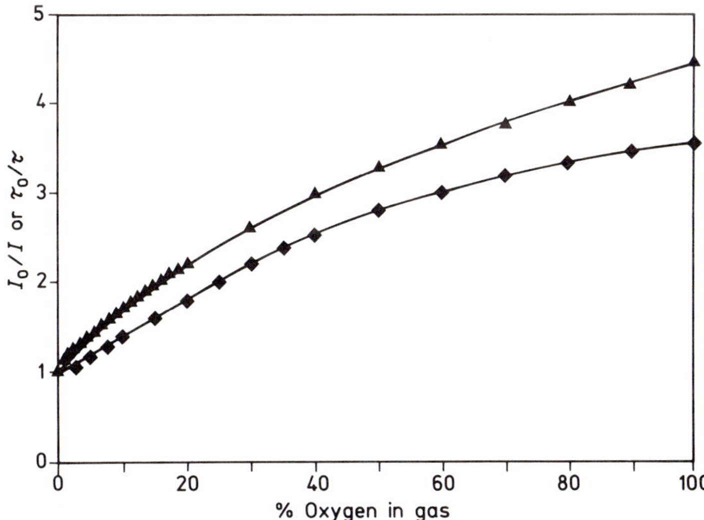

Figure 17-16. Stern–Volmer plots of the quenching of a silicone-entrapped ruthenium complex by oxygen. Upper curve based on intensity I_0/I easurements, lower curve based on lifetimes τ_0/τ [120]. Note the distinctly better linearity of the lifetime-based sensor at oxygen levels up to 30%.

17.2.1.5   Other Sensing Schemes

Various other technical modifications of sensor components have been proposed. Apart from conventional excitation, it has been suggested that a radioactive material could be used as an excitation source [121] or to produce electroluminescence [122]. Kautsky's method can be improved upon by covering the support with a thin layer of a water-repellant polymer [123]. Poor quenching of dyes in rigid polymer solution can be improved by addition of plasticizers [124], and the effect of photobleaching can be compensated for by making use of a reference indicator [18]. When two sensing layers are placed beneath each other at the fiber end, a sensor for oxygen and carbon dioxide is obtained [52].

17.2.1.6   Biomedical Applications

Oxygen sensors have been used in various kinds of biomedical instrumentation. Burkhard and Barnikol [123] utilized the quenching of polystyrene-adsorbed dyes to monitor oxygen in respiratory gases. Some of the oxygen sensors have been shown to exhibit excellent in vivo performance. Thus, an in vivo evaluation of a fiber-optic oxygen sensor [113] was performed in the blood stream of a 31-kg ewe after the sensor had been inserted through a PTFE catheter. Figure 17-17 shows the continuous oxygen record of the fiber sensor during a 3-h experiment in comparison with the blood values.

A fluorosensor for monitoring blood gases and pH in an extracorporal loop is commercially available [125]. Arterial or venous oxygen and carbon dioxide pressure, pH, and temperature can be determined continuously during cardiopulmonary bypass surgery. The system consists of a microprocessor-based instrument, bifurcated fiber-optic cables, and a disposable sensor head with fluorescent spots sensitive to the respective analytes.

The most advanced technology at present appears to be a small invasive catheter of 1 mm thickness suitable for on-line in vivo measurement of pH, pO_2, and pCO_2 along with temperature [50]. The device consists of a sensing rod 1 mm in diameter and composed of three fibers, each of which is sensitive to one of the above parameters. It can be inserted into blood vessels, eg, during operations, and gives a continous recording over periods of typically 12 h. Figure 17-18 shows a cross-section of the 1-mm tip of the sensor bundle that fits into a standard gage needle.

17.2.1.7   Blood Oximetry

Blood oximetry is another important field of application of fiber optics. In contrast to oxygen sensors, which yield information on *oxygen partial pressure,* oximetry is used to determine the amount of *hemoglobin-bound oxygen.* It can be performed with plain fibers utilizing the difference in the absorption spectra of oxyhemoglobin and hemoglobin (Figure 17-19). Because of the high absorbance of blood at analytical wavelengths below 600 nm, almost all work has been performed at wavelengths close to or beyond the red end of the visible spectrum. The advent of red and infrared LEDs together with red-sensitive photodiodes has contributed considerably to the stimulation of these experiments. Generally, the analytical wavelength is set between 600 and 750 nm. Internal standardization is achieved by measuring reflection at ca. 930 nm, which is one of the isosbestic wavelengths in the absorption spectra of hemoglobin and oxyhemoglobin [126].

The theory, design, and evaluation of an optical sensor for measurement of blood oxygen saturation and hematocrit was presented [127] that meets the requirement of a large dynamic range even in the deep-lying vessels, high reliability and stability over extended periods of time, and biocompatibility. It is based on the measurement of diffuse reflectance by a ratio method using three LEDs (one red, two infrared) as light sources placed on an implantable 6 × 1.6 mm area.

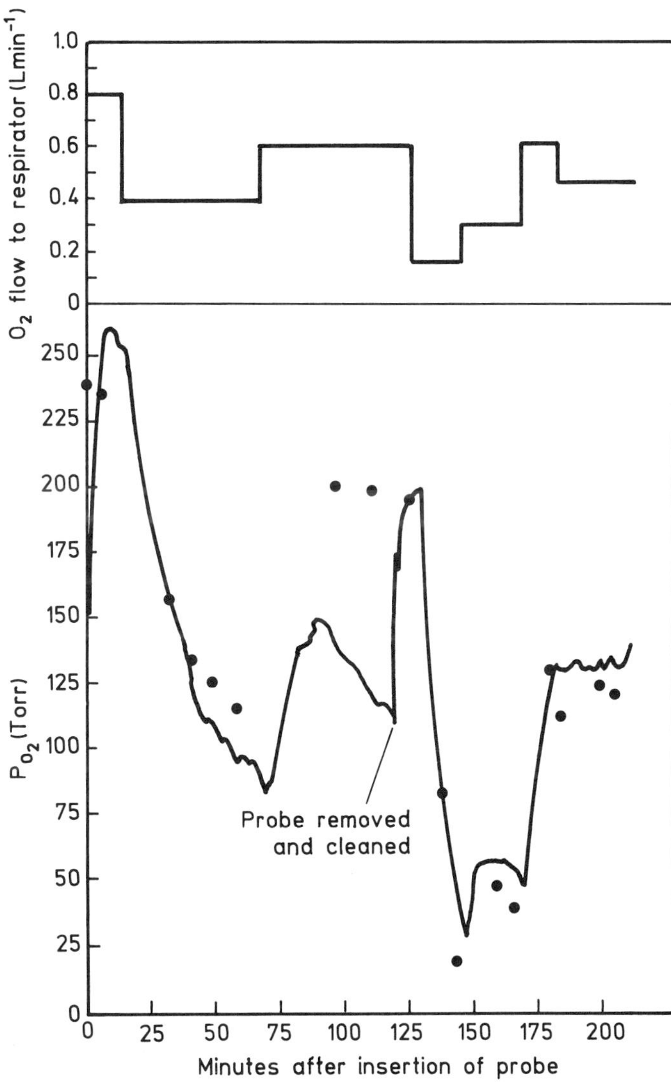

Figure 17-17. In vivo performance of the fiber-optic oxygen sensor shown in Figure 17-13 in a ewe [114]. Upper trace: variation of oxygen fraction in respiratory gas. Circles: data obtained ex vivo with an oxygen electrode.

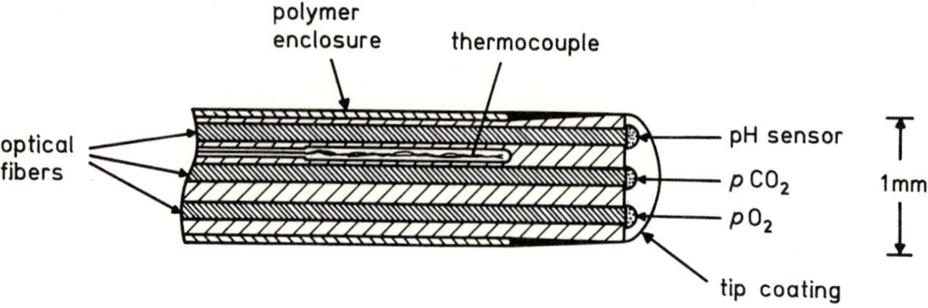

Figure 17-18. Cross-section through the tip of an invasive catheter for on-line measurement of pH, oxygen, carbon dioxide, and temperature [50].

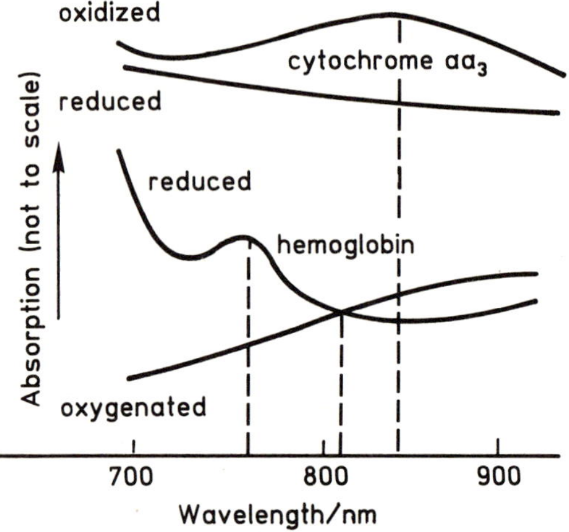

Figure 17-19. Absorption spectra of hemoglobin and oxyhemoglobin showing the isosbestic wavelength at above 800 nm which usually serves as the reference wavelength.

A solid-state linear diode-array spectrometer has been applied to the simultaneous determination of relative amounts of hemoglobin, carboxyhemoglobin, and oxyhemoglobin in whole blood [128]. The fiber is incorporated in a hypodermic syringe needle and the analytical wavelength is set between 520 and 620 nm. Derivative spectra, signal averaging, and least-squares calculations were applied to extract the maximum possible information from the spectra.

17.2.2   Carbon Dioxide Sensors

Carbon dioxide can be assayed via infrared absorption, or electrochemically by measuring the changes in the pH of a buffer solution as a result of varying CO_2 partial pressure above the solution. The latter principle can also be applied to optical sensors.

The response of a pH sensor to CO_2 occurs as a direct result of the proton concentration in the sensitive layer, which is related to the concentration of CO_2 through the following series of chemical equilibria:

(1) CO_2 (aq.) + H_2O $\rightleftharpoons$ H_2CO_3 (hydration)
(2) H_2CO_3 $\rightleftharpoons$ H^+ + HCO_3^- (dissociation, step 1)
(3) HCO_3^- $\rightleftharpoons$ H^+ + CO_3^{2-} (dissociation, step 2)

These are governed by the following equilibrium constants:

$$K_h = [H_2CO_3]/[CO_2] \text{ (aq.)} = 0.0026 \tag{17-52}$$
$$K_1 = [H^+] [HCO_3^-]/[H_2CO_3] = 1.72 \cdot 10^{-4} \tag{17-53}$$
$$K_2 = [H^+] [CO_3^{2-}]/[HCO_3^-] = 5.59 \cdot 10^{-11} \tag{17-54}$$

In most studies on CO_2 sensors, the total analytical concentration of carbon dioxide, ie, $[CO_2 \text{ (aq.)}] + [H_2CO_3]$, has been related to the response.

Lübbers and Opitz [109] followed the changes in fluorescence of a solution of 4-methyl umbelliferone in 1 µM hydrogencarbonate buffer as a function of CO_2 partial pressure. The indicator solution was covered with a 6-µm PTFE membrane and the ratio of fluorescence intensity at 445 nm measured under excitation at 318 and 357 nm was related to pressure.

The optode showed good reproducibility in the 0.1–10 kPa CO_2 range, with a response time of 3–4 s for 90% of the final value. It can be accelerated by addition of the enzyme carbonic anhydrase, which catalyzes the establishment of equilibrium (1). The short response is in striking contrast to the 0.5–2.0 min for the CO_2 electrode. When coupled to a fiber-optic system it allows the transcutaneous measurement of CO_2 pressure [128].

The same principle was applied to construct a compact instrument (5 × 6 × 14 cm in size) for CO_2 consisting of a blue LED as a light source, a long-wave-length-absorbing indicator (HPTS) dissolved in hydrogencarbonate and covered with a PTFE layer, two optical filters, and two photodiodes for light detection [129]. Linearized calibration graphs for this sensor type are shown in Figure 17-20. The slopes of the graphs can be governed by the pK_a of the

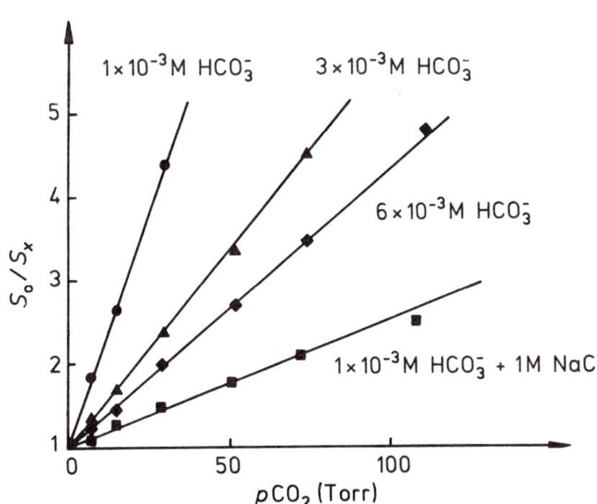

Figure 17-20.
Effect of buffer composition of the internal filling solution in a sensor for carbon dioxide [129].

indicator, ionic strength, and buffer concentration. Unfortunately, high buffer concentrations result in very long response times. These are further prolonged with increasing thickness of the sensor layer (typically 20 µm), PTFE covers (6 µm), and slow kinetics of the hydration and, in particular, dehydration of CO_2. For the system described above, the response for a change from 0 to 5% CO_2 was 15 s for 90% of the final value. The reverse change required ca. 34 s.

Zhujun and Seitz [130] used HPTS in hydrogencarbonate, covered with silicone, to sense CO_2. Fluorescence is measured with a bifurcated fiber system. The equation that was used to relate the CO_2 partial pressure to hydrogen ion concentration is

$$[H^+]^3 + N[H^+]^2 - (K_1 C + K_w)[H^+] = K_1 K_2 C \qquad (17\text{-}55)$$

where N is the internal hydrogencarbonate concentration, K_1, K_2, and K_w are the dissociation constants of carbonic acid (steps 1 and 2; see Equations (17-53 and 17-54) and water, respectively, and C is the analytical CO_2 concentration including both hydrated and unhydrated CO_2. It was shown that, within a limited range, there is linearity between CO_2 pressure and $[H^+]$ according to

$$[H^+] = K_1 / CN \, . \qquad (17\text{-}56)$$

In practice, the internal HCO_3^- concentration should be such that the CO_2 concentrations of interest yield pH changes between 6.5 and 8.0, where, for instance, a pH sensor based on HPTS with its pK_a of 7.3 is most sensitive.

Heitzmann and Kroneis [131] have prepared CO_2-sensitive fluorescent membranes by soaking cross-linked polyacrylamide beads with a solution of HPTS in hydrogencarbonate and embedding them in silicone rubber (Figure 17-21). The response to CO_2 was varied by adding different amounts of hydrogencarbonate, carbonate, and HPTS, all of which act as buffers. The polyacrylamide beads may be omitted, so that an emulsion of the HPTS–carbonate solution in silicone-rubber is obtained.

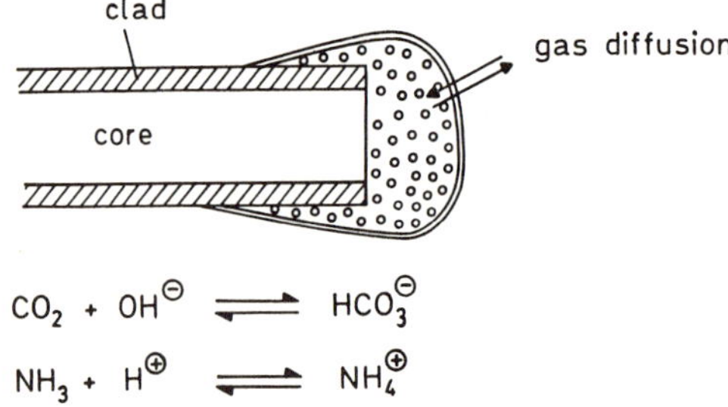

Figure 17-21.
Typical working chemistry for sensing acidic or basic gases such as carbon dioxide or ammonia. The gas penetrates through the silicone matrix into the aqueous droplets which contains a buffer plus a pH-sensitive dye. The resulting change in pH is indicated by the dye.

A CO_2 sensor with a fiber-immobilized indicator has also been described [132]. When two sensing layers with different optical properties and selectivity of one for O_2 and the other for CO_2 are attached to the end of a fiber, a single sensor for both species can be obtained [52]. A cross-section through an O_2/CO_2 sensor layer is shown in Figure 17-22 and a typical response curve in Figure 17-23.

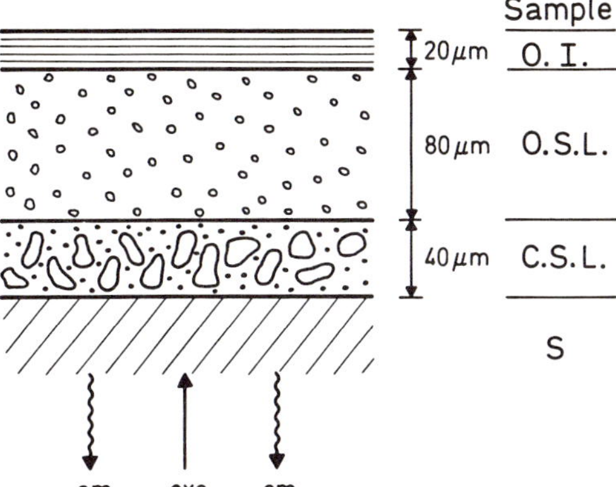

Figure 17-22
Cross-section through a sensor that responds to both oxygen and carbon dioxide [52]. S: solid support; CSL: carbon dioxide-sensitive layer (having green fluorescence); OSL: oxygen-sensitive layer (having a red fluorescence); OI: optical isolation. The directions of exciting and emitted light are also shown.

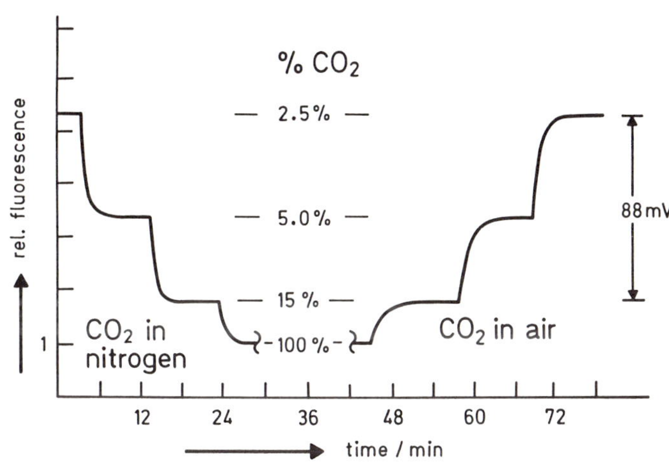

Figure 17-23. Response time, relative signal change, and reproducibility of a CO_2 sensor [52].

The principle of the fiber-optic pH sensor led Vurek et al. [133] to devise a CO_2 sensor. Instead of coupling the pH indicator dye to an insoluble polymer, a simple isotonic solution of salt, hydrogencarbonate, and dye was used, which was covered with a CO_2-permeable silicone-rubber membrane. The sensor's performance was demonstrated in vivo. Similarly, a fluorescein-based CO_2-sensitive system was reported by Hirschfeld et al. [134].

17.2.3 Sensors for Other Gases

17.2.3.1 Ammonia

The well known reaction between ninhydrin and ammonia, producing a purple coloration, was utilized to develop an ammonia probe. The basis for the analytical technique is the partial attenuation of a light signal traversing a cylindrical optical waveguide by successive total internal reflections. A fiber rod was covered with a solution of ninhydrin in polyvinylpyrrolidone and the changes in the absorption of the evanescent wave were followed [135]. Concentration and relative humidity both determine the slope of the transmission curves. The probe works irreversibly and is able to detect ammonia down to the 50 ppb level.

Another ammonia sensor specifically designed for use in bioliquids is based on the evanescent wave technique and can be applied to the vapor-phase determination of ammonia above blood and serum [136]. It utilizes the ninhydrin reaction occurring in the polymer coating of the fiber, and the resulting color change is monitored by total internal reflection. The probe is applicable to clinical determinations normally carried out in the vapor phase, but works irreversibly. A linear relationship exists between absorbance and ammonia concentration in the clinically useful range of 0–4.0 µg mL^{-1}. Comparison with the reference method showed a correlation coefficient of 0.92.

A reversible optical waveguide sensor for ammonia vapor was introduced more recently [137], consisting of a small capillary glass tube fitted with a LED and a phototransistor detector to form a multiple reflecting optical device. When the capillary was coated with a thin solid film composed of a pH-sensitive oxazine dye, the instrument was capable of reversibly sensing ammonia. Vapor concentrations from 100 to below 60 ppm were easily and reproducibly detected. A preliminary qualitative kinetic model was proposed to describe the vapor–film interactions.

Ammonia sensors based on the same principle as electrochemical ammonia sensors (viz, the change in pH of an alkaline buffer solution caused by ammonia) have been reported by various groups. Arnold and Ostler [138] followed the changes in the absorption of an internal buffer solution, to which *p*-nitrophenol was added, as ammonia passes by and gives rise to an increase in pH which causes a color change of the indicator. Wolfbeis and Posch [139] used a fine emulsion of an aqueous solution of a fluorescent pH indicator which simultaneously may act as a buffer (Figure 17-20). Another type of ammonia-sensitive probe was obtained by immobilizing Bromothymol Blue in a hydrophilic polymer and measuring ammonia-induced changes in reflectance [140]. All these kinds of sensors have a slow response and even slower recovery [141 a]. Much faster membranes have been obtained by incorporating both an ammonium carrier and a colored proton carrier in plasticized PVC membranes [141 b].

17.2.3.2 Sulfur Dioxide

Continuous determination of SO$_2$ in air can be performed by fluorimetry owing to its strong intrinsic fluorescence at about 330 nm. In complex samples the method is not very specific. Advantage can be taken of the observation that the fluorescence of benzo [*b*] fluoranthene and related polynuclear aromatic hydrocarbons (PAHs) is strongly quenched by SO$_2$; 80 ppm of SO$_2$ in gases can be detected [142], and the useful range is from 0.01 to 6.0% (v/v), as can be seen in Figure 17-24. Other gases likely to occur in air were found to be inert, except for oxygen, which also acts as a dynamic quencher. Its interference is neglibible for SO$_2$ levels below 6% in air at constant oxygen pressure, because the quenching efficiency of SO$_2$ is about 26 times higher than that of oxygen. For varying oxygen levels, a two-sensor technique was suggested.

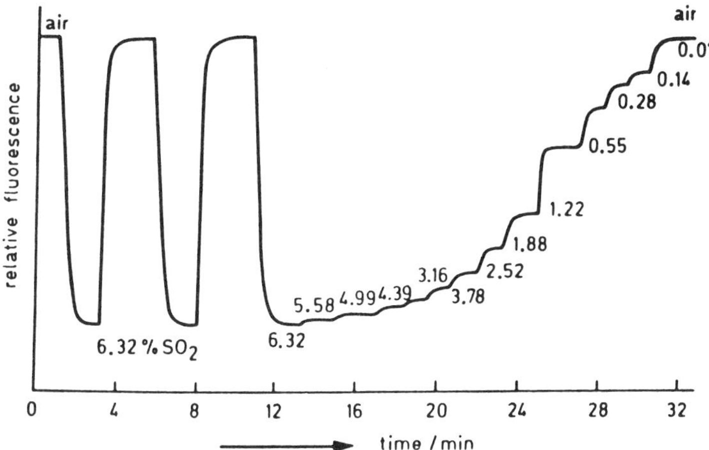

Figure 17-24. Response time, relative signal change, and reproducibility of an SO_2 sensor [142].

17.2.3.3 Hydrogen Sulfide

When paper is impregnated with lead acetate, it assumes a dark color when in contact with H_2S. This, in turn, changes its reflectance (as measured at 580 nm). Levels as low as 50 ppb of H_2S could be measured with high reproducibility [143]. Other methods for sensing H_2S are based on the reducing properties of sulfide or hydrogen sulfide [94, 96]. These devices are, however, probes rather than sensors because they act irreversibly.

In the only reversible approach [144], quaternized acridinium was immobilized on a solid support such as cellulose. In slightly alkaline medium, hydrogen sulfide adds to the strongly fluorescent dye and renders it nonfluorescent. Interferences by ionic quenchers may be eliminated by covering the sensor with a 4-µm silicone-rubber membrane which is permeable to gases but not to ions. The detection limit is of the order of 0.1 mmol L^{-1}.

17.2.3.4 Hydrogen

An optical fiber has been used as the sensing element for hydrogen [145]. It was coated with palladium, which expands on exposure to hydrogen because of the formation of PdH_x. The expansion depends on the partial pressure of hydrogen and stretches the fiber in both the axial and radial directions. This results in a change in the effective optical path length of the fiber, which can be detected by Mach–Zehnder interferometry. A large dynamic range is reported (1–30000 ppm), and the effect is fully reversible. Further information can be found in the literature [88, 146–150].

17.2.3.5 Flammable Gases

The formation of alkane hydrates by interaction with water molecules has been exploited to detect alkanes via the technique of multiple total internal optical scattering [151]. The hydrates are formed at a glass surfaces, and the change in refractive index of the coating causes a change in reflectance characteristics. The instrumental system employed utilizes LED light sources (560 and 660 nm) and different scattering intensities. For methane, concentrations up to ca. 1000 ppm may be detected at atmospheric pressure and room temperature. When platinum-coated fibers are exposed to hydrocarbons, an exothermic reaction of the hydrocarbon with oxygen will occur at the surface. The resultant heat of reaction can be transduced to a phase retardance in a fiber-guided light beam and detected by interferometry. Hydrocarbon gases at ca. 300 Pa appear to be detectable [149]. Various reviews on fiber-optic gas sensors have appeared [147, 148, 150].

Methane has also been detected with an evanescent wave sensor using the optical absorption by methane of the 3.39-µm line of a He–Ne laser [152]. The single fiber is used as both the sensor and the transmission line. When a fiber of diameter 1.8 µm and length 10 mm is used, the minimum detectable concentration of methane is less than 5% in air. Possible interferents are ethane, propane, and water. Alternatively, the absorption by methane and other hydrocarbons in the overtone range (1.3 and 1.65 µm) may be measured, since the fiber transmission is much better in this range. Thus, the 1644.9-nm absorption band of methane coincides with the emission line of the erbium : YAG laser [153]. Similarly, the absorption of carbon monoxide at 4.7 µm may be used by analogy with the tunable diode laser technique, which exhibits a fast response and good sensitivity [154]. Fluoride fibers are required in this wavelength range [148].

Vapors of polar solvents can be detected on-line by virtue of their decolorizing effect on certain blue indicator papers [155] and triphenylmethane dyes [156]. Typical vapors that can be detected are those of ethers, alcohols, esters and ketones, whereas hydrocarbons and chlorinated hydrocarbons are inert. Detection limits (with a device consisting of a yellow LED light source and a phototransistor as the detector) vary from 10 to 1000 ppm and both flow-through cell sensing and fiber-optic sensing (via reflectometry) were found to be applicable. Response times vary from 0.5 to 4 min.

17.2.3.6 Anesthetics

Halothane, a widely used inhalation narcotic, can be detected owing to its quenching effect on the fluorescence of polycyclic aromatic hydrocarbons [118]. Figure 17-25 shows the response of such a sensor to halothane in air, nitrogen, and oxygen. Interference by oxygen can be taken into account by a second sensor sensitive to oxygen only. The two-sensor technique allows the determination of halothane and/or oxygen with a precision of ±5%. No interferences are reported for CO, CO_2, N_2O, and fluorans (mixtures of compounds used in anesthesia). Potential applications include monitoring of halothane in blood (using plastic fiber catheters), oxygen in vitro in the presence of halothane, and anesthetic gases in the breathing circuit.

17.2.3.7 Humidity

Continuous monitoring of air moisture plays an important role in airports and similar locations to allow the timely prediction of fogging and icing. A simple version for a humiditiy sensor makes use of the color change of cobalt(II)chloride in gelatine cast as films on a 600-µm

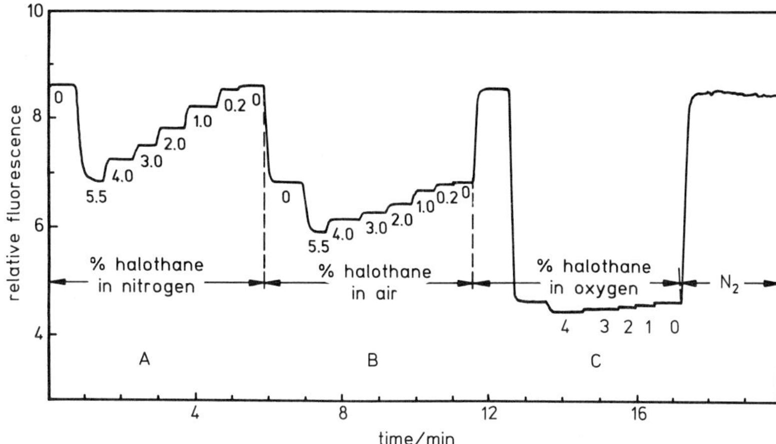

Figure 17-25. Response time, relative signal change, and reproducibility of a halothane sensor [118].

fiber when exposed to humidity [157]. The absorption is measured at 680 nm through the fiber by internal reflection, and relative humidity can be determined at levels between 40 and 80%.

Another sensor type [158] utilizes the effect of water vapor on the fluorescence quantum yield of adsorbed fluorophores. Thus, a silica-gel-adsorbed perylene dye, when excited with an LED at 490 nm, shows a ca 90% reduction in fluorescence on going from 0 to 100% relative humidity. Although this sensor covers the whole humidity range, the response is not linear, and oxygen interferes to some extent.

17.3 Optical Fibers in Process Control

The transmission of data representative of a physical or chemical value of a "process", using optical fibers, has aroused the interest of instrumentation manufacturers. The new concepts of fiberoptic sensors (FOS), the possibility of optical multiplexing, and the organization of networks has swollen the ranks of users demanding these new techniques. Thus, process control and monitoring are spreading in many branches of industry such as the nuclear, gas, oil, agricultural, and food areas where the advantages of fibers can be clearly identified in comparison with tried and proven electrical systems.

The most common application of FOS in process control makes use of simple on/off devices and electro-optical converters coupled directly with standard transducers. However, the latest research and development have focused on measurements of pressure, temperature, flow rate, and level, the four physical parameters for which demand is strongest. An excellent review of these sensors was published recently [159]. This review, focusing on industrial objectives, followed on the First International Conference on Optical Techniques in Process Control [160].

Japanese (Optical Measurement and Development System) and American (ITT) experiments on plant control, Norwegian works [161] on the monitoring of offshore installations, projects for gas installations with optical fibers [162] and experiments in Purex and Chemex nuclear plants [163] revealed the importance attached by industry to developments in real-time remote monitoring in an undisturbed electromagnetic environment. Recent sessions of the International Conference of Optical Engineering (Hamburg, 1988) spotlighted the opening for new markets in process control, both for physical parameters and for chemical analysis by spectrometric techniques [147, 164].

Fiber-optic applications in the chemical industry were first tried over 15 years ago with colorimetry. However, concepts related to remote spectrometry and refractometry with optical fibers are of much more recent date. In the former area, it was necessary to overcome problems due to coupling of "telecommunication" fibers with spectrometric analyzers for absorption [165], fluorescence [166], and Raman scattering [167]. The last area was studied at the same time as the intrinsic FOS, with the measurement of the variation in amplitude of an optical signal as a function of the refractive index of the optical fiber cladding which consists of the liquid to be measured [168]. A characteristic example is the vehicle battery acid concentration indicator [169].

Recent developments are related to both batch and continuous processes, including environmental monitoring (see Chapter 18). Applications in relatively inaccessible zones such as explosive, nuclear or high-temperature containments require new specific components and a control organization, revealing the considerable repercussions on the structure of future plants. This section attempts to summarize the most significant research of the past few years on remote control of chemical processes. The new concepts, multi-point measurement techniques, associated components, and aspects of real-time measurement techniques are also examined.

17.3.1   Concepts for Remote Process Control

17.3.1.1   Architecture of Remote Control

Spectrometry

Sampling is an essential function of spectrometric analytical control. The sample, which is as small as possible, must be representative of the medium to be analyzed. The analysis must be rapid in order to be able to take any required action on the process. The general characteristics of sampling require a knowledge of the fluid flow rates, mass transfer, and temperature variations, especially in compiling a material balance.

Three architectures are used for on-line control with optical fibers:

— In a first arrangement the process control system includes automatic sampling in the "hot" zone such as an explosion-proof container, the manual or automatic transfer of the sample to the laboratory, a robot to prepare the analysis, and transfer of the solution to be analyzed to an fiber-optic measurement cell [163]. This architecture is most commonly found today if the laboratory is habitually and directly involved in controlling the various steps in the process. It is justified if the samples must also be prepared in a "hot" zone, or if the analysis requires semiautomatic operations such as dissolution of solid samples, or sophisticated techniques such as chromatography, mass spectrometry, or emission spectrometry. An example is given with DC arc spectrometry for the determination of alkali metals [170].

— Optical fibers that can transmit light data over a long distance allow for an even more advantageous configuration, shown schematically in Figure 17-26 [164]. The sample itself need no longer be transferred to the laboratory. The optical fiber gathers the data in a zone close to the process and transmits it to the laboratory. The samples are transferred automatically after the addition of chemical reagents, either to a fiber-optic circulation cell or to a tank in which an optode is immersed. Samples are taken directly from the main stream or from a side stream. The analytical laboratory manages the analyzer, the optical multiplexing station, and the handling of results. Automatic systems for sampling, dilution, and reagent additions and the robot can be managed by the process personnel or, depending on the plant organization, directly by the laboratory itself. In a nuclear environment, and for safety reasons, the laboratory is itself located in the "hot" zone.

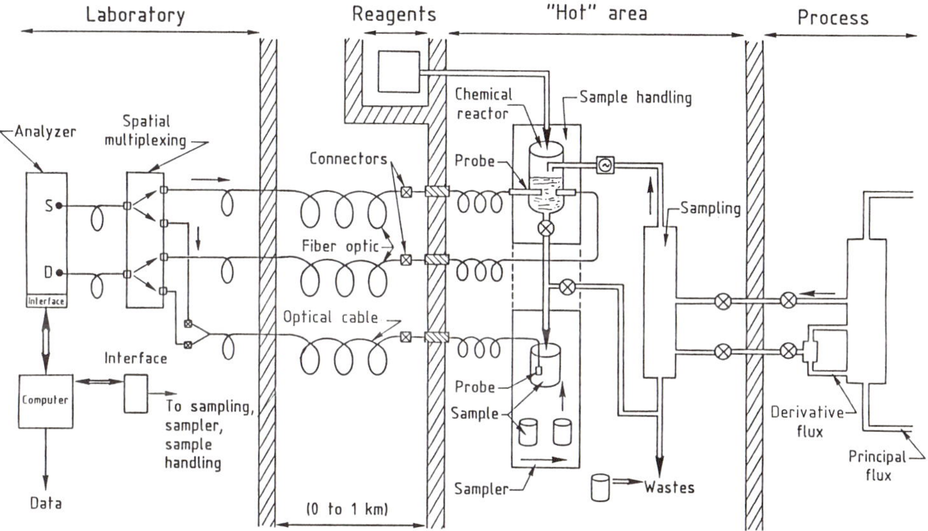

Figure 17-26. Schematic architecture of remote analytical on-line control with automatic sampling and fiber-optic chemical sensors.

— A better configuration avoids or limits sample transfers. It allows in situ measurement either on the main stream in a continuous process or in batchs, either on a side stream specially designed if a slight "hold-up" is acceptable. In this way, the advantage of remote real-time measurement is preserved. This architecture is shown in Figure 17-27. Circulation cells or small-diameter immersion probes are employed. The small fiber cross-section can thus be exploited to minimize the sample volume. The elimination of all mobile systems, including that of the optical multiplexer, is also a considerable asset for the representativeness of the measurement. When the service life may raise problems, the interchangeability of the circulation cells and probes must be considered from an external action zone using optical connectors. This configuration is designed individually for the dif-

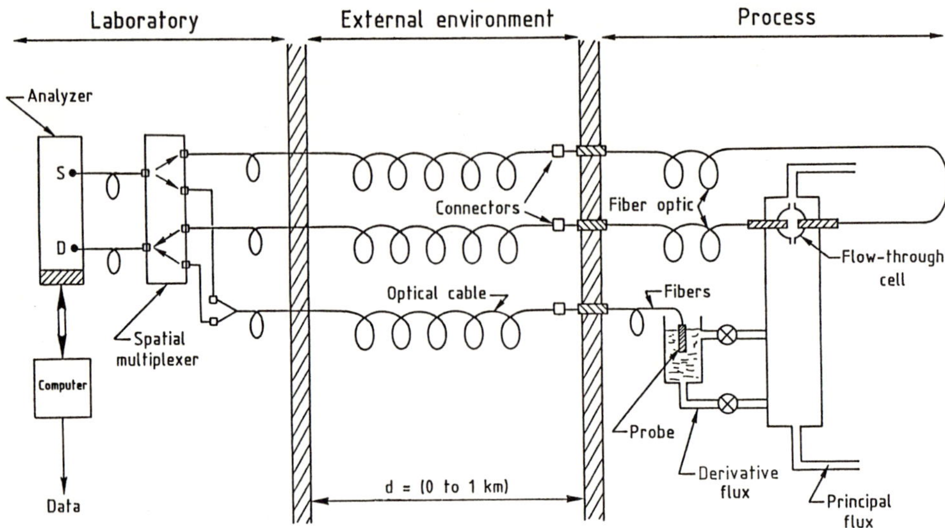

Figure 17-27. Schematic architecture of fiber-optic on-line control without sample handling.

ferent plant situations. It depends on the existing facilities. It is often difficult to achieve today because production plants were generally built before the advent of the concepts of remote analysis with optical fibers.

The advantages of these configurations are the electrical insulation, no electromagnetic interference, small cells, multi-point measurement for all architectures, no sample transfer, little holdup and real-time information in the third architecture. The main drawbacks are associated with the lack of familiarity with the use of optical fibers in a sometimes unsuitable "plant" configuration, and with the youth of a technology that is still not fully under control with optical multiplexing and instrumentation connections and costly (lacking a "mass public").

Reflectometry

Reflectometry uses the principle of light backscattering at the fiber end. A light pulse transmitted in the fiber is fed back as an echo. The signal propagation time depends directly on the speed of light in the waveguide, and hence on the length of the fiber. The attenuation is determined in relation to this length. The variation in attenuation along the fiber can be identified in amplitude and in transit time. Figure 17-28 illustrates the type of measurement, which helps to determine the attenuation of several sensors in series on the same optical line. This technique is highly advantageous for the location and remote measurement of refractive index or evanescent waves variations in a waveguide under the influence of a chemical species, for example. It justifies the many developments of optical time domain reflectometry (OTDR) and the derivative methods POTDR (polarization OTDR) and FOTDR (frequency OTDR) for physical measurements [147] and, more recently, in chemistry with oil-leakage sensors [171].

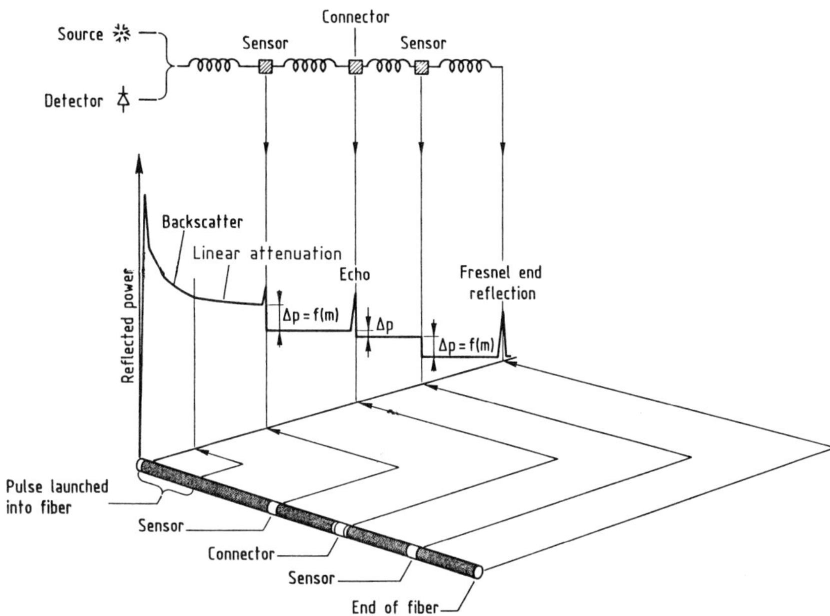

Figure 17-28. Multi-point measurements by optical time domain reflectometric detection.

17.3.1.2 *Analytical Spectrometric Concepts*

Absorption

It is mainly in absorption spectrometry that the concepts of on-line remote control by optical fibers have been defined [172]. As a starting hypothesis, they all use the "in situ" on-line measurement over long periods and they can be summarized as follows:

– Simultaneous measurement at $n + 1$ wavelengths, where n is the number of independent species to be determined in solution. The simplest case is the measurement at two wavelengths [161, 173]. One wavelength is specific to the product to be analyzed, and the other serves as an internal spectral reference as shown in Figure 17-29a. The variation in background at the absorption peak (λ_p) is assumed to be the same as for valley (λ_v). Thus the absorbance A is proportional to the concentration of species the M to be analyzed:

$$A = (A_{\lambda_p} - A_{\lambda_v}) = K\,[\mathrm{M}] \ . \tag{17-57}$$

– Simultaneous measurement on the sides of the absorption spectrum. This is justified if the identification of the wavelengths is absolute and the observation window $(\Delta\lambda)$ is constant. In principle, these conditions exclude wavelength-scanning devices in favor of photodiode-array spectrometers. These measurements on the sides allow the extension of the measurement to $n + x + 1$ wavelengths in accordance with the complexity of the background and

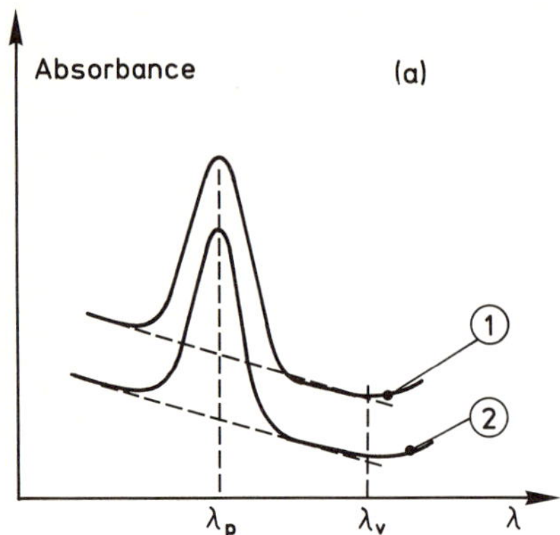

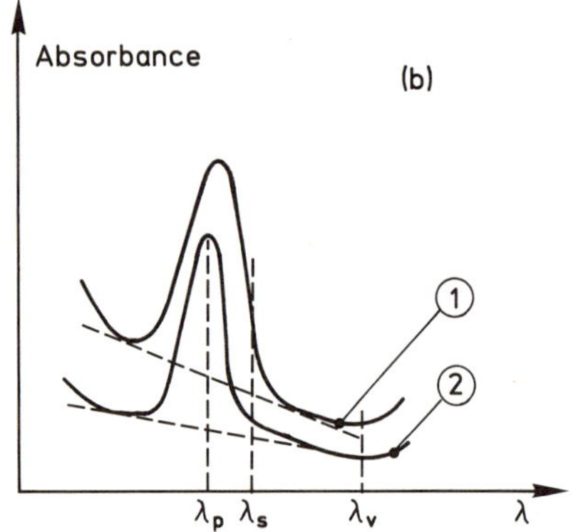

Figure 17-29.
Schematic spectral representation of spectrophotometric measurement for one species in process control. The equal slopes absorbance = f(λ) indicate: (a) an equal background compensation between λ_p (peak wave length) and λ_v (valley wavelength) for two noises ① and ② ; (b) an unequal background not compensated easily when noises ① and ② vary with the wavelength.

noise correction ($x \geqslant 0$). At three wavelengths ($n = 1$ and $x = 1$), according to Figure 17-29b, the absorbance function thus becomes

$$A = \log(I_{\lambda_p}/I_{\lambda_v}) + K \log(I_{\lambda_p}/I_{\lambda_s}) = K \, [\mathrm{M}] \tag{17-58}$$

where I is the intensity measured at the peak (p) and valley (v) or side (s) wavelength (λ). For strong concentrations of the species M, this measurement on the sides serves to extend artificially the dynamic range of the measurement by a factor of 100–1000, while avoiding equipment distortions.

– Simultaneous measurement at $N + n + 1$ wavelengths, where N is the number of interfering functions (or species) on the n species investigated. For $n = 1$ and $N = 3$ (acidity H, ionic strength I, *temperature* θ), the spectrum of the species must be analyzed at five wavelengths, or treated by an algorithm. Indeed, the spectral distorsions must be analyzed before allowing on-line control. Significant examples have been described in nuclear environments such as the $U^{(VI)}$ measurement in the presence of interacting species such as $U^{(III)}$ and $U^{(IV)}$, or for monitoring of uranium and plutonium [34] in media of variable acidities, total nitrates and ionic strength.

– Simultaneous measurement as iso-absorbance on the sides of optical spectra as seen in Figure 17-30. One species (2) may be measured relative to another (1) considered as a standard and which serves as a reference. Traces of (2) can thus be detected even if species (1) is highly absorbent at the measurement wavelengths λ_{s_1} and λ_{s_2}. This concept was demonstrated in the Chemex process for the chemical isotopic separation of uranium [34, 163].

– Simultaneous detection of multiple interfering species of which the optical spectra have first been identified (Figure 17-30b). This detection is mainly useful for studies of chemical kinetics and spectral distorsions resulting from the formation of complexes in solution and from the influence of the medium. It is indispensable in the near-infrared (NIR) region, where the spectra may be complex.

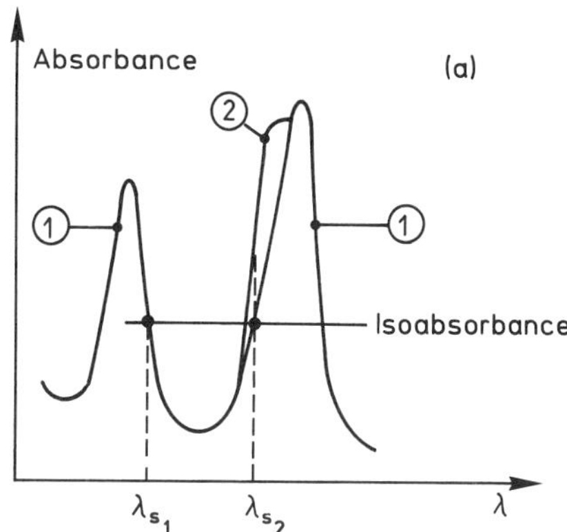

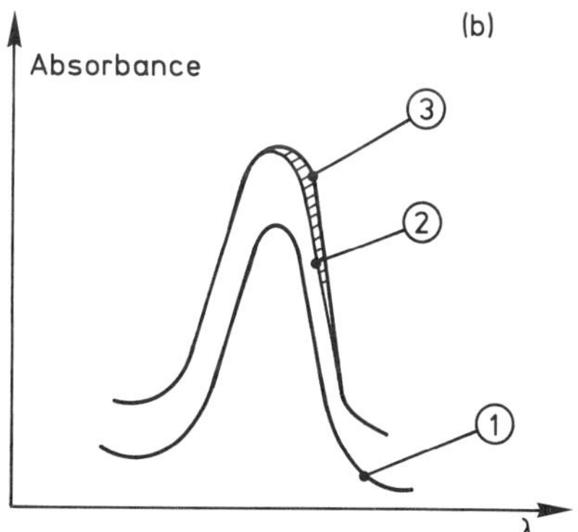

Figure 17-30.
Spectrophotometric measurement of two species (I and II) in process control. (a) Determination with iso-absorbance technique (λ_s = slide wavelength); (b) representation of an equal background variation only (curves ① and ②). For a multi-component determination (curve ③) the spectral complementary variation is represented by the hatched zone.

Thus the number of wavelengths for on-line control is $n + N + x + 1$. However, the simplest instruments have a capacity of two ($N = 0$, $x = 0$) or three ($N = 0$, $x = 1$) wavelengths.

In considering an "anion" effect, an experimental function of the following type can be generalized:

$$A = \sum_{i=0}^{n} [\mathrm{M}_i]\,(a_i\,[\mathrm{S}_i] + b_i) \tag{17-59}$$

where A is the absorbance, S is the anionic species, M is the cation, and a and b are calibration coefficients for the i species present. Sometimes it is necessary to use a second-degree equation:

$$A = \sum_{i=0}^{n} [M_i]\,(a_i\,[S_i]^2 + b_i\,[S_i] + c_i)\,. \tag{17-60}$$

Examples were given for $U^{(VI)}$, $U^{(IV)}$ and $Pu^{(III)}$ in Purex process control [34]. Also, the spectral operating method usually requires reliance on algorithms and chemometrics [174].

Fluorescence and Raman Emissions

The selectivity of fluorescence generally helps in avoiding the need for multi-wavelength measurements. However, a knowledge of the intensity ratio at two wavelengths is often recommended. The identification of the half-life of the species to be analyzed is also a factor in selelctivity. Thus, time spectrometric instruments with pulsed lasers and photodiode arrays are ideal for process control. A recent example was given for the in-line control of traces of uranium in nuclear plants [175].

The major drawbacks of fluorescence in on-line control result from the variation in quenching or enhancement of the fluorescence under the influence of interferents such as impurities of the medium and uncontrolled species. In addition, interference fluorescence due to the fibers is significant and can be increased considerably under ionizing radiation (see Chapter 12). Although the use of a single fiber is possible [166], sometimes it is preferable to use multi-fiber optodes when determinations are made at long distances and also when using shielding windows in industrial environments; in this way, the undesirable reflections of the laser at the fiber end are avoided.

In Raman spectrometry, the scattering of the incident photon may gain in energy (anti-Stokes rays) or give up energy (Stokes rays) from the change in vibration or spin energy of the molecules. A photon is re-emitted at a different energy (and hence wavelength). The emission intensity is directly proportional to the concentration of the species and specific to chemical bonds. Its application in process control is only beginning to be considered for the in situ analysis of solutions [167, 176] and the cartographic observation of powdered solids [177].

17.3.2   Multi-Component Measurement Techniques

In addition to remote measurement, another major asset of optical fibers in process control derives from the possibility of multi-point measurements in a network architecture. Spatial multiplexing was briefly described in Chapter 12, Figures 12-11 a, b and c. This is the main technique usable in spectrometry with extensive spectral observation [82]. In fact, the construction of an optode network requires a proper balancing of the various channels. This condition is only satisfied after determining the connection balance of each channel (see Section 17.3.4.2). This multiplexing, which uses a single source and a single detector, is only applicable for the same determination for all the measurement points. An example was recently described for the measurement of flammable gases with a monitoring unit for 30 measurement points [178].

Wavelength division multiplexing (WDM) has also been considered. The spectrum of the source, which may be a light-emitting diode (LED), is divided into sections. Each of these sections corresponds to an elementary source. This subdivision is obtained by means of a grating. In a system with a concave mirror combined with a reflective diffraction grating [179] or with a reflective curved grating [180], the width of the spectral band is directly related to the arrangement of the receiving fibers. The system can operate as a multiplexer and a demultiplexer. The subdivision of an LED from 810 to 870 nm into nine wavelengths with a resolution of less than 1 nm and a cross-talk of $-18\,$dB m has been obtained for process control applications [181]. Further, a GRIN rod/prism grating was used with eleven input/output fibers covering a range of 90 nm [182]. It was coupled with a 35-element self-scanned photodiode array. The whole system is a compact spectrometer with multi-channel analysis capability.

Two other architectures with multiple sources have been suggested [183]. They offer the advantage of allowing for simultaneous determinations of different types of measurement such as absorption, fluorescence, and evanescence. Figure 17-31a shows the architecture with an

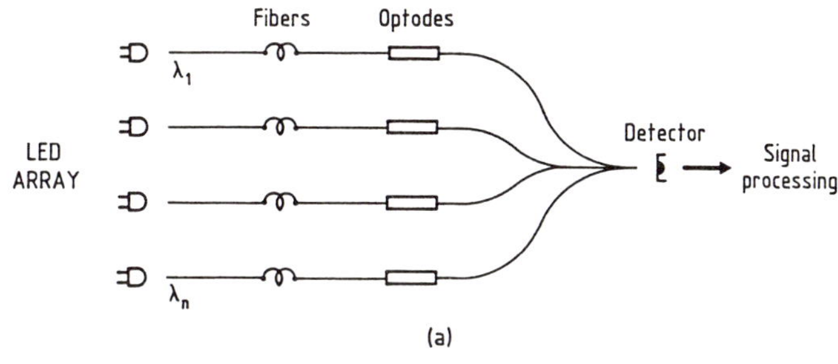

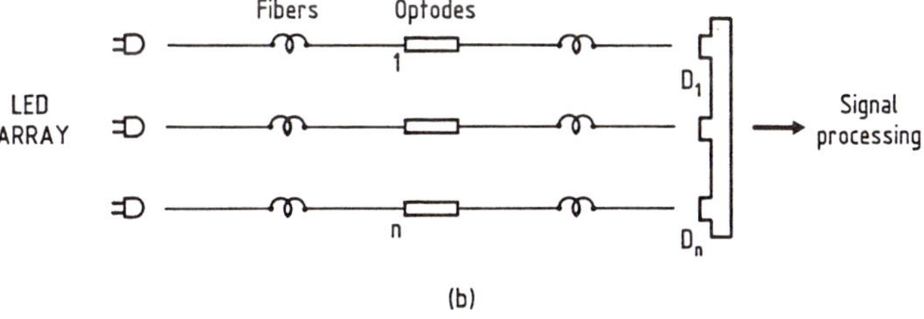

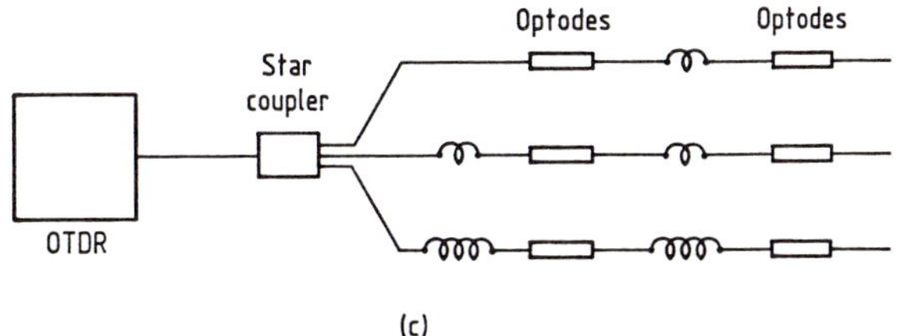

Figure 17-31. Multiplexing. (a) Multiple sources and one detector; (b) multiple sources and multiple detectors; (c) spatial and time-division multiplexing OTDR with delayed lines.

array of eight LEDs and a single detector. The brightness of each LED is adjusted. The channels are selected by on/off switch times of 1–2 ms using an assembly language algorithm. The system is simple and inexpensive. Its limitations are attributed to the chemical sensitivity and selectivity of each sensor. The second architecture, shown in Figure 17-31b, features several transmitters and several receivers. It uses the concepts of chemometrics [174]. Using several detectors, each of which can respond to at least two components of a mixture, the data are analyzed. The measurements obtained from various sensors in line or in parallel are related to algorithms or to combinatorial analysis.

Time multiplexing on a line using the OTDR technique can be coupled with space multiplexing on several lines, as shown in Figure 17-31c. Fiber-optic delay lines serve to individualize each of the sensors, or preferably to identify a local malfunction over the entire line length. This principle was demonstrated by Wada et al. [171] for monitoring oil leaks, by measuring the refractive index of the surface of a waveguide. The limits of this system are generally given by the connecting balance of the optical line with the limits of the OTDR technique.

17.3.3 Cells and Optodes

Many research projects in on-line process control have been limited to the use of passive optodes. Among the flow-through cells for absorption spectrometry the simplest consists of an optical adaptation between the light-emitting fiber and the receiving fiber (see Figure 17-10). The optical path in solutions may vary from 1 mm to 0.5 m for linear cells [34]. An additional fiber forming a loop serves to have the source and detection points on the same side according to the arrangement of Freeman et al. [184] for the remote determination of copper in industrial electroplating baths.

Two models of long optical path cells, shown in Figure 17-32, have been investigated. One is an adaptation of the White cell [185] for measuring hydrocarbons [173, 186]. The second, used in solutions for nuclear plants, was made specially with optical fibers for maximum improvement of the optical path/dead volume ratio of the cell [187]. The best performance ob-

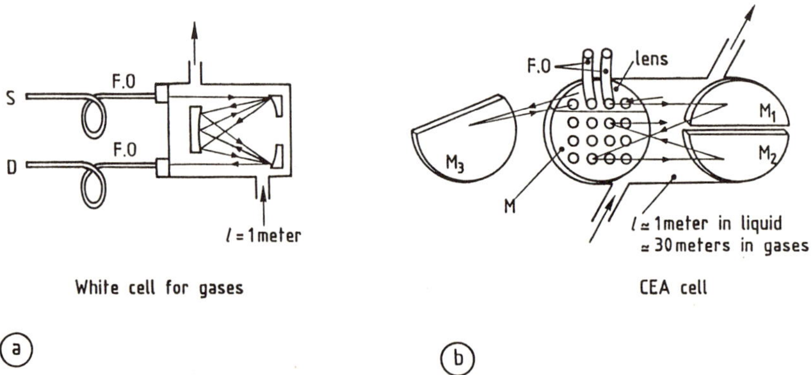

Figure 17-32. Flow sensing in absorption with long optical path cells. (a) White cell [185]; (b) CEA cell [187].

tained with plastic cladding silica (PCS) fibers of diameter 1 mm is about 3 mL in real length for a 75-cm optical path in solution and 300 mL for a 30-m optical path in gases.

Passive immersion probes use a mirror for absorption measurements, either with two monofibers [34] or with multi-fibers [188], as shown in Figure 17-33. The drawback of these mirrors is the possible decrease in their reflectance owing to corrosion or dirt. Their interchangeability, the knowledge of their natural drift, and the use of an internal wavelength reference are necessary in process control. For reflectance on solid products, arrangements of fibers in a ring around a central light-emitting fiber (Figure 17-33 c) have been proposed for infrared measurements in foodstuffs [189]. The measurement of particles in suspension illuminated by a laser beam is considerably facilitated by a fiber-optic array arrangement [190], as shown in Figure 17-33 d.

Today, no active optode is capable of application in industrial environments for process control. Indeed, the required properties such as reversibility, durability, and reliability are very difficult to obtain. Nevertheless, various approaches have been made in the laboratory, eg, pH measurements in acidic or basic media [34], the detection of uranium in phosphate medium [166], and the continuous measurement of the concentration of vapors in polar organic solvents using blue thermal paper placed in a flow-through cell [155]. Moreover, the features of the flow-injection analysis (FIA) technique [69, 70] are being adapted in the laboratory to process control.

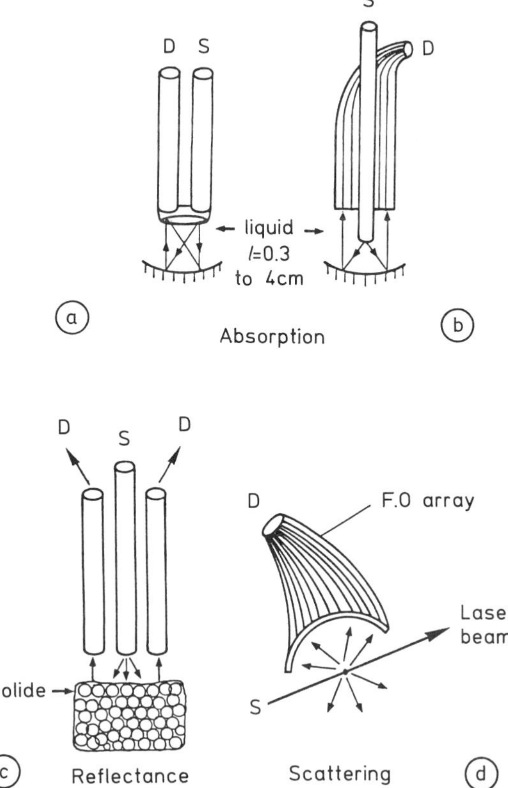

Figure 17-33.
Passive optodes in (a and b) absorption (c) reflectance [188] and (d) scattering [190].

Passive optodes shown in Figure 17-34 are used for Raman or fluorescence spectrometry. Two types of monofiber optode have been described [166], as shown in Figure 17-34a. They require the use of a pierced mirror and suitable optics to separate the source laser and fluorescence emission beams at the fiber–measuring instrument junction. With a single fiber the observation of fluorescence is divided into four zones in order to calculate the effective optical path (ℓ_e). For a fiber with radius r_0 and numerical aperture A_n and observation in a medium with refractive index n_m, this path was evaluated by Deaton [191] as

$$\ell_e = 1.303 \; r_0 \, ((n_m/A_n)^2 - 1)^{1/2} \,. \tag{17-61}$$

In Raman spectrometry, the use of a microsphere and capillary tube at the fiber tip increases the received signal [167]. This observation was also reported recently for a multi-fiber system [192] for which the signal (S) satisfies the equation

$$S = A'/2k \, (1 - e^{-2kz}) \tag{17-62}$$

where z is the length of the capillary, k an attenuation factor due to light losses through the walls and A' a characteristic constant of the capillary and of the experimental conditions. The sensitivity is enhanced by a factor of five compared with a probe without a capillary. A thick-walled glass tube was found to be more efficient than a thin-walled or a stainless-steel tube [193].

The problems caused by interfering light due to the fiber under the influence of the laser excitation have led to multi-fiber configurations. Figure 17-34 shows an arrangement of two fibers, one for excitation (λ_e) and the other for fluorescence emission (λ_m). The acceptance cones show that the optode is more effective if the fibers make an optimum angle to each other [194]. However, this architecture is difficult to achieve industrially. Hence, in fluorescence and

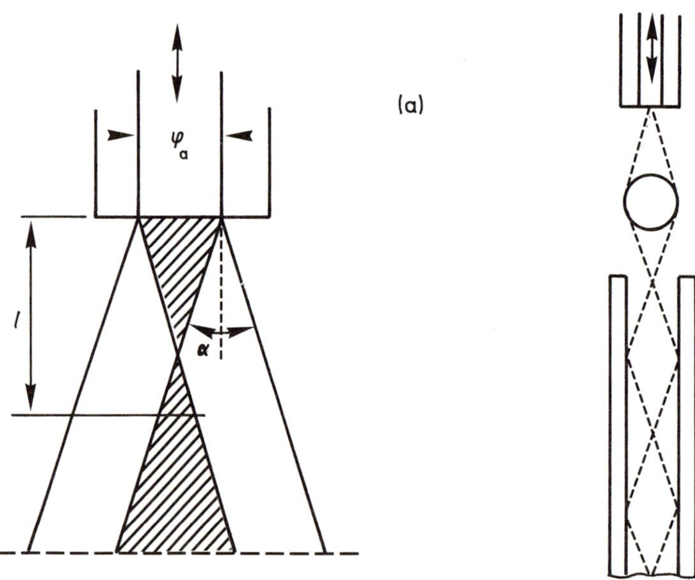

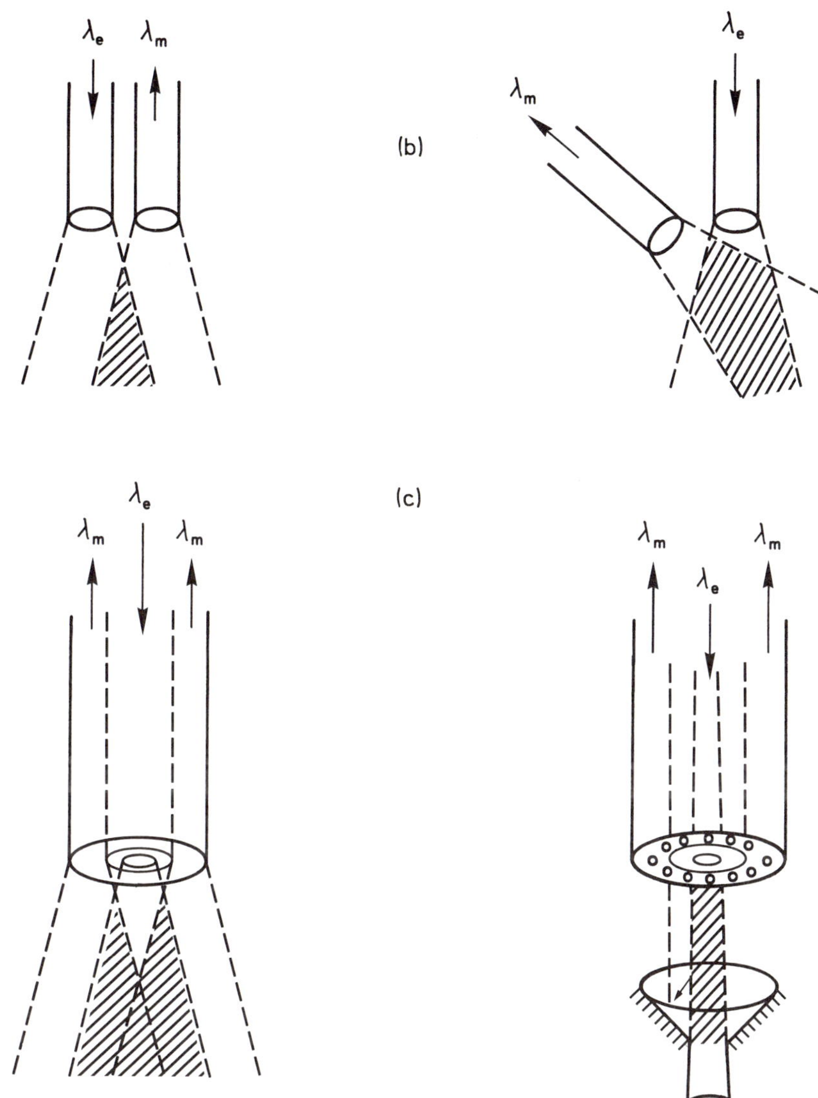

Figure 17-34. Passive optodes in fluorescence and Raman spectrometry for process control. (a) Monofiber; (b) bi-fiber; (c) fibers in crown [198] and crown with drilled mirror [199].

Raman spectrometry for process control, the usual structures of optodes have a central fiber (excitation) and receiver fibers (emission) laid out in a ring over one or more diameters. Various suggestions have been made recently for determining stray light rejection in fiber-optic probes [195], for calculating the fluorescence modulation and the effect of the absorbent species on measurements [196], and for evaluating the relative power received by the adjacent fibers in various configurations [193, 194, 197a and b].

Two proposals for optodes, shown in Figure 17-34c, have also been made. One consists of a structure of encased fibers [198] with a collector fiber in a ring containing the excitation fiber, and the other uses a pierced mirror with a multi-fiber arrangement or encased fibers [199].

All these optodes are still in the development stage. They are indicative of the new requirements in spectrometry for process control and for remote environmental measurements (geology, etc.). Specific products are likely to emerge in the years to come.

17.3.4   Aspects of Real-Time-Techniques

17.3.4.1   Instruments

Typically, a process control instrument must be reliable over long periods of time in spite of the absence of recalibration or even checkups. A minimum time is required between the sampling and transmitting the analytical information to the process. Minimum constraints and maintenance can be imposed in a plant environment, and few analyzers satisfy these conditions today. Current developments represent a preliminary approach, with four types of instruments:

— "Monitoring" devices, generally with discrete wavelengths, such as the Telephot (Commissariat à l'Energie Atomique (CEA), France) [34] or the modular units of Oriel (Stamford, CT, USA). These generally perform on-line in situ remote control functions with optical fibers. However, they need light sources with long lifetimes, typically 10 000 h, without spectral drift. LEDs as sources and PIN photodiodes as detectors are preferable owing to their stability.
— Commercial analyzers mainly intended for laboratory use have been adapted with optical fibers to most spectrophotometers (Photonetics Society). Assemblies of fibers on spectrofluorimeters [82, 199] and Raman [167, 177, 193, 200], infrared (IR), near-IR [189] mid-IR [201], and Fourier transform (FT) IR [202, 203] spectrometers have already proved their value for remote spectroscopic sensing. However, these units are only partially adequate for continuous in situ process control because of their weak dynamic range after coupling of fibers and a maladjusted numerical aperture with respect to optical fibers. Photodiode-array instruments, without spectral scanning, are nevertheless the best for on-line control. This class includes the series of Hewlett-Packard HP 8450–8452 spectrophotometers and the Fluo 2001 spectrometer (Dilor Society) [175] for fluorimetric and Raman determinations.
— New spectrophotometers are designed around optical fibers with wide spectral ranges (0.2–1 or 3 µm) and modular. The spectrometer from Guided Wave is already used in process control (batch). However, its design with spectral scanning and a photomultiplier makes it more similar to an analytical instrument than to a unit for plant process control. It can be associated with twelve optically multiplexed channels. The DTC 1000 spectrophotometer [34] called Spectrofip (Photonetics Society) was recently developed for remote control of nuclear materials (0.4–0.95 µm) with a resolution of 0.6 nm. It is of the video spectrometer type with photodiode arrays (1728). A spectrum is obtained in 10 ms and the minimum time between each measurement is about 0.5 s. It is an ideal device for

on-line monitoring or for analyzing chemical kinetics in process control. A new UV-visible version (DTC 2000) with a Reticon photodiode array is in production (CEA). A Zeiss MCS 21 multi-channel spectrometer (0.2–1 µm) with 512 photodiode elements has been described recently for the analysis of liquid streams in industrial product control [204].
- Developments are based on the possibilities offered by LEDs, laser diodes as sources, and solid-state photodetectors directly coupled by optical fibers. Typical examples of multi-point measurements with optodes have been proposed by Smardzewski [183] (see Section 17.3.2). These developments should make the instruments more rugged and cheaper, and hence more accessible to industrial users.

17.3.4.2 Limitations

Apart from the constraints inherent in the process, the limitations to the use of optical fibers generally concern the following:

- The spectral extent of the fibers in the UV or IR region, and the natural absorption of the material (see Chapter 12).
- The short dynamic range available in measuring instruments; this is especially true in spectrophotometry. A test example for in situ control of the dissolution of medicinal drugs [205] is given in Figure 17-35. The dynamic range of an HP 8450 instrument, initially 30 dB, coupled by fibers drops and ultimately becomes zero in the UV region below 250 nm.

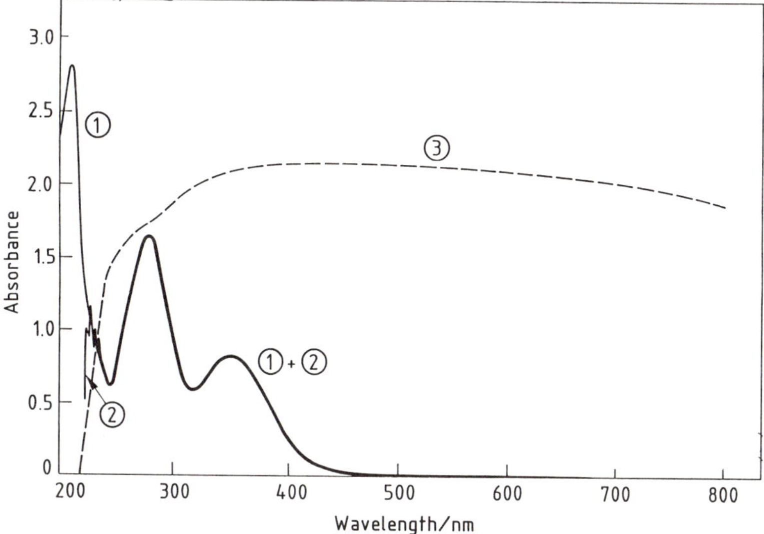

Figure 17-35. Spectrum of *o*-nitrophenol with HP 8450 spectrophotometer. 1: Measurement without fibers; 2: with fibers (2 × 6m length); 3: determination is possible below the disposal dynamic range represented by the dotted line [205].

- The connection balance of the optical line: the overall loss is the sum of several losses, viz, those due to the fiber-source coupling, to Rayleigh scattering and the absorption in the fibers, the optical losses due to the optode geometry, to the connectors, the multiplexing, and the fiber–detector coupling. In process control, all losses must remain constant to avoid disturbing the measurement.
- The use of probes or optodes with a geometry fixed for a wide range of concentration implies operating limits in process control. A technique has been used successfully in absorption spectrometry [34]. As shown in Figure 17-36, its consists in shifting the measurement wavelength according to the concentration zones to be measured, in order to remain within the limits of the signal-to-noise ratio corresponding to the performance of the measuring instrument.
- The use of an internal reference is usually necessary in process control. Several techniques have been suggested: a spectral reference in wavelength [172, 186], a reference optical line

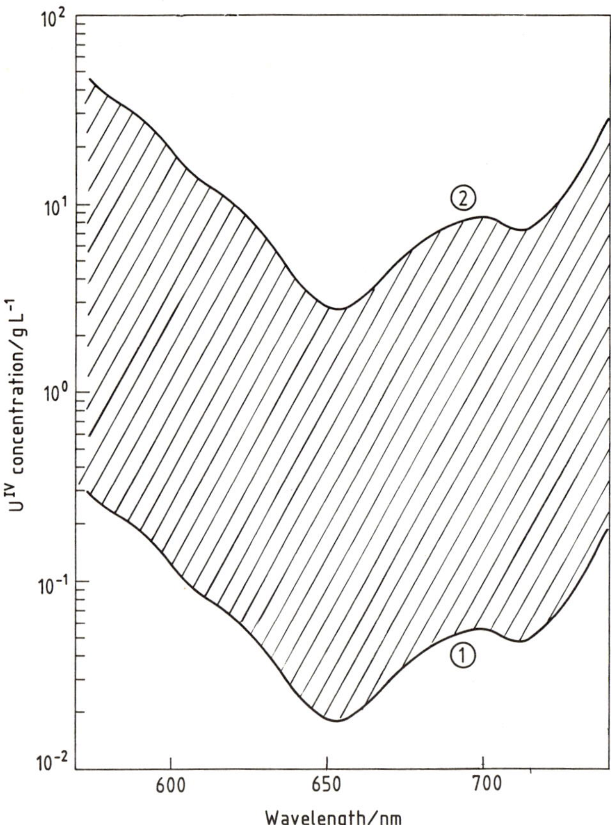

Figure 17-36. Usable zone (hatched) in on-line control of uranium(IV) with a 4-cm optical path length probe. 1: Minimum for 10^{-2} absorbance; 2: maximum for 1.6 absorbance. Example: the 600-nm wavelength on the side of the absorption spectrum can be used for a concentration range between 0.1 and 20 gL^{-1}. The reference wavelength for uranium(IV) is at 530 nm.

[206], the Raman spectrum of the fiber [82], the addition of a compound at the end of optode for an intensity reference [207], the fluorescence lifetime [208], or using a matrix lifetime technique for compensating for the effects of quenching in fluorescence [209]. This reference is necessary in process control, eg, for following the spectral losses of the optical line.

17.3.5 Applications

The main applications of optical fibers in monitoring and process control are very diversified. In particular, we can mention the following.

- Nuclear control for reprocessing plants [163] with spectrophotometry [34, 206, 210–212] and fluorescence [166, 175, 207] for gaseous isotopic separations [213, 214] or in the environmental sciences (see Chapter 18). The results in pilot plants [34, 206, 210] indicate a significant future for these new techniques.
- Remote control of flammable gases, methane, propane, etc. [161, 173, 178, 215].
- Optical measurement of oil leaks with planar waveguides [171] or fibers [168].
- Monitoring of some gases such as NO_2 [186, 216] and combustion gases [217] and measurements in the mid-IR region with fluoride glasses for CO_2 at 4.5 μm [201] or anesthetic agents [203].
- Remote determinations in industrial electroplating baths [184].
- Turbidity monitoring [190, 218].
- Foodstuffs [189] for in situ determination of moisture (flour, cakes), proteins, and sugar.
- Remote sensing in the near-IR region for solvent mixtures [219].
- Raman spectrometry [167, 177, 192–194, 197a], which is also potentially suitable for remote process control.

17.4 References

[1] Pietrzyk, D. J., *Analytical Chemistry,* 2nd ed.; London: Academic Press, 1979.
[2] Bishop, E., *Indicators;* Oxford: Pergamon Press, 1972.
[3] Meites, L., *Handbook of Analytical Chemistry;* New York: McGraw-Hill, 1963.
[4] Charlot, G., *Les Méthodes de la Chimie Analytique;* Paris: Masson, 1961.
[5] Kolthoff, J. M., Elving, B. J., *Treatise in Analytical Chemistry, Part I, II*; New York: Wiley, 1975.
[6] Janata, J., *Anal. Chem.* **59** (1987) 1351–1356.
[7] Edmonds, T. E., Flatters, N. J., Jones, C. F., Miller, J. N., *Talanta* **35** (1988) 103–107.
[8] Vogel, A. I., *A Textbook of Quantitative Analysis,* 4th ed.; London: Longman, 1979.
[9] Albert, A., Serjeant, E. P., *The Determination of Ionization Constants,* 3rd ed.; London: Chapman and Hall, 1984.
[10] Peterson, J. I., Goldstein, S. R., Fitzgerald, R. V., Buckhold, D. K., *Anal. Chem.* **52** (1980) 864–869.
[11] Goldstein, S. R., Peterson, J. I., *U.S. Pat. 4200100,* 1980.
[12] Boisdé, G., Perez, J. J., *Proc. SPIE, Int. Soc. Opt. Eng.* **798** (1987) 238–245.
[13] Narayanaswamy, R., Sevilla, III F., *Anal. Chim. Acta* **189** (1986) 365–369.
[14] Zhujun, Z., Seitz, W. R., *Anal. Chim. Acta* **160** (1984) 47–55.
[15] Cramp, J. H., Ferguson, R. F., Brian, D. B., *Eur. Pat. 61884,* 1982.

[16] Kirkbright, G. F., Welti, N. A., *Eur. Pat. 126600*, 1984.

[17] Boisdé, G., Perez, J. J., *Eur. Pat. 284513*, 1988.

[18] Lübbers, D. W., Opitz, N., *Ger. Off. 2508637*, 1976; *Chem. Abstr.* **85** (1976) 173867; *Ger. Off. 2720370*, 1979; *Chem. Abstr.* **90** (1979) 214720.

[19] Lübbers, D. W., Völkl, K. P., Opitz, N., *Ger. Off. 3001669*, 1981; *Chem. Abstr.* **96** (1982) 14814.

[20] Wolfbeis, O. S., Kroneis, H., Offenbacher, H., *Ger. Off. 3343636, 3343637*, 1984.

[21] Marsoner, H. E., Kroneis, H., Wolfbeis, O. S., *Eur. Pat. Appl. 109959*, 1984; *Chem. Abstr.* **101** (1984) 65205; *U.S. Pat. 4587101*, 1987.

[22] Wolfbeis, O. S., Leiner, M. J. P., Posch, H. E., *Aust. Pat. Appl. 19A 755/87*, 1987.

[23] Cramp, J. H. W., Mortier, R. M., Murray, R. T., Reid, R. F., *Br. Pat. 2103786A*, 1981.

[24] Peterson, J. I., *U.S. Pat. 4194877*, 1980.

[25] Goldstein, S. R., Peterson, J. I., Fitzgerald, R. V., *J. Biomech. Eng. Trans. ASME* **102** (1980) 141–146.

[26] Watson, R. M., Markle, D. R., Ro, Y. M., Goldstein, S. R., McGuire, D. A., Peterson, J. I., Patterson, R. E., *Am. J. Physiol. Heart Circ. Physiol.* **246** (1984) H232-H238.

[27] Taite, G., Young, R. B., Wilson, G. J., Stewart, D. J., McGregor, D. C., *Am. J. Physiol. Heart Circ. Physiol.* **12H** (1982) H1027-H1031.

[28] Abraham, E., Markle, D. R., Fink, S., Ehrlich, H., Tsang, M., Smith, M., Meyer, A., *Anesth. Analg.* **64** (1985) 731–736.

[29] Suidan, J. S., Young, B. K., Hetzel, F. W., Seal, H. R., *Clin. Chem.* **29** (1983) 1566.

[30] Goldfinch, M. J., Lowe, C. R., *Anal. Biochem.* **138** (1984) 430–436.

[31] Kirkbright, G. F., Narayanaswamy, R., Welti, N. A., *Analst* **109** (1984) 15–17.

[32] Kirkbright, G. F., Narayanaswamy, R., Welti, N. A., *Analyst* **109** (1984) 1025–1028.

[33] Edmonds, T. E., Ross, I. D., *Anal. Proc.* **22** (1985) 206–207.

[34] Boisdé, G., Blanc, F., Perez, J. J., *Talanta* **35** (1988) 75–82.

[35] Boisdé, G., *unpublished results*, 1988.

[36] Moreno, M. C., Martinez, A., Millan, P., Camara, C., *J. Mol. Struct.* **143** (1986) 553–556.

[37] Jones, T. P., Porter, M. D., *Anal. Chem.* **60** (1988) 404–406.

[38] Saari, L. A., Seitz, W. R., *Anal. Chem.* **54** (1982) 821–823.

[39] Hirschfeld, T., Deaton, T., Milanovich, F., Klainer, S., *Opt. Eng.* **22** (1983) 527–531.

[40] Milanovich, F. P., Hirschfeld, T., Wang, T., Klainer, S. M., Walt, D., *Proc. SPIE, Int. Soc. Opt. Eng.* **494** (1984) 18–24.

[41] Fuh, M.-R. S., Burgess, L. W., Hirschfeld, T., Christian, G. D., Wang, F., *Analyst* **112** (1987) 1159–1163.

[42] Munkholm, C., Walt, D. R., Milanovich, F. P., Klainer, S. M., *Anal. Chem.* **58** (1986) 1427–1430.

[43] Kawabata, Y., Tsuchida, K., Imasaka, J. I., Ishibashi, N., *Anal. Sci.* **3** (1988) 7–9.

[44] Jordan, D. M., Walt, D. R., Milanovich, F. P., *Anal. Chem.* **59** (1987) 437–439.

[45] Lübbers, D. W., Opitz, N., Speiser, P. P., Bisson, H. J., *Z. Naturforsch., Teil C* **32** (1977) 133–134.

[46] Opitz, N., Lübbers, D. W., *Sensors Actuators* **4** (1983) 473–479.

[47] Wolfbeis, O. S., Fürlinger, E., Kroneis, H., Marsoner, H., *Fresenius' Z. Anal. Chem.* **314** (1983) 319–324.

[48] Offenbacher, H., Wolfbeis, O. S., Fürlinger, E., *Sensors Actuators* **9** (1986) 73–84.

[49] Wolfbeis, O. S., Offenbacher, H., *Sensors Actuators* **9** (1986) 85–91.

[50] Gehrich, J. L., Lübbers, D. W., Opitz, N., Hansmann, D. R., Miller, W. W., Husa, J. K., Yafuso, M., *IEEE Trans. Biomed. Eng.* **33** (1986) 117–132.

[51] Miller, W. W., Yafuso, M., Yan, Chen. F., Hui, H. K., Arick, S., *Clin. Chem.* **33** (1987) 1538–1542.

[52] Wolfbeis, O. S., Weis, L. J., Leiner, M. J. P., Ziegler, W. E., *Anal. Chem.* **60** (1988) 2028–2030.

[53] Attridge, J. W., Leaver, K. D., Cozens, J. R., *J. Phys. E* **22** (1987) 548–553.

[54] Benaim, N., Grattan, K. T. V., Palmer, A. W., *Analyst* **111** (1986) 1095–1097.

[55] Grattan, K. T. V., Mouaziz, Z., Selli, R. K., *Proc. SPIE Int. Soc. Opt. Eng.* **798** (1987) 230–237.

[56] Guthrie, A. G., Narayanaswamy, R., Welti, N. A., *Talanta* **35** (1988) 157–159.

[57] Wolfbeis, O. S., Schaffar, B. P. H., Kaschnitz, E., *Analyst* **111** (1986) 1331–1334.

[58] Wolfbeis, O. S., Baustert, J. H., *J. Heterocycl. Chem.* **22** (1985) 1215–1218.

[59] Wolfbeis, O. S., Marhold, H., *Fresenius' Z. Anal. Chem.* **327** (1987) 347–350.

[60] Wyatt, W. A., Poirier, G. E., Bright, F. V., Hieftje, G. M., *Anal. Chem.* **59** (1987) 572–576.

[61] Kadin, H., *Anal. Lett.* **17** (1984) 1245–1257.

[62] Laitinen, H. A., Harris, W. E., *Chemical Analysis,* 2nd ed.; New York: McGraw-Hill, 1975.

[63] Hammett, L. P., *Physical Organic Chemistry,* New York: McGraw-Hill, 1970.

[64] Opitz, N., Lübbers, D. W., *Adv. Exp. Med. Biol.* **169** (1984) 907.

[65] Christian, L. M., Seitz, W. R., *Talanta* **35** (1988) 119-122.

[66] Monici, M., Boniforti, R., Buzzigoli, G., De Rossi, D., Nannini, A., *Proc. SPIE, Int. Soc. Opt. Eng.* **798** (1987) 294-300.

[67] Trettnak, W., Leiner, M. J. P., Wolfbeis, O. S., *Biosensors* **4** (1988) 15-26.

[68] Arnold, M. A., *Anal. Chem.* **57** (1985) 565-566.

[69] Růžička, J., Hansen, E. H., *Anal. Chim. Acta* **173** (1985) 3-21.

[70] Woods, B. A., Růžička, J., Christian, G. D., Charlson, R. J., *Anal. Chem.* **58** (1986) 2496-2502.

[71] Seitz, W. R., *CRC Crit. Rev. Anal. Chem.* **19** (1988) 135-173.

[72] Schultz, J. S., Mansouri, S., Goldstein, I. J., *Diabetes Care* **5** (1982) 245-253.

[73] Zhujun, Z., Mullin, J. L., Seitz, W. R., *Anal. Chim. Acta* **184** (1986) 251-258.

[74] Adler, J. D., Ashworth, D. C., Narayanaswamy, R., Moss, R. E., Sutherland, I. O., *Analyst* **112** (1987) 1191-1192.

[75] Wolfbeis, O. S., Schaffar, P., *Anal. Chim. Acta* **198** (1987) 1-12.

[76] Wolfbeis, O. S., Hochmuth, P., *US Pat. 4892640,* 1990.

[77] Schaffar, B. P., Wolfbeis, O. S., Leitner, A., *Analyst* **113** (1988) 693-697.

[77b] Morf, W. E., Seiler, K., Lehmann, B., Behringer, Ch., Hartmann, K., Simon, W., *Pure & Appl. Chem.* **61** (1989) 1613.

[78] Saari, L. A., Seitz, W. R., *Anal. Chem.* **55** (1983) 667-670.

[79] Saari, L. A., Seitz, W. R., *Analyst* **109** (1984) 655-657.

[80] Zhujun, Z., Seitz, W. R., *Anal. Chim. Acta* **171** (1985) 251-258.

[81] Dowling, S. D., Seitz, W. R., *Spectrochim. Acta, Part A* **40** (1984) 991-993.

[82] Hirschfeld, T., *J. Instrum. Soc. Am.* **85** (1985) 305-317.

[83] Saari, L. A., Seitz, W. R., *Anal. Chem.* **56** (1984) 810-813.

[84] Bright, F. V., Poirier, G. E., Hieftje, G. M., *Talanta* **35** (1988) 113-118.

[85] Kawabata, Y., Tahara, R., Imasaka, T., Ishibashi, N., *Anal. Chim. Acta* **212** (1988) 267-271.

[86] Urbano, E., Offenbacher, H., Wolfbeis, O. S., *Anal. Chem.* **56** (1984) 427-429.

[87] Hirschfeld, T., Deaton, T., "Remote Fiber Optic Fluorimetry: Specific Analyte Optrodes", invited paper, Pittsburg Conference Atlantic City, NJ, March 1982.

[88] Wolfbeis, O. S., (ed.), *Fiber Optic Chemical Sensors and Biosensors,* Vol. 1 & 2; Boca Raton: CRC Press, 1991.

[89] Fujiwara, K., Fuwa, K., *Anal. Chem.* **57** (1985) 1012-1016.

[90] Wyatt, W. A., Bright, F. B., Hieftje, G. M., *Anal. Chem.* **59** (1987) 2272-2276.

[91] Narayanaswamy, R., Russel, D. A., Sevilla, F., III, *Talanta* **35** (1988) 83-88.

[92] Hardy, E. E., David, D. J., Kapany, N. S., Unterleitner, F. C., *Nature (London),* **257** (1975) 666-667.

[93] Hirschfeld, T., Deaton, T., Milanovich, F., Klainer, S., *The Feasibility of Using Fiber Optics for Monitoring Groundwater Contaminants, EPA Report AD-89-F-2A074,* 1983.

[94] Narayanaswamy, R., Sevilla, III, F., *Analyst* **111** (1986) 1085-1088.

[95] Narayanaswamy, R., Sevilla, III, F., *Int. J. Opt. Sensor,* **1** (1986) 403-413.

[96] Martinez, A., Moreno, M. C., Camara, C., *Anal. Chem.* **58** (1986) 1877-1881,

[97] Benaim, N., Grattan, K. T. V., Palmer, A. W., *Analyst* **111** (1986) 1095-1097.

[98] Wolfbeis, O. S., Hochmuth, P., *Mikrochim. Acta* **III** (1984) 129-148.

[99] Wolfbeis, O. S., Schaffar, B. P., Chalmers, R. A., *Talanta* **33** (1987) 867-870.

[100] Dittmer, T., Nuyken, O., Pask, S., *J. Chem. Soc., Perkin Trans. II* (1988) 151-155.

[101] Halwani, J., Dexpert-Ghys, J., Piriou, B., Caro, P., *Analusis* **15** (1987) 229-305.

[102] Trettnak, W., Wolfbeis, O. S., *Fresenius' Z. Anal. Chem.* **326** (1987) 547-550.

[103] Kautsky, H., Hirsch, A., *Z. Anorg. Allg. Chem.* **222** (1935) 126; a brief summary on this and related work was given by Kautsky, H., *Trans. Faraday Soc.* **35** (1939) 216.

[104] Pollak, M., Pringsheim, P., Terwoord, D., *J. Chem. Phys.* **12** (1944) 295.

[105] Zakharov, I. A., Grishaeva, T. I., *Zh. Anal. Khim.* **35** (1980) 481; **36** (1981) 112; *Zh. Prikl. Spektrosk.* **31** (1979) 703.

[106] Freeman, T. M., Seitz, W. R., *Anal. Chem.* **53** (1981) 98.

[107] Hendricks, H. D., *Mol. Phys.* **20** (1979) 189.

[108] Bergman, I., *Nature (London)* **218** (1968) 396.

[109] Lübbers, D. W., Opitz, N., *Z. Naturforsch. Teil C* **30** (1975) 532.

[110] Lübbers, D. W., Opitz, N., *Sensors Actuators* **4** (1983) 641.

[111] Lübbers, D. W., Opitz, N., *Anal. Chem. Symp. Ser.* **17** (1983) 609.

[112] Wolfbeis, O. S., Offenbacher, H., Kroneis, H., Marsoner, H., *Mikrochim. Acta* **I** (1984) 153.

[113] Kroneis, H. W., Marsoner, H. J., *Sensors Actuators* **4** (1983) 587.

[114] Peterson, J. I., Fitzgerald, R. V., Buckhold, D. K., *Anal. Chem.* **56** (1984) 62.

[115] Cox, M. E., Dunn, B., *Appl. Opt.* **24** (1985) 2114.

[116] Cox, M. E., Dunn, B., *Proc. SPIE Int. Soc. Opt. Eng.* **576** (1985) 60.

[117] Opitz, N., Lübbers, D. W., in: *Oxygen Transport to Tissue, IV,* Bruley, D., Bicher, H. I., Reneau, D. (eds.); New York: Plenum Press, 1984, p. 261.

[118] Wolfbeis, O. S., Posch, H. E., Kroneis, H. K., *Anal. Chem.* **57** (1985) 2556.

[119] Bacon, J. R., Demas, J. N., *Anal. Chem.* **59** (1987) 2780.

[120] Lippitsch, M. E., Pusterhofer, J., Leiner, M. J. P., Wolfbeis, O. S., *Anal. Chim. Acta* **205** (1988) 1.

[121] Stevens, B., *U.S. Pat. Appl. 3612866,* 1971; *Chem. Abstr.* **76** (1972) 20945.

[122] Stanley, C. C., Kropp, J. L., *U.S. Pat. 3725658,* 1973, TRW Inc.

[123] Burkhard, O., Barnikol, W. K. R., *Prax. Klin. Pneumol.* **37** (1983) 805.

[124] Marsoner, H., Kroneis, H., *Eur. Pat. Appl. 109958,* 1984; *Chem. Abstr.* **101** (1984) 65204.

[125] Lübbers, D. W., Gehrich, J., Opitz, N., *Life Support Syst.* **4** (1986) 94.

[126] Landsman, M. L. J., Knop, N., Kwant, G., Mook, G. A., Zijlstra, W. G., *Eur. J. Physiol.* **373** (1978) 273, and references cited therein.

[127] Schmitt, J. M., Meindl, J. D., Mihn, F. G., *IEEE Trans. Biomed. Eng.* **33** (1986) 98.

[128] Milano, M. J., Kim, K. Y., *Anal. Chem.* **49** (1977) 555.

[129] Opitz, N., Lübbers, D. W., in: *Proceedings of International Symposium on Oxygen Transport to Tissue,* Vol. IV, Bruley, D., Bicher, H. I., Reneau, D. (eds.); New York, London: Plenum Press, 1984, p. 757.

[130] Zhujun, Z., Seitz, W. R., *Anal. Chim. Acta* **160** (1984) 305.

[131] Heitzmann, H. A., Kroneis, H. K., *U.S. Pat. 4557900,* 1986.

[132] Munkholm, C., Walt, D. M., Milanovich, F. P., *Talanta* **35** (1988) 109.

[133] Vurek, G. G., Peterson, J. I., Goldstein, S. W., Severinghaus, J. W., *Fed. Proc. Fed. Am. Soc. Exp. Biol.* **41** (1982) 1484.

[134] Hirschfeld, T., Miller, F., Thomas, S., Miller, H., Milanovich, F., Gaver, R., *J. Lightwave Technol.* **5** (1987) 1027.

[135] David, D. J. Wilson, M. C., Ruffin, D. S., *Anal. Lett.* **9** (1976) 389.

[136] Smock, P. L., Orofino, T. A., Wooten, G. W., Spencer, W. S., *Anal. Chem.* **51** (1979) 505.

[137] Giuliani, J. F., Wohltjen, H., Jarvis, N. L., *Opt. Lett.* **8** (1983) 54.

[138] Arnold, M. A., Ostler, T. J., *Anal. Chem.* **58** (1986) 1137.

[139] Wolfbeis, O. S., Posch, H. E., *Anal. Chim. Acta* **185** (1986) 321.

[140] Caglar, P., Narayanaswamy, R., *Analyst* **112** (1987) 1285.

[141a] Shariari, M. R., Zhou, Q., Sigel, G. H., *Opt. Lett.* **13** (1988) 407.

[141b] Ozawa, s., Hauser, P. C., Seiler, K., Tan, S. S. S., Morf, W. E., Simon W., *Anal. Chem.* **63** (1991) 640–644.

[142] Wolfbeis, O. S., Sharma, A., *Anal. Chim. Acta* **208** (1988) 53.

[143] Narayanaswamy, R., Sevilla, F., *Fresenius' Z. Anal. Chem.* **329** (1988) 789.

[144] Wolfbeis, O. S., "Fiber Optic Fluorosensors in Analytical and Clinical Chemistry", in: *Molecular Luminescence Spectroscopy. Methods and Applications,* Vol. 2, Schulman, S. G. (ed.); New York: Wiley, 1988, Chap. 3, p. 240.

[145] Butler, M. A., *Appl. Phys. Lett.* **45** (1984) 1007.

[146] Inaba, H., Chan, K., Ito, M., *Proc. SPIE, Int. Soc. Opt. Eng.* **514** (1984) 211.

[147] Culshaw, B., Dakin, J., *Optical Fiber Sensors: Systems & Applications,* Norwood (MA): Artech House, 1989.

[148] Saito, M., Takizawa, M., Ikegawa, K., Takami, H., *J. Appl. Phys.* **63** (1988) 269.

[149] Farahi, F., Leilabady, P.A., Jones, J. D. C., Jackson, D. A., *J. Phys. E.* **20** (1987) 435, and references cited therein.

[150] Chan, K., Ito, H., Inaba, H., *J. Lightwave Technol.* **5** (1987) 1706.

[151] Guiliani, J. F., Jarvis, N. I., *Sensors Actuators* **6** (1984) 107.

[152] Tai, H., Tanaka, H., Yoshino, T., *Opt. Lett.* **12** (1987) 437.

[153] White, K. O., Watkins, W. R., *Appl. Opt.* **14** (1975) 2812.

[154] Sachse, G. W., Hill, G. F., Wade, L. O., Perry, M. G., *J. Geophys. Res.* **92** (1987) 2071.

[155] Posch, H. E., Wolfbeis, O. S., Pusterhofer, J., *Talanta* **35** (1988) 89.

[156] Dickert, F. L., Lehmann, E. H., Schreiner, S. K., Kimmel, H., Mages, G. R., *Anal. Chem.* **60** (1988) 1377.

[157] Russell, A. P., Fletcher, K. S., *Anal. Chim. Acta* **170** (1985) 209.

[158] Posch, H. E., Wolfbeis, O. S., *Sensors Actuators* **17** (1988) 77.

[159] Pitt, G. D., Extrance, P., Neat, R. C., Batchelder, D. N., Jones, R. E., Barnett, J. A., Pratt, R. H., *IEE Proc.* **132**, Part J (1985) 214-248.

[160] Stephens, H. S., Stapleton, C. A. (ed.), *Optical Techniques in Process Control, The Hague 14-16th June 1983;* Cranfield, Bedford, U.K.: *BHRA Fluid Engineering,* (1983).

[161] Stueflotten, S., Christensen, T., Iversen, S., Hellvik, J. D., Almas, K., Wien, T., Graav, A., *Proc. SPIE, Int. Soc. Opt. Eng.* **514** (1984) 87-90.

[162] Place, J.D., Pinchbeck, D., *Fiber Optic Sensors for the Gas Industry, 51st Autumn Meeting, London;* London: Institution of Gas Engineers, No. 1269, 1985.

[163] Boisdé, G., Blanc, F., Mauchien, P., Perez, J. J., in: *Fiber Optic Chemical Sensors and Biosensors,* Wolfbeis, O. S. (ed.); Boca Raton, FL: CRC Press, 1991, Chap. 14.

[164] Boisdé, G., Perez, J. J., *Proc. SPIE, Int. Soc. Opt. Eng.* **1012** (1988) 58-65.

[165] Perez, J. J., Boisdé, G., Goujon de Beauvivier, M., Chevalier, G., Isaac, M., *Analusis* **8** (1980) 344-351.

[166] Milanovich, F. P., Hirschfeld, T., *Adv. Instrum.* **38** (1983) 404-418.

[167] MacGreery, R. L., Fleischmann, M., Hendra, P., *Anal. Chem.* **55** (1983) 146-148.

[168] Kawahara, K. F., Fiuten, R. A., Silvus, H. S., Newman, F., Frazar, J. F., *Anal. Chim. Acta* **15** (1983) 315-327.

[169] Harmer, A. L., *IEE, Proc.* **221** (1983) 104-108.

[170] Faires, L. M., Bieniewski, T. M., Apel, T. M., Niemczyk, T. M., *Appl. Spectrosc.* **39** (1985) 9-13.

[171] Wada, H., Okuda, E., Yamasaki, T., in: *4th International Conference on Optical Fiber Sensors, Tokyo, Proceedings of OFS 86;* Tokyo: Optics Industry Technical Developments Association, 1986, pp. 109-112.

[172] Boisde, G., Perez, J. J., *Proc. SPIE Int. Soc. Opt. Eng.* **514** (1984) 227-232.

[173] Chan, K., Ito, H., Inaba, H., *Appl. Opt.* **23** (1984) 3415-3420.

[174] Hirschfeld, T., Callis, J. B., Kowalski, B. R., *Science* **226** (1984) 312-318.

[175] Decambox, P., Kirsch, B., Mauchien, P., Moulin, C., in: *Proceedings of Opto 88, ESI Public.;* Paris: Masson, 1988, pp. 357-361.

[176] Gomy, C., Jouan, M., Nguyen Quy Dao., *Anal. Chim. Acta* **215** (1988) 211-221.

[177] Nguyen Quy Dao, Plaza, P., Joyeux, M., *Analusis* **14** (1986) 334-343.

[178] Azakawa, Y., Fukunaga, H., Inaba, H., in: *4th International Conference on Optical Fiber Sensors, Tokyo, Proceedings of OFS 86;* Tokyo: Optics Industry Technical Develpment Association, 1986, pp. 135-138.

[179] Laude, J. P., *Eur. Pat. 37787 B1,* 1984.

[180] Tagonan, G. L., *U.S. Pat. 4274706,* 1981.

[181] Laude, J. P., Lerner, J. M., *Proc. SPIE, Int. Soc. Opt. Eng.* **503** (1984).

[182] Fuh, M.-R. S., Burgess, L. W., *Anal. Chem.* **59** (1987) 1780-1783.

[183] Smardzewski, R. R., *Talanta* **35** (1988) 95-101.

[184] Freeman, J. E., Childers, A. G., Steele, A. W., Hieftje, G. M., *Anal. Chim. Acta* **177** (1985) 121-128.

[185] White, J. U., *J. Opt. Soc. Am.* **32** (1942) 285.

[186] Inaba, H., Kobayasi, T., Hirama, M., Hamza, M., *Electron. Lett.* **15** (1979) 749-751.

[187] Boisdé, G., Boissier, A., *U.S. Pat. 4188126, 4225232,* 1980.

[188] Mathisen, E. S., *IBM Tech. Disclos. Bull.* **18** (1976) 3757-3758.

[189] Bellon, V., Boisdé, G., *Proc. SPIE, Int. Soc. Opt. Eng.* **1055** (1989) 350-358.

[190] Extrance, P., Pitt, G. D., in: *Optical Techniques in Process Control, The Hague 14-16th June, 1983,* Stephens, H. S., Stapleton, C. A. (Eds.); Cranfield, Bedford, UK: BHRA Fluid Engineering, 1983, pp. 43-54.

[191] Deaton, T., "Instrumentation and Methodology for Remote Fiber Fluorimetry", *Thesis;* University Microfilms International, 1984, p. 115.

[192] Bonnaventure, B., Lucas, M., *Analusis* **17** (1989) 185–187.

[193] Hendra, P. J., Ellis, G., Cutler, D. J., *J. Raman Spectrosc.* **19** (1988) 413–418.

[194] Plaza, P., Nguyen Quy Dao, Jouan, M., Fevrier, H., Saisse, H., *Appl. Opt.* **25** (1986) 3448–3454.

[195] Skogerboe, K. J., Yeung, E. S., *Anal. Chem.* **59** (1987) 1812–1815.

[196] Ping Huan, Walt, D. R., *Anal. Chem.* **59** (1987) 2391–2394.

[197a] Louch, J., Ingle, J. D., *Anal. Chem.* **60** (1988) 2540–2542.

[197b] Myrick, M. L., Angel, S. M., Desiderio, R., *Appl. Opt.* **29** (1990) 1333–1344.

[198] Plaza, P., Nguyen Quy Dao, Jouan, M., Fevrier, H., Saisse, H., *Analusis* **15** (1987) 504–507.

[199] Boisdé, G., Kirsch, B., Mauchien, P., Rougeault, S., in: *Proceedings of Opto 88, ESI Public.;* Paris: Masson, 1988, pp. 294–299.

[200] Newby, K., Reichert, W. M., Andrade, J. D., Benner, R. E., *Appl. Opt.* **23** (1984) 1812–1815.

[201] Pruss, D., Dreyer, P., Koch, E., *Proc. SPIE Int. Soc. Opt. Eng.* **799** (1987) 117–122.

[202] Archibald, D. D., Lin, T. T., Honigs, D. E., *Appl. Spectrosc.* **42** (1988) 468–472.

[203] Pruss, D., *Mater. Sci. Forum* **32–33** (1988) 321–330.

[204] Gerlinger, H., Schlemmer, H., *GIT Fachz. Lab.* **3** (1988) 167–170.

[205] Boisdé, G., Rougeault, S. Perez, J. J., in: *Proceedings of Opto 86, ESI Public.;* Paris: Masson, 1986 pp. 71–82.

[206] Groll, P., Romer, J., Roder, I., Pershon, M., *Anal. Chim. Acta* **190** (1986) 265–269.

[207] Malstrom, R. A., *Report DP-MS 85-76,* 1985, E. I. DuPont de Nemours, Savannah River Laboratory.

[208] Klainer, S., Hirschfeld, T., Bowman, H., Milanovich, F., Perry, D., Johnson, D., *Report LBL-11891,* 1981, Lawrence Berkeley, Laboratory, Berkeley, CA.

[209] Hieftje, G. M., Haugen, G. R., *Anal. Chim. Acta* **123** (1981) 225–261.

[210] O'Rourke, P. E., Van Hare, D. R., *Report DP-MS 87-100,* 1987, E. I. DuPont de Nemours, Savannah River Laboratory.

[211] Baldwin, D. L., Stromatt, R. W., *Report PNL-SA 15318,* 1988, Pacific Northwest Laboratory, Richland.

[212] Brown, M. L., Mills, C. L., Kyffin, T. W., *Report NDR-1266 (D),*1985, UKAEA Northern Division.

[213] Cates, M. R., Allison, S. W., Marshall, B., Davies, T. J., Franks, L. A., Nelson, M. A., Noel, B. W., *Proc. SPIE Int. Soc. Opt. Eng.* **506** (1984) 64–69.

[214] Saturday, K. A., *Anal. Chem.* **55** (1983) 2459–2460.

[215] Chan, K., Ito, H., Inaba, H., *Appl. Phys. Lett.* **45** (1984) 220–222.

[216] Kobayasi, K., Hirama, M., Inaba, H., *Appl. Opt.* **20** (1981) 3279–3280.

[217] Kychakoff, G., Kimball-Linne, M. A., Hanson, R. K., *Appl. Opt.* **22** (1983) 1426–1428.

[218] Extrance, P., Pitt, G. D., Scott, B. J., Verrells, M. V., *IEE Proc.* **221** (1983) 109–113.

[219] Schirmer, R. E., Zetter, M., *Adv. Instrum.* **42** (1987) 173–178.

18 Applications of Optochemical Sensors for Measuring Environmental and Biochemical Quantities

Wolfgang Trettnak, Michael Hofer, Otto S. Wolfbeis, Joanneum Research, Graz, Austria

18 Contents

18.1   Environmental Sensors

Optical fibers offer the unique possibility of performing spectroscopy at the site of interest. In other words, the sample no longer has to come to the spectrometer, but rather the spectrometer goes to the sample. Hence all problems associated with sampling, surface deposition of analyte, delay in analysis, work-up procedures, and decomposition of samples and analytes during the work-up can, in principle, be avoided. It is for these reasons that tremendous expectations are associated with fiber sensor technology in environmental sciences.

18.1.1   Monitoring Water Flow and Quality

18.1.1.1   Water Flow

The use of fluorescent tracers in groundwater studies is important for following the route of the water flow from specific injection sources. In recent work, [1] tracers such as Rhodamine 6G have been shown to be detectable after argon laser excitation at 488 nm over a distance of typically 100 m at 10 ppb (10 μg L^{-1}) concentrations. By increasing the laser power to a few milliwatts, tracers can be detected at ca 1 ppm over distances as long as ca 1 km.

18.1.1.2   Mineral Oils and Polycyclic Aromatic Hydrocarbons

Since these polutants have a strong intrinsic fluorescence, they may be monitored by fluorimetry. There are two ways to perform this: in one, light from an intense light source is directed onto the sample of interest, eg, in airborne sensing of oil spills on the surface of the sea, and in the other, a fiber-optic light guide is used to direct the light beam to the frequently inaccessible site of analysis. Polycyclic aromatic hydrocarbons and oil spills can now be detected with airborne fluorosensors using UV excitation, preferentially from a laser source. The field performance of several laser fluorosensors for the detection of oil and dye spills has been described [2–4].

In the second method, remote sensing is performed via fibers. One of the first fiber sensors for oil pollution monitoring [5] consisted of a fiber without cladding, inserted through a stainless-steel capillary, and coated with an organophilic material (a long-chain alkyl group). The input radiation is at 632.8 nm from a low-power helium–neon laser. The normal total internal reflection is changed, as the pollutant is absorbed by the organophilic cladding because of a change in its refractive index. The sensor can detect diesel oil (17 mg L^{-1}) and crude oil (3mg L^{-1}) in water. Of course, it is not very selective.

The performance of a field instrument for the detection of automotive fuel (gasoline) in the ppm range has also been reported in detail [6]. The sensor consists of a fluorescent dye attached to the tip of a fiber, which acts as a monochromatic light source, and a 2-cm long coating of a refractive index-matched material possessing selective high affinity for gasoline. Figure 18-1 shows the working principle of the sensor. In the absence of gasoline, the returning light has a high intensity because air (or water) has a smaller refractive index than the core

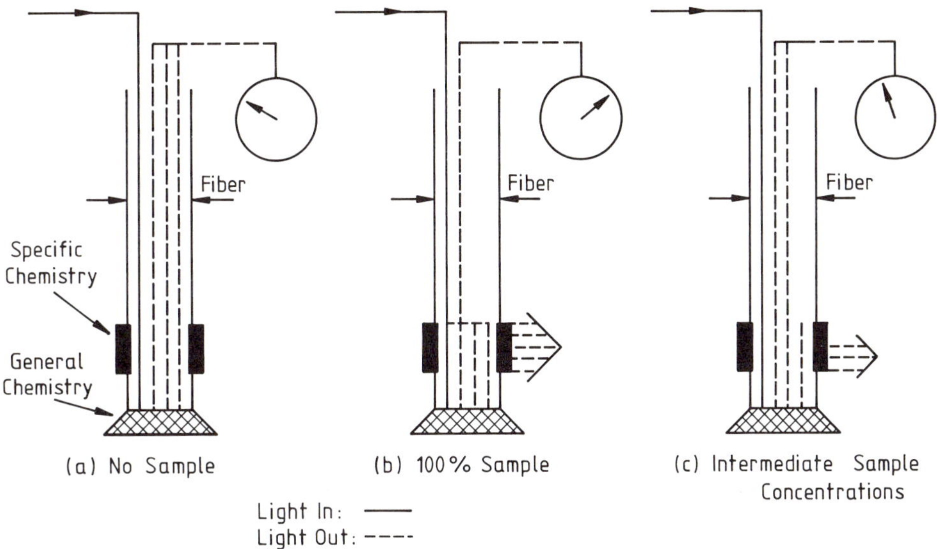

Figure 18-1. Working principle of a sensor based on analyte-induced changes in refractive index of the cladding. From [6].

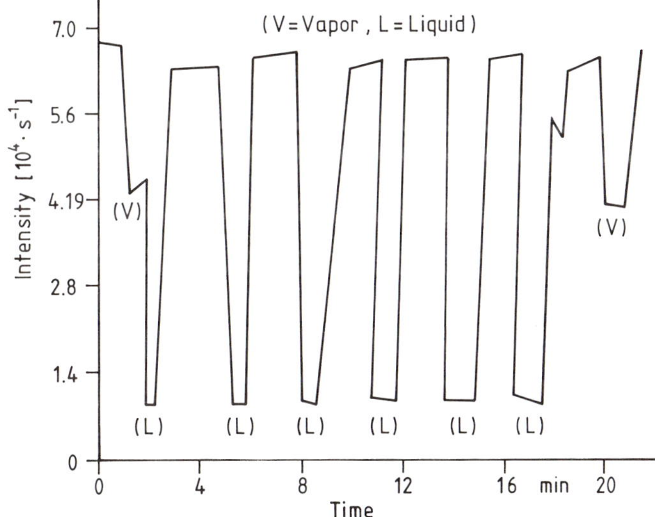

Figure 18-2. Response time and reversibility of a gasoline sensor exposed to gasoline in vapor (V) and as a liquid (L). From [6].

and thus acts as a cladding. Under these circumstances, all of the light is propagated through the fiber. When gasoline is present, it is attracted through the high-affinity material coated on the surface of the core, and there is an increase in refractive index at the surface of the core. As the refractive index of the cladding changes and approaches that of the core, light leaks out from the sides. The return signal decreases proportionally to the amount of gasoline present. The sensor is fast and completely reversible (Figure 18-2) and responds to all brands of gasoline. This concept is of particular interest because it may be applied to any other species for which a selective polymer coating is found. A sensor for detection of ppm levels of gasoline in water is available now from FiberChem Inc., Las Vegas (NV).

18.1.1.3 Phytoplankton and Chlorophyll

Airborne laser fluorosensing of chlorophyll (Chl) and phycoerythrine promises to be a valuable technique for the determination of phytoplankton abundance and distribution in water [7, 8]. To enhance and control photosynthesis, most free drifting algae contain a number of so-called primary absorbing pigments. The absorbed excitation energy is transferred to Chl A, whose emission (centered at 685 nm) is predominantly observed. Alternatively, the emission of the strongly fluorescent photoreceptor phycoerythrine may be measured.

Compared with passive radiometric techniques, the fluorosensor methods for phytoplankton have several attractive features which are unique. These include relative insensitivity to daylight or cloudiness, specifity of response to Chl, and the potential for depth profiling using range-gating techniques. On the other hand, they are sensitive to other factors such as energy transfer, water composition, algae species, and sediments [8, 9].

A method which has been developed to circumvent the lack of understanding physical processes in remote Chl sensing is the normalization of fluorescence signals by the Raman water OH stretch signal occurring at a frequency shift of 3418 cm^{-1} [8]. Nearly linear relationships between normalized fluorescence and in situ Chl concentration (0.2–20 µg L^{-1}) were observed by Brislow et al. [10] with 470-nm excitation and by Hoge and Swift [11] with 532-nm excitation.

Linear vertical gradients in Chl concentration may be evaluated by the same method by applying a combination of stochastic and analytical techniques. This makes it possible to model LIDAR systems with realistic geometric constraints and reduced computer resources [12]. In this approach the media are respresented by combinations of layers with constant Chl concentrations within each layer, but varying Chl concentration in different layers. Statistically significant differences can be seen under certain conditions in the Raman water-normalized fluorescence signals between homogeneous and nonhomogeneous cases, when the Chl concentrations vary linearly with depth in the water medium.

A simple and sensitive in situ algae fluorosensor based on fiber optics was described by Lund [13]. It was used for in situ studies of natural sweet water. Excitation spectroscopy of four algae species revealed differences in their excitation spectra, which therefore may yield more specific and more accurate information than data obtained with single-wavelength excitation.

18.1.1.4 Groundwater and Seawater Quality

Natural groundwater and creek water is known to exhibit blue emissions owing to the presence of fluorescent humic and fulvic acids. Lignin sulfonate pollution manifests itself in an increase in the fluorescence of waste water. In many cases its strong enough to be of use in the continuous fluorimetric sensing of pollutants. Lakshman [14] analyzed a number of soil and run-off samples for the nature and stability of the intrinsic fluorescence. A strong correlation exists between intensity and water quality parameters such as total carbon, total organic carbon, and total inorganic carbon. It can be used in the quantitative measurement of agricultural pollution.

Chudyk et al. [15] reported results obtained in a test of remote fluorescence analysis of groundwater contaminants using UV laser light sources and fiber optics. Several priority pollutants such as phenols, toluene, and xylenes and also naturally occuring humic acid, all of which display UV fluorescence, were readily detected over a distance of 20–25 m. Typical detection limits over this distance are 10 ppb for phenol, 1 ppb for *o*-cresol, and 0.07 ppb for xylenes. A prototype instrument for monitoring phenolic groundwater contaminants has been described, and its suitability demonstrated by using phenol as a model contaminant [16].

Chlorinated hydrocarbons are frequently encountered in ground-, drinking, and waste waters. A method for the detection of organic chlorides has been developed [6]. It makes use of the Fujiwara color reaction, which is a general assay for organic compounds that contain two or more chlorine atoms. Reaction of alkaline pyridine with polychlorinated organic compounds yields a red fluorescent chromophore whose formation is detected. The formation of the fluorophore is irreversible, so the device is a probe rather than a sensor.

The sensor consists of a single 200-μm glass core fiber to transmit exciting light (in initial experiments the argon laser 488-nm line) to the optode, and to carry the fluorescence back to the spectral sorter. The sensing unit (Figure 18-3) is fixed to the tip of the fiber with a 400-μm glass capillary. Pyridine is kept in the optode by a Mylar membrane which retains the

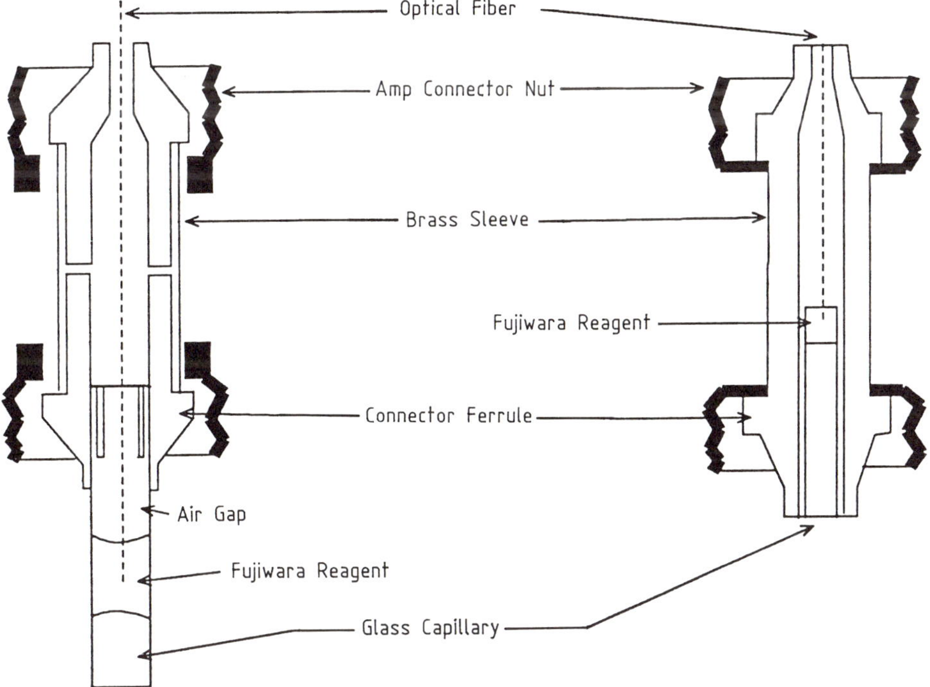

Figure 18-3. Comparison of two configurations for a fiber-optic probe for chlorinated hydrocarbons based on the Fujiwara reaction. From [6].

pyridine and water, but passes the organic chloride. A 10 mol/L quaternary ammonium hydroxide solution keeps the pyridine basic, since the Fujiwara reaction yields a nonfluorescent product when it turns acidic. Field tests [17, 18] with a portable instrument which is shown diagrammatically in Figure 18-4 resulted in sensitivities of less than 1 ppm for chloroform and related pollutants.

Mercury is one of the most toxic pollutants in waste waters. It can be detected by virtue of its static quenching effect on the fluorescence of indoles. When indole-3-acetic acid is covalently linked via long spacer groups to quartz beads welded to the tip of a quartz fiber or onto a quartz slide, its fluorescence (occurring at 350 nm) is quenched by mercury(II). Thus, the emitted light provides a continuous and reversible signal for the mercury(II) concentration of the solution in contact with the fiber end [19]. Mercury(II) concentrations down

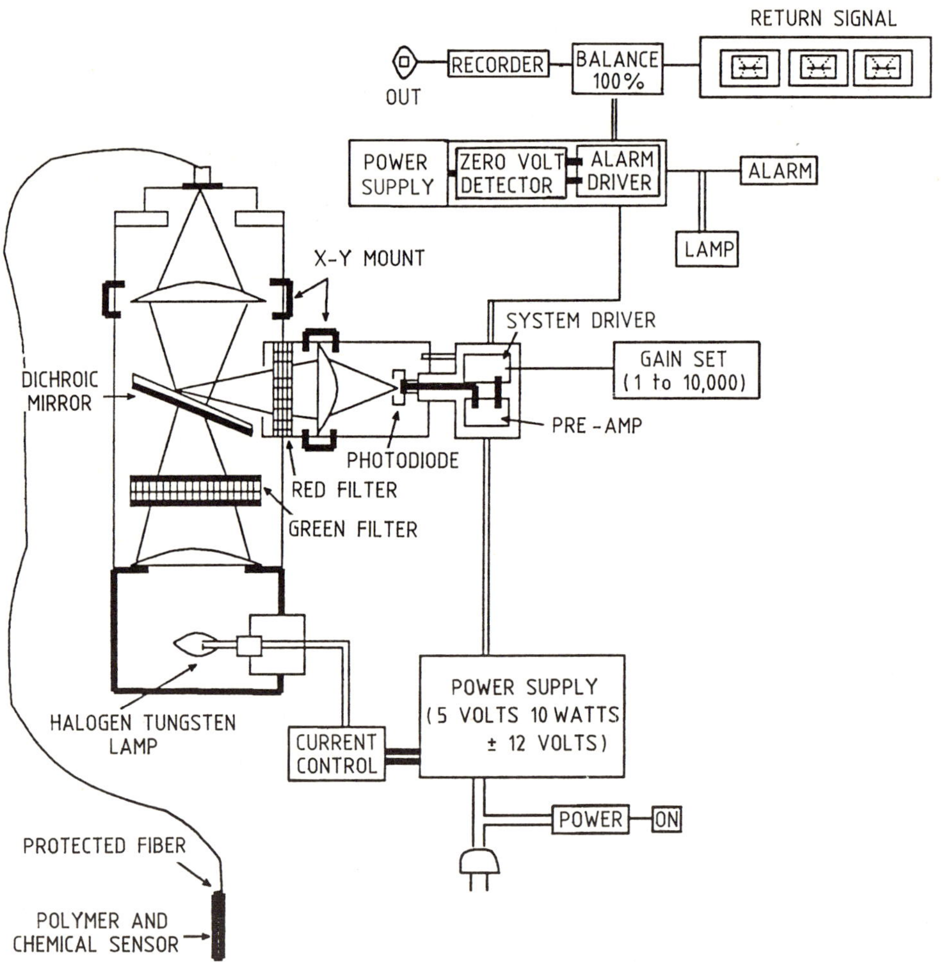

Figure 18-4. Schematic diagram of a portable fiber-optic probe for detection of chlorinated hydrocarbons in groundwater. From [6].

to 1 µmol/L are detectable, but the strong background from both fiber and solution under UV excitation (295 nm) frequently limit the sensitivity.

A promising field of application of fiber optodes is in seawater analysis. In order to sense electrolytes such as zinc [20], a probe was developed in which a fluorogenic ligand (8-hydroxyquinoline-5-sulfonate) was pushed through an ultrafiltration membrane into the optical viewing volume of the fiber at a rate of ca 1 µL min^{-1}. The formation of the fluorescent metal-ligand complex is seen be the fiber. Since this ligand also complexes magnesium(II), a more specific ligand was subsequently used. The sensor responds reversibly within a few seconds to 20–200 ppb concentrations of zinc. Work has also been performed with oxygen optodes to measure the depth profiles of oxygen pressure in seawater [21].

A great deal of work [22, 23] has been dedicated to the development of sensors for actinide ions such as UO_2^{2+} (uranyl). These are encountered in groundwater around nuclear fuel-processing facilities and in soil after underground tests. Sensing radioactive analytes is now possible using fiber optics because the UO_2^{2+} ion shows an intrinsic fluorescence with maxima at 518 and 546 nm, which can be detected via fibers. Its intensity is strongest in acidic solution at pH 1.6. In 8% phosphoric acid, the emission is particularly intense, being 150 times stronger than in water, probably by virtue of the combined effect of lowering the pH and dehydration by formation of the phosphate complex. The lifetime is 197 µs in 8% phosphoric acid, but is strongly reduced by external quenchers such as iron(III). Unfortunately, the quantum yields are strongly affected by other factors, eg, lowering the temperature causes an exponential increase and added salts a decrease [22].

The Purex process is another process that requires continuous monitoring of uranyl ion [24]. This has been achieved by exciting the fluorescence of uranyl ion in nitric acid medium at 337 nm using a nitrogen laser. Phosphoric acid is additionally added because it complexes the ion, enhances its fluorescence, and thereby lowers the detection limit.

The extension of the technique to the remote sensing of other fluorescers such as plutonium(III) is obvious. Detection of 10^{-14} mol/L Pu(III), which in solution is nonfluorescent, was said to be a realistic expectation when a special technique is applied which, however, does not work on-line [22]. Remote fiber sensing of Pu^{III} has obvious safety advantages, but the absence of a fluorescent plutonium species in *solution* is a difficulty in remote fiber spectrometry. Pu^{III} can be determined [22] because it has an energy level nearly coincident with one of the highly fluorescent terbium(III) ions, allowing the measurement of Pu^{III} by the quenching of a known amount of added Tb^{III}. To enhance the effect, the lifetime of the Tb^{III} fluorescence is lengthened by using D_2O as a solvent and adding complexing agents such as acetate or trifluoroacetate.

Plutonium and uranium can be assayed during the oxidation and reduction steps in the extraction processes using a remote photometer equipped with a dye laser light source [25]. Silica fibers with 200-µm cores served as light guides, so to have access to a highly radioactive environment. However, given the detection limits of the order of 0.14 g L^{-1} for plutonium and 3.0 g L^{-1} for uranium, the sensitivities are still too low for environmental applications.

18.1.1.5 *Detection of Enzyme Inhibitors*

Although many pesticides (eg, organophosphates) are fluorescent by themselves (albeit weakly), the spectra are spread over a wide range. A number of measurements would therefore be required to detect these species with a plain fiber. A more general way to detect pesticides is to group them according to their biological effect, since they act as inhibitors of the enzyme acetylcholinesterase (AChE). Other inhibitors include heavy metal ions such as mercury and lead.

A reservoir sensor has been developed [23] in which the AChE-inhibiting properties of pesticides are exploited. In this sensor type (Figure 18-5), cholinesterase is immobilized on a solid support and brought into contact with a chromogenic synthetic enzyme substrate. The enzyme converts the yellow substrate into a blue product. The resulting absorption is measured through a fiber using a yellow light-emitting diode (LED) as a light source. The presence of inhibitors retards reaction and reduces the absorbance. The method can be used to detect both

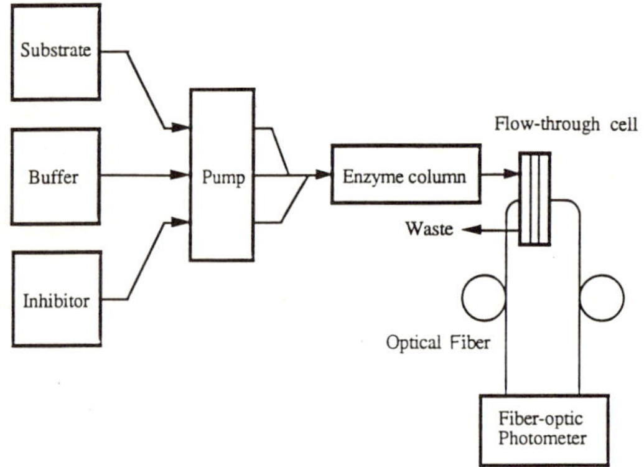

Figure 18-5.
Experimental arrangement for continous monitoring of inhibitors of acethyl-cholinesterase (AChE) in drinking water. The enzyme converts the yellow substrate into a blue product. If the AChE is inhibited, the blue dye no longer is formed and this is measured via the fiber-optic photometer.

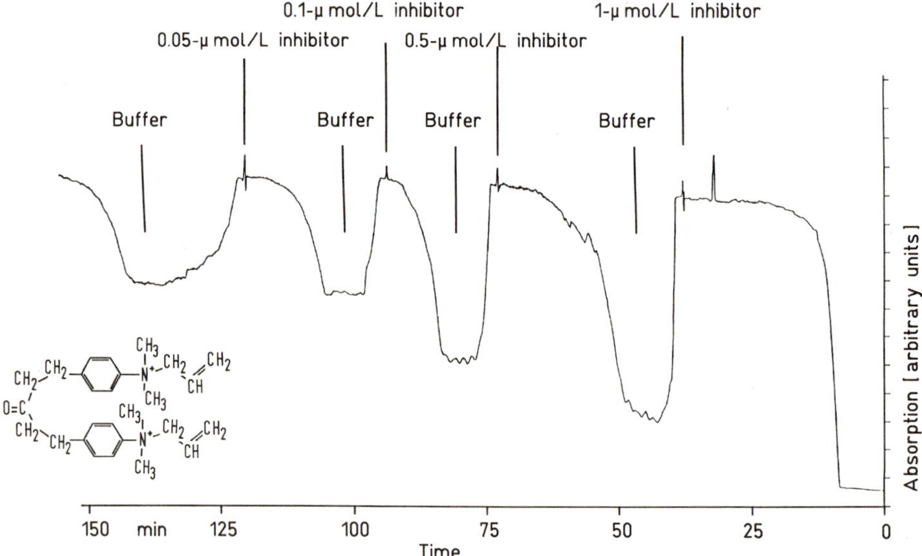

Figure 18-6. Reversible inhibition of acetylcholinesterase by nanomolar quantities of anticholinesterase (BW 284 C51, a reversible allyl-ammonium type inhibitor) as monitored with the arrangement shown in Fig. 18-5.

pesticides in drinking water [26] and other AChE inhibitors [27]; 20 nmol/L of a pesticide are easily detectable (Figure 18-6).

18.1.2 Air Pollution

Although it is generally agreed that fiber sensors will greatly extent the range of instrumentation useful for air pollution studies, there is a certain lack of methods that work with real samples. Most papers describe principles rather than applications. Thus, the sensors developed for sulfur dioxide, hydrogen sulfide, carbon monoxide, and other species (as described in Section 17-2) await their application to practical environmental studies.

Reports on sensors for nitrogen oxides (NO_x) are scarce, although there are various observations as to the quenching of the fluorescence of polycyclic aromatic hydrocarbons by NO_x. This may be utilized for sensing purposes. The chemiluminescence produced in the reaction between NO and ozone is another attractive possibility that might be exploited for this purpose.

An extensive literature exists on the measurement of primary gaseous pollutants via their intrinsic absorption using optical fibers [28]. Typical examples include methane [29], NO_2 [30], SO_2 [31], propane [32], and fluorinated compounds [33]. The whole wavelength range that is compatible with standard fibers has been covered. Table 18-1 lists the species that have been studied so far, together with the wavelengths of operation.

Hydrogen cyanide can be detected in air by oxidation of an atmospheric sample with chloramine-T impregnated on an ion-exchange resin [37] to yield cyanogen bromide, which reacts with picoline and barbituric acid to give a color change at ca 545 nm. HCN levels down to 1 ppm are detectable with this probe. Toxic gases such as 1 ppm of chlorine, hydrogen chloride, and NO_2 have been detected by virtue of the quenching of fluorescent porphyrin films incorporated into Langmuir–Blodgett films [38]. The assays were performed without the use of fibers, but they could easily have been used. City gas has been sensed via fibers through

Table 18-1. Gaseous species that can be detected via fiber absorptiometry or fluorimetry in the 0.2–5-μm wavelength range.

Species	Analytical wavelength / μm	Reference
Carbon dioxide	ca. 4.9	34
Alkanes	ca. 3.6 and 1.68	32
Methane	1.66, 1.33	29, 32, 35
Ammonia	1.28–1.32	36
HCl	ca. 1.1, 3.46	–
Carbon monoxide	ca. 2.2 or 4.7	–
Nitrogen dioxide	3.44, 3.09, 0.48 (broad)	–
Nitrogen oxide	5.32, 2.69	–
Chlorine	0.4–0.5	–
SO_2, NO, NO_2, O_3	0.22–0.25	31
Hydrogen sulfide	4.13, 3.83, 3.73, 2.64	–
SO_3	3.61	–

methane absorption employing a low-loss optical fiber link and an InGaAsP LED at ca 1.33 μm [39]. The status of optical fiber sensors up to 1986 has been briefly reviewed [40]. Other kinds of optical pollution sensors are described in Section 17.3.

In addition to the fiber sensors, there have been numerous reports on remote fiberless optical detection of pollution from stationary sources. Mobile remote sensors can have cost advantages over on-site instruments and also are much more versatile. Pollutants in smokestack emissions have been identified by irradiation with a laser from a mobile unit equipped with a telescope, a monochromator, and low-noise detection electronics. The gases and particles in the plume scatter the laser light in various directions. A fraction is scattered back to the receiver and analysed to detect the amount and type of gas in the plume. Many of these methods await their application to fiber-optic sensing schemes which are inherently safer than direct laser spectroscopic schemes.

18.2   In Vivo Sensors for Chemical Species

An attractive feature of fiber sensors is the possibility of performing in vivo tests and monitoring. Numerous fiber-optic sensors have already been described that measure physical parameters of the human body [41]. Pressure, temperature, physiological flow, strain, motion, displacement, or flow velocity can be monitored by optical methods such as variable reflection, laser Doppler velocimetry, optical holography, or diffraction. In this section the application of optosensing methods to the determination of molecular species encountered in clinical and biomedical analysis is described.

One may differentiate between two types of sensors. In one type, a plain fiber is used to monitor the intrinsic properties of an analyte such as hemoglobin with its intense red color, or reduced nicotine adenine dinucleotide (NADH) with its strong blue fluorescence under UV excitation. In the other type, the indicator-mediated sensors for pH, pO_2, pCO_2, electrolytes, and other uncolored low molecular weight species discussed in previous sections may also be used in vivo. In fact, many of those have been designed especially for biomedical applications. Some of them have found their first commercialization in blood analyzers.

Two serious problems are encountered in the design, manufacture, and performance of in vivo sensors: the lack of biocompatibility of the materials used and the poor long-term stability. The latter, however, plays only a minor role in the case of disposable optodes, which are in use only for the duration of a particular operation or test. Disposable sensor heads for clinical analytes seem to be the most promising candidates for practical use at present. Another problem results from the need for sterilization, which is difficult to solve in the case of biosensors with their thermally labile components such as enzymes.

Current problems with photobleaching and signal drift do not severely limit the performance of these sensors. In fact, they are sometimes not intended to provide extremely precise absolute analytical data. Rather, it is the relative change of a parameter that has to be recognized as soon as possible during the course of an operation on a critically ill patient, or a fairly quick indication of whether abnormal clinical values require a more reliable assay in a clinical laboratory.

Fiber-optic sensors may also be combined to form a bundle of fiber applicable to the simultaneous sensing of, eg, physiological pH, oxygen, electrolytes, anesthetics, glucose, creatinine, temperature, and flow rate. This is possible because of the minute size of the fibers. There is considerable industrial activity in this direction.

18.2.1 Plain Fiber Sensors

Plain fiber sensors are simpler in design and manufacture than are indicator-phase sensors (Section 18.2.2). However, they are mostly less selective and suffer from interferences by any substance that adsorbs at the same analytical wavelength, even small changes in the optical parameters such as refractive index at the sensing tip, and ambient light. They therefore are usually operated at at least two wavelengths and/or with background subtraction.

Indicator-mediated optrodes frequently have an optical isolation at the fiber tip to prevent ambient light and sample from interfering in the optical system, and a fairly constant tip chemistry with its almost invariable refractive index (see Figure 18-11). However, as with plain fiber sensors, additional discriminations such as pulsed excitation plus electronic background subtraction or sequential excitation are useful techniques for improving selectivity.

Plastic, glass, and quartz fibers can be used, depending on the analytical wavelength applied. Plastic fibers have a larger aperture, are flexible, and are easy to work with, but have poor transmission below 420 nm and do not tolerate multiple heat sterilization. Glass fibers are available in small size and have low attenuation, but have a small aperture and are suitable only down to ca 380 nm. Quartz fibers transmit in the UV region, but the aperture is not better than that of glass. Both glass and quartz fibers are fragile and present a risk of breaking in vivo. However, plain fiber sensors usually have sufficient biocompatibility because a wide variety of materials with proven biocompatibility are available. This contrasts with the indicator-mediated sensors, where material constraints often limit compatibility. One way to improve the in vivo performance is to cover the sensor surface with covalently bound heparin.

18.2.1.1 Fiber Oximeters

The fact that hemoglobin (Hb) and oxyhemoglobin (OxyHb) have different absorption spectra has resulted in the development of a widely known class of sensors called oximeters. These devices measure the *oxygen saturation* (OS) of blood by exploiting the difference in the spectra of Hb and OxyHb. This contrasts with the oxygen optodes described in Section 17.2, which measure *oxygen partial pressure* pO_2. OS is defined as the amount of oxygen carried by the Hb in erythrocytes relative to the maximum carrying capacity. Typical arterial blood is 95% saturated and venous blood may be 75% saturated. Another substance whose absorption of light in the near-infrared region changes according to its reduced or oxidized state is cytochrome aa_3, the terminal member of the respiratory chain.

Fiber oximeters are commercially available from Oximetric (1212 Terra Bella Ave., Mountain View, CA 94034, USA) and BTI (4765 Walnut Street, Boulder, CO 80301, USA). The Oximetrix 3 instrument uses three LEDs, emitting at different wavelengths, to send light down a single fiber in a disposable catheter. The reflected light returns through a receiving fiber to the instrument, is detected, and then analyzed by a microprocessor in real time. The state of the art up to 1983 has been reviewed [42].

18.2.1.2 *Miscellaneous Plain Fiber Sensors for Chemical and Biochemical Species*

The intensity of reflected or transmitted light at specific wavelengths in the visible and near-infrared regions can provide useful information concerning the metabolic state of tissue [43]. Fiber optics can play an important role in this research area. Metabolism can also be studied at an early stage in the respiratory chain by monitoring the redox state of NADH. The redox potential E_h of the NAD$^+$/NADH system at pH 7.0 is known to be $(0.320 \pm 0.03) \times \log ([NAD^+]/[NADH])$ Volts, and reflects the redox status of the tissue. NADH displays strong fluorescence at 455 nm when excited at ca 340–350 nm through a quartz fiber. NAD$^+$, in contrast, is nonfluorescent. Very small areas of tissue can be examined so that metabolism at a localized level can be followed [44–46].

NADH is favorably excited by a nitrogen laser system in combination with single-strand fibers [46]. In a typical experiment, the 337-nm nitrogen laser output is split into two beams: one beam is focused onto the incident end of the optical fiber and serves to excite NADH, and the other is used to pump a dye laser whose output at 805 nm is also focused onto to incident end of the fiber. The intensity of the reflected 805-nm line (and, in later work, the 586-nm line) can be used to correct for blood-induced disturbances of the rat heart. The metabolic state of the moving heart was determined in a similar way. Applications of this procedure include pharmacological studies by monitoring the protective effect of drugs (such as pentobarbital) in myocardial ischemia, evaluation of revascularization procedures by monitoring NADH for estimation of the reversibility of ischemic injury, and myocardial protection during cardiopulmonary bypass.

A fiber sensor for continuous monitoring of NADH in culture broths of bioreactors is available from Ingold (Switzerland). Another class of substances whose absorption of light in the near-infrared region changes according to its reduced or oxidiced state are cytochromes, the terminal members of the respiratory chain and the oxidized and reduced flavoproteins.

An elegant way to suppress stray light in plain fiber fluorimetry of biological samples is to make time-resolved measurements in the nanosecond time regime. In a typical arrangement [47], light pulses from a spark source are selected in an optical filter and passed through a beam splitter. The exciting beam then enters a light guide, the distal end of which contacts the sample. Fluorescence, scattered light, and reflected light are conveyed back to the beam splitter where part of it is reflected into the detector and registered after selection in a secondary filter. The time difference between light source and detector pulses is measured with a delayed coincidence system. The method is not applicable to the determination of the concentrations of fluorescent biomolecules of fluorescent biological species such as NADH or FAD, but can be used to probe minor changes in the microenvironment of such species.

Plain fibers also have been used to perform in vivo measurements of drugs in minute volumes of body fluid by using conventional methods of absorptiometry [48] and fluorimetry [49], but also more sophisticated methods such as two-photon-excited fluorimetry and sequentially excited fluorimetry [49]. Detection limits of ca 0.5 µmol/L were determined for the drug adriamcin (doxorubicin) in interstitial fluid and whole blood and are comparable for the three techniques. In constructing the sensor, the protective coating and optical cladding of a 200-μm quartz fiber was stripped off to about 2 cm from its terminus. Then, a piece of close-fitting glass capillary tubing was slid over the stripped fiber and attached to a modified 20-gauge needle with epoxy resin. The available capillary volume was about 200 nL. Using this sensor, virtually all fluorescence emanated from the first 1 mm of the solution in front of the fiber tip. In an alternative version, the fiber may be threaded into a bluntended 26-gauge needle of a syringe. When the fiber is withdrawn a certain distance, a defined sample chamber is created within the needle. Figure 18-7 shows a schematic diagram of the experimental arrangement. The sensor has also been used for pharmacokinetic studies.

Similarly, bilirubin and methotrexate can be determined in serum with a fiber-optic system terminated in a 19-gauge hypodermic needle and a reflective cap at its end so to produce a small absorbance cell [50]. Although these methods usually do not yield absolute analyte concentrations owing to background absorption or fluorescence of serum, they do reflect relative concentration changes sufficiently correctly. This kind of sensor was expected to be applicable also to the photometric determination of other important clinical analytes such as drugs, toxins, and biomolecules, but the limited selectivity and sensitivity of spectrophotometry will possibly also limit the scope of the method when applied directly to serum or whole blood.

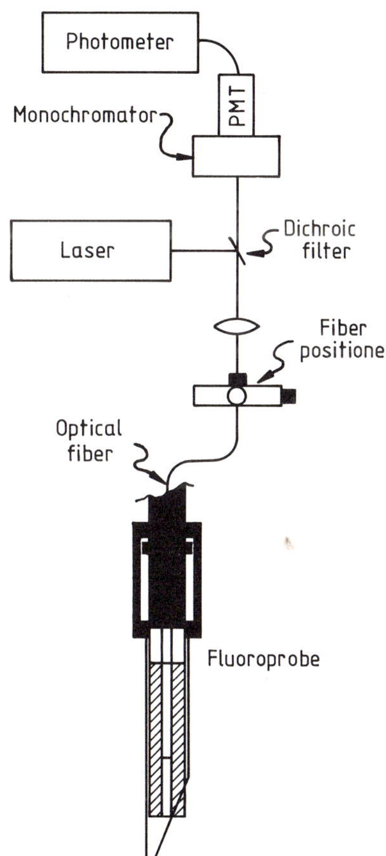

Figure 18-7.
Experimental arrangement for performing in vivo fluorimetry with a plain single-strand fiber inserted into the body through a standard needle catheter. The volume "seen" by the fiber in ca. 200 nL. From [49].

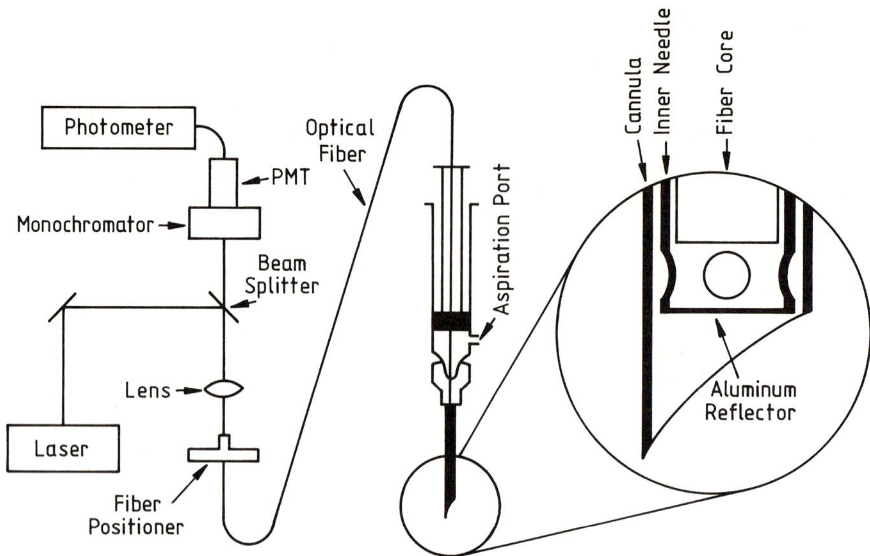

Figure 18-8. Fiber-optic catheter for invasive absorbance measurement and associated instrumentation. From [50].

A concept of the fiber-optic-based absorbance sensor is shown in Figure 18-8. Again, the fiber is threaded in a standard catheter, thus allowing its insertion into tissue or body fluids. A piece of aluminium foil is attached to the end of the inner needle (which contains the optical fiber). Fluids can be drawn into the sample irradiation cavity by aspiration, the volume between foils and fiber being filled through the hole shown. Typical pathlengths (twice the distance from the fiber tip to the foil) are 0.5–4.3 mm.

An optical fiber probe has been used to both excite and collect fluorescence from a suspension of cells [51]. The configuration of the probe in the flow cytometer is such that one or a few cells are sensed at a time, with a convenient cell concentration. With fluorescently labelled antibodies to cellular antigens, the fiber cytometer is able to identify the presence of a specific set of cells with high sensitivity.

18.2.2   In Vivo Sensors for pH, Oxygen, Carbon Dioxide, and Electrolytes

Most pH sensors described in Section 17.1 are, in principle, suitable for measuring physiological pH values. They exhibit sufficient precision to allow a resolution of ± 0.02 pH unit or better in the near-neutral pH range. Some have already been tested with respect to in vivo or ex vivo performance.

18.2.2.1   Early Experiments in Continuous in Vivo Measurement of Blood Gases and pH

An evaluation of the pH sensor [52] in sheep, developed by Peterson et al. demonstrated its aptitude for in vivo blood pH measurement. Although not identical with the values obtained with an invasive microelectrode and with data obtained on withdrawn blood, the agreement between numerical data and trends is very good. It was not possible to say which of the measurements was the most correct. This demonstrates that the fiber-optic method is generally applicable for blood pH measurements in vivo, and gives as good an indication of pH levels as electrode methods. The pH of ewe blood as determined by an electrode, a blood gas analyzer, and a fiber-optic device is shown in Figure 18-9.

An even more precise version of this sensor type was specially designed for in vivo application and consists of a 25-gauge hypodermic needle with an ion-permeable side window and 75-μm fibers. It had a response time of 3 s. Together with computerized signal processing and three-point calibration, a precision of 0.001 pH unit in the pH range 7.0–7.4 was achieved. The sensor has been used in studies of the transmural pH gradient in myocardial ischemia [53] and for continuous measurement of intravascular pH [54].

Some of the oxygen sensors described in Section 17.2 have been shown to exhibit excellent in vivo performance. They have been used successfully in in vivo studies although a major problem, namely a lack of biocompatibility, does not seem to have been resolved yet.

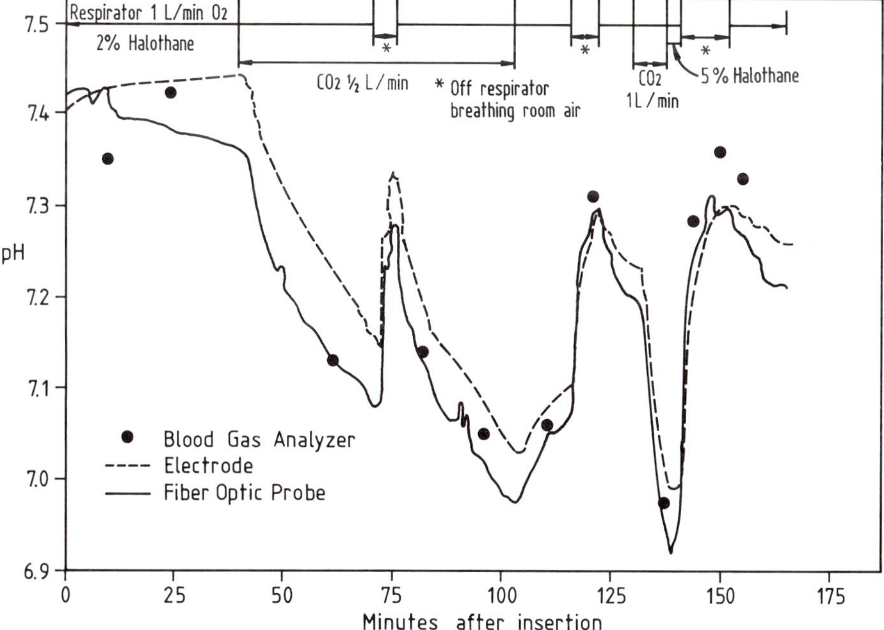

Figure 18-9. In vivo evaluation of a fiber-optic pH sensor in sheep undergoing varying respiration conditions. The blood pH analyzer data were obtained ex vivo and the electrode data with an inserted microelectrode. After [52].

The design and performance of optical sensors for blood gas and blood pH analysis has been described [55]. The sensors consist of small glass cylinders onto which a sensing layer is placed. The oxygen sensor is based on the quenching of the fluorescence of a polycyclic aromatic hydrocarbon. pH is measured with fluorescent pH indicators covalently bound to the carrier surface. pCO_2 is measured through the carbonic acid equilibrium (see Section 17.2) in an aqueous buffer entrapped in microvesicles in a silicone membrane (Figure 18.2). The standard deviations were 0.003 for pH (ionic strength range 60 ± 40 mM), less than 1 Torr for pO_2 (0–150 Torr range), and less than 1 Torr for pCO_2 (4–150 Torr range). A 10-µl blood sample is required for complete blood gas analysis and the response time is below 1 min. However, for whole blood the standard deviation of pH determination was 0.02 unit, probably because of the interfering effect of varying ionic strength.

It was only in 1984 that the first in vivo study using a fiber-optic oxygen sensor was reported [56]. The sensor is based on the quenching of the fluorescence by molecular oxygen of a dye adsorbed on hydrophobic particles and kept in position at the fiber and with porous polyethylene tubing (see Section 17.2). The dye is excited at ca 470 nm and fluoresces maximally at 515 nm with an intensity that depends on oxygen partial pressure. The ratio of green fluorescence to scattered blue excitation light (which serves as an internal reference) is the op-

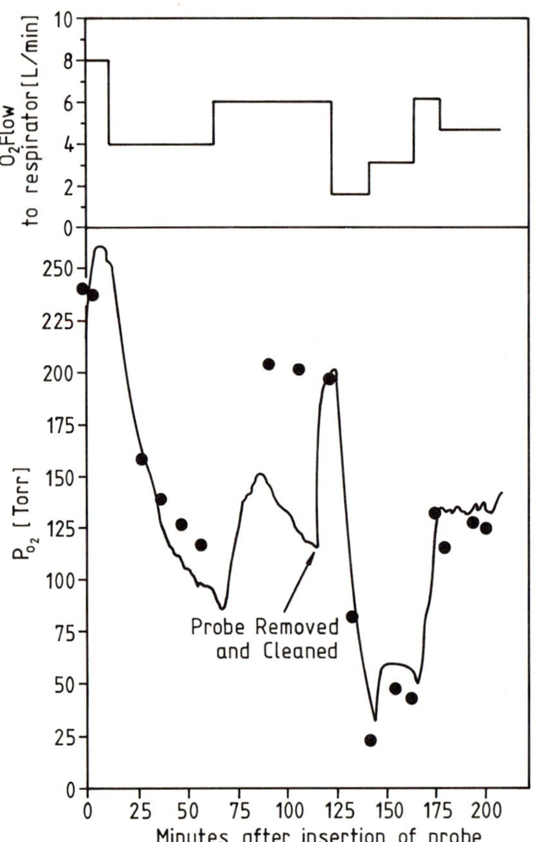

Figure 18-10.

In vivo test of an oxygen optode in an ewe. The breath gas composition was varied as shown on top, and the change in pO_2 was followed with the inserted fiber sensor. The data points are values obtained ex vivo with a blood gas analyzer. From [56].

tical information. A resolution of ± 1 Torr up to 150 Torr oxygen is reported. Figure 18-10 shows the trace of a continuous recording of oxygen pressure in the bloodstream of a ewe after the sensor had been inserted through a PTFE catheter.

The theory, construction, and performance of a catheter tip optical pCO_2 probe have been described [57]. The sensor, called the Opticap, was made from plastic fiber optics. One fiber carriers light to the sensitive tip, which is a silicone-rubber tube, 0.6 mm diameter $\times$ 1.0 mm long, filled with a Phenol Red–$KHCO_3$ solution. Ambient pCO_2 controls the pH of the solution, which influences the optical transmittance of the Phenol Red. A second fiber carriers the transmitted signal to a receiver; the resulting electrical signal is linearly related to pCO_2 over the range 2,7–10.7 kPa (20–80 Torr). The probe was tested as a tissue pCO_2 sensor on the cerebral cortex of the cat and as an arterial pCO_2 sensor. The drift over 1 day's use was 0.6 kPa (4.5 Torr) or less and individual probes have been used as long as 12 weeks.

A similar fiber-optic sensor has been devised [58] to measure and monitor the partial pressure of carbon dioxide in blood via fluorescence or phosphorescence changes in sensing dyes. The pCO_2 optode uses a nanoliter-size droplet of a concentrated fluorescein derivative in a buffered sodium ascorbate solution as a pH indicator. The pH of the droplet equilibrates with the volatile constituents of blood across an air bubble trapped at the tip of the capillary surrounding the fiber. Ascorbic acid is used to reduce photobleaching, which is further minimized by working on the appropriate portion of the concentration quenching curve. Signal levels of 300000 counts per second are obtained at 0.5 μW using the 488-nm line of an argon ion laser. A response range of 25% was obtained when the sensor was evaluated from 1.97 to 7.40% CO_2 in air at controlled humidity and temperature.

18.2.2.2 On-Line Measurement of Blood Gases and pH in an Extracorporeal Loop

CDI-3M Healthcare market a triple sensor (GasStat) for continuous blood gas and blood pH analysis in an extracorporeal loop during cardiopulmonary bypass (CPB) [59]. Blood is in contact with three sensing membranes whose fluorescence is monitored via fiber-optic cables. Figure 18-11 shows an exploded view of the disposable sensor and how it is connected to the loop. Two permeable membranes separate the bloodstream from the sensor chemistry placed on the disposable sensor, which is connected to the fiber-optic cable and also has an electrical contact for temperature measurement using a thermistor. Light from a xenon lamp is guided through the fibers to the sensor spots (two of which are red at the sample side because of a red layer that serves as an optical isolation). Under blue light excitation, the sensor spots display green fluorescence.

The pH-sensitive chemistry consists of a cellulosic material to which hydroxypyrene trisulfonate (HPTS) is covalently bonded. The CO_2-sensitive material is a fine emulsion of a hydrogen-carbonate buffer (plus HPTS) in a two-component silicone. The oxygen-sensitive chemistry is simply a solution of chemically modified decacyclene (which is strongly quenched by oxygen) in a one-component silicone. To make it insensitive toward halothane (an inhalation narcotic), it is covered with a thin layer of black PTFE, which also serves as an optical isolation. The fluorescence intensities of the three sensing spots can be related to pO_2, pH, and pCO_2 via modified Stern-Volmer or Henderson–Hasselbalch algorithms.

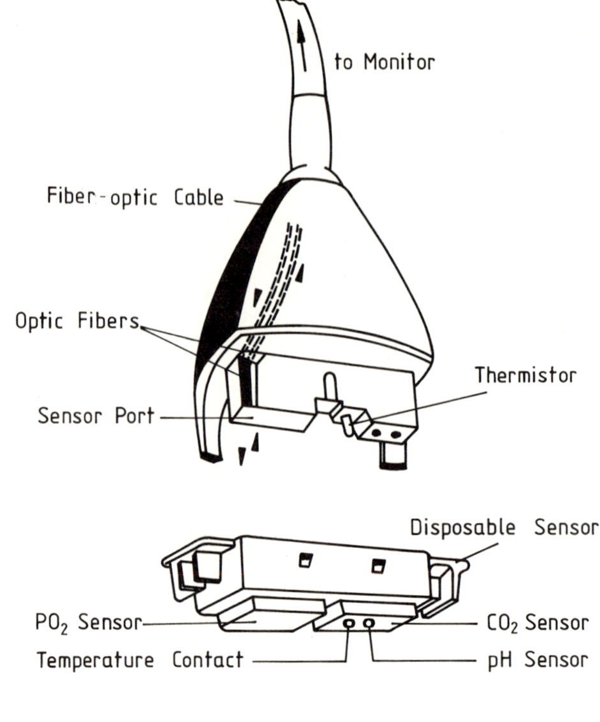

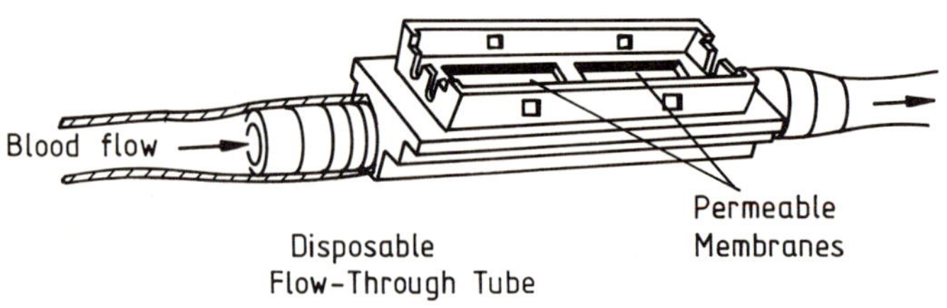

Figure 18-11. Exploded view of a disposable sensor for on-line blood gas analysis in an extracorporeal loop during cardiopulmonary bypass. After [59].

The device has two channels, one for arterial and the other for venous blood, and is capable of recording data for both the actual temperature and the corresponding 37 °C data. Before use, the sensors are two-point calibrated using a calibrating device supplied with the instrument. This device contains two disposable gas cylinders with tonometered gas levels of pO_2 and pCO_2. During calibration, the gases bubble through a 100-mmol/L sodium hydrogen carbonate buffer solution in the sensor cuvette for 7–10 min at room temperature to establish the two-point gas calibration graph. The pH calibration points are determined by the levels of carbon dioxide and the buffer composition in the cuvette which are, for gas 1, 9.87 kPa pO_2 and 2.47 kPa pCO_2. This gas adjusts to a pH of 7.61. For gas 2, the respective values are 29.7, 7.87 and 7.10. The data refer to atmospheric pressure (101.325 kPa) and 22 °C and have to be corrected to the actual barometric pressure for precise measurements. Both sensors are calibrated simultaneously prior to their insertion in the flow-through chamber. The response time (t_{90}) is of the order of 3 min. The device is said to be stable for up to 12 h in operation.

Various evaluations of the performance of the GasStat have appeared. Hill et al. [60] determined blood gas values in 50 patients undergoing CPB using a conventional off-line blood gas machine (BGM) and both the GasStat and the CardioMet 4000 on-line monitors (see Section 18.2.2.3). The latter two achieved agreement within 7.5% for pO_2 values over the entire range of temperatures. pH values from the GasStat were within 5% of the corrected values from the BGM. The pCO_2 values varied within 11% during hypothermia. The GasStat agreed with the CardioMet over varying temperatures and ranges of pO_2, while the BGM again demonstrated inaccuracy above 100 Torr pO_2 and below 37 °C. It was concluded that, owing to the increased cost of on-line monitoring, the cost/benefit ratio should be considered, but that the advantage of on-line, real-time parameters provides a much safer and precise perfusion management system giving the essential information for making rapid accurate decisions.

The GasStat was also used in another series of on-line measurements of blood gases during rewarming on hypothermic CPB [59]. In 20 patients, blood samples were obtained for 25, 30, 34 and 36 °C, and analyzed with both the continuous sensor and with a BGM, for determination of oxygen content and calculation of whole-body oxygen consumption (V_{O_2}). On-line measurement of venous pO_2 provided a reasonably accurate reflection of the changing metabolic requirements during the rewarming period, more so than arterial pO_2. The data suggested that the continous measurement of oxygen tension, particularly pV_{O_2}, provides a simple and effective method of monitoring the adequacy of oxygen delivery, particularly during the rewarming period. The results were consistently within 5% of the values obtained with a BGM in the blood gas laboratory.

An authorative study on the performance of the GasStat has been published by Sigaard-Andersen et al. [61]. The system was evaluated by reference to in vitro measurements using a BGM. GasStat measurements were performed at the actual temperature of the blood in the extracorporeal circuit, while reference measurements were performed at two fixed temperatures (25 and 37 °C) with interpolation of the values to the actual temperature of the GasStat. Ten patients undergoing coronary artery bypass grafting during hypothermic extracorporeal circulation with hemodilution were monitored in both the venous and the arterial lines with the GasStat with 6–9 samplings of arterial and venous blood from each patient, giving a total of 136 samples.

18.2.2.3 In Vivo Blood Gas Analysis

A fluorescence-based fiber-optic measurement system has been developed for monitoring pH, pCO_2, and pO_2 through a 20-gauge (1 mm) radial artery catheter without compromising capabilities for monitoring arterial pressure or for blood withdrawal [62–65]. The measuring probe consists of three optical fibers to which the sensing chemistries are attached, and a thermocouple that measures temperature (Figure 17-18). The probe does not yet meet the in vivo biocompatibility and reliability requirements for a one-time use of up to 72 h. The components that are in contact with the patient's blood are nontoxic, nonhemolytic, fairly nonthrombogenic, and sterilizable. Blood compatibility is enhanced by including covalently bound heparin.

The optoelectronic system consists of a microprocessor-based monitor where excitation light originates from a xenon lamp (operated at 20 Hz) and is focused into the respective quartz fibers. The returning fluorescence signal is collected in a subunit called the patient interface module (PIM) near the patient after signal demultiplexing and pre-amplification. The three active and reference signals are normalized against lamp intensity, filtered, and referenced, and then used to calculate actual values according to modified Henderson–Hasselbalch and Stern–Volmer equations, respectively. The calibration device contains two gas bottles and is similar to that described for the extracorporeal CDI system (Section 18.2.2.2).

The heart of the instrument lies within the PIM. Each of the three channels is composed of two highly collinear quarter-pitch graded-index ("grin") lenses separated by a 5-mm cube beam splitter, as shown in Figure 18-12. The grin lens on the right is part of the disposable unit. Above and below the cube is a photodiode plus filter stack, for purposes of fluorescence emission detection and excitation normalization, respectively. This arrangement combines the three functions of a reproducible low-loss optical connector, a means of picking off some excitation light for lamp energy normalization purposes, and a means of picking off most of the fluorescence emissions returning from the fiber tips. The PIM is connected to the xenon light source via three 250-μm fibers. When a few microjoules per pulse of light are delivered into each fiber, about 1 nJ of fluorescence is collected by each signal photodiode.

The three sensing fibers are bonded together over their last 6 cm and then chemically sterilized, and are designed to reside within a standard 20-gauge (1-mm) radial artery catheter. The probe is thin enough to allow easy blood withdrawal and undistorted propagation of the arterial pressure signal to a pressure transducer. The PIM is clamped at the bedside and connected to the monitor located elsewhere.

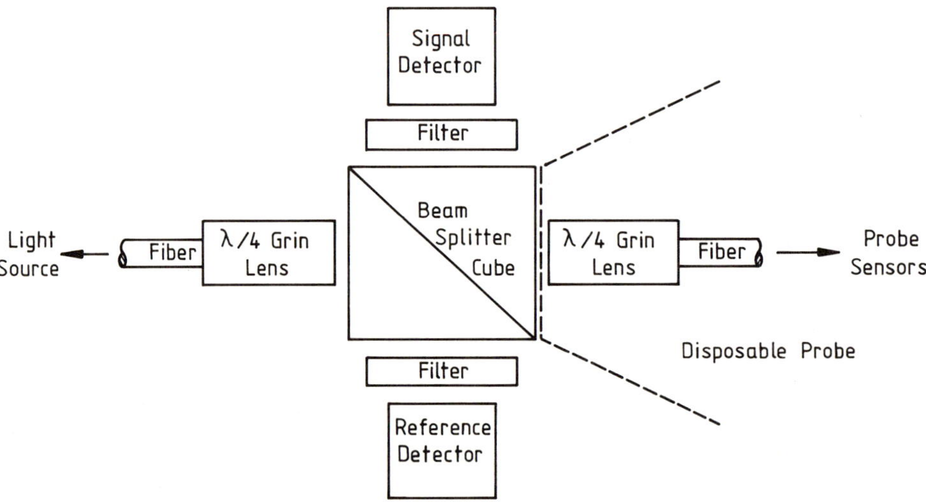

Figure 18-12. Optical components of the patient interface module. From [63].

Before use, the sensors are calibrated by a standard method: after attachment of the disposable unit to the PIM, the calibration cuvette containing the sensor tip is placed in the calibration device and an automatic cycle is started. During the calibration process the temperature of the solution is maintained at 37 °C while two primary standard calibration gases are sequentially bubbled through the solution until equilibrium is attained. The pH is adjusted via a defined pCO_2, which establishes a defined pH in a 100-mmol/L sodium hydrogen carbonate buffer. At the end of the calibration (which usually takes 20–30 min) the tip is removed from the calibration cuvette and threaded through the catheter into a blood circuit as described above. Figure 18-13 demonstrates the technique of inserting the catheter into the blood vessel.

The in vitro accuracy of the system has been tested against commercial blood-gas measurement instruments; comparison with tonometry and blood gas values gave a correlation coefficient $r \geq 0.98$ for all three sensors. The standard error for all sensors was within the College of American Pathologists' accuracy guidelines for measuring blood gas. Instrumental drift was minimal, indicating that system performance characteristics should not be the limiting factor in obtaining clinically useful information for up to 72 h. The response times of the sensors in animal and in vitro studies were < 2 min, suitable for monitoring physiological changes in blood gas values. Figure 18-14 shows the response function during an on-line measurement of pH and a comparison with data obtained by intermittent blood sampling. However, for reasons that cannot be discussed within the frame of this book, the reliability of the sensor is not good enough for medical application.

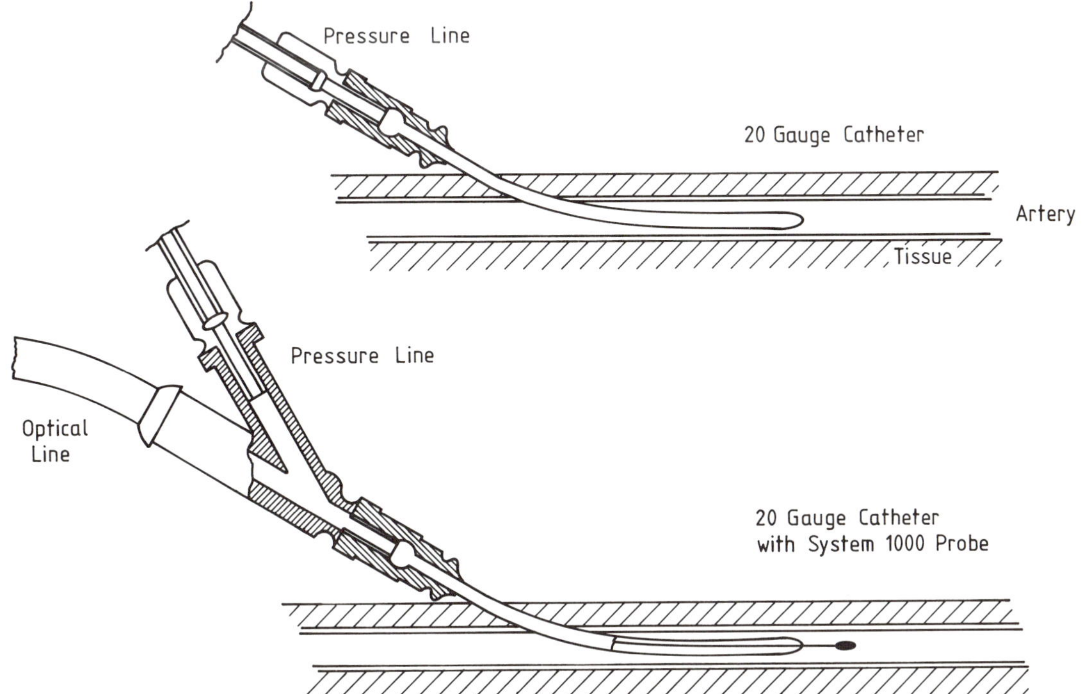

Figure 18-13. Radial artery catheter before and after insertion of the optical sensor. From [65].

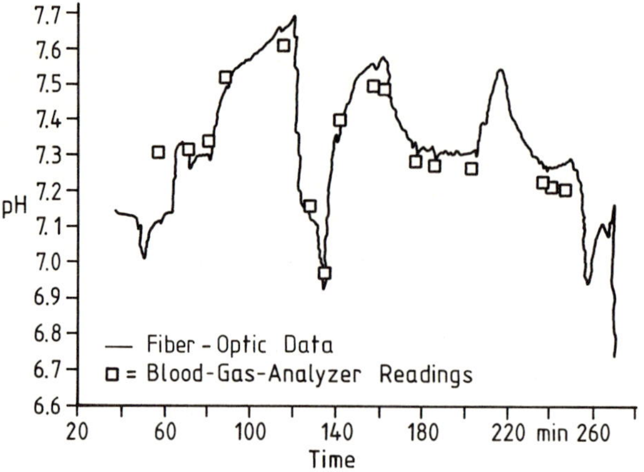

Figure 18-14.
Sensor performance in the animal model: pH via fiber optics versus laboratory electrochemical blood pH analyzer data.

18.3   Biosensors

Biosensors, in contrast to in vivo sensors discussed in Section 18.2, incorporate a biological component such as an enzyme, an antigen, or an antibody. They can therefore be divided into enzyme-based sensors (Section 18.3.1) and immunosensors (Section 18.3.2). There are three serious problems encountered in the design, manufacture, and performance of in vivo sensors and biosensors. One is the lack of biocompatibility of the materials used, and a second the poor long-term stability. The latter, however, plays no role in the case of disposable optodes. Disposable sensor heads for clinical analytes seem to be the most promising candidates for practical use at present. The third problem is the need for sterilization of in-vivo sensors which is difficult to solve in cases of biosensors with their thermally labile components. Specifically in the case of immunosensors, the binding strength between antigen and antibody is often such that it prevents the sensor from being reversible, so that it will become a probe good for a single assay only.

18.3.1   Enzyme-Based Biosensors

18.3.1.1   Glucose Biosensors

Glucose is probably the most frequently assayed nonionic analyte in clinical chemistry, but only recently have reliable sensors been developed for this species [66, 67]. At present, the lack of a suitable glucose sensor can be regarded as the rate-limiting step in the development of an artificial pancreas.

A fully reversible optical sensor for glucose was first described in 1980 and is based on the measurement of oxygen produced during its enzymatic oxidation by glucose oxidase (GOD) in the presence of catalase [68]. The enzymes were covalently immobilized on cross-linked bovine serum albumin, spread on 12-μm Cellophane, and covered with a 12-μm PTFE membrane. These reagent layers were covered with an oxygen-sensitive fluorescent layer, viz, a solution of pyrenebutyric acid (PBA) in a viscous solvent. When glucose solution is brought into contact with the sensing multilayer, a decrease in fluorescence intensity is observed as a result of oxygen consumption. The response to glucose is proportional to its concentration, provided that the oxygen supply for the first enzymatic reaction remains constant, and glucose does not saturate the enzyme. Since the system is based on diffusional processes, the response also depends on factors such as thickness, viscosity, enzyme concentration, and diffusion characteristics of all layers. The performance is most sensitive to oxygen supply from outside the sensor. The effects of substrate saturation, glucose oxidase concentration, and layer thickness were reported and found to be in agreement with a model of slow diffusion that determines the response. This type of sensor is, however, not compatible with standard fiber optics because of the UV excitation that is required for the indicator system used.

Therefore, the sensor was later improved and adapted to the fiber-optic technique by using an indicator (decacylene) with spectral properties more compatible with plastic fibers [69]. A cross section through the sensing membrane is shown in Figure 18-15, and typical response curves are given in Figure 18-16. The response time of the sensor was further improved by attaching a more active and thinner enzyme layer [70]. It has been used to determine glucose in wine and juice, using a flow-injection manifold [71].

Since the oxidation of glucose by GOD also yield protons, a pH optode is another candidate for transducing a biochemical reaction into a measurable parameter [72]. This type of biosensor is, however, more susceptible to external interferences, since the relative pH change (which is related to the glucose concentration) also depends on the buffer capacity and ionic strength of the sample. Like the glucose sensor with the oxygen optode as the transducer, it also depends on the oxygen supply and must not be saturated with substrate.

When glucose is oxidized by glucose dehydrogenase, NADH is simultaneoulsy formed, whose fluorescence can be detected via fiber optics [73]. This concept will be discussed in

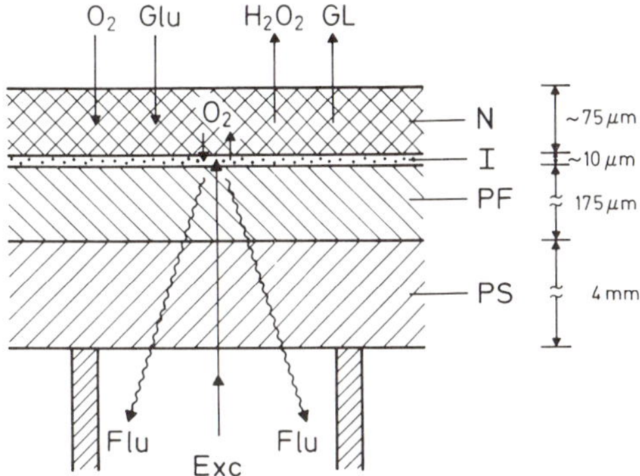

Figure 18-15. Cross section through a typical enzyme-based glucose biosensor membrane with an oxygen optode as the transducer. The membrane can be attached to the end of a fiber. The directions of exciting light (Exc) and fluorescence (Flu) are shown. PS, polymer support; PF, polymer foil needed for production of thin sensing membranes; I, oxygen indicator layer; N, nylon net with immobilized glucose oxidase plus catalase. The diffusional processes of oxygen, glucose (Glu), gluconolactone (GL), and hydrogen peroxide are also shown. From [69].

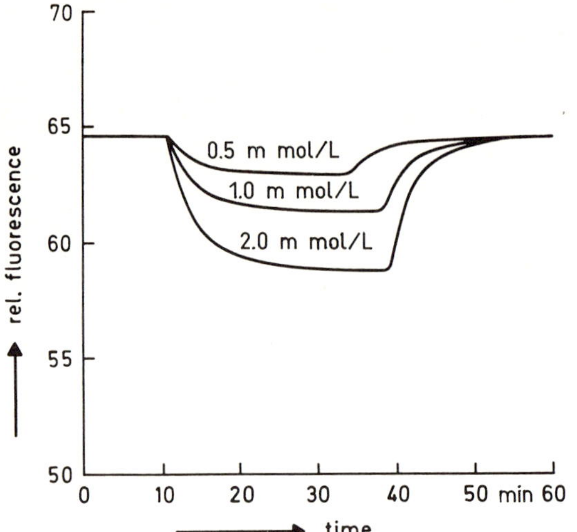

Figure 18-16.
Response of glucose sensor with an oxygen optode as the transducer to various concentrations of glucose passing by. From [69].

more detail in the case of the lactate sensor [74] (Section 18.3.1.2). A newer type of glucose sensor utilizes the intrinsic fluorescence of glucose oxidase (GOD), which has a flavine adenine dinucleotide (FAD) as a coenzyme [75]. The coenzyme has an intrinsic fluoresence with a maximum at 520 nm with 450-nm excitation. When fully reduced by an excess of glucose, the spectra are almost unchanged, but the quantum yield increases by about 20%. A typical trace, showing the sharp increase at the point of saturation, is given in Figure 18-17. This kind of sensor has several advantages: unlike NADH-based optodes, its working wavelengths are in the visible region so they are compatible with standard fibers; they are simple in construction because the enzyme (which is both the recognizing element and the transducer element) has only to be entrapped at the fiber end (Figure 18-18); and it is fully and rapidly reversible, since the reduced coenzyme is rapidly converted into FAD by oxygen. On the other hand, the dynamic range is fairly small (Figure 18-17).

A different kind of glucose sensor exploiting the phenomenon of competitive binding of substrates to an enzyme has also been described [76]. Fluorescein-labelled dextran competes with glucose for binding to concanavalin A, which is immobilized on Sepharose. Figure 18-19 shows a schematic diagram of the device. The end of a bifurcated fiber optic fits into a hollow fiber with a plug on the end. The substrate is fixed on the walls of the fiber and thus in a position out of its numerical aperture; therefore, it cannot be seen by the fiber. Glucose can diffuse freely through the hollow fiber or a dialysis membrane, which is impermeable to the large dextran molecules. Increasing glucose concentration displaces the labelled dextran, causing it to diffuse into the illuminated solution volume. Thus, fluorescence intensity as seen by the fiber follows the glucose concentration. This sensor principle is of particular interest because the design idea has broad potential for application to any analytical problem for which a specific competitive binding system can be devised. However, response times are extremely long, and the selectivity for glucose is poor.

Co-immobilization of pH-sensitive dyes and proton-consuming or proton-generating enzymes to a thin transparent membrane sandwiched between an LED light source and a silicon photodiode–amplifier detector system permits the characteristic color change observed on contact with a transparent substrate to be measured. Goldfinch and Lowe [77] used Bromocresol Green immobilized on Cellophane to measure the decrease in pH produced by the catalytic action of immobilized glucose oxidase on the oxidation of glucose. The glucose optoelectronic sensor responds linearly to glucose over the range 0–70 mmol/L and displays a half-life of 7–8 days.

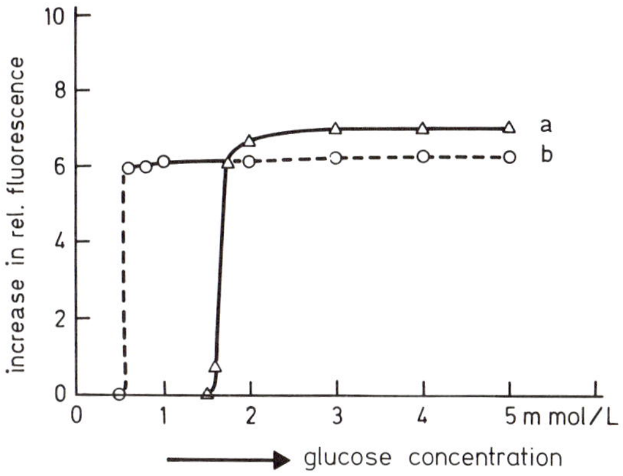

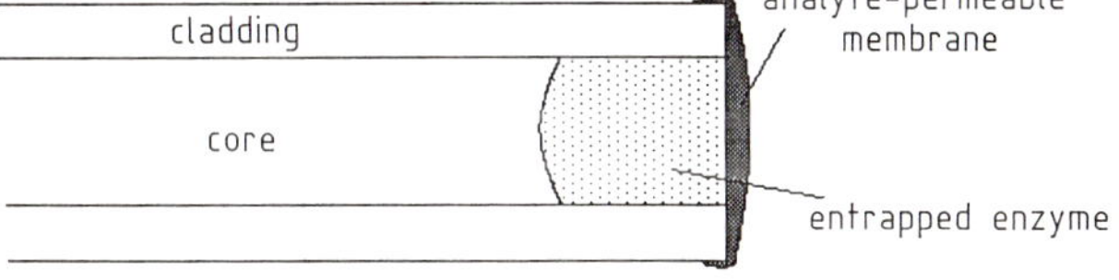

Figure 18-17.
Calibration graph for the intrinsic glucose sensor based on measurement of the fluorescence of glucose oxidase when glucose solutions pass by. (a) In air-saturated samples; (b) glucose in solutions containing 7.88% oxygen at 727 Torr. From [75].

Figure 18-18.
Schematic diagram of a single-fiber biosensor based on measurement of the intrinsic fluorescence of an oxidase. From [79].

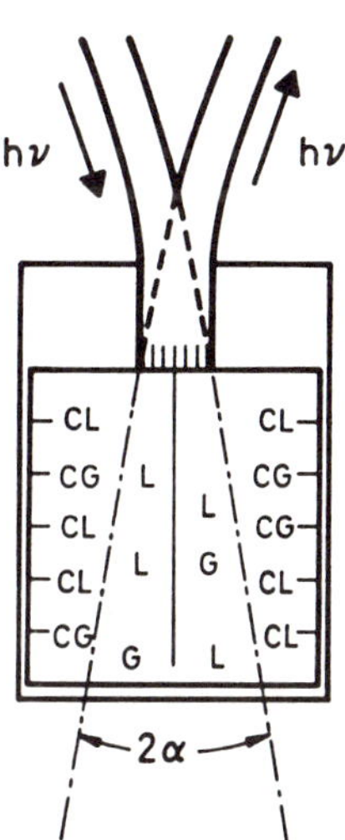

Figure 18-19.
Schematic diagram of a glucose sensor based on competitive binding. The fiber "sees", within its numerical aperture 2α, only the labelled dextran in solution, and not the dextran bound to the walls. G, glucose; L, labelled dextran; C, concanavalin A immobilized on the wall of the hollow fiber. Redrawn from [76].

18.3.1.2   *Lactate Biosensors*

Transducer devices similar to those used for glucose have also been used to determine lactate. Immobilized lactate oxidase converts lactate into acetate (plus CO_2 and H_2O_2) and consumes 1 equivalent of oxygen. Two oxygen-sensing layers were applied [78] to measure the difference in the oxygen partial pressure that is produced enzymatically within the complex sensor membrane. Two oxygen-sensitive indicators with different spectral properties (pyrenebutyric acid and perylene) were applied. Their excitation and emission maxima (342/395 nm and 436/490 nm, respectively) are sufficiently separated that both sensor layers can be submitted to simultaneous fiber fluorimetry. An almost linear ratio between lactate concentration and normalized fluorescence intensity was found in the $0-1$ mg mL^{-1} concentration range.

Another class of lactate biosensors was described by Wangsa and Arnold [74]. It is based on both the immobilization of a dehydrogenase at the sensing tip of a fiber-optic probe and the fluorimetric detection of consumed or generated NADH, which, in contrast to nicotine adenine dinucleotide (NAD$^+$), is highly fluorescent (350/450 nm). The lactate sensor employs a layer of covalently immobilized lactate dehydrogenase. Although this is a kinetic method and is not reversible unless additional reagents are added, it is a most general approach because of the many available NADH-coupled bioreactions and the stability of the corresponding enzymes. Response times ranged from 5 to 12 min, and a reproducible sensor response was maintained for at least 7 days.

Lactate can also been assayed by monitoring, via fibers, the intrinsic fluorescence of lactate monooxygenase, which, in fact, is a highly fluorescent enzyme [79]. Its fluorescence varies with the lactate concentration in a solution passing by, as shown in Figure 18-20. In order to compensate for the small analytical range (similar to that shown in Figure 18-17 for glucose), the sensor has been exposed to the sample for only a limited time in a fashion similar to flow-injection analysis. The dynamic range can be expanded considerably by this technique (Figure 18-21).

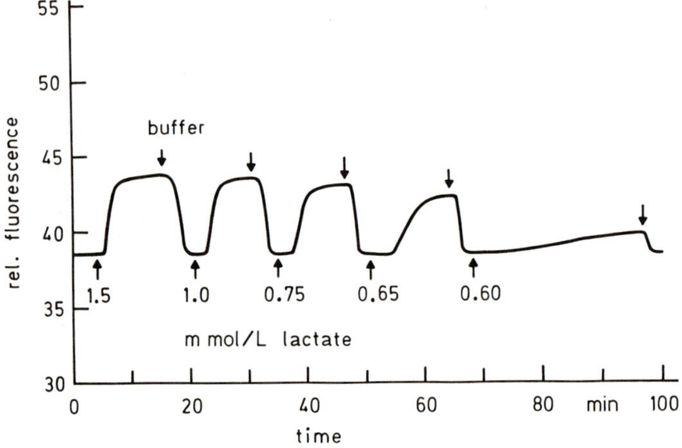

Figure 18-20. Typical response curve, response time, and analytical range of an intrinsic lactate sensor. From [79].

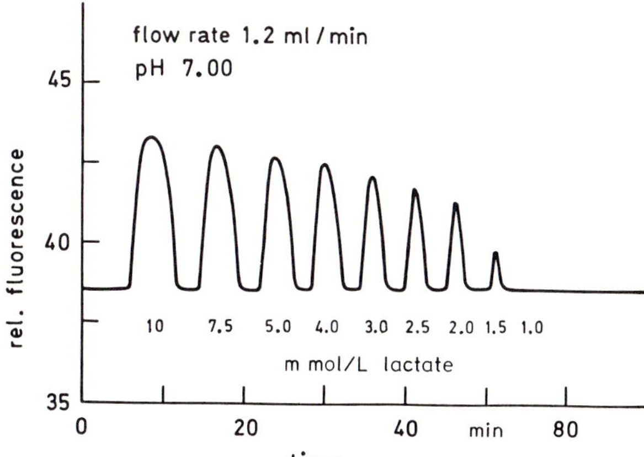

Figure 18-21.
Typical time response curve of an intrinsic lactate biosensor exposed to air-saturated lactacte solutions for 55 s, followed by flushing with citrate buffer. From [79].

18.3.1.3 Biosensors for Other Species

The principle of detecting pH changes caused by enzymic action has been applied to sense urea [77]. Immobilized Bromothymol Blue changes its color because of an increase in pH that is produced by the activity of the enzyme urease. The sensor responds to urea concentration in the range 0–40 mmol/L linearly up to 10 mmol/L and with an output voltage change of ca 125 mV min^{-1} at 10 mmol/L urea. The sensor is completely reversible and is regenerated with pH 7.0 buffer. The half-life is largely conditioned by the nature of the enzyme and the sensor preparation, being of the order of 3–4 weeks.

Oxygen is consumed when alcohol oxidase (AOD) oxidizes alcohol. This permits the continuous assay of ethanol by measurement of the oxygen gradient across a sensing membrane [80]. As oxygen is consumed during the reaction, quenching of the indicator by oxygen becomes less efficient. The resulting increase in fluorescence intensity can be related to the ethanol concentration, [EtOH], by the following modified Stern–Volmer equation

$$F_0/F = 1 + K_d [Q] - K' [EtOH] \tag{18-1}$$

where F_0 and F are the fluorescence intensities in the absence and presence of oxygen, respectively K_d is the quenching constant (which depends on temperature, viscosity, and fluorophore lifetime), [Q] is the oxygen concentration in the *sample* (equal to apO_2), and K' is a parameter specific for the enzymatic layer. This relationship only holds, however, if oxygen is continuously available from the substrate solution side, the enzyme is not overloaded, a diffusion equilibrium is established, and acetaldehyde and hydrogen peroxide (formed during the reaction) do not retard the enzyme activity. Unfortunately, AOD is a labile enzyme, and both the lifetime and detection limits are poor. Figure 18-22 shows a typical calibration plot and the dynamic range which was specifically designed for use in biofermenters.

A different type of ethanol bioprobe was designed on the basis of an NAD$^+$-dependent dehydrogenase reaction [81]. Alcohol dehydrogenase plus NAD$^+$ was retained at the end of a fiber by a microporous PTFE membrane, and the rate of formation of fluorescent NADH

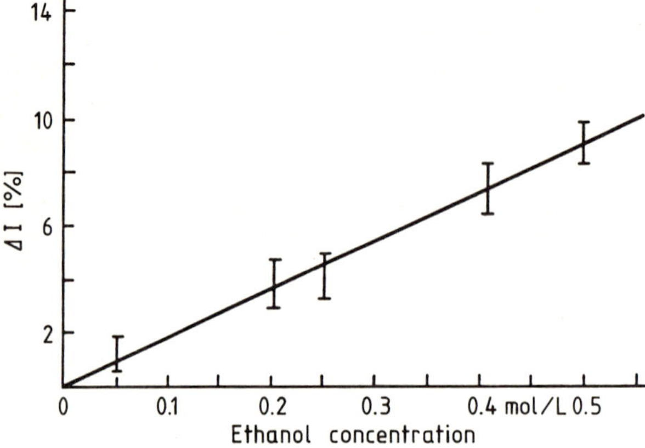

Figure 18-22.
Calibration graph and analytical range of an enzyme-based ethanol sensor with an oxygen optode as the transducer. From [80].

Table 18-2. Selection of biochemical analytes for which enzyme-based optical sensors have been described, and respective transducers[a].

Analyte	Enzyme	Transduction via
Glucose	Glucose oxidases	pH, O_2
	Glucose dehydrogenase	NADH[b]
Lactase	Lactate mono-oxygenase	O_2, CO_2
	Lactate oxygenase	O_2
	Lactate dehydrogenase	NADH[b]
Pyruvate	Lactate dehydrogenase	NADH[b]
Ascorbate	Ascorbate oxidase	O_2
Bilirubin	Bilirubin oxidase	O_2
Ethanol	Ethanol oxidase	O_2
	Alcohol dehydrogenase	NADH[b]
Urea	Urease	NH_3, NH_4^+, pH
Creatinine	Creatinine iminohydrolase	NH_4^+
Uric acid	Uricase	O_2
Penicillin	Penicillinase	pH
Glutamate	Glutamate decarboxylase	CO_2
	Glutamate oxidase	O_2
	Oxalate decarboxylase	CO_2
Xanthine	Xanthine oxidase	O_2

[a] Data from ref. [67] where the respective references may be found, and unpublished results of the author's group.

[b] Through formation of fluorescent NADH but without regeneration or recycling.

was followed as ethanol entered the reaction volume. A linear relationship between rate and ethanol concentration was observed over the alcohol concentration range 0.9–9 mmol/L.

The pH-sensitive membrane as described by Goldfinch and Lowe [77] was applied to determine penicillin and related antibiotics by virtue of a proton-producing reaction catalyzed by penicillinase. The optoelectronic probe consists of a cellulose membrane containing the enzyme and co-immobilized Bromocresol Green, which responds to penicillin G in the concentration range 0–20 mmol/L. The change in output voltage of the silicone diode light detector was approximately linear within the range 0.5–5.0 mmol/L, but displayed increasing saturation at concentrations above 10.0 mmol/L. Response of the probe to ampicillin and cephaloridine was also observed. Subsequent to each determination, the membrane was regenerated by flushing the cell with 5 mmol/L phosphate buffer (pH 7.0).

Penicillin also has been assayed with a single-fiber enzyme-based sensor [82]. The enzyme penicillinase, which converts penicillins to penicilloic acid, was immobilized in a thin glutaraldehyde/bovine serum albumin membrane on a porous glass bead to which fluorescein isothiocyanate was immobilized. The decrease in pH is seen by the fluorescein, which is excited by an argon laser. A response time of 20–45 s is reported, with a detection limit of 0.1 mmol/L penicillin G and V. Kulp et al. [83] reported on a similar optode for penicillin with immobilized penicillinase at the tip of a fiber, 1-hydroxypyrene trisulfonate or fluorescein as the pH indicator, and a tungsten–halogen light source. Both kinds of sensors, like all others exploiting a pH optode as the transducer, suffer from the fact that the relative signal change depends strongly on the sample pH, the buffer capacity and, much less so, on interference by ionic strength. Table 18-2 summarizes the species for which optical biosensors have been described, and the respective transducers that were employed.

18.3.1.4 Determination of Enzyme Activities

When a chromogenic or fluorogenic enzyme substrate is immobilized at the fiber tip, which then is immersed in an enzyme solution, it will be hydrolyzed, which results in an increase in absorbance or fluorescence. This offers the possibility of remote (or invasive) determination of enzyme activities [84]. Specifically, the activity of carboxylesterases was determined.

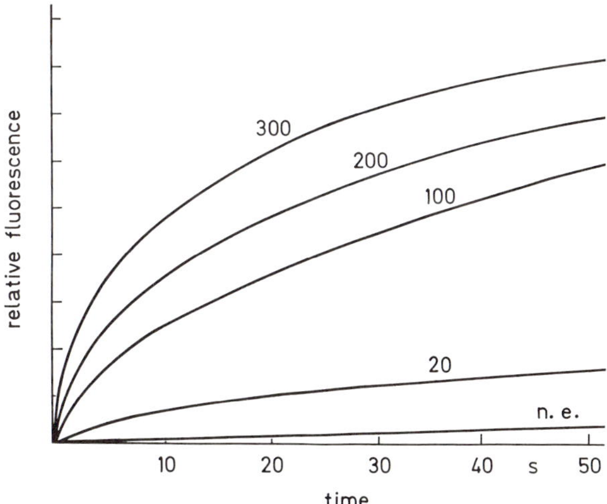

Figure 18-23. Increase in fluorescence when a non-fluorescent substrate for carboxylesterase immobilized on the end of a fiber is brought into contact with increasing amounts of carboxylesterase in solution. The numbers refer to mg of enzyme in solution. From [84].

Figure 18-23 shows a typical plot obtained by immersing a fiber-immobilized substrate in an enzyme solution. Note, however, that the device is a probe rather than a sensor because the enzyme substrate on the fiber is consumed and cannot be reactivated during the sensing process.

The method was designed for direct determination of hydrolases such as esterases, phosphatases, sulfatases, and amylases. The synthetic substrates used in this method do not present a health risk when applied in vivo, since it is the acid or sugar component that is released. Various modifications of the design of the fiber tip and the remote gathering of optical information have been suggested [85].

18.3.2   Immunosensors

Immunological reactions are known to proceed with outstanding selectivity. This, combined with the sensitivity of fluorimetry, has led to the development of various kinds of fluoroimmunoassay. Notwithstanding their commercial success, these detection principles are hardly suitable for continuous sensing purposes because the assay is of the non-homogeneous type and the binding of antigen by antibody is so strong that it is virtually irreversible. Moreover, the intrinsic colour and fluorescence of most biomatter can severely hamper conventional spectroscopic measurements. Most of the sensing techniques in waveguide and fiber immunoassay are therefore performed by total internal reflection. Figure 18-24 shows how the evanescent wave hits the antibody on the surface of the waveguide. Excellent reviews on immunosensors exist [86].

Optical fibers without cladding can be used to monitor the adsorption of chemical species such as proteins on an optical fiber core [87]. Fluorescence is induced by the evanescent wave field of the light propagating in the core. For example, the intrinsic fluorescence of the amino acid tryptophan in immunoglobulin (IgG) and the fluorescence of fluorescein-labelled IgG were detected and used to calculate the amount of protein adsorbed on hydrophilic glass and quartz surfaces. Fluorescent, nonadsorbing dextran was used as a calibration molecule which approximates the diffusion properties of the protein. Remote spectroscopic sensing of the adsorption of Rhodamine B-labelled IgG at the tip of a 600-μm fiber optic has also been described [88, 89].

In a typical application for immunosensing, the emission of fluorescein-labelled antibody bound to a hapten–protein conjugate adsorptively immobilized on a quartz plate in contact

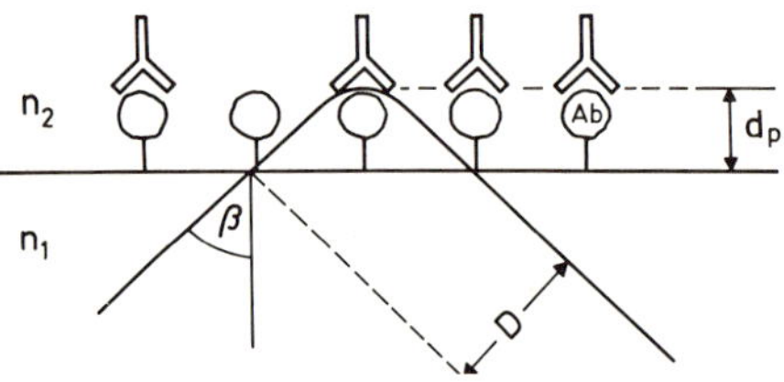

Figure 18-24. Schematic diagram of the evanescent wave which, when light is totally reflected at a dielectric interface of refractive indices n_1 and n_2, penetrates into phase 2 where it can excite the fluorescence of an antibody (Ab); d_p is the depth of penetration and D the so-called Goos–Hänchen shift.

with the antibody solution was monitored [90]. The quartz plate acts as a waveguide. The presence of any free hapten in solution reduces the amount of antibody free to bind to the surface and thus reduces the fluorescence signal. Following the decrease in fluorescence gives a measure of the concentration of free hapten present. Because of its low penetration depth, the evanescent wave mainly interacts with the immobilized material rather than with the bulk solution. It therefore reports the binding processes occurring at the interface. The technique is simple, fast, and has high intrinsic sensitivity and specifity. Free morphine at a concentration of 2.10^{-7} mol/L is readily detected.

Recently, an examination of the binding kinetics of Rhodamine G-labelled antibodies to dinitrophenol, also immobilized to a quartz light guide, was reported [91]. In a similar experiment, Rhodamine G-labelled IgG and insulin were shown to adsorb non-specifically to serum albumin-coated fused silica with both reversible and irreversible components. The characteristic time of the most rapidly reversible component measured was ca 5 ms and was limited by the rate of bulk diffusion. The binding was followed by fluorescence which, collected by a microscope from a ca 5-μm^2 surface area, spontaneously fluctuates as the solute molecules randomly bind to, unbind from, and/or diffuse along the surface in chemical equilibrium.

The general characteristics of these immunoassay systems include kinetic monitoring of the immunological reactions without major interferences from the bulk solution and without a formal separation step of antibody-bound from free antigen. The approaches were extended by attaching the specific antiserum to the surface of the waveguide and monitoring the sequestration of antigen by the labelled antibody. This, however, is an almost irreversible process, and these devices are therefore better named probes or dosimeters rather than sensors [92].

The theory, instrumentation, and preliminary application of the evanescent wave probing technique specifically for the case of human IgG have been described in some detail [93]. The antibody is covalently immobilized onto the surface of either a planar microscope slide or a cylindrical waveguide made of fused quartz. The reaction of fluorescein-labelled antibody with an antigen in solution is detected by the evanescent wave component of the light beam. Specifically, methotrexate (MTX, a cancer drug) was detected by the reduction in light transmission as as the molecule was bound to anti-MTX serum immobilized on the liquid waveguide interface. Flushing the slide with 0.01 mol/L hydrochloric acid disrupts the antibody–antigen binding and makes the probe reusable.

IgG has been measured by a sandwich immunofluoroassay. First, immobilized sheep antiserum was incubated with antigen to form one half of the sandwich. After washing out excess of IgG, the solid phase complex was reacted with fluorescein-isothiocyanate (FITC)-labelled antiserum and the reaction was monitored for 15 min. Figure 18-25 shows the detection of the binding process on the surface of a slide. Following the injection of FITC-labelled anti-IgG, a rapid increase in fluorescence intensity (X) is observed. This is mainly due to free molecules fluorescing within the penetration depth of the evanescent wave. Second, a slower increase in binding over the next 500 s is evident, reaching a plateau after ca 600 s. The curve obtained when no antigen was present shows fluorescence because of the presence of free molecules, but lacks the rapid binding event.

A no-label, homogeneous, optical immunoassay for human IgG has been developed and tested [94] (Figure 18-26). Following adsorption of antiserum to the surface of an optical waveguide, the immobilized antibody was then reacted with a solution containing antigen. The

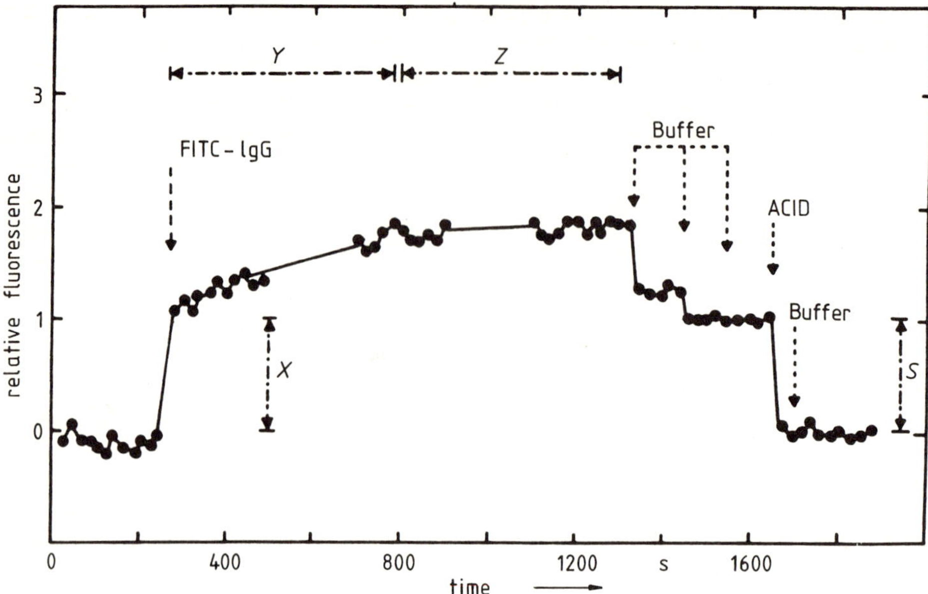

Figure 18-25. Specific binding of fluorescein-labelled immunoglobulin G (IgG) by antiserum immobilized on a glass waveguide. The binding process is monitored by the increase in fluorescence excited by the light beam propagating in the glass. When IgG is added, there is an increase in fluorescence (X) due to free molecules fluorescing close to the waveguide surface. Over distance Y a plateau level Z is reached. After removing supernatant IgG with buffer, an intensity S remains which is an absolute measure of specific binding. Treatment with strong acid (0.01 mol/L hydrochloric acid) removes bound IgG and makes the probe reusable. Redrawn from [94].

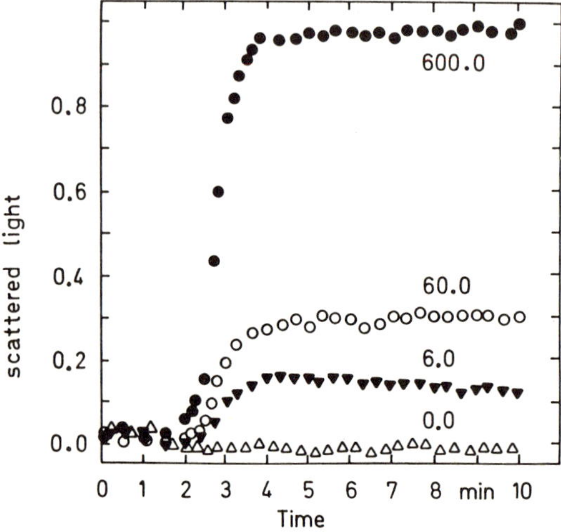

Figure 18-26. Generation of light-scattering signal at the surface of an optical waveguide bearing immobilized antibody. IgG was added after 2 min at concentrations ranging from 0.0 to 60.0 μg mL^{-1}, and scattering was monitored at 440 nm. Redrawn from [94].

reaction was detected utilizing the evanescent wave component of a light beam totally internally reflected within the waveguide. The growing antigen–antibody layer resulted in an increase in scattered light which was monitored kinetically.

Tromberg et al. [95] described the development of a fiber-optic chemical sensor based on a competitive-binding fluorescence immunoassay. The binding of labelled rabbit IgG to unlabelled anti-IgG was monitored via the fiber. Alternatively [96], the fiber-optic approach may be coupled with an enzyme-linked immunosorbent assay (ELISA), or the evanescent-wave technique may be exploited to monitor binding process occurring at the quartz/sample interface [97, 98]. In another sensing approach, lanthanide chelates with their extremely long fluorescence decay time and large Stokes' shift have been used for the design of a new generation of non-isotopic immunoassays [99]. Measuring fluorescence after a lag phase, during which the emission intensity of virtually all other molecules has dropped to zero, allows an immunoassay to be performed with great sensitivity (Figure 18-27). In this assay, an antibody is labelled with a europium chelate and its binding to fiber-immobilized immunoglobulin is monitored. Detection limits are of the order of 0.1 µg mL $^{-1}$ anti-IgG.

A fiber-optic flow-injection analysis in which a fluorescent reporter molecule is utilized to monitor a competition reaction between a labelled and an unlabelled antigen molecule for an immobilized antibody has been reported [100]. It is based on the evanescent wave technique.

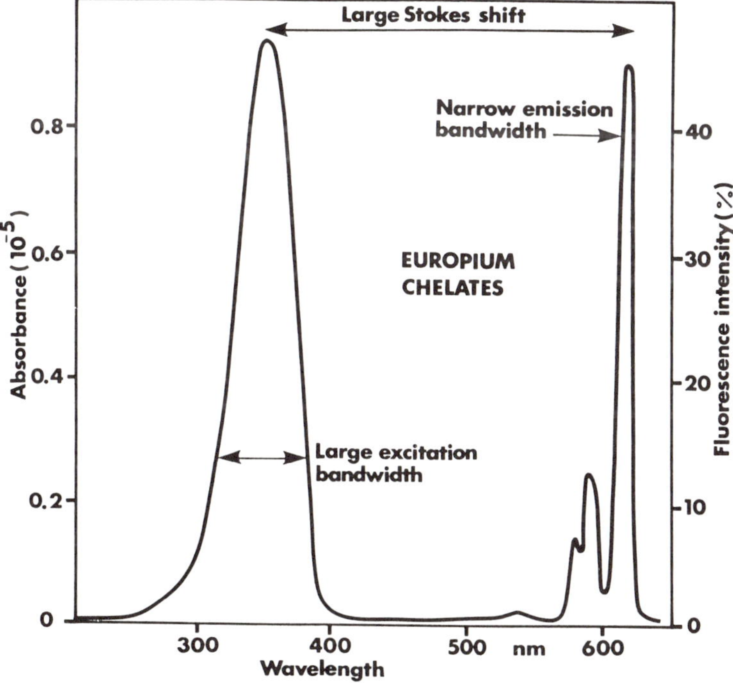

Figure 18-27. Fluorescence decay profile of an europium chelate as used in a time-resolved fluorescence immunoassay. Background fluorescence disappears after a few nanoseconds, whereas the chelate decays in the millisecond time range. Reproduced by permission, courtesy of LKB Produkter, Bromma, Sweden.

Although this method makes any washing steps superfluous, it is difficult to apply it in vivo because of the need for labelled antigens. Vo-Dinh and co-worker have described a portable fiber-optic monitor for use in combination with enzyme-linked immunosorbent assay [86] and a fiber-optic immunosensor for benzo[*a*]pyrene [101].

A fiber-optic sensor based on a homogeneous fluorescence energy transfer immunoassay operates in a continuous, reversible manner for the determination of the anticonvulsant drug phenytoin [102]. Phycoerythrin-labelled phenytoin and Texas Red-labelled anti-phenytoin were sealed inside a short length of cellulose dialysis tubing, which was cemented to the distal end of a fiber. When the sensor was alternately placed into various solutions with various concentrations of free phenytoin, the drug crossed the dialysis membrane and displaced a fraction of the labelled phenytoin from the antibody. The resulting change in fluorescence signal was measured with a fiber fluorimeter. Typical analytical ranges were 5–500 μmol L^{-1} of phenytoin. Response times ranged from 5 to 30 min for different sensors. The chemical kinetics of the two competitive binding processes in the tubing were modeled mathematically [103].

18.4   Reference

[1] Hirschfeld, T., Deaton, T., Milanovich, F., Klainer, S., *Opt. Eng.* **22** (1983) 527.

[2] O'Neil, R. A., Buja-Bijunas, L., Rayner, D. N., *Appl. Opt.* **19** (1980) 863.

[3] Capelle, G. A., Franks, L. A., Jessup, D. A., *Appl. Opt.* **22** (1983) 3382.

[4] Franks, L. A., Capelle, G. A., Jessup, D. A., *Appl. Opt.* **22** (1983) 1717.

[5] Kawahara, F. K., Fiutem, R. A., Silvus, H. S., Newman, F. M., Frazar, T. H., *Anal. Chim. Acta* **151** (1983) 315.

[6] Klainer, S., Goswami, K., Dandge, D. K., Simon, S. J., Herron, N. R., Eastwood, D., Eccles, L., in: *Fiber Optic Chemical Sensors and Biosensors,* Wolfbeis, O. S. (ed.); Boca Raton, FL: CRC Press, 1991, Chap. 13.

[7] Kim, H. H., *Appl. Opt.* **12** (1973) 1454.

[8] Hoge, F. E., Swift, R. N., "Airborne Mapping of Laser-Induced Fluorescence of Chlorophyll and Phycoerythrin in a Gulf Stream Warm Core Ring" in: *Mapping Strategies in Chemical Oceanography,* Zirino, A. (ed.), *Adv. Chem. Ser.* **209** (1985) 353.

[9] Poole, L. R., Esaias, W. E., *Appl. Opt.* **21** (1982) 3756; **22** (1983) 380.

[10] Brislow, M., Nielson, D., Bundy, D., Furtek, R., *Appl. Opt.* **20** (1981) 2889.

[11] Hoge, F. E., Swift, R. N., *NASA Conf. Publ. 2188,* US GPO Washington, DC, 1981, p. 349.

[12] Venable, D. D., Punjabi, A. R., Poole, L. R., *Appl. Opt.* **23** (1984) 970, and references cited therein.

[13] Lund, T., *IEEE Proc.* **131 H** (1984) 49, and references cited therein.

[14] Lakshman, G., *Water Resour. Res.* **11** (1975) 705.

[15] Chudyk, W. A., Carrabba, M. M., Kenny, J. E., *Anal. Chem.* **57** (1985) 1237.

[16] Kenny, J. E., Jarvis, G. B., Chudyk, W. A., Pohlig, K. V., *Anal. Instrum.* **16** (1987) 423.

[17] Milanovich, F., Hirschfeld, T., Miller, H., Garvis, D., Anderson, W., Miller, F., Klainer, S. M., *Lawrence Livermore Natl. Lab. Report UCID-1977/4,* Vol. 2, 1985.

[18] Milanovich, F. P., Garvis, D. G., Angel, S. M., Klainer, S. M., Eccles, L., *Anal. Instrum.* **15** (1986) 347.

[19] Wolfbeis, O. S., "Fiber Optic Fluorosensors in Analytical and Clinical Chemistry", in: *Molecular Fluorescence Spectrocopy. Methods and Applications,* Vol. 2, Schulman, S. G. (ed.); New York: Wiley, 1988, Chap. 3.

[20] Lieberman, S. H., Inman, S. M., Stromvall, E. J., "Fiber Optic Fluorescence Sensors for Remote Detection of Chemical Species in Seawater", in: *Chemical Sensors,* Turner, D. R. (ed.); London: Electrochemical Society, 1987.

[21] McFarlane, R., Hamilton, M. C., *Proc. SPIE* **798** (1987) 324.

[22] Malstrom, R. A., Hirschfeld, T., *Anal. Chem. Symp. Ser.* **19** (1984) 25; in this paper the legends of Figure 4 and 5 should apparently be interchanged.

[23] Boisdé, G., Blanc, F., Mauchien, P., Perez, J. J., in: *Fiber Optic Chemical Sensors and Biosensors,* Wolfbeis, O. S. (ed.); Boca Raton, FL: CRC Press, 1990, Chap. 14.

[24] Boisdé, G., Kirsch, B., Mauchien, P., Rougeault, S., in: *Proceedings of Opto 88;* ESI Publications, Paris: Masson, 1988, pp. 294–299.

[25] Groll, P., Römer, J., Röder, L., Persohn, M., Schlosser, B., *Anal. Chim. Acta* **190** (1986) 265.

[26] Trettnak, W., unpublished results, 1991.

[27] Koller, E., Wolfbeis, O. S., in: *Proceedings of NATO Advanced Workshop on Detection and Characterization of Chemical and Biological Species;* Salamanca, Spain, April 1–3, 1989. Reidel Publ., Dordrecht (NL), A. Sanz-Medel (ed.), p. 15.

[28] Inaba, H., *Springer Ser. Opt. Sci.* **39** (1983) 288.

[29] Stueflotten, S., Christensen, T., Iversen, S., Hellvik, J. O., Almas, K., Wien, T., Graav, A., *Proc. SPIE* **514** (1984) 87.

[30] Kobayashi, T., Hirama, M., Inaba, H., *Appl. Opt.* **20** (1981) 3279.

[31] Matthews, T. G., Hawthorne, A. R., Gammage, R. B., *Oak Ridge Natl. Laboratory Report TM 8007,* 1982.

[32] Chan, K., Ito, H., Inaba, H., *Appl. Phys. Lett.* **45** (1984) 220.

[33] Cates, M. R., Allison, S. W., Marshall, B., Davies, T. J., Franks, L. A., Nelson, M. A., Noel, B. W., *Proc. SPIE* **506** (1984) 64.

[34] Mazé, G., Cardin, V., Poulain, M., *Proc. SPIE* **576** (1983) 10.

[35] Chan, K., Ito, H., Inaba, H., *Appl. Opt.* **22** (1983) 3802.

[36] Inaba, H., *Jpn. Pat. Appl. 59/183,348,* 1984.

[37] Jawad, S. M., Alder, J. F., *Anal. Chim. Acta* **246** (1991) 259.

[38] Beswick, R. B., Pitt, C. W., *J. Colloid Interface Sci.* **124** (1988) 146.

[39] Alarcon, M. C., Ito, H., Inaba, H., *Appl. Phys.* **B 43** (1987) 79.

[40] Narayanaswamy, R., *Proc. SPIE* **798** (1987) 249.

[41] Neuman, M. R., Fleming, D. G., Cheung, P. W., Ko, W. H., *Physical Sensors for Biomedical Applications;* Boca Raton, FL: CRC Press, 1980.

[42] Schweiss, J. F. (ed.), *Continuous Measurement of Blood Oxygen Saturation in the High Risk Patient;* Mountain View, CA: Oximetric, 1987.

[43] Crowe, J. A., Rea, P. A., Wickramasinghe, Y. A. B. D., Rolfe, P., "Towards Non-Invasive Monitoring of Cerebral Metabolism", in: *Fetal and Neonatal Physiological Measurements,* Rolfe, P. (ed.), Vol. 2; London: Butterworths, 1987, p. 150.

[44] Chance, B., Legallais, V., Sorge, J., Graham, N., *Anal. Biochem.* **66** (1975) 498.

[45] Mayevsky, A., Chance, B., *Science* **217** (1982) 537.

[46] Renault, G., Duboc, D., Degeorges, M., *J. Appl. Cardiol.* **2** (1987) 91.

[47] Alvager, T., Branham, M., *Proc. Indiana Acad. Sci.* **87** (1977) 365.

[48] Tromberg, B. J., Eastham, J. F., Sepaniak, M. J., *Appl. Spectrosc.* **38** (1984) 38.

[49] Sepaniak, M. J., *Clin. Chem.* **31** (1985) 671, and references cited therein.

[50] Coleman, J. T., Eastham, J. F., Sepaniak, M. J., *Anal. Chem.* **56** (1984) 2246.

[51] Briggs, J., Fisher, M. L., Gazharossian, V. E., Becker, M. J., *J. Immunol. Methods* **81** (1985) 73.

[52] Peterson, J. I., Goldstein, S. R., Fitzgerald, R. V., Buckhold, D. K., *Anal. Chem.* **52** (1980) 864.

[53] Watson, R. M., Markle, D. R., Ko, Y. M., Goldstein, S. R., McGuire, D. A., Peterson, J. I., Patterson, R. E., *Am. J. Physiol. Heart Circ. Physiol.* **15** (1984) H 232.

[54] Abraham, E., Markle, D. R., Fink, S., Ehrlich, H., Tsang, M., Smith, M., Meyer, A., *Anesth. Analgesia* **64** (1985) 731.

[55] Marsoner, M. J., Kroneis, H. W., Offenbacher, H., Karpf, H., in: *Physiology and Methodology of Blood Gases and pH,* Vol. 6, Maas, A. H. J., Boink, F. B. T. J., Saris, N. E. L., Sprokhelt, R., Wimberley, P. D. (eds.); Copenhagen: International Federation for Clinical Chemistry, 1986.

[56] Peterson, J. I., Fitzgerald, R. V., Buckhold, D. K., *Anal. Chem.* **56** (1984) 62.

[57] Vurek, G. G., Feustel, P. J., Severinghaus, J. W., *Ann. Biomed. Eng.* **11** (1983) 499.

[58] Hirschfeld, T., Miller, F., Thomas, S., Miller, H., Milanovich, F., Gaver, R., *J. Lightwave Technol.* **5** (1987) 1027.

[59] Philbin, D. M., Inada, E., Sims, N., Philbin, D. M., Misiano, D., Schneider, R. C., *Perfusion* **2** (1987) 127.

[60] Hill, A. G., Groom, R. C., Vinansky, R. P., Lefrak, E. A., *Proc. Am. Acad. Cardiovasc. Perfusion* **6** (1985) 148.

[61] Siggaard-Andersen, O., Gothgen, I. H., Wimberley, P. D., Rasmussen, J. P., Fogh-Andersen, N., *Scand. J. Clin. Lab. Invest.* **48**, Suppl. 189 (1988) 77.

[62] Gehrich, J. L., Lübbers, D. W., Opitz, N., Hansmann, D. R., Miller, W. W., Tusa, J. K., Yasufo, M., *IEEE Trans. Biomed. Eng.* **33** (1986) 117.

[63] Tusa, J., Hacker, T., Hansmann, D. R., Kaput, T. M., Maxwell, T. P., *Proc. SPIE* **713** (1986) 137.

[64] Miller, W. W., Yafuso, M., Yan, C. F., Hui, H. K., Arick, S., *Clin. Chem.* **33** (1987) 1538.

[65] Hansmann, D. R., Gehrich, J. L., *Proc. SPIE* **906** (1988) 4.

[66] Turner, T. E., Karube, I., Wilson, G. (eds.), *Biosensors;* London: Oxford University Press, 1988.

[67] Wolfbeis, O. S. (ed.), *Biosensors & Bioelectronics;* Boca Raton, FL: CRC Press, 1991.

[68] Völkl, K. P., "Die Enzym-Optode", *PhD Thesis;* University of Bochum, 1980.

[69] Trettnak, W., Wolfbeis, O. S., *Analyst* **113** (1988) 1519.

[70] Schaffar, B. P. H., Wolfbeis, O. S., *Biosensors & Bioelectronics,* **6** (1990) 137.

[71] Dremel, B. A. A., Schaffar, B. P. H., Schmid, R. D., *Anal. Chim. Acta* **225** (1989) 293.

[72] Trettnak, W., Leiner, M. J. P., Wolfbeis, O. S., *Biosensors,* **4** (1989) 15.

[73] Narayanaswamy, R., Sevilla, F., *Anal. Lett.* **21** (1988) 1165.

[74] Wangsa, J., Arnold, M. A., *Anal. Chem.* **60** (1988) 1080.

[75] Schultz, J. S., Mansouri, S., Goldstein, I. J., *Diabetes Care* **5** (1982) 245.

[76] Trettnak, W., Wolfbeis, O. S., *Anal. Chim. Acta* **221** (1989) 195.

[77] Goldfinch, M. J., Lowe, C. R., *Anal. Biochem.* **138** (1984) 430.

[78] Lübbers, D. W., Völkl, K. P., Grossmann, U., Opitz, N., in: *Progress in Enzyme and Ion Selective Electrodes,* Lübbers, D. W., Acker, H., Buck, R. P., Eisenmann, E., Kessler, M., Simon, W. (eds.); Berlin: Springer, 1981, p. 67.

[79] Trettnak, W., Wolfbeis, O. S., *Fresenius' Z. Anal. Chem.* **334** (1989) 427.

[80] Wolfbeis, O. S., Posch, H. E., *Fresenius' Z. Anal. Chem.* **332** (1988) 255.

[81] Walters, B. S., Nielsen, T. J., Arnold, M. A., *Talanta* **35** (1988) 151.

[82] Fuh, M. R., Burgess, L. W., Christian, G. D., *Anal. Chem.* **60** (1988) 433.

[83] Kulp, T. J., Camins, I., Angel, S. M., Munkholm, Ch., Walt, D. R., *Anal. Chem.* **59** (1987) 2849.

[84] Wolfbeis, O. S., *Anal. Chem.* **58** (1986) 2874.

[85] Wolfbeis, O. S., "The Development of Fibre Optic Sensors by Immobilization of Fluorescent Probes", in: *Proceedings of the NATO Advanced Workshop on Analytical Uses of Immobilized Biological Compounds,* Guilbault, G. G., Mascini, M. (eds.); Dordrecht: Reidel, 1988, p. 219.

[86] Vo-Dinh, T., Griffin, G. D., Sepaniak, M. J., "Fiber Optic Immunosensors", in: *Fiber Optic Chemical Sensors and Biosensors,* Wolfbeis, O. S. (ed.); Boca Raton, FL: CRC Press, 1991, Chap. 18.

[87] Rockhold, S. A., Quinn, R. D., VanWagenen, R. A., Andrade, J. D., Reichert, W. M., *J. Electroanal. Chem.* **150** (1983) 261.

[88] Newby, K., Reichert, W. M., Andrade, J. D., Benner, R. E., *Appl. Opt.* **23** (1984) 1812.

[89] VanWagenen, R. A., Rockhold, S., Andrade, J. D., "Probing Protein Adsorption: Total Internal Reflection Intrinsic Fluorescence", in: *Interfacial Phenomena and Applications,* Cooper, S. L., Pappas, N. A. (eds.); New York: Wiley, 1982, p. 351.

[90] Kronick, M. N., Little, W. A., *J. Immunol. Methods* **8** (1975) 235.

[91] Thompson, N. L., Axelrod, D., *Biophys. J.* **43** (1983) 103.

[92] Andrade, J. D., Lin, J. N., Herron, J., Reichert, M., Kopecek, T., *Proc. SPIE* **718** (1987) 280.

[93] Sutherland, R. M., Dähne, C., Place, J. F., Ringrose, A. R., *J. Immunol. Methods* **74** (1984) 253.

[94] Sutherland, R., Dähne, C., Place, J., *Anal. Lett.* **17** (1984) 43.

[95] Tromberg, B. J., Sepaniak, M. J., Vo-Dinh, T., Griffin, G. D., *Anal. Chem.* **59** (1987) 1226.

[96] Vo-Dinh, T., Griffin, D. G., Ambrose, K. R., *Appl. Spectrosc.* **40** (1986) 696.

[97] Stoker, R. S., VanWagenem, R. A., Andrade, J. D., Mosher, D. F., *J. Colloid Interface Sci.* **106** (1985) 459.

[98] Andrade, J. D., VanWagenem, R. A., Gregonis, D. E., Newby, K., Lin, T. N., *IEEE Trans. Electron Devices* **32** (1985) 1175.

[99] Petrea, R. D., Sepaniak, M. J., Vo-Dinh, T., *Talanta* **35** (1988) 139.

[100] Looyman, R. P. H., De Bruijn, H. E., Greve, T., *Proc. SPIE* **798** (1987) 290.
[101] Vo-Dinh, T., Tromberg, B. J., Griffin, G. D., Ambrose, K. R., Sepaniak, M. T., Gardenhire, E. M., *Appl. Spectrosc.* **41** (1987) 735.
[102] Anderson, F. P., Miller, W. G., *Clin. Chem.* **34** (1988) 1417.
[103] Miller, W. G., Anderson, F. P., *Anal. Chim. Acta* **227** (1989) 135.

19 Sensors and Monitoring Systems in Environmental Control

Konrad Colbow, Karen L. Colbow, Simon Fraser University, Burnaby, B.C., Canada

Contents

In a reference work on sensors like the present one, the reader would not like to miss a contribution on applications in environmental control. On the other hand this field is extremely dependent on the legal aspects in different countries, what makes the inclusion of this topic to our series difficult. Nevertheless we decided to present at least a brief survey which lists some example information about gas sensors and monitoring systems in environmental control. The description is confined to the situation in North America and Canada.

Further aspects, eg, measurement of toxic substances in water and soil, are not dealt with here. The reader is referred to an extensive literature available for the particular situation in individual countries around the world.

The editors

19.1   Introduction

Combustible and toxic gas sensors and monitoring systems are available for a wide variety of gases. Different sensors and sensor systems have advantages in different applications. Data for six important gases are collected in Table 19-1. Information from N. O. Korolkoff's review [1] and a collection of information available from the Government of British Columbia, Canada [2-6] are extensively quoted in this article without further referencing.

Gas monitoring systems should be chosen to suit their intended use since significant differences exist between available sensing technologies, price, and performance. There are basically two types of gas detection technologies. With instrumentation monitors, a sample of air is drawn through a piece of tubing, the sample draw line, to an analysis instrument, which is most commonly based on mass spectrometry, flame emission spectrometry, infrared spectrometry, or colorimetric (paper tape) response.

In the other detection technology the chemical/electrical response of a small sensor located at the desired site is monitored. One usually distinguishes here between solid state sensors and electrochemical sensor cells. Since the sensor output is transmitted electrically, the distance between it and the monitoring system is usually not a limitation.

With instrumentation monitors detection capabilities are in the low ppb range for some gases. Such systems may be used to monitor a very large variety of gases but, unless the instrument is placed quite close to the monitoring point, the length of the sample draw line increases the overall system response time. Electrochemical and solid state sensors often do not have the high sensitivity required for some gases, but they do have fast response times and are well suited for monitoring large areas from a central location.

The sensitivity and selectivity of both types of monitoring systems are affected to varying degrees by interferences arising from the presence of other gases. A positive or negative response, along with a possible false alarm may be produced in the presence of an interfering gas or varying humidity. A sensor may respond to its target gas directly, or indirectly by detecting a byproduct created in a reaction with something in the environment. In the latter case one must ensure that the same environmental conditions exist as during the calibration run. For example, an indirect detection of arsenic pentafluoride would involve the monitoring of hydrogen fluoride, a byproduct created when arsenic pentafluoride reacts with moisture in the air. The result in this case is higher sensitivity for the indirect detection, but the requirement for more stringent environmental control if quantitative measurements are desired.

Table 19-1. Kind of gases and their standard concentration to be detected for preservation of the environment on the earth.

		NO_x	SO_2	CO	O_3	CO_2	Chlorofluorocarbon (CFC)
Request of sensors (concentration)		air pollution (0.01–0.3 ppm) boiler, combustor (10–1000 ppm) automotive exhaust (10–1000 ppm) heater, water heater (10–500 ppm)	air pollution (0–2 ppm) coal or oil burner (1–1000 ppm) incinerator (10–100 ppm)	air pollution (0.1–10 ppm) misfire (10–500 ppm) fire and gas-poisoning (5–50 ppm) automotive exhaust (0–10%) metal or coal mine (1–50 ppm)	air pollution and photochemical smog (0–0.5 ppm)	air pollution (200–400 ppm)	air pollution (200 ppt)
Standard concentration	environmental standard of air pollution	below 0.04–0.06 ppm (daily average)	below 0.04 ppm (daily average) below 0.1 ppm (daily average)		below 0.06 ppm (1 hourly average) as an oxidant concentration		
	Standard level at exhaust sources — fixed facility	60–800 ppm depend upon kind and scale of facility	0.1–5 m^3/hr depend upon height of exhaust or location of facility				
	Standard level at exhaust sources — auto-mobile	0.48–0.98 g/km (gas, LPG) 350–520 ppm (diesel)		2.70–17.0 g/km (gas, LPG) 980 ppm (diesel)			
	general allowable concentration	3 ppm (NO_2) 25 ppm (NO)	2 ppm	50 ppm	0.1 ppm	5000 ppm	

There exist both low-volume and high-volume air samplers. The former consists of a sample inlet, collection media, gas moving device and a gas flow meter. The collection media which consists of impinger solutions and/or filters isolate the sample; the gas flow meter measures the total volume of gas sampled. The temperature and barometric pressure at the meter are also recorded. The sampling periods typically vary from 24 hours for ammonia and mercury to 7 days for fluoride and particulate. The impinger for asbestos, coal, and particulate consists of silica gel. A 0.01-mol/L-H_2SO_4 impinger solution followed by a silica gel impinger is used to isolate ammonia. Fluoride is collected via a 0.01-mol/L-NaOH impinger and silica gel filter, and a solution of 1% $KMnO_4$ and 10% H_2SO_4 and silica gel is used for mercury. The maximum flow rate is usually approximately 5 L/min.

High-volume samplers, used to sample particulates, typically utilize a high flow-rate blower to draw air into a covered housing and through a glass fiber filter at a typical rate of 1–2 m^3/min. The flow rate of the sampled air is measured with a mass flow probe mounted in the throat section of a 20 cm × 25 cm filter holder of the high-volume sampler. The electrical output of the probe and associated circuitry is used as the control signal for a motor speed controller. Thus, as external conditions change, the controller adjusts the power to the motor so that a constant mass flow rate is maintained. The filters are weighed before and after the sampling period to determine the total mass of collected particulate. The total volume sampled is determined by the sampling time (usually 24 h) and the flow rate. The calculated mass concentration of suspended particulates in the ambient air is expressed in $\mu g/m^3$.

Inertial impactors are used to separate particles based on their size. These devices operate by impinging the air sample at high velocity against a surface so as to produce a sharp change in the direction of the air stream. The larger particles with greater inertia are thrown onto the impaction surface while the smaller particles with less inertia escape impaction and continue with the air stream. In this manner, particles which have inertia greater than a certain critical value are selectively removed from the air stream. Selective removal from the air stream is defined in terms of the 50% cut-point which refers to the particle size for which collection efficiency is 50%. For example, if the cut-point of an impactor is quoted as 15 μm then ideally 50% of all particles having an aerodynamic size of 15 μm and 100% of all particles having an aerodynamic size of less than 15 μm will be collected. All particles regardless of shape and density which have the same settling velocity will have the same aerodynamic size.

19.2   Commonly Used Gas Sensing Methods

The six most common gas sensing methods are discussed in some detail in this section. Minimum detection limits, typical upper concentration range, and response times are presented in Table 19-2.

Flame Emission Spectrometry (FES)

In flame emission spectrometry, air samples are mixed with hydrogen in a reaction chamber and activated through combustion. Optical filters are used to isolate the characteristic wavelengths of the emitted light as the gas molecules in the reaction chamber are raised to excited states and emit light in returning to their ground state. A photomultiplier converts the light intensity to an electronic signal. Compensating optical filters correct the output signal for contributions from other light sources.

Specific chemical scrubbers which are switched into the sample stream to remove the target gas may be used as double checks for possible gas leaks before activating an alarm. Flame-emission-spectrometry monitors are not affected by humidity, air flow, or temperature. Commercially available systems have been perfected to the point where they can respond in only 2 s of sampling time. Calibration of such systems is recommended every six months. The latter procedure takes approximately 30 min with standards provided. (The primary manufacturer of this type of system is Telos Labs.)

Table 19-2. Detection limits, range, and response times for: flame emission spectrometry (FES), Fourier transform infra red (FTIR) spectrometry, mass spectrometry (MS), paper tape (PT), solid state (SS) sensors, and electrochemical (EC) sensors. Threshold limit value (TLV): maximum exposure in 8 h period in 40 h work week [7].

Gas	TLV (ppm)	Minimum detection limit and range (ppm) and TLV response time (second row: in s)					
		FES	FTIR	MS	PT	SS	EC
Ammonia (NH$_3$)	25	0.2–1000 8	1.0–1000 30	200.0	2.8–250 30	2.5–100 15–60	1–100 ppm 30–300 s
Arsenic Penta-fluoride (AsF$_5$)		0.01–30 8	0.5–1000 30			0.1–1 60	0.1–9 30
Arsine (AsH$_3$)	0.05 8	0.01–30	0.5 15	0.05 15	0.007–1 15–60	0.05–0.5 30	0.02–1 30
Boron Trichloride (BCl$_3$)		0.02–30 8	1.0–1000 30	1.3–2.6 2.5		25–500 15–60	0.1–15 30
Boron Trifluoride (BF$_3$)	1	0.02–30 8	0.5–1000	1.0–1 7.5		2.0	0.2–9 30
Carbon Monoxide (CO)	50		5.0–1000		5–150 60	2.5–200 15–60	1–500 30
Chlorine (Cl$_2$)	1			0.25–1 8.7	0.1–10 15	0.5–10 15–60	0.1–10 30
Diborane (B$_2$H$_6$)	0.1	0.02–30 8	1.0	0.5	0.025–0.1 30	0.5–10 60	0.02–0.2 30
Dichlorosilane (H$_2$SiCl$_2$)		0.5–1000 8	1.0–1000	2.5–5 2.3		2.5–50 15–60	0.5–15 30
Freon (Family)	1000		1.0–1000	10–1000 1.6		10–1000 5–60	0.5–150 30
Hydrogen (H$_2$)				98.0–200 3		2.5–100 15–60	0.1%–5% 30
Hydrogen Chloride (HCl)	5		1.0–1000	1.3–5 2.5	0.2–50 10	2.5–50 15–60	0.1–25 60
Hydrogen Fluoride (HF)	3		0.5–1000	40.0	0.7–30 60	2.5–50 15–60	0.1–25 60

Table continued on next page

Table 2. (continued)

Gas	TLV (ppm)	Minimum detection limit and range (ppm) and TLV response time (second row: in s)					
		FES	FTIR	MS	PT	SS	EC
Hydrogen Sulfide (H_2S)	10	0.02–100 8	10.0–1000 30	15.0	0.1–100 7	0.5–50 15–60	0.1–50 30
Nitric Oxide (NO)	25	0.2–100 8	1.0–1000 30	16.0 3.3	2.5–125 60	1–50 15–60	1–100 30
Nitrogen Dioxide (NO_2)	3	0.2–100 8	1.0–1000 30	1.0–3 11		1–20 15–60	0.1–15 30
Oxygen Deficiency (O_2)	19.5%						0.1–25% 30
Phosphine (PH_3)	0.3	0.01–30 8	0.5	1.2	0.010–3 10	0.3–10 60	0.01–1 <30
Phosphororus pentafluoride (PF_5)		0.01–30 8	0.5–1000 30			3–50	0.1–9 30
Silane (SiH_4)	5	0.5–30 8	1.0–1000 30	1.2–5 2.8	1.3–50 30	0.5–10 15–60	0.5–15 30
Silicon tetra-chloride ($SiCl_4$)		0.5–30 8	1.0–1000 30	1.3–10 2.5		50–1000 15–60	0.1–15 30

Infrared Spectrometry (FTIR)

Gases can be identified by measuring their characteristic infrared absorption wavelengths due to their molecular chemical bonds. Analysis by the absorption of infrared light is based on the Beer-Lambert Law. Fourier transform infrared spectrophotometers (FTIR) have been the most widely used IR instrument for monitoring gases and chemicals. Some FTIR analyzers, operating with a laser spectral reference, provide excellent selectivity and sensitivity for the identification of gases and chemical vapors, even if the chemicals are similar. Molecules that do not absorb IR radiation, such as diatomic nonpolar Cl_2 and F_2 are not detectable. There is usually no need for periodic recalibration after initial instrument setup. (A primary manufacturer of FTIR gas detection systems is Telos Labs.)

Mass Spectrometry (MS)

Monitors using mass spectrometry require a controlled sample draw system in conjunction with the mass spectrometer analyzer. A capillary inlet allows a small controlled gas sample to enter the ionization region of the mass spectrometer, where the sample molecules are bombarded by electrons, ionized, and formed into an ion beam. Electrostatic and magnetic fields then separate the ion species according to energy and mass/charge ratio. The ions are directed to either a Faraday-cup collector or a high-sensitivity electron multiplier detector. The measured beam intensities correspond to the gas concentrations being analyzed and the mass spectra may be used to identify sample constituents with a high degree of specificity.

One disadvantage of mass spectrometry is its susceptibility to interference gases. For example, the detection of hydrides, such as phosphine and diborane, is susceptible to interference from oxygen and nitrogen. When interference occurs due to the presence of a known compound, it may be corrected to a certain degree by software subtraction. (Perkin-Elmer is a primary manufacturer of this type of system.)

Paper Tape (PT)

Paper tape detectors are chemically treated papers which react only with a particular gas or family of gases. A continually fed tape, when exposed to air from a sample draw line, changes color in direct proportion to the amount of target gas present. A stain becomes darker at higher concentrations of the target gas. Color changes are continuously monitored by an electro-optical detection system which determines the amount of light reflected from the tape. Accuracy and reproducibility of the paper tape system depend on the quality of the tape, as well as on the sophistication of the electro-optical detection system. At a maximum sample draw length of 90 m, TLV (threshold level value) response times using advanced detectors can range from 7 s for HCl gas to 60 s for HF. Minimum detection limits range from 7 ppb for AsH_3 to 5 ppm for CO. Paper tape monitors are calibrated and verified at the factory and require only simple cleaning and reverification on a weekly or monthly basis. The cost of cleaning is minimal, the verification card is supplied with the unit, and the time required is 5–10 min. The sensors are affected only by extremes in humidity, air flow, and temperature. They can be located in adverse environments and outdoors, but some conditions may require a custom enclosure. When detection tape is used continuously during monitoring, a one-month supply currently costs about $140 for a 16-point system. However, detection tapes are not available for all gases commonly used in industry. (A primary U.S. manufacturer of gas detecting paper tape and tape detection systems is MDA Scientific.)

Solid State Sensors (SS)

Solid state sensors are a rapidly growing field. These sensor may be expected to assume growing importance for gas sensing in the near future due to their small size, low cost, and improving reliability. A recent review has been written by Madou and Morrison [8].

Of prime commercial importance at this time are the metal-oxide semiconductor sensors such as tin oxide. The sensor consists of a heater and a collector imbedded in a metal oxide material. Under operation at constant temperature, the sensor has a fixed resistance between two electrodes. When the target gas is present, it causes a resistance change proportional to its concentration. An electronic package converts this resistance change to a readout of gas concentration. Gas selectivity and reaction to interference gases and humidity are controlled by the manufacturer through selection of the oxide material, the use of catalysts, and by maintenance of the proper surface temperature with the heater. Although various metal oxides have been used as the basis of semiconductor gas sensors, tin oxide (SnO_2) is by far the most popular. It has been used in both thin-film and sintered-powder form. The sensors typically operate at 350 °C, where humidity interference is considerably reduced. Reducing gases in the atmosphere remove chemi-adsorbed oxygen (mainly in the form O^-) and thus increases the conductivity of the sample. The lack of gas selectivity [9] may be partially overcome by enhancing the response to certain gases through catalysts such as platinum, palladium, and ruthenium. Gas selectivity may also be achieved through thermal cycling of the sample with a cycling time of about 20–50 s [10–12]. Different reducing gases produce conductivity changes in the tin oxide sensor during different time intervals in the heating-cooling cycle. (Taguchi gas sensors (Figaro Engineering Inc. Japan) have been the leading tin oxide based gas sensors.)

In many cases the sensitivity of a solid state monitor is not adequate for efficient TLV detection, as required by newer legislation. Response times average between 15 and 60 s.

Metal-oxide monitors are relatively inexpensive ($ 550 to $ 1600). Calibration, needed every 60 to 90 d, takes about 2 min and costs about $ 30. Monitors have a 5 to 10 year life with an element replacement cost of $ 165–$ 275. (Primary manufacturers are International Sensor Technology and Matheson Gas Products.)

Electrochemical Sensors (EC)

Electrochemical gas sensors consist of two electrodes immersed in a common electrolyte, which may be liquid, a gel, or an impregnated porous solid which is isolated from the environment. The isolation barrier may be a gas permeable membrane or a capillary diffusion barrier. When a polarizing voltage is applied between the electrodes and a gas is introduced into the sensor, an oxidation-reduction reaction generates an electrical current that is linearly proportional to the gas concentration. Each unit is designed for maximum sensitivity for a given gas with minimum interferences from other gases. Electrochemical sensors are usually affected by humidity and air flow. A slow loss of fluid (usually water) through the barrier is often a problem and requires periodic additions of water or electrolyte to the sensor. Manufacturers use a variety of methods to minimize this problem, such as a two-membrane configuration and a gel-like electrolyte (Enterra Instrumentation), a capillary design (City Technologies), or an electrolyte bound in a matrix (Interscan). Bionics uses an organic electrolyte, which according to the company requires replenishment only every 6 months.

The response time of EC sensors is typically 30 s and minimum detection limits range from 0.02 to 15 ppm. EC sensors range in price from $ 1,000 to $3,000. Calibration and maintenance are usually required every 3 months and sensor life averages 18 to 36 months. Replacement cells cost $ 175 to $ 300 depending on sensor type. (The primary manufacturers of EC monitors are Bionics Instrument Co., Ltd., Enterra Instrumentation Technologies, Gas Tech Inc., Interscan Corp., MDA Scientific Inc., National Draeger, Inc., Sierra Monitor Corp. and Telos Labs.)

19.3   Principle Air Pollutants

Suspended particulates are airborne particles such as smoke, dust, fly-ash, and pollen. The degree to which they are a health hazard depends on the size and chemical characteristics of the particulate matter. Particles smaller than 10 μm in diameter pose the greatest threat to health because they penetrate deep into the lungs und cause respiratory problems.

Lead, originating primarily from the use of leaded gasoline in motor vehicles, may be absorbed into the body and accumulated in bone and soft tissues to the point where blood-forming, nervous and kidney systems are affected.

Sulphur dioxide is produced through the combustion of fossil fuels containing sulfur, such as oil and coal. It is a colorless gas with a pungent odor and can undergo reactions in the atmosphere which contribute to acid rain. It can cause upper respiratory tract irritation and increase the occurrence of respiratory disease.

Carbon monoxide gas is produced through the incomplete combustion of carbon containing fuels, thus a major source is automobile exhausts. It is colorless and odorless and at lower doses may impair perception and reflexes, while at higher concentrations it may cause asphyxiation.

Nitrogen dioxide is present in the combustion discharges from industrial furnaces, building heating systems, and motor vehicles. It is a reddish-brown gas with an irritating odor, and has been associated with respiratory disease. It is also involved in the production of smog.

Ozone is produced as a result of chemical reactions of hydrocarbons and nitrogen oxides in the air in the presence of sunlight. It is a pungent-smelling, faintly bluish gas. Ozone can irritate the respiratory system and impair lung function. Other oxidants that accompany ozone are strong eye irritants.

Continuous Ambient Air Analyzers — Some Specific Examples

Coulometric titration of SO_2, H_2S: In a cell containing Br-/Br_2 in an acidic medium, electrolytically generated bromine is titrated with ambient levels of either SO_2 or H_2S. The voltage perturbation seen by the analyzer as a result of the change in the redox potential when SO_2 (or H_2S) is introduced into the cell will be proportional to the SO_2 concentration in the sample stream.

Pulsed fluorescence of SO_2: A sample cell containing SO_2 is exposed to incident UV radiation delivered in pulses in the 230–190 nm range. This incident radiation is absorbed producing electronically excited SO_2 molecules which revert back to the ground state by an emission process known as fluorescence. A photomultiplier tube is used to measure the fluorescence intensity, which is proportional to the SO_2 concentration.

Nondispersive infrared photometry of CO: Equivalent chopped sources produce IR radiation which is incident upon a reference cell and a sample cell containing CO. The radiation emerging from the reference cell will be equivalent to the incident radiation, such that the intensity differential between the reference cell and the sample cell will be proportional to the CO concentration.

Gas chemiluminescence measurements of NO, NO_2, ozone: In a reaction chamber nitric oxide in the presence of excess ozone produces photons, the emission intensity of which is proportional to the original NO concentration. The emission intensity is measured by a photomultiplier tube and amplified by an electrometer. The same technique is applicable for NO_2 if it is first converted to NO via a catalytic converter. The same method is used to measure ozone concentrations; the only difference is that the reaction chamber contains excess CH_2 rather than ozone.

19.4 Selection Criteria for Monitors

Selection criteria of gas monitor systems involve additional decisions than those based solely on minimum detection capability, upper range, and response time listed in Table 19-2. Of prime importance are the sensitivity to interference gases, air flow, temperature, and humidity. Also of concern are long-term stability, and monitor maintenance and calibration.

Some monitors are prone to drift in their zero point and are affected by external environmental conditions. Thus, they require sophisticated maintenance by skilled personnel or even by the manufacturer. Calibrating a monitor to a known concentration of a target gas is the most accurate method. However, in general it is desirable to have a linear repeatable response over the widest possible gas concentration range. Some sensors require a stabilization period before calibration can be performed following maintenance. This results in a longer overall calibration time. The cost of materials used for calibration is another factor to consider.

Regulations and court cases have put pressure on industry not only to monitor hazardous materials for the welfare of employees and the community but also to insure that accurate records of those monitoring efforts are kept. Thus, sophisticated monitoring systems with a variety of capabilities have been developed. Manufacturers offer systems that can count from 1 to 1024 points, allowing coverage of a large area. Generally, the greater the number of points, the longer the response time. Monitors with an electrical signal from the sample point have the advantage of faster response times when compared to systems using sample draw lines. An ideal monitoring/data transmission system should be capable to interface readily with new sensors.

In the event of system failure, relay logic may be used for shut down of heating, ventilation, air conditioning and gas supply, or the activation of emergency exhaust systems, audible and visual alarms, security notification, etc. Internal relays may be logically addressed so that some are activated under low level alarm conditions while others respond to high-level conditions. The ability to set alarm levels in response to specific user safety requirements is a useful feature in a gas monitor system. By setting an alarm level below TLV a window of protection for personnel is provided before dangerous levels of exposure occur. A system with remote notification capability can communicate a dangerous condition. Beeper annunciation is one example, but information through phone lines or modems may allow safety personnel to react faster. A PC-based system allows the flexibility of modem use, networking, and system upgrading as new software is available or as operating requirements change.

19.5 References

[1] Korolkoff, N. O., "Survey of Toxic Gas Sensors and Monitoring Systems", *Solid State Technology,* Dec. 1989, pp. 49–64.
[2] McQuaker, N. R., Habooseh, H., *Calibration and Operation of Continuous Air Monitors;* Ministry of Environment, Vancouver, B.C., Canada, April 1989.
[3] McQuaker, N. R., Sandberg, D. K., Pengilly, D., *Calibration and Operation of Measurement Systems used in Air Quality Monotoring;* Ministry of Environment, Vancouver, B.C., Canada, April 1989.

[4] McQuaker, N. R., *A Laboratory Manual for the Chemical Analysis of Ambient Air, Emissions, Precipitation, Soil and Vegetation;* Ministry of Environment, Vancouver, B.C., Canada, April 1989.

[5] *Industrial Hygiene: Industrial Hazards and their Evaluation;* Workers Compensation Board of British Columbia, Vancouver, B.C., Canada, 1989.

[6] *Laboratory Analytical Methods;* Workers Compensation Board of British Columbia, Vancouver, B.C., Canada, 1989.

[7] ACGIH, *TLVs and Biological Indices for 1988/1989,* March 1989.

[8] Madou, M. J., Morrison, S. R., *Chemical Sensing with Solid State Devices;* San Diego: Academic Press, 1989.

[9] Morrison, S. R., "Selectivity in Semiconductor Gas Sensors"; *Sens. Actuators* 12 (1987) 425-440.

[10] Sears, W. M., Colbow, K., Consadori, F., "General Characteristics of Thermally Cycled Tin Oxide Gas Sensors"; *Semiconductor Science and Technology* 4 (1989) 351-359.

[11] Sears, W. M., Colbow, K., Slamka, R., Consadori, F., "A Restricted Flow Thermally Cycled Gas Sensor"; *Sens. Actuators* B1 (1990) 62-67.

[12] Wlodek, S., Colbow, K., Consadori, F., "Signal Shape Analysis of a Thermally Cycled Tin Oxide Gas Sensor"; *Sens. Actuators* B3 (1991) 63-68.

20 Humidity Sensors

HIROMICHI ARAI, TETSURO SEIYAMA, Kyushu University, Fukuoka, Japan

Contents

20.1   Introduction

Recently the use of humidity sensors has greatly increased in a range of industries such as in the production of electronic devices, precision instruments and foodstuffs. As humidity is a permanent environmental factor, its measurement and/or control are of importance not only for human comfort but also for many industries and technologies. In recent years, major domestic applications of humidity sensors have been: eg, automatic humidity control in air conditioners, automatic cooking by microwave ovens, and dew sensing on the cylinder heads of VTRs [1, 2].

Humidity is defined as the concentration of water molecules in the atmosphere. In practice, parameters important in the measurement of humidity are partial pressure of water, mixing ratio, specific humidity, absolute humidity, mole fraction of water vapor, relative humidity (r. h.), and dew-point temperature. Of these measurements, relative humidity, which is the ratio of the actual water pressure to the saturated pressure, is widely used. The concentration of water molecules in the air is low and, moreover, the effects of water are very complicated not only chemically but also physically. The measurement of humidity is difficult compared to the measurement of temperature. In many industries from the electronic industry to agriculture, there is a demand for humidity control. For example, a dry atmosphere is required for the pro-

Table 20-1. Applications of Humidity Sensors

Industry	Applications	Operation range
Community safety	Nuclear power reactor	above 80 °C, 80% r. h.
	Thermoelectric power plant	above 100 °C
	Humidity in boiler	100–400 °C
	Water in gas	city gas atmosphere
Domestic electric appliance	Dessiccator	50–100% r. h.
	Air conditioner	30–100% r. h.
	Control of cooking	5–200 °C, humidity
	Various kinds of air conditioner	5–40 °C, 40–70% r. h.
	Reprographic machine	150 °C, humidity
Medical equipment	Sterilizer	humidity above 100 °C
	Medical equipment for respiratory organs	20–30 °C, above 90% r. h.
	Incubator	10–300 °C, 50–80% r. h.
Industry	Paper manufacture	10–30 °C, 50–100% r. h.
	Oven	100–500 °C, above 50% r. h.
	Film dessiccation	20–60 °C, below 20% r. h.
	Humidity control in factory	5–40 °C, 0–50% r. h.
	Humidifier for industry	30–300 °C, above 50% r. h.
	Refrigerated show case	below 8 °C, below 30% r. h.
	Printing	20–25 °C, 90% r. h.
Agriculture	Observation robot for agriculture	Open-air
	Green house	5–40 °C, 0–100% r. h.
	Protection of plantations	−10–60 °C, 50–100% r. h.

duction of electronic devices such as LSI or IC, but a wet atmosphere is required for the germination or growth of plants. Because of the growing demands for humidity control, the development of humidity sensors is of increasing importance, not only from the point of view of humidity measurement but also to provide a detecting device for air conditioning systems.

A humidity sensor has to satisfy the following practical requirements: (1) high sensitivity over a wide humidity range, (2) quick response, (3) good reproducibility and no hysteresis, (4) robustness and long life, (5) resistance to contaminants, (6) insignificant dependence on temperature, and (7) simple structure and low cost. For particular applications, other requirements should be satisfied, such as low power, low weight, or microprocessor compatibility.

As summarized in Table 20-1, the applications of humidity sensors are widespread, covering domestic appliances, automobiles, medical equipment, various industries, and agriculture. The purposes of humidity sensors and their operating conditions are quite different depending on the field of application. Various types of humidity sensors have been investigated and developed in response to varying demands. Many physcial and chemical phenomena are utilized, and a classification of humidity sensors which have been proposed is shown in Table 20-2. The sensing materials, principles, operating temperatures, humidity ranges and response times for each of these are also summarized. There have been many reports describing humidity sensors using electrolytes, organic polymers and metal oxides. They are classified into three types: ionic conductivity, electronic conductivity, and capacitance types. However, a high proportion of humidity sensors, as shown in Table 20-2, utilize the physical adsorption or absorption of water molecules. In this chapter, the principles and areas of application of several types of humidity sensors will be described and some recent advances will be discussed.

20.2 Mechanism of Water Adsorption on Solid Surfaces

Water vapor chemisorbs on an oxide surface by a dissociative mechanism to form two surface hydroxyls for each water molecule, i.e., the hydroxyl group adsorbing on a surface metal cation and the proton forming a second hydroxyl group with an adjacent surface O^{2-} ion [3]. On the other hand, the first layer of physically adsorbed water vapor on the hydroxylated substrate is localized by double hydrogen bonding of a single water molecule, as will be shown by arguments based on energy, thermodynamics, and dielectrics. The chemisorption of water vapor on de-hydroxylated α-Fe_2O_3 has been investigated by measuring the heat during the immersion of this α-Fe_2O_3 into water at 25 °C. Figure 20-1 shows the net differential heat of water adsorption as a function of coverage on α-Fe_2O_3. The initial net differential heat of chemisorption is at least -50 kcal/mol of water vapor, which is consistent with the dissociative mechanism of chemisorption discussed above. The net differential heat of adsorption decreases sharply as the water coverage increases until the surface of α-Fe_2O_3 is completely hydroxylated, then it slowly decreases with the physical adsorption of the first layer of water. The differential heat of physical adsorption as multilayers is very small.

The dielectric constant from 70 Hz to 300 Hz of physically adsorbed water is constant within the BET monolayer, but rises sharply with the onset of the second layer [3]. This behavior suggests that the first layer of physisorbed water is immobile, but that the succeeding

Table 20-2. Characteristics of Humidity Sensors

Materials	Type	Sensing element	Principle	Operation Range	Response Time	Application	References
Electrolyte	Lithium chloride humidity sensor	LiCl + Organic binder LiCl + Botanical fiber	Ion conductivity change of LiCl	20–90% r.h. 0–60 °C	2–5 min.	Humidity measurement	[7]
Polymer	Swelling-type humidity sensor	Hygroscopic resin + Carbon	Conductivity change depending on carbon particle distance changing by swelling	20–90% r.h. −30–40 °C	<1 min.	Humidity measurement	[46]
		Conductive polymer	Conductivity change of polymer by water adsorption	30–90% r.h. 0–50 °C	<1 min.	Measurement and control of humidity	[45]
	Organic polymer humidity sensor	Hydrophilic polymer	Capacitance change of polymer by water adsorption	0–100% r.h. −40–80 °C	<1 min.	Humidity measurement and control	[52]
Ceramics	Proton-type ceramic humidity sensor	$MgCr_2O_4$-TiO_2 TiO_2-V_2O_5 $ZnCr_2O_4$-$LiZnVO_4$ Amorphous SiO_2 + Metal Oxide	Conductivity change by physical adsorption of water on the porous ceramic surface	10–100% r.h. 0–150 °C	<10 min.	Humidity measurement and control	[16] [18] [20], [21] [57], [58]
	Semiconductor-type humidity sensor	Perovskite-type oxide ZrO_2-MgO	Conductivity change by chemisorption of water on oxide semiconductor	10^2–10^5 ppm 300–600 °C	2–3 min.	Air conditioner Control of food cooking Drier system	[29[, [31] [30]
	Capacitance-type humidity sensor	Al_2O_3 thin film	Conductivity change by physical adsorption of water on the porous Al_2O_3 film	1–2000 ppm 25 °C	<10 s	Detection or measure-ment of humidity in IC or LSI package	[33]
Dew con-densation	Resin dispersion-type dew sensor	Hygroscopic resin + Carbon Cr_2O_3-glass slit	Conductivity change depending on the carbon particle distance changing by swelling of resin	94–100% r.h. −10–40 °C	<10 s	Dew detector or VTR cylinder	[56] [53]
Others	FET-type humidity sensor	Organic polymer film	Change of threshold voltage caused by the capacitance change of polymer films depen-ding on adsorption or desorp-tion of water	0–100% r.h. 0–40 °C	<30 s	Air conditioner Humidity measurement	[52]

Table 20-2. Continued.

Materials	Type	Sensing element	Principle	Operation Range	Response Time	Application	References
	Thermal conductive-type humidity sensor	Thermistor	Difference of thermal conductivity between dry air and wet atmosphere	0–100% r.h. 0–40 °C	10 s	Humidity measurement	
	Microwave moisture sensor	Dielectric substance	Attenuation of microwave transmitting along the dielectric substrate by moisture in sample	0.3–70% r.h. 0–35 °C	few seconds	Detection of moisture in grain, wood and paper	

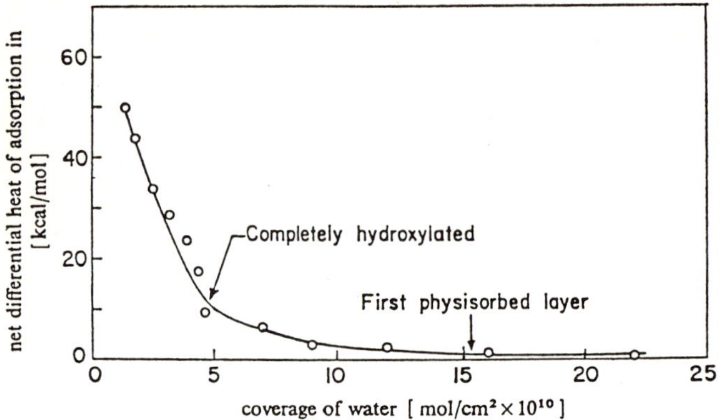

Figure 20-1. Heat of adsorption of water vapor on a-Fe$_2$O$_3$.

layers are mobile. Characteristic relaxation frequencies of adsorbed water, which are obtained from the dielectric constant and the dielectric loss, increase smoothly with increasing coverage. The physisorption of water as coverages from three to six layers seems to have a hydrogen-bonded ice-like structure. The smooth increase in characteristic frequency with increasing coverage suggests that the structuring process is a gradual one toward on ice-like order. The first physisorbed layer is localized by the hydrogen bonding of a single water molecule to two surface hydroxyls as shown in Figure 20-2. Thus the adsorbed water of the first layer has a restricted freedom of rotation. The next few layers develop an ordered ice-like structure leading to the formation of a liquid-like adsorbate.

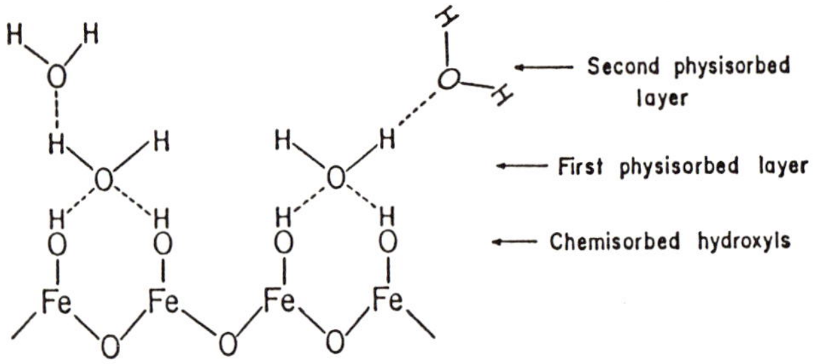

Figure 20-2. Schematic representation of water adsorption on a-Fe$_2$O$_3$.

The polarization effects, isotopic effects (exchange of D_2O for H_2O), and non-ohmic behavior of adsorbed water on silica gel (surface area 660 m^2/g) are similar to those of an electrolyte solution [4]. The resistivity was found to decrease exponentially with water content and to increase roughly exponentially with OH content; this was determined from a simultaneous measurement of resistivity and the near-infrared spectra of adsorbed water and surface hydroxyl groups. The adsorbed water on silica gel is immobile because the water molecule is hydrogen-bonded to two substrate hydroxyls, but the conductivity of silica gel increases as the amount of adsorbed water increases.

The physisorption changes from monolayer to multilayer as water vapor pressure (P) increases. When the pores are cylindrical with one end closed, condensation occurs in all pores with radii up to r_K given by the Kelvin equation:

$$r_K = \frac{2\,\gamma M}{\rho\,R\,T\,\ln\,(P_S/P)}$$ (20-1)

where r_K is the Kelvin radius, γ, ρ, and M are the surface tension (72.75 dyn/cm at 20 °C), density, and molecular weight of water, respectively, and P_S is the water vapor pressure at saturation. Figure 20-3 shows the Kelvin radius as a function of relative humidity from 20 to 100 °C. The condensation occurs in all the pores with radii up to r_K, and under a constant P; the smaller the r_K or the lower the temperature, the more easily it takes place [5]. The occurrence of capillary condensation is important in proton-type humidity sensors [6].

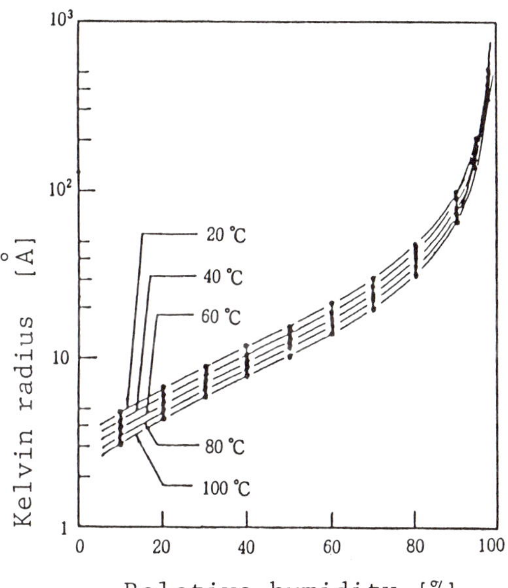

Figure 20-3.
Kelvin radius as a function of relative humidity.

The conduction mechanism may depend on the surface coverage of adsorbed water as shown in Table 20-3. Hopping of H^+ dominates on the surface when no H_2O is present [4]. When water is present but surface coverage is not complete (fractional coverage, $\theta < 1$), H_3O^+ diffusion dominates. When molecular water is abundant ($\theta > 1$), the proton-transfer process dominates. As the amount of hydroxyl groups on the surface increases, the water molecules are doubly hydrogen-bonded to surface hydroxyls and cannot move or rotate freely. On the other hand, when the amount of hydroxyl groups on the surface is low, the adsorbed water molecules are singly hydrogen-bonded to the sublayers and have more freedom to rotate. This results in the change of the dielectric constant and the increase of electrical conductivity. Adsorption of water on other metal oxides also appears to proceed by the same mechanism as in the α-Fe_2O_3 or the silica gel.

Table 20-3. Transport Mechanism as a Function of H_2O Coverage

Fractional H_2O surface coverage	Favored state of protons	Transport mechanism
$\theta = 0$	H^+	H^+ hopping on SiOH surface
$0 < \theta < 1$	H_3O^+	H_3O^+ diffusion on SiOH surface and H^+ transfer between adjacent H_2O molecules in clusters
$\theta > 1$	$H_3O^+(aq)$	H^+ transfer between adjacent H_2O molecules within a continuous film of water

20.3 Electrolyte Humidity Sensors

Water evaporates from or condenses onto an electrolyte solution depending on the relative humidity of the surrounding atmosphere. Electrolyte humidity sensors utilize this phenomenon. Lithium chloride is a typical material [7]. An electric hydrometer using lithium chloride was developed in 1939 by Dunmore [7] and is still widely used. In a typical lithium chloride humidity sensor the lithium chloride solution is impregnated into a plant pith substrate (10 mm × 4 mm × 0.2 mm) [8] and Pt electrodes are applied to both faces of the element. The plant pith possesses a fine reticulate structure and is therefore a suitable porous binder for the electrolyte. Lithium chloride solution trapped in such a porous binder is too stable to flow out even under very humid conditions. Since the humidity range covered by one such unit is about 30% r.h., a wider humidity range from 10 to 100% can only be measured by using a number of elements with different characteristics.

As shown in Table 20-2, lithium chloride humidity sensors have rather slow responses to humidity compared to other types of humidity sensors. This type of humitidy sensor cannot, therefore, be used in automatic control systems which require a rapid response. They are, however, widely used for meteorological observations such as radiosonde because of their reproducibility, long term stability, and low cost [7]. Care should be taken when using this type of sensor in very humid environments because the accuracy and lifetime are lowered. Thin films of potassium metaphosphate [9] and barium fluorite [10] are also used as humidity sensors. The film of potassium metaphosphate is deposited on a glass substrate by vacuum sublimation and Ag paste is used as the electrodes. Such sensors show a response to humidity within 2 s, but are not stable [9].

20.4 Ceramic Humidity Sensors

Ceramic materials have an advantage in terms of mechanical strength and thermal and chemical stability. A number of investigations have been carried out on humidity sensors utilizing porous ceramic elements. There have been a number of reviews of ceramic humidity sensors [1, 5], and some examples of recent developments are introduced in this chapter.

Alumina film sensors have been used in various industries such as medicine and foodstuffs. Figure 20-4 (a) shows a schematic view of an alumina film sensor which is formed by the anodic oxidation of an aluminum plate in an acid solution and then the deposition of a Au film as an electrode. The equivalent circuit of this sensor element is shown in Figure 20-4 (b). The resistance and capacitance are changed by the adsorption and desorption of water. This alumina film sensor has fast response times, and even a few ppm of water in the atmosphere can be detected. However, the use of this element is restricted to dry atmospheres because of the hysteresis and the change of sensor characteristics which occur during operation [11].

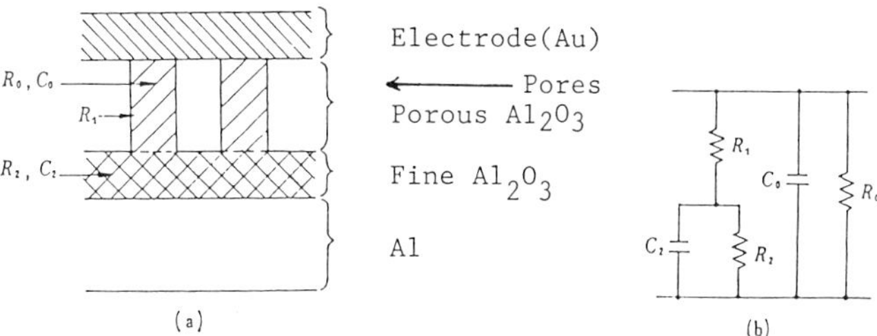

Figure 20-4. Alumina film sensor (a) and its equivalent circuit (b).

The resistivity of ZnO with added Li_2O and V_2O_5, following calcination at 800–900 °C, depends on the relative humidity. This sensor has the advantages of high sensitivity, little hysteresis, and good long term stability. The high sensitivity is retained even after 2000 h of operation. Since the detection mechanism of this sensor depends on the adsorption or desorption of water, the sensor structure is simple but the response time is rather long [12].

A humidity sensor using colloidal Fe_3O_4 can be used to measure a wide range of relative humidities. It is also very reproducible. Figure 20-5 is a schematic view of the sensor element. Fe_3O_4 power (100–250 A) is applied (30 µm thick) on to a comb shaped electrode on an alumina substrate. The resistance of the element decreases as the relative humidity increases over the whole humidity range as shown in Figure 20-6. This sensor has the advantages of low cost, long lifetime, and good accuracy [13].

The addition of K_2CO_3 to α-Fe_2O_3 has a beneficial effect on the humidity sensing characteristics as shown in Figure 20-7. The resistance of this sensor decreased with an increase in operation temperature and the sensor became sensitive below 50% r.h. The response and recovery times of this sensor are about 10 and 20 s, respectively [14].

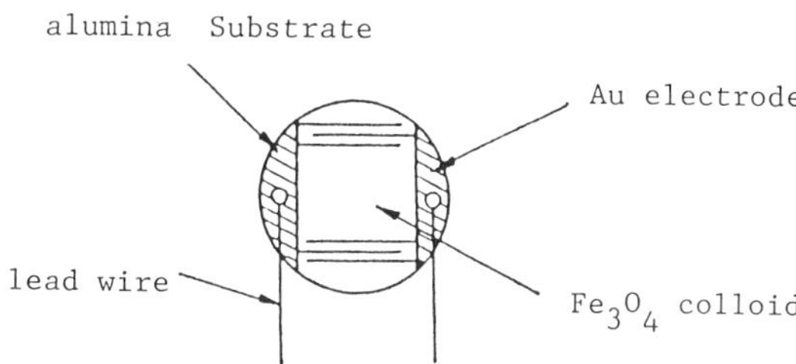

Figure 20-5. Schematic view of a colloidal Fe_3O_4 sensor.

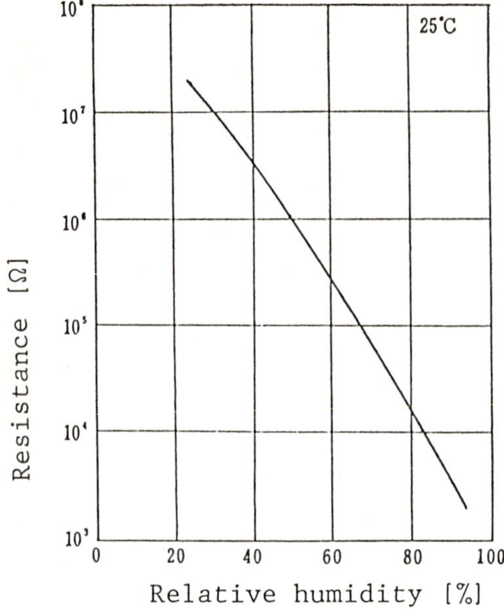

Figure 20-6.
Resistance-humidity characteristic of a colloidal Fe_3O_4 sensor.

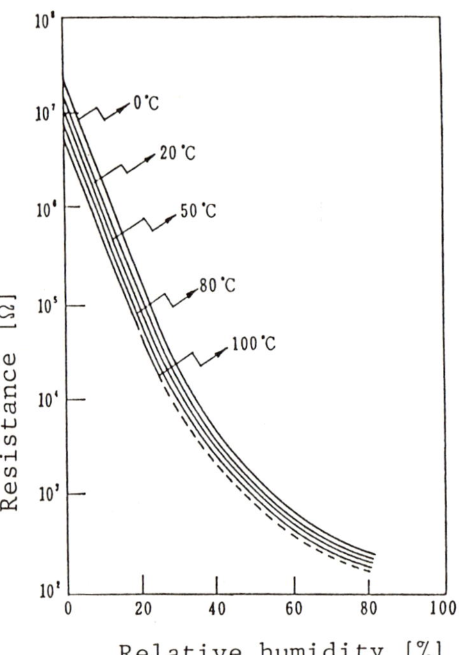

Figure 20-7.
Sensor characteristics of the Fe_2O_3-K_2O ceramic sensor.

The mixed oxides of TiO_2-SnO_2 precipitated onto Al_2O_3 substrates and calcined in air at 1300 °C, respond to a relative humidity in the range from 10 to 90% as shown in Figure 20-8. Increasing the TiO_2 content enhanced not only the sensitivity but also the intrinsic resistance of the element. Following a sudden change in humidity from 90% r. h. to 50% r. h., a steady level of signal was attained within 30 s. The resistance of the element is unaffected by temperature in the range of 20 to 100 °C, but decreases above 100 °C showing semiconducting properties. The intrinsic resistance of TiO_2-SnO_2 was reduced by the addition of Sb_2O_5, but increased by raising the firing temperature. It was also noted that the sensitivity of this sensor depended on the structure of the electrode as shown in Figure 20-9 [15].

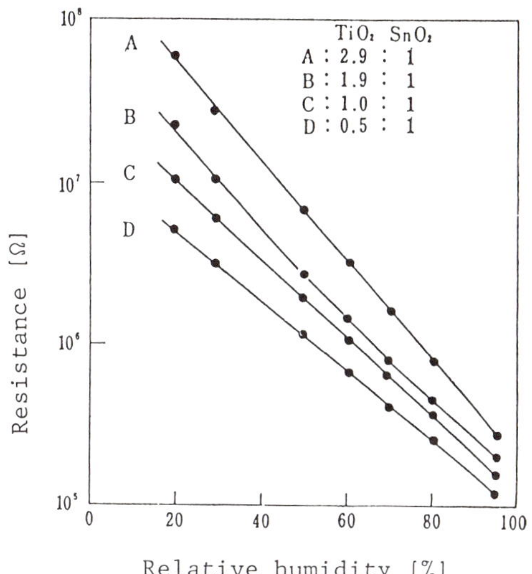

Figure 20-8.
Resistance-humidity characteristics of the TiO_2-SnO_2 ceramic sensor at 40 °C.

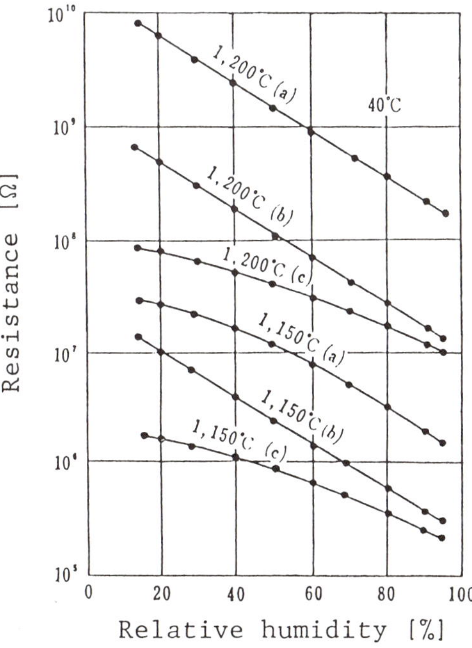

Figure 20-9.
Effects of element structure and sintering temperature on sensor characteristics:
a) external electrode type, b) internal electrode type, c) sandwich type.

A humidity sensor using $MgCr_2O_4$-TiO_2 was developed by Nitta and coworkers for use in microwave ovens [16]. The sensing material is a porous ceramic with a pore size distribution from 1000 to 3000 Å and 25–35% porosity. The water molecules adsorb within the pores resulting in a decrease of resistance. The resistance is unaffected by changes in operating temperatures below 150 °C. A porous RuO_2 was applied to both sides of the element; this is used as an electrode and does not disturb the adsorption or desorption of water. The resistance of the element decreases as the relative humidity increases over the whole humidity range as shown in Figure 20-10. This sensor responded within 20 s to the introduction of water [17]. The base resistance of the element increased gradually during operation. This drift appears to result from the transmutation of the surface of the element or by contamination from other adsorbed gases. In order to obtain a reproducible resistivity, the element is cleaned by heating it at 360 °C before each operation via one of the RuO_2 electrodes.

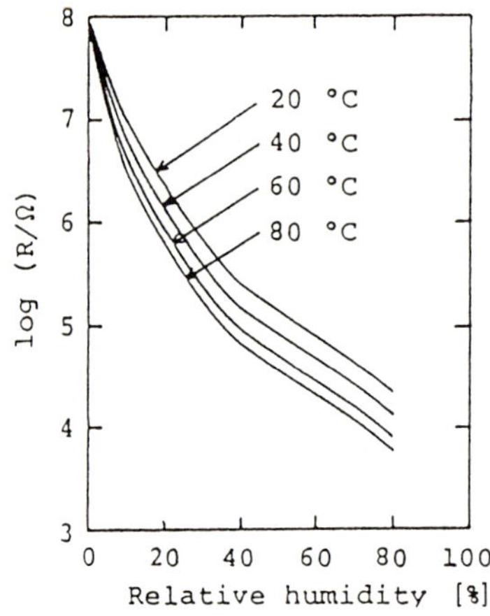

Figure 20-10.
Resistance (R)-humidity characteristics of the $MgCr_2O_4$-TiO_2 ceramic sensor at several temperatures.

A porous TiO_2-V_2O_5 sensor developed by Katayama and coworkers is also based on ionic conduction [18]. A small amount of V_2O_5 additive (1 mol%) is effective in controling the intrinsic electrical conductivity, pore size distribution, and average particle size of the elements. A fairly good humidity sensitivity with a response time of less than 10 s has been reported. The effects of the amount of V^{5+} in TiO_2 have been emphasized by Sudo et al. [19]. This element measures humidity over a wide temperature range from 5 to 90 °C as shown in Figure 20-11. With an increasing amount of V_2O_5 additives, the mechanical strength and the degree of sintering of the element was increased, but the intrinsic resistance decreased as the amount of V^{5+} increased. The sensor characteristics appear to depend on the intrinsic resistance, pore size distribution, and average particle size of metal oxides. The highest sensitivity was obtained in the element with a pore radius of 0.2–0.5 μm and a 45% porosity. This sensor responds within 10 s, and shows a fairly good sensitivity. The conduction mechanism seems to be the Grotthuss-type proton conduction similar to $MgCr_2O_4$-TiO_2 [19].

A ceramic sensor using $ZnCr_2O_4$-$LiZnVO_4$, developed by Yokomizo and coworkers [20, 21], consisted of $ZbCr_2O_4$ grains 2–3 μm in size, which were covered with a thin glazed film of $LiZnVO_4$. In this case, hydrated Li^+ ions seem to contribute to charge transport and sensitivity. This ceramic sensor hardly shows a drift in resistance and can be operated without heat cleaning, unlike the proton conduction-type sensors. Figure 20-12 shows the temperature dependence of this sensor. When temperature increases, relative resistance characteristics shift to lower resistance. The temperature coefficient between 10 °C and 40 °C is about 0.7% r. h./ °C.

An improvement effected by alkali metal oxide additives is also reported on H_3PO_4-$ZrSiO_2$ [22] and zirconium phosphate [23, 24]. The addition of alkali metal ions, therefore, appears to be effective in several respects. Maintaining the alkali metal ions on the surface in humid

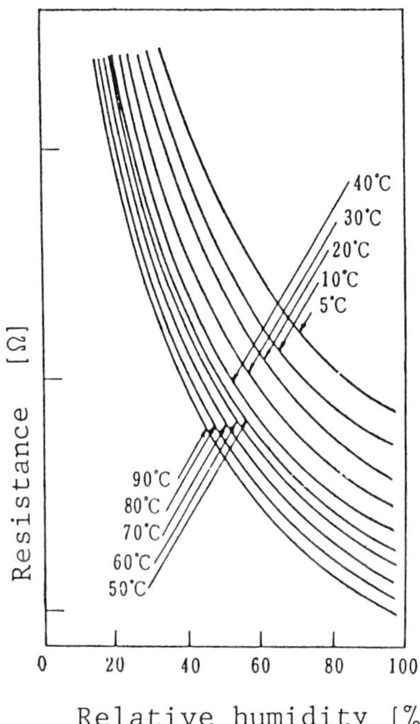

Figure 20-11.
Temperature dependence of sensor characteristics on TiO_2-V_2O_5.

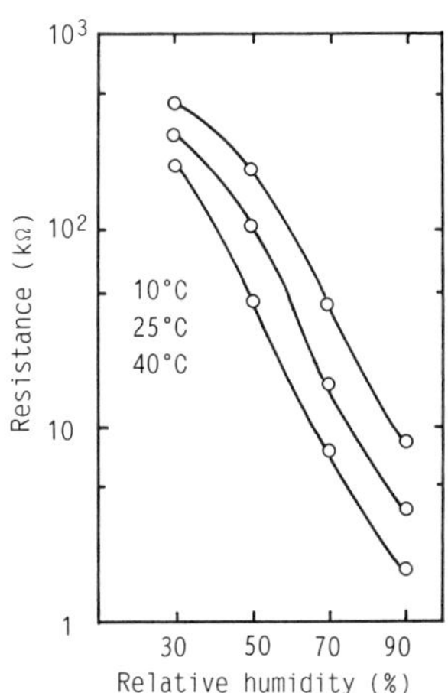

Figure 20-12.
Temperature dependence of sensor characteristics on $ZnCr_2O_4$-$LiZnVO_4$ ceramics.

atmospheres is, however, a critical problem in these sensors. In an attempt to improve sensor sensitivity using additives other than alkali metal ions, a number of different metal oxides have been added to $PbCrO_4$. The humidity-sensitive characteristics of $PbCrO_4$ was strongly dependent on the additives. The effects in this case seem to be due to a change from $PbCrO_4$ to $PbCrO_5$ [25].

It has been recognized that the sensing characteristics are strongly dependent on surface properties such as whether the surface is hydrophilic or hydrophobic, on the microstructure in terms of parameters such as pore size distribution, and on intrinsic properties such as electrical resistance. Controlling the microstructure of the element seems to be particularly important in enhancing the sensing characteristics. Simulation analysis of impedance-humidity characteristics was carried out using the Kelvin equation (Eg. (20-1)). A simple sensor element structure as shown in Figure 20-13 is assumed when calculating the impedance-humidity characteristics. The following assumptions are made:

1) The specimen is a disc with an electrode on both sides and its dimensions are 1.0 cm in diameter and 0.05 cm in thickness.

2) All pores are aligned perpendicular to the electrode and the porosity is 30%.

3) The impedance of the specimen model is 10^7 Ω in dry air.

4) The specific conductivity of condensed water is 8.2×10^7 $(\Omega^{-1}\ cm^{-1})$.

5) The maximum radius r_i of pores, which are filled with water by capillary condensation, is given by the Kelvin equation.

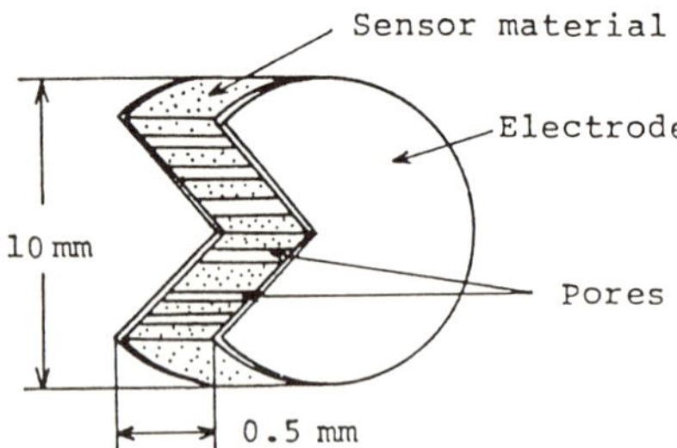

Figure 20-13. Structure of the model sensor.

The impedance-humidity characteristics of model specimens (a), (b), (c), and (d) with small, large, uniform, and more complicated pore size distribution, respectively, are shown in Figure 20-14. With the small pore size distribution (a), the impedance, as shown in Figure 20-15, changes only in a narrow humidity range below 10% r.h., because capillary condensation of water vapor occurs in all the pores up to this value. In contrast, the impedance of model (c) has a marked sensitivity only above 80% r.h. Model (b) shows low sensitivity over the humidity range from 10% to 80% r.h. and cannot be used as a humidity sensor. On the other hand, the more complicated distribution (d), which spreads from about 5 to 90 Å, shows excellent characteristics Figure 20-15. It is obvious that control of pore size distribution is important in fabricating a humidity sensor with the required characteristics. This calculation has been applied to two sensor elements: $MgAl_2O_4$ with a large volume of micropores and $MgFe_2O_4$ with a small volume of micropores, as shown in Figure 20-16. Reasonable agreement between the theoretical and experimental results is obtained if the dissolved ions from the sensor element are washed out with distilled water as shown in Figure 20-17. [5].

The microstructure of the La_2O_3-TiO_2 glass ceramic system is sensitive to heat treatment and subsequent leaching conditions, so that the humidity sensitivity of the element can be favorably controlled by a phase separation. Elements, heat-treated at 600 °C for 12 h followed by leaching at 85 °C for 34 h, exhibit excellent humidity sensitive characteristics in a number of La_2O_3-TiO_2 type porous glass ceramic humidity sensors as shown in Figure 20-18. Controlling the pore size distribution, selecting the intrinsic impedance of the element, and evacuating the condensed water seem to be the most important factors in achieving high sensitivities [26].

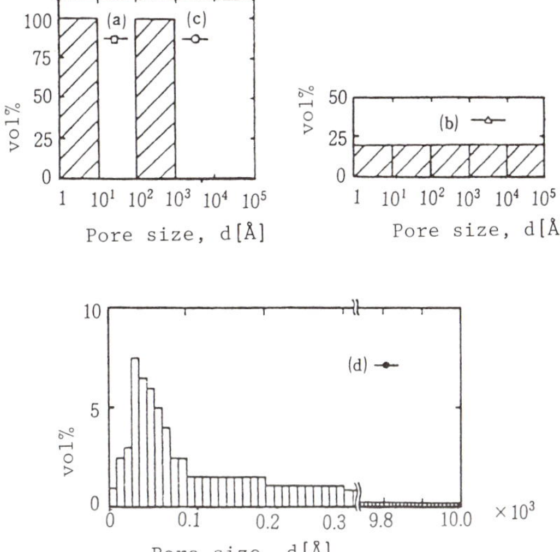

Figure 20-14.
Pore size distribution of the model
elements.

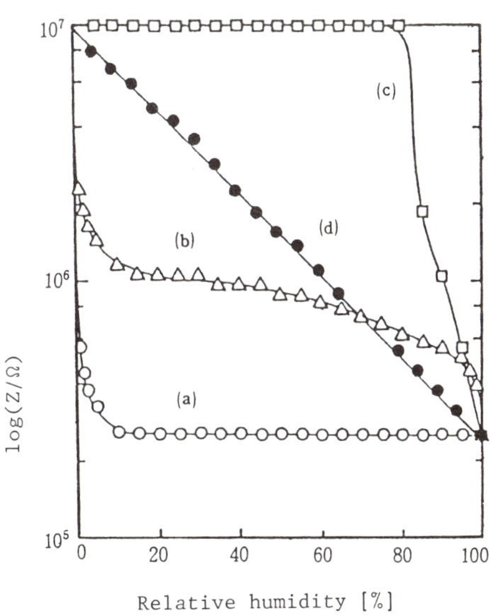

Figure 20-15.
Impedance (Z)-humidity characteristics of the
model sensor with pore size distribution as
shown in Figure 14 (a) to (d), respectively.

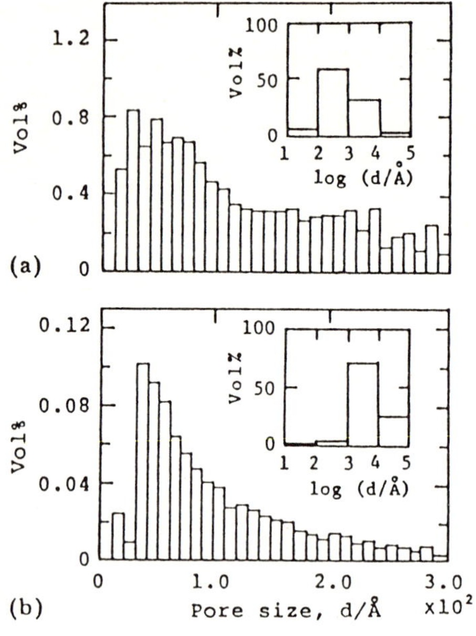

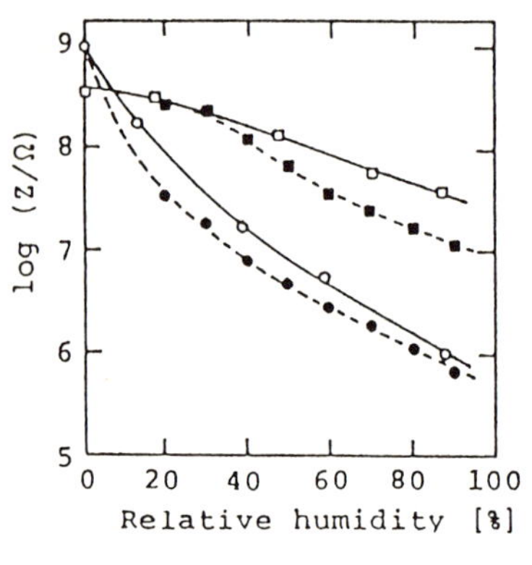

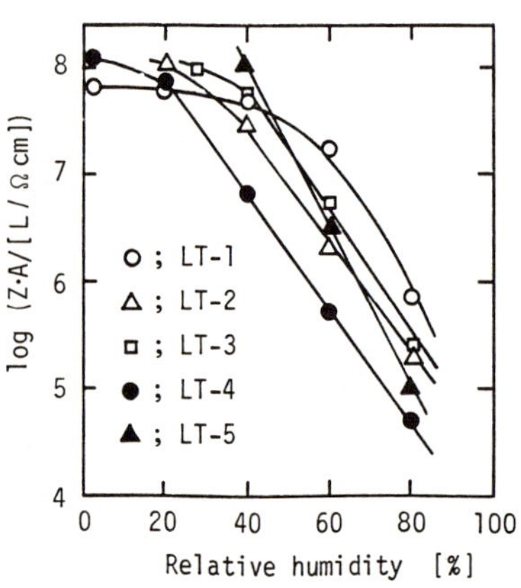

Figure 20-16.
Pore size distribution of the elements:
(a) $MgAl_2O_4$ and (b) $MgFe_2O_4$.

Figure 20-17.
Impedance (Z)-humidity characteristics of $MgAl_2O_4$ and $MgFe_2O_4$ elements.
○ $MgAl_2O_4$ observed   ● $MgAl_2O_4$ calculated
□ $MgFe_2O_4$ observed   ■ $MgFe_2O_4$ calculated

Figure 20-18.
Impedance (Z)-humidity characteristics of La_2O_3-TiO_2 porous glass-ceramics. The heat treatment was performed at 600 °C for 3h (LT-1), Gh (LT-2), 9h (LT-3), 12h (LT-4) and 15h (LT-5) with a leach period of 34h for each sample.

Capillary condensation of the water vapor affects not only the ionic conduction but also the capacitance of porous ceramics. The capacitance changes caused by the adsorption of water are also utilized for the detection of humidity. The capacitance of tantalum oxide films prepared by anodic oxidation shows a sensitivity to changes in relative humidity. The structure of this element, (Figure 20-19), shows that a part of the Ta_2O_5 film surface (part B in Figure 20-19) is in contact with a porous MnO_2 layer. The capacitance at 0% r.h. reflects that of Ta_2O_5 films in contact with MnO_2, but the capillary condensation of water in the pores of MnO_2 (part A in Figure 20-19) enlarges the effective area of the electrode so that the capacitance of the element increases as the relative humidity increases as shown in Figure 20-20 [27].

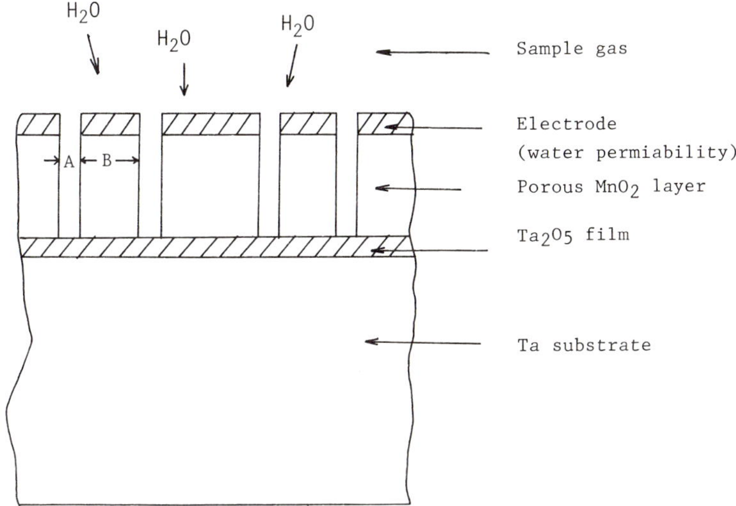

Figure 20-19. Schematic view of a capacitance-type humidity sensor consisting of Ta_2O_5 film.

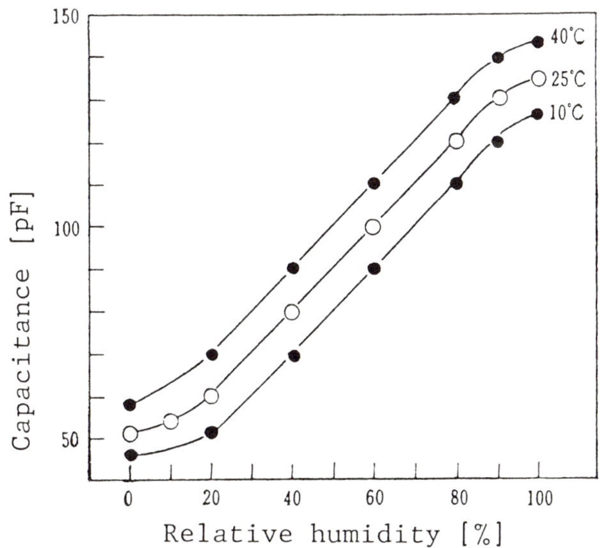

Figure 20-20.
Capacitance-humidity characteristics of a Ta_2O_5 film capacitor.

20.5   Semiconductor-type Humidity Sensors

When water molecules are adsorbed on semiconducting oxides, the conductivity increases or decreases according to whether the oxides are n- or p-type. In response to demands for operation at elevated temperatures, semiconductor humidity sensors using metal oxides, such as perovskite-type oxide, have been proposed as shown in Table 20-2.

The resistance of the $Ni_{1-x}Fe_{2+x}O_4$ humidity sensor increased with a rise in water vapor pressure as shown in Figure 20-21. The maximum sensitivity was obtained at $x = 0.4$. This sensor exhibits a fairly good linearity with humidity and reproducibility but lacks interchangeability and productivity [28].

The conductivity of $SrSnO_3$ (n-type) increases with an increase in water vapor pressure, while that of $SrTiO_3$ (p-type) decreases as shown in Figure 20-22. The adsorption of electron

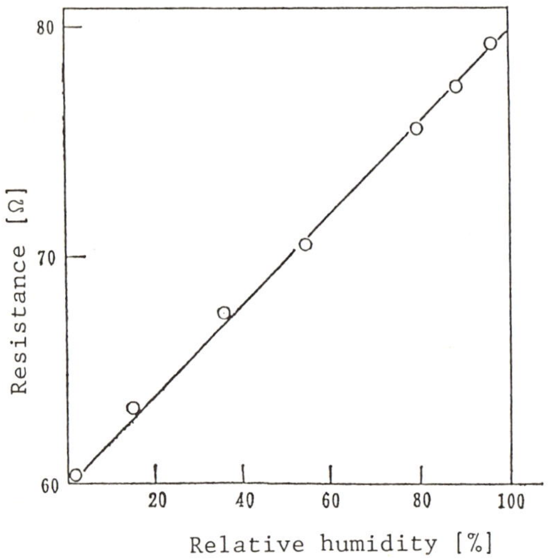

Figure 20-21.
Sensor characteristics of a $Ni_{1-x}Fe_xO_4$ ceramic sensor.

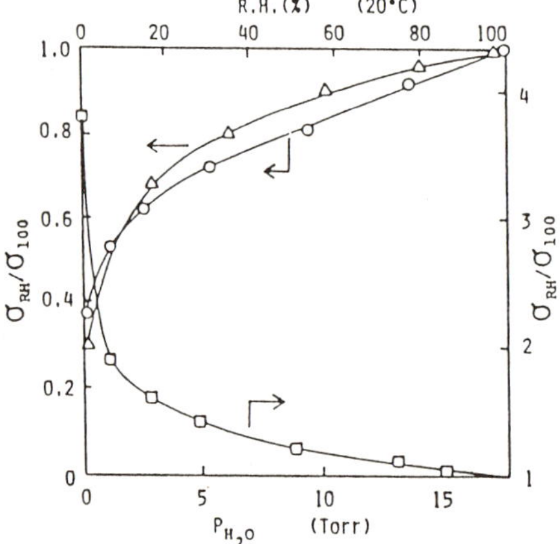

Figure 20-22.
Conductivity ratio-humidity characteristics at 400 °C:
○ $SrSnO_3$,   □ $SrTiO_3$,
△ $Sr_{0.9}La_{0.1}TiO_3$.

donor molecules results in a conductivity change which is larger for n-type semiconducting oxides than for p-type oxides. This suggests that n-type oxides are more desirable than p-type oxides as materials for humidity sensing [29].

In the case of contact with reducing gases such as hydrocarbons, hydrogen and carbon monoxide, the reaction of these gases with adsorbed oxygen or lattice oxygen at the oxide surface causes electron donation to the oxide. In water adsorption, on the other hands, no such electron-donating reactions are conceivable. The following mechanisms may, however, be suggested to account for the apparent electron donation from water molecules to oxides:

(1) non-dissociative adsorption with one electron donation:

$$H_2O \text{ (g)} \overset{K_1}{\rightleftarrows} H_2O^+\text{(ad)} + e^- \tag{20-2}$$

if $[H_2O^+] = [e^-]$

$$\sigma \propto P_{H_2O}^{1/2} \tag{20-3}$$

(2) dissociative adsorption with one electron donation:

$$H_2O \text{ (g)} + O_0^{2-} + V_0^- \overset{K_2}{\rightleftarrows} 2OH_0^-\text{(ad)} + e^- \tag{20-4}$$

V_0^- : oxygen vacancy trapping one electron

$$\sigma \propto P_{H_2O}^{1/3} \tag{20-5}$$

(3) dissociative adsorption with two electron donation:

$$H_2O \text{ (g)} + O_0^{2-} + V_0^{2-} \rightleftarrows 2OH_0^-\text{(ad)} + 2e^- \tag{20-6}$$

V_0^{2-} : oxygen vacancy trapping two electrons

$$\sigma \propto P_{H_2O}^{1/4} \ . \tag{20-7}$$

In the respective cases, the conductivity should be proportional to the one-half, one-third, and one-fourth power of the water vapor pressure (Eqs. 20-3, 20-5, 20-7).

The conductivity-humidity characteristics of the partially substituted perovskite oxide specimens at 400 °C are shown in Figure 20-23. The log-log plots of the conductivity (σ) versus water vapor pressure (P_{H_2O} [Torr]) give straight lines with slopes (a) of 0.12 to 0.03. Since the observed values for the perovskite-type oxides were one-third and less, mechanisms (20-2) or (20-3) are considered to be influential. The largest value was found in $Sr_{0.9}La_{0.1}SnO_3$, suggesting that it is the best sensor material. The a value strongly depends on operating temperature and the highest value is obtained at 400 °C.

A composite ceramic ZrO_2-MgO with porous structure and n-type semiconductivity is a promising material [30]. When the ZrO_2-MgO ceramic at high temperature between 400 °C and 700 °C is exposed to an ambient atmosphere containing water vapor, reversible chemisorption becomes dominant and the electrical conduction changes with gas chemisorption.

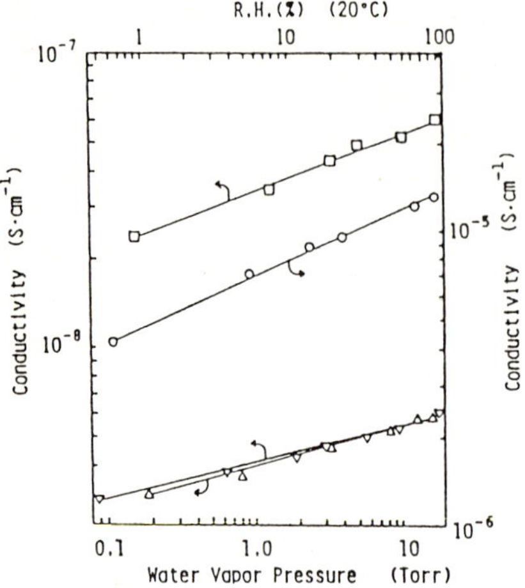

Figure 20-23. Dependence of electric conductivity of perovskite-type oxides on the water vapor content in air at 400 °C:

○ $Sr_{0.9}La_{0.1}SnO_3$ (α = 0.23),   □ $Ca_{0.9}La_{0.1}SnO_3$ (α = 0.18),
△ $Sr_{0.9}La_{0.1}TiO_3$ (α = 0.13),   ▽ $Ca_{0.9}La_{0.1}TiO_3$ (α = 0.12).

Figure 20-24 shows the resistance of ZrO_2-MgO as a function of water vapor content (ppmw). The resistance decreases rapidly with an increase in water vapor from 10^{12} to 10^5 ppmw. Compared with the ionic-type humidity sensor, the response of the semiconductor-type is rather slow because of the slow rate of chemisorption or the subsequent electron transfer process on the oxide surface. The microstructure of the elements as defined by surface area and average particle size, has a less pronounced effect on sensing characteristics than is the case in the ionic-type humidity sensors [31].

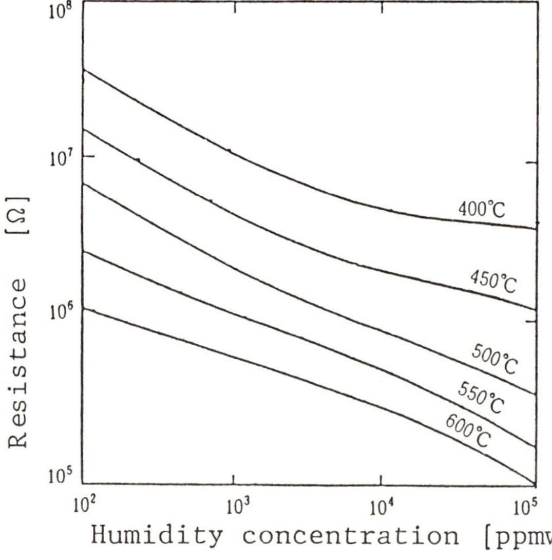

Figure 20-24.
Humidity concentration (ppmw)
Sensor characteristics of ZrO_2-MgO ceramics.

Tsurumi et al. recently reported that the rectifying I-V characteristics of the Pd-ZnO diode exhibited humidity sensitivity. With increasing humidity, the current under the forward bias increases remarkably by physisorption of water at the junction interface. The forward current change shows a linear relation with relative humidity from 0 to 90% r.h. [32].

An integrated MOS capacitor was developed for a humidity sensor by using a photolithographic technique as shown in Figure 20-25. A porous, thin alumina layer of 1 μm thickness was prepared by anodic oxidation on to silicon dioxide. The porous upper electrode of gold was fabricated by a photolithographic technique. The humidity-dependent capacitance can be detected from a variation of the drain current as shown in Figure 20-26 [33]. Such an approach is important because the progress in integration techniques will enable us to realize sophisticated or intelligent sensor in future.

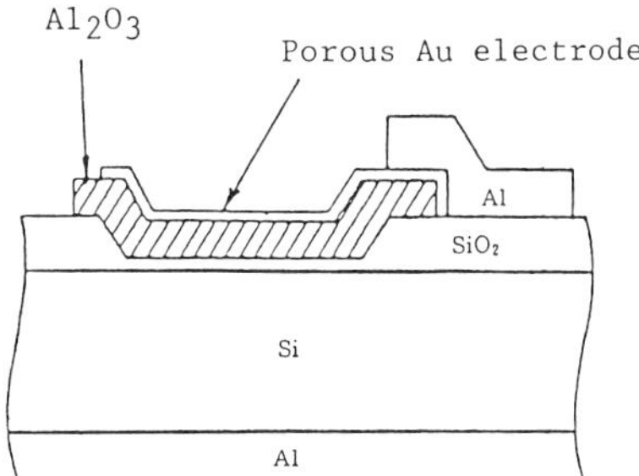

Figure 20-25.
Schematic view of a humidity-sensitive MOS capacitor.

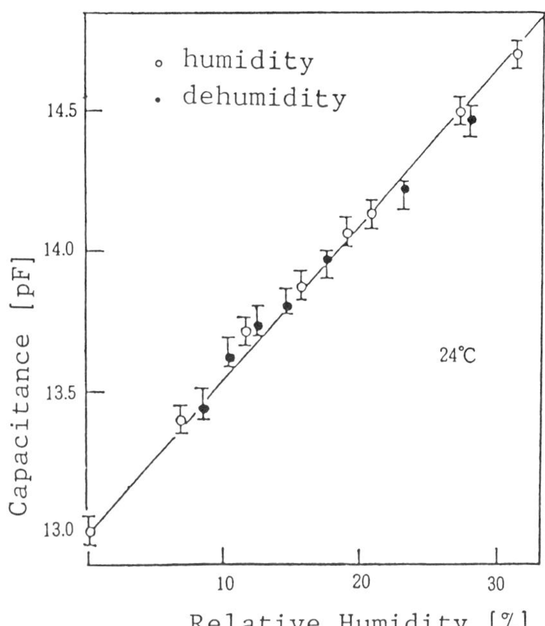

Figure 20-26.
Sensor characteristics of a MOS capacitor.

20.6   Polymer Film Humidity Sensors

The use of organic polymers for humidity sensing is well known, and various devices incorporating organic polymer humidity sensing elements have been reported [34–39]. Because the water content of the polymer is directly related to relative humidity at any given temperature, the value of the detected property at the particular temperature will provide a direct indication of the relative humidity. In spite of the variety of polymers available and the many improvements made during recent years, polymeric humidity sensors still suffer from a variety of disadvantages including hysteresis, non-linearity, instability and short life, swelling and oxidation [40]. These disadvantages become increasingly severe at elevated temperatures and high relative humidities. Although polymer film humidity sensors will probably be used less frequently in practical applications than ceramic sensors, their potential has been increased through developments in solid-state transducer technology. Polymer films are well suited to standard IC processing techniques for the fabrication of small, low cost sensors. Sensors which utilize polymer films can be classified into impedance-and capacitance-types; the impedance-type being further subdivided into ionic and electronic conduction types.

20.6.1   Impedance-Type Humidity Sensors

The resistance of ion-exchange resin decreases by the absorption of water because the ionic pair of the polymer electrolytes is dissociated according to the following equation.

$$R-SO_3H + H_2O \rightleftarrows R-SO_3^- + H_3O^+ . \tag{20-8}$$

Humidity sensors using cross-linked styrene-sulfonate have been developed by Pope [41] using this mechanism. Cross-linked copolymers prepared from styrene-sulfonate are fabricated on polystyrene substrates furnished with Ag electrodes. This sensor responds to water vapor at temperatures below 100 °C and rinsing the element is enough to restore the sensor characteristics in case of contamination. A similar sensor which consists of polystyrene-sulfonate containing 4–10% divinylbenzene, has been developed by Musa [42]. This sensor exhibits extremely low hysteresis, 3% r.h., but the long term stability is poor.

A polymer film sensor, with good stability under various conditions, has been developed by Hijikigawa et al. [43]. This sensor consists of an Al_2O_3 substrate, a pair of interdigitated Au electrodes, a humidity-sensitive polymer, and a protective film prepared by coating cellulose ester. The humidity sensing membrane is a cross-linked copolymer consisting of sodium styrenesulfonate, N,N'-methylene-bis-alkylamide, and polyvinyl alcohol. This sensor has good durability against water, examined by the cycling test, as shown in Figure 20-27. Furthermore, this sensor is stable in the presence of alcohols as well as that of water [43].

The impedance-type humidity sensor has the advantages of simple structure and rapid response. Moreover, long term operation becomes possible when the element has been stabilized by coating with a resin film. This type of sensor is also easily compatible, depends on the chemical structure of the polymers, and, in principle, enhancement of the water sensitivity can be achieved by cross-linking or copolymerisation.

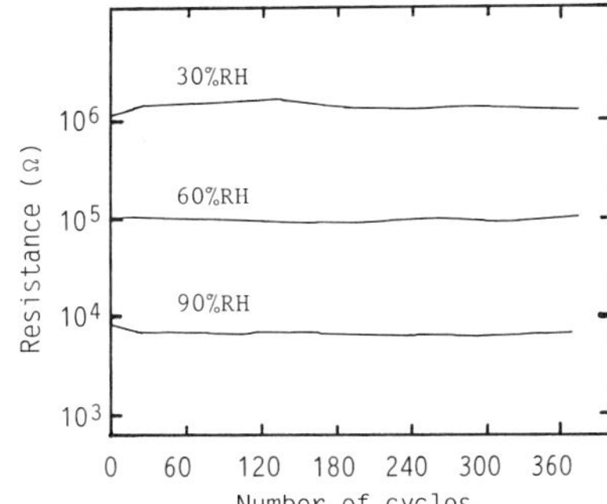

Figure 20-27.
Durability against the cycling test: the sensor was immersed in water for 1 minute and then was dried for 59 minutes.

A thin film humidity sensor has been developed by Otsuka et al. for use in air conditioning systems. This element, which is illustrated schematically in Figure 20-28, was put into practical use in 1978. The humidity-sensitive copolymer thin film is fabricated on an alumina substrate with patterned comb shaped Au electrodes. This humidity sensor exhibited good reproducibility and low hysteresis. The electrical resistance of the sensor decreases with increasing relative humidity (Figure 20-29). Following an abrupt change in humidity from 30% to 90%, a steady signal was obtained within 1 min [44].

The impedance of organosilicon compounds possessing hydrophilic groups varies linearly with the relative humidity. The sensitivity of these materials depends strongly on the type of hydrophilic group and increases as follows:
$OH > NH_2 > N^+(CH_3)_3Cl^- > SO_3H$. Since the amount of adsorbed water increases in the following order, $N^+(CH_3)_3Cl^- > SO_3H > NH_2 > OH$, the effects of adsorbed water on the electric conduction seem to depend on the acid-base property of the hydrophilic groups. On the other hand, the activation energy of the element with $N^+(CH_3)_3Cl^-$ decreases as the amount of adsorbed water increases. This suggests that the dissociation of $N^+(CH_3)_3Cl^-$ is promoted by adsorbed water to increase the ionic conduction of the element [45].

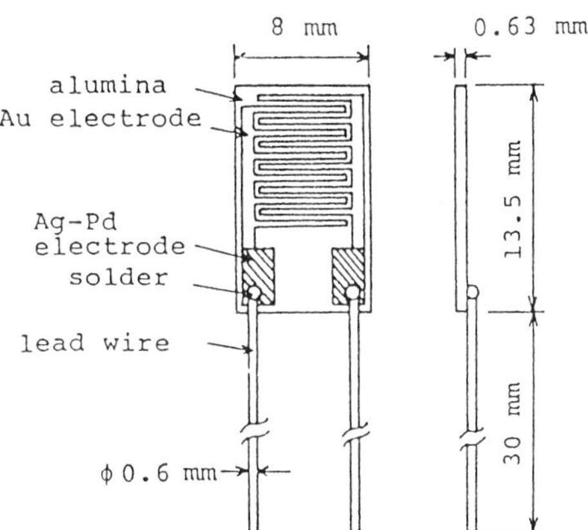

Figure 20-28.
Schematic view of the „Hument HPR"-type humidity sensor.

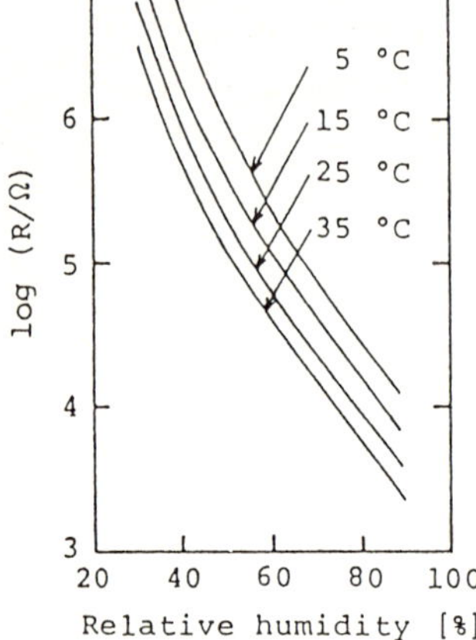

Figure 20-29.
Resistance-humidity characteristics of the „Hument" sensor.

20.6.2   Swelling-type Humidity Sensors

Some types of organic polymers such as polyethyleneoxide-sorbitol swell, i.e., change volume by absorption of water. The devices, obtained by spreading carbon or metal powder on films of these materials, are sensitive to water vapor. Many such polymers have been studied as humidity sensors, but the successful ones are polyethyleneoxide-sorbitol [13] and hydroxyethyl-cellulose [46]. Since swelling of the polymer disturbs the ohmic contact between carbon or metal particles dispersed in organic films, the resistance of the element increases with an increase in humidity as shown in Figure 20-30 [13]. Generally, the swelling-type sensor is operationally superior but the hysteresis is severe. A humidity sensor prepared by dispersing

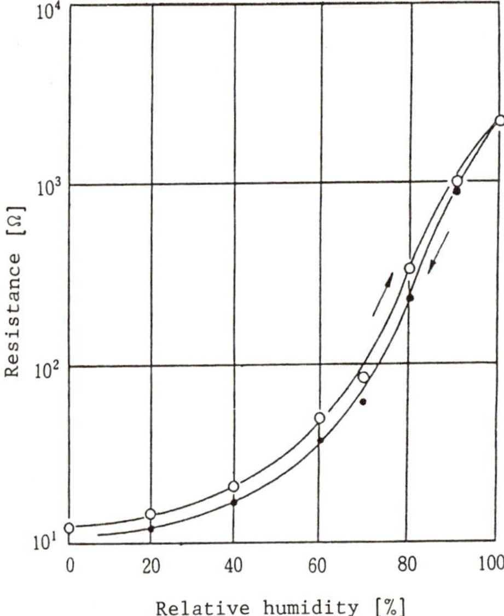

Figure 20-30.
Resistance-humidity characteristics of a swelling-type sensor consisting of polyethylenoxide-sorbitol.

3.5 µm carbon powder on the membrane of cellulose-ester, polyethyleneoxide-sorbitol, and alkylallyl-polyethylene-alcohol has been put into a practical use in radiosonde. The humidity-sensitive characteristics of this sensor are independent of operating temperatures between 0 and 40 °C.

20.6.3 Capacitance-type Humidity Sensors

A capacitance-type humidity sensor developed by Vaisala consists of a comb shaped Au electrode and cellulose acetate dissolved in ethylene dichloride as humidity-sensitive materials. A schematic view of this sensor is shown in Figure 20-31 [47]. This sensor is now widely used in meteorological observations and in many other humidity measuring instruments. As illustrated in Figure 20-32, the capacitance-humidity characteristics show a linear relation from 0 to 100% r. h. [48]. This sensor has the advantages of good accuracy, low hysteresis, and fast response time.

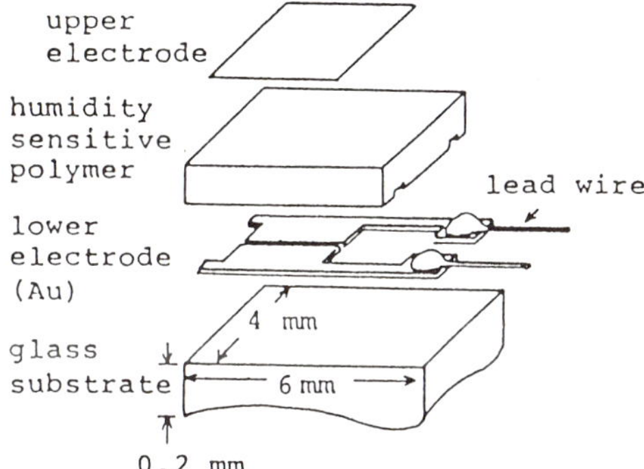

Figure 20-31.
Configuration of the „Humicape" humidity sensor.

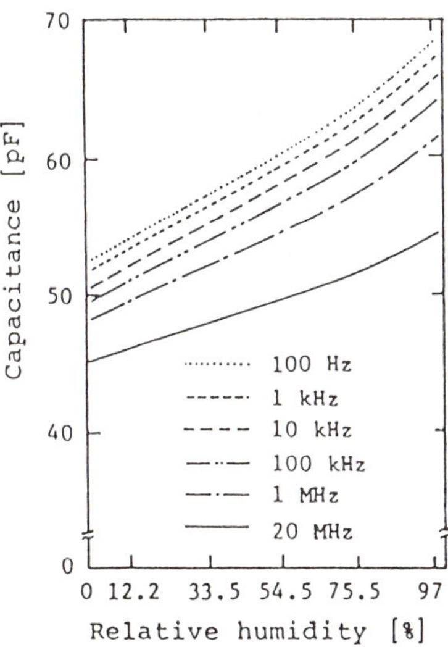

Figure 20-32.
Frequency dependence of capacitance-humidity characteristics of „Humicape".

The fabrication of humidity sensors on silicon chips has recently become possible using IC production technology [37, 49]. This realizes a small, low cost humidity sensor, and makes it possible to integrate the humidity sensor with other sensors or signal-handling circuitry on the same chip. A new integrated temperature and humidity sensor developed by Yamamoto et al., consists of a polymer capacitor on the p-n diode of a temperature sensor [50] as illustrated in Figure 20-33. A thin film of polyimide is used as the moisture-sensitive material

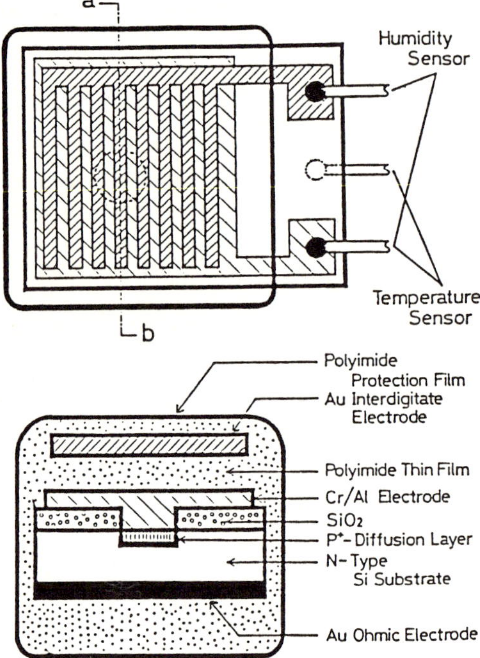

Figure 20-33.
Structure of the integrated temperature humidity sensors. Top view (above) and cross sectional view at a–b (below).

and a comb shaped gold electrode is vacuum evaporated on to the polyimide film. This sensor shows a good linear response to humidity (Figure 20-34) and a rapid response. Some hysteresis was apparent in the high humidity region, but was less than 2% r. h. The sensor characteristics are independent of the temperature between 5 and 50 °C.

A microchip humidity sensor based on an insulated-gate field effect transistor (IGFET) was recently developed by Hijikigawa et al. [51, 52]. A temperature sensing diode is fabricated on the same chip. The structure of this FET is shown schematically in Figure 20-35. A hydroscopic polymer, cross-linked cellulose acetate butylate (CAB), whose electric capacitance changes with relative humidity, is used as the humidity sensing membrane. The upper-and the under-gate electrodes are porous gold (100–200 Å thick) and gold-titanium alloy (3000 Å thick), respectively. The equivalent circuit of this FET sensor is shown in Figure 20-36. Both the constant DC voltage V_0 which drives the host IGFET and a small AC voltage $\tilde{v}_0$ are applied to the upper-gate electrode. The upper- and under-gate electrodes are connected through a resistor R_B which is assembled outside the chip. The output voltage V_{out} is theoretically given by the following equation:

$$V_{\text{out}} = V_0 R_L g_m / (1 + C_i / C_s) \tag{20-9}$$

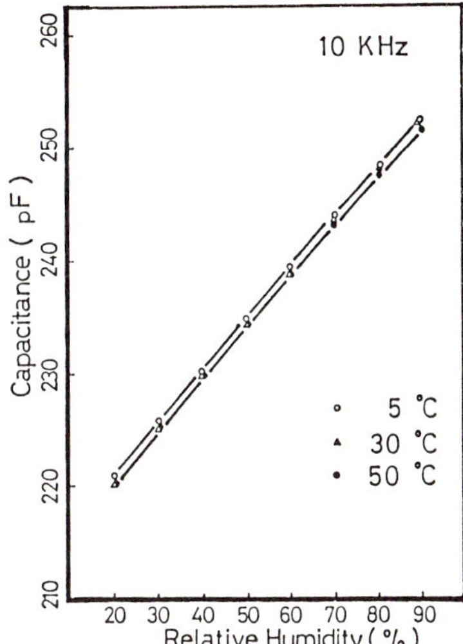

Figure 20-34.
Capacitance-humidity characteristics at various
temperatures of the temperature-humidity sensor.

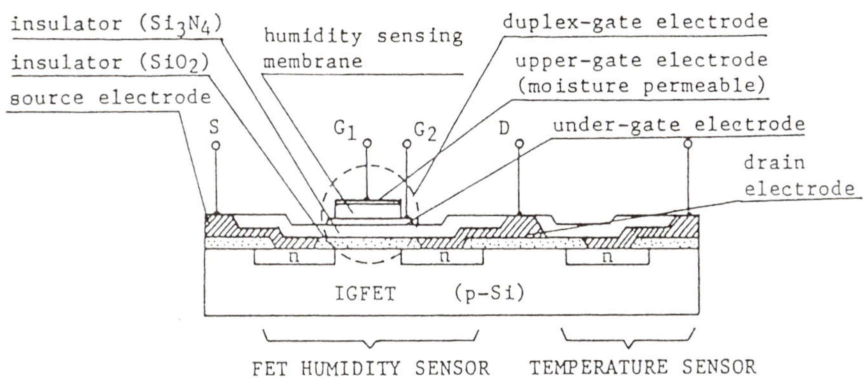

Figure 20-35. Schematic cross-section of an FET humidity sensor.

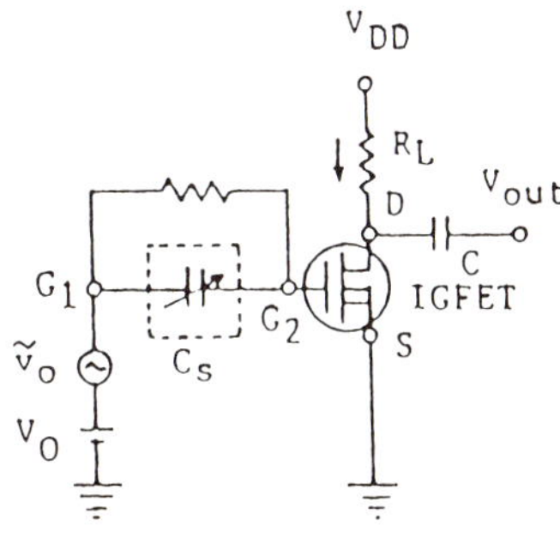

Figure 20-36. Equivalent circuit of an FET humidity sensor.

sensor one electrode is formed by a comb-like etched, n-doped polycrystalline silicon. The other electrode is diffused in a high-sensitivity p-substrate. The two electrodes thus have a very short spacing of about 1 μm. The capacitance change of each type of sensor is shown in Figure 20-40 as a function of the surface coverage that is occupied by water. The capacitance of sensor (I) increased exponentially with an increase in dew coverage, and that of sensor (II) shows a linear dependence on coverage. Sensor (III) has the largest sensitivity for the initial drops.

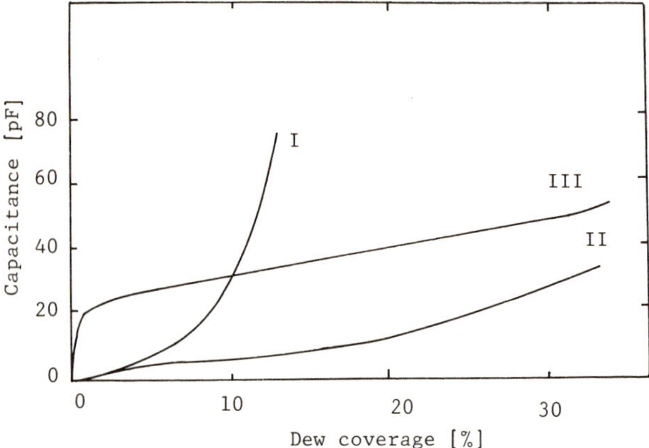

Figure 20-40.
Dew sensitivity of the three types of MOS capacitors.

An electronic conduction-type humidity sensor was developed by Ishida et al. [56]. The sensor structure is very similar to the one illustrated in Figure 20-41. A cross-linked hydrophilic polymer in which carbon particles are dispersed is used as the humidity-sensitive film. The swelling of the polymer disturbs the ohmic contacts between dispersed carbon particles and thus the electronic resistance of the element increases sharply as the relative humidity approaches 100%, as shown in Figure 20-41. This element was put into practical use as a dew detector for the VTR cylinder in 1978.

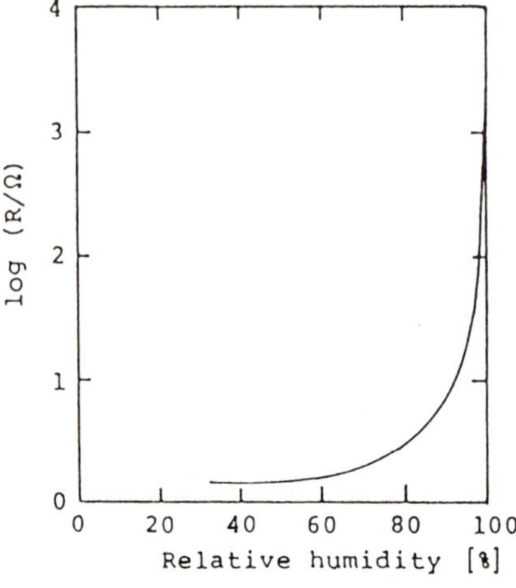

Figure 20-41.
Resistance-humidity characteristics of a polymer dew-point sensor.

20.8 References

[1] Seiyama, T., Yamazoe, N., Arai, H., *Sens. Actuators* **4** (1983) 85.

[2] Yamazoe, N., Shimizu, Y., *Sens. Actuators* **10** (1986) 379.

[3] McCaffery, E., Zettlemoyers, A. C., *Dis. Far. Soc.* **52** (1971) 239.

[4] Anderson, J. H., Parks, G. A., *J. Phys. Chem.* **72** (1968) 3662.

[5] Shimizu, Y., Arai, H., Seiyama, T., *Sens. Actuator* **7** (1985) 11.

[6] de Bore, J. H., *The Structure and Properties of Porous Materials;* London: Butterworth 1958, pp. 68.

[7] Dunmore, F. W., *J. Research National Bureau Standards* **23** (1939) 701.

[8] Yamada, Y., *Denshi-Gijyutsu* **21**, No. 9 (1979) 26.

[9] Wexler, A., *J. Res. Natl. Bur. Standards* **55** (1955) 71.

[10] Nakamura, N., *Electroanalytical Chem. Interfacial Electro Chem.* **47** (1974) 175.

[11] Kovac, M. G., *Solid State Technology* **2** (1978) 35.

[12] Ichinose N., Yokomizo, Y., Katsura, K., *Denshi-Zairyo* **16**, No. 2 (1977) 69.

[13] Kanou, R., *Sensa Gijyutsu Nyumon;* Tokyo: Kogyo Chosakai, 1978, pp. 131.

[14] Matsuura, M., Matsuoka, M., *Funtai oyobi Funmatsuyakin* **23**, No. 7 (1976) 242.

[15] Ichinose, N., *Kinou-Zairyo* **1** (1981) 19.

[16] Nitta, T., Terada, Z., Hayakawa, S., *J. Am. Ceram. Soc.* **63** (1980) 295.

[17] Nitta, T., Terada, Z., Fukushima, F., *IEEE Trans. Electron Devices* **ED-29**, No. 1 (1982).

[18] Katayama, K., Akiba, T., *Proc. Int. Meet. Chemical Sensors;* Fukuoka, 1983, pp. 433.

[19] Sudo, Y., *Denshi-Zairyo* **19**, No. 9 (1980) 74.

[20] Yokomizo, Y., et al., *Sens. Actuators* **4** (1983) 599.

[21] Uno, S., et al., *Proc. Int. Meet. Chemical Sensors;* Fukuoka, 1983, pp. 375.

[22] Sadaoka, Y., Sakai, Y., *J. Mater. Sci. Lett.* **5** (1986) 656.

[23] Sadaoka, Y., et al., *Proc. 4th Solid-State Sensors and Actuators (Transducers '87);* Tokyo, 1987, pp. 637.

[24] Sadaoka, Y., Sakai, Y., Mitsui, S., *J. Mater. Sci. Lett.* **6** (1987) 377.

[25] Miura, N., Yashima, I., Yamazoe, N., *Nippon Kagaku Kaishi* **1985** (1985) 1644.

[26] Shimizu, Y., Okada, H., Arai, H., *Proc. 2nd Int. Meet. Chemical Sensors;* Bordeaux, 1986, pp. 380.

[27] Nishino, A. et al., *Nat. Tech. Rep.* **26** (1980) 442.

[28] Kanou, R., *Erekutoroniku Seramikusu* **1** (1974) 15.

[29] Arai, H. et al., *Proc. Int. Meet. Chemical Sensors;* Fukuoka, 1983, pp. 393.

[30] Nitta, T., Fujishima, F., Matsuo, Y., *Proc. Int. Meet. Chemical Sensors;* Fukuoka, 1983, pp. 387.

[31] Shimizu, Y. et al., *Chem. Lett.* **1985** (1985) 917.

[32] Tsurumi, S., Mogi, K., Noda, J., *Proc. 4th Solid-State Sensors and Actuators (Transducers '87);* Tokyo, 1987, pp. 661.

[33] Inagaki, S., Kawano, Y., Kodato, S., *Proc. 1st. Sensor Symposium;* Fukuoka, 1981, pp. 115.

[34] Senturia, S. D., Garverick, S. L., Togashi, K., *Sens. Actuators* **2** (1981) 59.

[35] Delapierre, G., et al., *Sens. Actuators* **4** (1983) 97.

[36] Hermans, E. C. M., *Sens. Actuators* **5** (1984) 181.

[37] Denton, D. D., Senturia, S. D., *Proc. 2nd Solid-State Sensors and Actuators (Transducers '85);* Philadelphia, 1985, pp. 202.

[38] Huang, P. H., *Sens. Actuators* **8** (1985) 23.

[39] Sakai, Y., Sadaoka, Y., *Porc. 2nd Int. Meet. Chemical Sensors;* Bordeaux, 1986, pp. 353.

[40] Crawk, J., Park, G. S., *Diffusion in Polymers;* New York: Academic Press, 1968, pp. 259.

[41] Pope, M., *U.S. Patent 2728831* 1955.

[42] Musa, D. C., Schnable, G. L., *Humidity and Moisture* **1**, 1965, 346.

[43] Miyoshi, S. et al., *Proc. Int. Meet. Chemical Sensors,* Fukuoka, 1983, pp. 451.

[44] Otsuka, K., Kinoki, S., Usui, T., *Denshi-Zairyo* **19**, No. 9 (1980) 68.

[45] Sakai, Y. et al., *Proc. 4th Solid-State Sensors and Actuators (Transducers '87);* Tokyo, 1987, pp. 677.

[46] Stine, S. L., *Humidity and Moisture* **1** (1965) 436.

[47] Vaisala KK, *Catalog, Ref. No. A9371.*

[48] Nakaasa Instrument Co. Ltd., *Catalog, C911-84001-15K-2D* 1984, p. 78.

[49] Glenn, M. C., Schuetz, J. A., *Proc. 2nd Solid-State Sensors and Actuators (Transducers '85);* Philadelphia, 1985, pp. 217.

[50] Yamamoto, T., Kurakami, K., Takai, T., *Proc. 4th Solid-State Sensors and Actuators (Transducers '87);* Tokyo, 1987, pp. 658.

[51] Hijikigawa, M. et al., *Proc. 4th Sensor Symp.;* Tsukuba, 1984, pp. 135.

[52] Hijikigawa, M., et al. *Proc. 2nd Solid-State Sensors and Actuators (Transducers '85);* Philadelphia, 1985, pp. 221.

[53] Nitta, K., Terada, J., *Nat. Tech. Rep.* **22,** No. 6 (1976) 885.

[54] Ziegler, H., Rolf, K., *Sens. Actuators* **11** (1987) 37.

[55] Regtien, P. P. L., *Sens. Actuators* **2** (1981) 85.

[56] Ishida, T., *Denshi-Gijyutsu 21,* No. 9 (1979) 38.

[57] Uchikawa, F. et al., *Ceram. Bull.* **63** (1984) 1043.

[58] Uchikawa, F., Shimamoto, K., *Am. Ceram. Soc. Bull.* **64** (1985) 1137.

21 Biosensors for Monitoring Pesticides in Water

PETRA KRÄMER, ROLF DIETER SCHMID, GBF, Braunschweig, FRG

Contents

21.1 Introduction

The need for environmental and pollution control is now greater than ever. The monitoring of pollutants in air and soil as well as the monitoring of water and waste water, particularly the control of pesticides in drinking water and groundwater, is of major importance. Regulations are now being very strictly enforced in Europe with a maximum allowable pesticide concentration of 0.1 µg/l per pesticide in drinking water [1]. The development of devices for fast, inexpensive on-line control is therefore especially interesting. Biosensors are such analytical tools [2–7].

A number of different possible biological materials are available for monitoring water quality, which act by sensing analytes:

- enzymes [8]
- antibodies [9]
- cell organelles
- whole cells, microorganism [10]
- chemorecepters [11, 12]
- flowers, blossoms [13, 14]
- plant tissue slices [15].

The biological material is combined with various sensing devices (transducers) such as potentiometric or amperometric electrodes, microcalorimeters, gas-sensing electrodes, fiber optic devices, field effect transistors (FETs), piezoelectric crystals, and devices for measuring light absorbance/fluorescence [16, 17].

The main use of enzymes and antibodies (Table 21-1) as sensing materials in biosensors is for the detection of pesticides. In addition, they have been incorporated into a device used to detect microbes [27]. Antibodies are specific to the components to be detected. Nevertheless, there are some cross reactivities with substances of similar structure. Enzymes are mostly used in an inhibition test. This means that special substances cause an inhibition of the enzyme when they are in the sample. These inhibitions can be caused by a variety of substances with different priorities. In general, this reaction is not as specific as the antibody reaction.

Biosensors containing cell organelles, whole cells, microorganisms or plant tissue slices (see Table 21-2) function by detecting substances that interfere in some way with these materials, for example, by the inhibition of photosynthesis, the enhancement of special catabolism, or others mechanisms.

21.2 Measuring Principles

Because of the great diversity of biosensors for environmental monitoring, in this chapter it is only possible to focus on a single application: the detection of pesticides in water. Three different principles are described with three different biological materials and transducers.

Table 21-1. Biosensors for pesticides using enzymes or antibodies as sensing material.

Biocatalyst Sensing material	Substance	Transducer	Range/detactable level	Reference
monoclonal antibodies	2,4-dinitrophenol	potentiometric membrane electrode	$10^{-6} - 10^{-5}$ M	[18, 19]
monoclonal antibodies	pyrethroid pesticides	optical fiber	priliminary studies	[20]
polyclonal antiserum	triazines atrazine, propazine simazine	fluorimeter	atrazine: 0.01 –10 µg/l propazine: 0.002–10 µg/l simazine: 0.1 –10 µg/l	[21]
polyclonal antiserum	parathione methyl parathione malathione	piezoelectric crystals	parathione: 36 ppb 36–680 ppb* (in gas phase) * different for the different pesticides	[22]
butyryl-cholin-esterase	organophosphorous pesticides and carbamates	pH electrode	3–15 ppb (µg/l) 5×10^{-7} M (ethylparaoxone)	[23, 24]
butyryl-cholin-esterase	organophoshorous pesticides and carbamates	potentiometric cell	0.1–20 ppm (mg/l) depending on the pesticide Sarin: 0.0001 µg/l	[25]
acetyl-cholin-esterase from bovine erythrocytes	organophosphorous pesticides and carbamates	– photometer – amperometric electrode	carbofuran: 0.5–500 µg/l	[26]

Table 21-2. Biosensors for pesticides using whole cells, cell organelles, or chemoreceptors as sensing material.

Biocatalyst Sensing material	Substance	Transducer	Range/detectable level	Reference
chemoreceptor intact (1989) *Callinectes sapidus*	pesticides: amitrol lindan fonofos	Ag/AgCl- reference action potential	amitrole: 10^{-2} mol/l lindane: 10^{-6} mol/l fonofos: 10^{-6} mol/l	[28]
Trichosporum cutaneum	*BOD* (biological oxygen demand) (organic pollution)	oxygen electrode	3–60 mg/l	[29, 30]
(C. butyricum)				[9]
Synechoccocus	herbicides (linurone, atrazine, metoxuron)	amperometric electrode (with mediator)	linurone 50–500 µg/l atrazine 50 µg/l–1 mg/l	[31]
Synechococcus (PCC 6301) *Chlorella sp.*	herbicides: DCMU, chlortolurone linurone	amperometric electrode (with mediator ferricyanide)	*Synechococcus:* DCMU: 1×10^{-6} (233 ppb) linurone: 2 ppb–2 ppm chlortolurone: 2 ppm	[32]
Synechococcus *Chlorella vulgaris*	chlortolurone linurone isoproturone	amperometric electrode (with mediator) O_2-electrode	chlortolurone: 10–500 µg/ml linurone: 150 µg/l isoproturone: 20–500 µg/ml	[33]
Synechococcus *algae* (PC 6301)	atrazine, linurone 2,4-dichlorophenoxy- acetic acid	amperometric electrode (with mediator hexa- cyano-ferrate (III))	atrazine: 50–500 µg/l linurone	[34]
Synechococcus (PC 6301)	linurone DCMU (3-(3,4-dichloro- phenyl)-NN'-dimethyurea	amperometric electrode (with mediator)	linurone: 20 ppb DCMU: 20 ppb	[35]
peroxisomal fraction	herbicides (metribuzin)	potentiometric ammonia-gas- sensing electrode	under development	[15]
yeast	BDO	amperometric electrode		[36]

a) The first principle is based on using antibodies as recognition elements in biosensors. This type of biosensor depends on the transfer of a competitive ELISA (enzyme-linked immunosorbent assay) to a flow injection immunoanalysis (FIIA [37]).

The device consists of a reactor with a membrane. A specific antibody is immobilized on this membrane for a specific pesticide. The pesticide of the standard or sample and the corresponding pesticide conjugated to an enzyme compete for the binding sites of the antibodies. This step is followed by washing with a buffer and the addition of a substrate for the enzyme. Depending on the law of mass action, there is a correlation between the concentration of the pesticide and the signal measured. The higher the product concentration of the enzyme reaction, the lower will be the concentration of the pesticide in the probe (Figure 21-1 [37]).

The antibodies are replaced after each test. With this device it is possible to measure pesticides of the triazine group at a very low level (0.01 μg/l).

Until now the system has only been used with standard solutions. An attempt is being made to apply FIIA to the monitoring of drinking water or industrial waste water.

Because of the specificity of the antibodies, every group of pesticides will require a specific antibody and the corresponding pesticide-enzyme conjugate.

b) The second concept for the detection of pesticides is based on the inhibition of an enzyme in the presence of phosphorous pesticides and carbamates. Kindervater et al. [26]. described a system using flow injection analysis (FIA) for rapid automated determination of pesticides in drinking water. The enzyme used in this device was acetylcholinesterase (ACHE) from bovine erythrocytes. It was immobilized on magnetic particles, so that it was possible to hold the enzyme during the assay and to replace the enzyme after it was inactivated (Figures 21-2 and 21-3) [26].

With an incubation time of one minute, the pesticide carbofuran caused a detectable inactivation of the enzyme at a concentration of the pesticide of 0.5 μg/l in 1 ml of a water sample without enrichment. Further investigations and improvements to increase sensitivity are in progress.

c) The third principle for the determination of pesticides in water involves the development of whole-cell biosensors [31, 32, 34].

For the screening of herbicides, Rawson et al. [31, 32] used mainly the cyanobacterium *Synechococcus* (PCC 6301). These bacteria were either immobilized on an alumina membrane or entrapped in alginate. An electrochemical mediator (K_3FeCN6) was used to cause an interaction between the electrode and the photosynthetical electron transport system of the cyanobacterium.

Using this biosensor, it was possible to detect herbicides of the urea, nitrile, and triazine families (eg, linurone, ioxynil, atrazine) with different sensitivities (125–350 μg/l). The sensitivity improved by 20–50 μg/l after fresh microbial cells were placed into the device.

The selectivity of whole-cell biosensors is poor for the most part compared to enzymatic or antibody-based biosensors. On the other hand, they often have a higher stability, are easier to prepare, and are generally less expensive.

This system can also be transferred to flow injection analysis, and work is in progress to develop such a device [38].

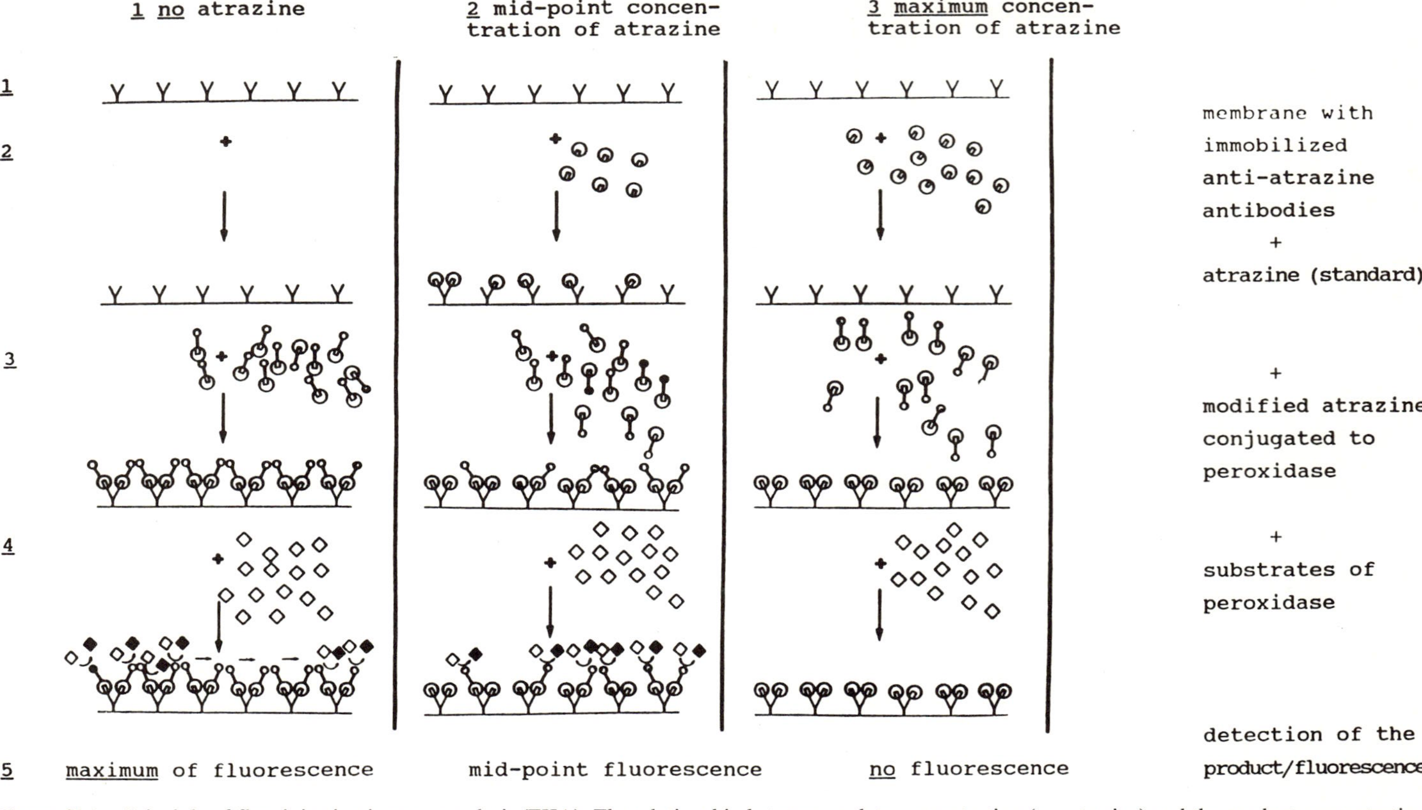

Figure 21-1.   Principle of flow injection immunoanalysis (FIIA). The relationship between analyte concentration (eg, atrazine) and the product concentration of the enzyme reaction is inversely proportional [37].

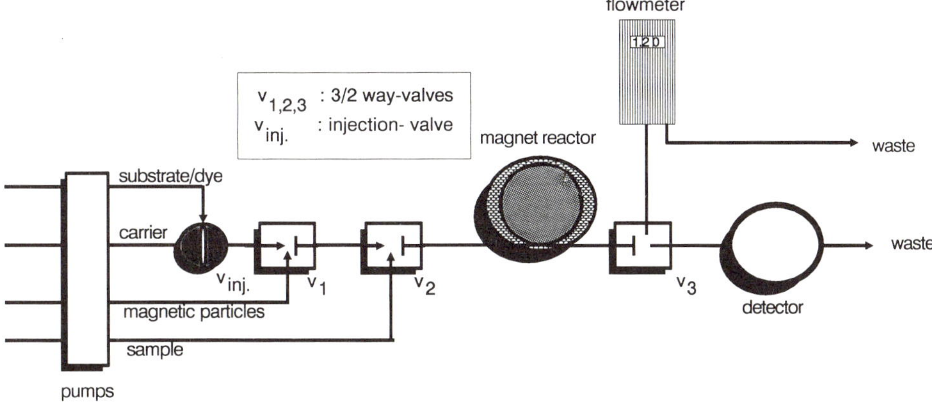

Figure 21-2. Arrangement of the flow injection system used for the determination of phosphorous pesticides and carbamates [26].

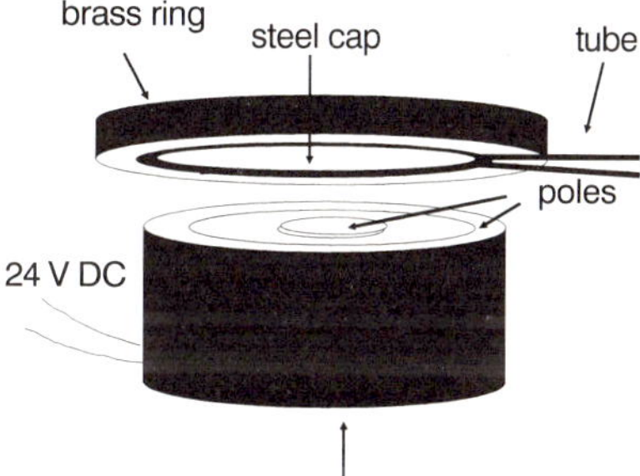

Figure 21-3. Enlargement of the magnet reactor used for exchanging inhibited acetylcholinesterase [26].

21.3 Conclusion

Compared to research and development for biosensors for clinical applications [39, 40], not much research has been carried out in the field of environmental control. Biosensors for pesticide control in water are still in a developmental stage, and no biosensors are commercially available to date for this field of application. Field tests are presently carried out with the mediated amperometric biosensor developed by Rawson et al. [31], but the sensitivity and specificy of this biosensor is not sufficient for the high demands of drinking water quality control.

Biosensors certainly have the potential to become useful environmental screening tools but a great deal of research has to be done to improve long time stability, reliability, and ease of application.

21.4   References

[1] Verordnung über Trinkwasser und über Wasser für Lebensmittelbetriebe (Trinkwasserverordnung – TrinkwV) *Bundesgesetzblatt, Jahrgang 1986,* Teil 1, 760–772.

[2] Gronow, M., "Biosensors", *Trends in Biosciences,* (1984) 336–340.

[3] Neujahr, H. Y., *Biosensors for Environmental Control, Biotechnology and Genetic Engineering Reviews,* Vol. 1 (1984) 167–186.

[4] Evans, G. P., Briers, M. G., Rawson, D. M., "Can Biosensors Help to Protect Drinking Water?" *Biosensors* 2 (1986) 287–300.

[5] Karube, I., Tamiya, E., "Biosensors for Environmental Control", *Pure and Applied Chemistry* 59 (1987) 545–554.

[6] Kindervater, R., Schmid, R. D., "Biosensoren zur Analytik von Wasser und Abwasser", *Z. Wasser-Abwasser-Forsch.* 22 (1989) 84–90.

[7] Schmid, R. D. et al. "Biosensoren zur Analytik von Pflanzenschutzmitteln in Wasser", *Z. Wass. Abw. Forsch.* (1991) in press.

[8] Karube, I., "Enzyme Sensors for Clinical, Process, and Environmental Analyses", In: *Enzyme Technology,* 7a (13) (1987) 686–708.

[9] Karube, I., Suzuki, M., "Novel Immunosensors", *Biosensors* 2 (1986) 343–362.

[10] Karube, I., "Micro-Organism Base Sensors", *Biosensors, Fundamentals and Application,* Turner, A. P. F., Karube, I., Wilson, G. S. (eds.); Oxford: Oxford University Press, (1987) pp. 13–19, ISBN 0-19-854724-2.

[11] Buch, R. M., Rechnitz, G. A., "Intact Chemoreceptor-Based Biosensors: Responses and Analytical limits", *Biosensors* 4 (1989) 215–230.

[12] Buch, R. M., Rechnitz, G. A., "Neuronal Biosensors", *Anal. Chem.* 61 (1989) 533A–542A.

[13] Uchiyama, S., Rechnitz, G. A., "Biosensors Using Flowers as Catalytic Material", *Anal. Lett.* 20 (1987/a) 451–470.

[14] Uchiyama, S., Rechnitz, G. A., "Biosensors Using Flower Petal Structures", *J. Electroanal. Chem.* 222 (1987/b) 343–346.

[15] Sidwell, J. S., Rechnitz, G. A., "Progress and challenges for biosensors using plant tissue materials", *Biosensors* 2 (1986) 221–233.

[16] Gronow, M., Kingdon, C. F. M., Anderton, D. J., *Biosensors, Spec. Publ. – R. Soc. Chem.* 54 (Mol. Biol. Biotechnol.) (1985) 295–324.

[17] Higgins, I. J., Lowe, C. R., "Introduction to the Principles and Applications of Biosensors", *Phil. Trans. R. Soc. London* B 316 (1987) 3–11.

18] Bush, D. L., Rechnitz, G. A., "Monoclonal Antibody Biosensor for Antigen Monitoring", *Anal. Lett.* 20, (1987) 1781–1790.

[19] Bush, D. L., Rechnitz, G. A., "Comparison of Antibodies as Molecular Recognition Elements for Biosensors Design", *Analytical Letters* 21 (11) (1988) 1947.

[20] Northrup, M. A., et al., "Development and Characterization of a Fiber Optic Immuno-Biosensor", *NATO ASI Ser., Ser. C,* Vol. 280 (1989) 229–241.

[21] Krämer, P., Schmid, R., „Der Nachweis von Pflanzenschutzmitteln aus der Triazin-Gruppe mit Hilfe der Fließinjektions-Immuno-Analyse", Poster zur GDCh-Tagung „Fachgruppe Wasserchemie", 1990, accepted.

[22] Ngeh-Ngwainbi, J., et al., "Parathion Antibodies on Piezoelectric Crystals", *J. Am. Chem. Soc.* 108 (1986) 5444–5447.

[23] El Yamani, H., et al., "Realisation d'une ensemble automatique pour la mesure de la toxicite des eaux de riviere", *J. Francais d'Hydologie* 18 (1987) 67–75.

[24] El Yamani, H., et al., "Automated System for Pesticide Detection", *Sensors and Actuators* 15 (1988) 193.

[25] Goodson, L. H., Jacobs, W. B., "Monitoring of Air and Water for Enzyme Inhibitors", *Methods in Enzymology* 44 (1976) 647–658.

[26] Kindervater, R., Künnecke, W., Schmid, R. D., "Exchangeable Immobilized Enzyme Reactor (EIMER) for Enzyme Inhibition Tests in FIA Using a Magnet Device. The Determination of Pesticides in Drinking Water", *Anal. Chim Acta* (1990) accepted.

[27] Muramatsu, H., et al., "Piezoelectric Immuno Sensor for the Detection of Candida Albicans Microbes", *Analytica Chim. Acta* **188** (1986) 257–261.

[28] Zink, P., Rechnitz, G. A., „Intakter Chemorezeptor als Biosensor zur Pesticiderfassung", *Fresenius Z. Anal. Chem.* **333** (1989) 645–656.

[29] Karube, I., "Microbial Sensors for Process and Environmental Control", *ACS Symposium Ser.* **309** (1986) 330–348.

[30] Karube, I., Suzuki, S., "Biosensor for Fermentation and Environmental Control", *Biotech.* **83** (1983) 625–632.

[31] Rawson, D. M., Willmer, A. J., Turner, A. P. F., "Whole-Cell Biosensor for Environmental Monitoring", *Biosensors* **4** (1989) 299–311.

[32] Rawson, D. M., Willmer, A. J., Cardosi, M. F., "The Development of Whole Cell Biosensors for On-Line Screening of Herbicide Pollution of Surface Waters", *Toxicity Assessment* **2** (1987) 325–340.

[33] Gaisford, W. C., Rawson, D. M., "Biosensors for Environmental Monitoring", *Measurement + Control,* **22** (1989) 183–186.

[34] Van Hoof, F. M., et al., "The Evaluation of Bacterial Biosensors for On-Line Screening of Water Pollutants", *Environmental Technol. Lett.* (1990) in press.

[35] Cardosi, M. F., "Development of Novel Biosensors", *Analytical Proceedings* **27** (1987) 143–145.

[36] Hikuma, M., et al., "Amperometric Estimation of BOD by Using Living Immobilized Yeasts", *Eur. J. Appl. Microbiol. Biotechnol.* **8** (1979) 289–297.

[37] Krämer, P., Stöcklein, W., Schmid, R., "Fließinjektions-Immuno-Analyse zum Nachweis von Pflanzenschutzmitteln", *Vom Wasser* **73** (1989) 345–350.

[38] Pastrik, K. H., Kärst, U., Schmid, R. D. "Mutanten von Cyanobakterien mit erhöhter Empfindlichkeit gegenüber Herbiziden", *Z. Wass. Abw. Forsch.* (1991) in press.

[39] Scheller, F. W., et al., "Biosensors: Trends and Commercialization", *Biosensors* **1** (1985) 135–160.

[40] Scheller, F. et al., "Research and Development of Biosensors", *A review Analyst* **114** (1989) 653–662.

22 Sensors in Biotechnology

Thomas-H. Scheper, Universität Hannover, FRG
Kenneth F. Reardon, Colorado State University, Fort Collins, USA

Contents

22.1   Introduction

Biotechnological processes can be considered to involve a biological system surrounded by chemical and physical environments. The activity and state of this biological component are often extremely sensitive to environmental influences (eg, the dependence of bacterial growth rates on temperature), and thus the monitoring of physical, chemical, and biological parameters is very important. A number of analytical techniques and sensors have been used for this task, some of which are indicated in Figure 22-1. Review articles dealing with various aspects of these measurement techniques are available [1-17]. Since fermentation is the most widely studied bioprocess, nearly all sensors have been developed for fermenters; however, the measurement principles and instrumentation can usually be applied to other operations with ease.

In this chapter, only brief descriptions of traditional probes of the physical and chemical environments will be presented. Instead, our focus will be on the on-line (and quasi on-line) monitoring of bioprocesses. This type of measurement is required for bioprocess development, control, and optimization and places special demands on the types of sensors involved. In addition, special attention will be given to biosensors. These new probes provide on-line measurements of compounds that would otherwise require off-line analysis and have greatly increased the power of automated bioprocess monitoring and control.

22.2   On-Line and In Situ Monitoring

For the purposes of this review, we will consider on-line sensors to be those that are directly integrated into the process, while quasi on-line systems are devices that are peripheral to the process. An in situ measurement is made in the location in which the substance to be quantified is normally found; i.e., the sensor is brought to the sample, rather than bringing the sample to the sensor.

In order to fully describe the state of a bioprocess at all times, whether for research or control purposes, one would want continuous, on-line monitoring of all possible parameters. Unfortunately, real on-line sensors are only available for a relatively small number of measurements; most of these probes are recent developments.

The reasons for the current state of limited on-line sensor development stem in large part from the rather challenging requirements placed on their design by (1) their biological/biochemical environment, (2) their on-line placement, and (3) the use of their output in process control schemes. Specifically, an on-line sensor for a biotechnological process must be (steam) sterilizable (if it is an in situ probe) or be attached to a sterile sampling device (if it is a quasi on-line sensor system). Furthermore, an in situ sensor must be unaffected by microbial growth and protein deposits on its surfaces, while the sampling device of a quasi on-line system must be resistant to clogging by high concentrations of microorganisms. Since one of the most important uses of these sensors is for process control, their signals must be easily interpretable and either continuously or frequently emitted, and the sensors themselves must be able to respond quickly to changes in the quantity that they measure. Of course, a sensor's cost must be reasonable with respect to the advantages it confers in order to be accepted by industrial users.

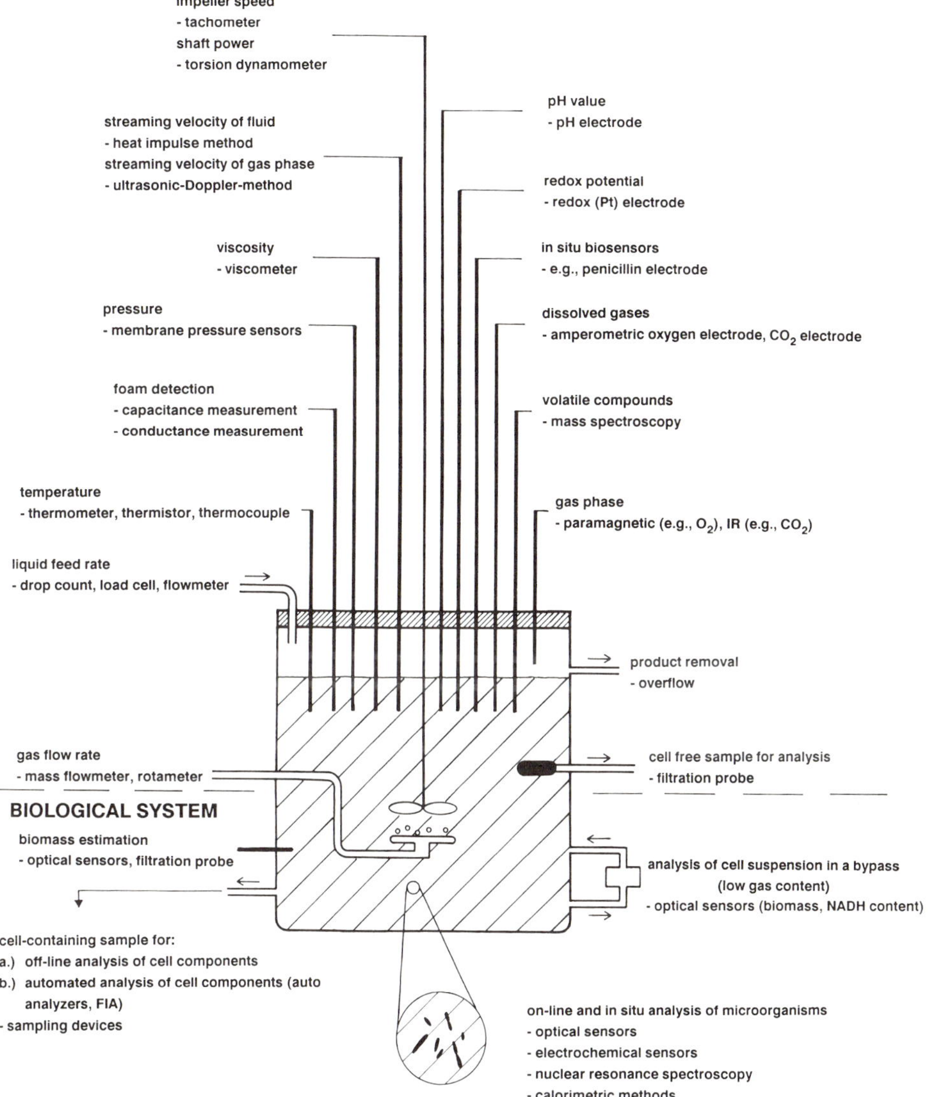

Figure 22-1. Different sensors used for fermentation monitoring in biotechnology (based on Reference [2]).

22.3   Types of Sensor Systems

22.3.1   In Situ Sensors

In situ sensors offer the possibility to monitor interesting parameters directly in a bioreactor without sampling-related time delays. Unfortunately, there are some problems associated with their use. One of these concerns the response time, which can differ drastically during the period of use. For example, the response time of a clean pH or pO_2 electrode is in the range of several seconds, but any deposition of medium components on the membrane surface might lengthen the response time to some minutes [18]. Also, fouling on pO_2 electrode membranes falsifies the data since microorganisms on the membrane consume oxygen. In addition to this effect, signal drifts may cause drastic changes in the sensor effectiveness. In situ recalibration would compensate for these problems, but calibration during a bioprocess is very difficult to achieve since stepwise changes in the analyte concentration (usually required for calibration) are normally not possible when the probe is in place. Nonetheless, some efforts have been made to build calibration systems. Kok and Hogan have described an in situ calibrator for oxygen electrodes [19], while Ingold Meßtechnik (Urdorf, Switzerland) offers a calibrator system for pH electrodes (see Figure 22-2). In both cases, the electrode can be withdrawn from the fermentation broth, recalibrated, and inserted into the medium again under sterile conditions. It is also possible to exchange the electrode.

a)

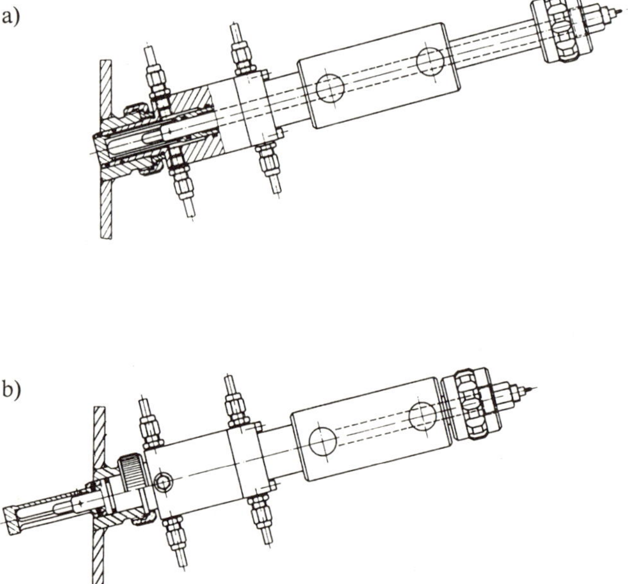

b)

Figure 22-2. Ingold InTrac™ system. a) position for calibration, rinsing, electrode exchange, and sterilization; b) measuring position. (Reproduced by permission of Ingold Meßtechnik, Urdorf, Switzerland).

Additional problems arise for in situ biosensors. These sensors are combinations of biological compounds for a specific detection assay and a suitable transducer (see Chapter 14). Since the biological part of such sensors cannot be sterilized, a membrane has to be used as sterile barrier between sensing element and the fermenter environment. The sensitive biological compounds (eg, enzymes) are brought into the sensor after the sterilization process. In Figure 22-3, an in situ version of a glucose sensor is shown [20-22]. The inner part of the glucose sensor with the immobilized enzymes is plugged into the outer part (with the sterile polycarbonate membrane) after the sterilization procedure. Buffer solution is injected to fill the compartment between membrane and sensor. Although the response time of this mediator-based amperometric glucose sensor increased from 1-2.5 min within 6 d, the signal was linear up to 5.5 g/L glucose, and the sensor was sufficiently stable for 14 d [20-22].

This example illustrates both the benefits and problems of in situ sensors. On the one hand, continuous measurements of the glucose concentration would be of great use in controlling a cultivation process. On the other hand, it is clear that changes in the membrane permeability for the substrate along with enzyme deactivation influence the sensor performance. Furthermore, the linear range of these sensors is small and sample dilution is impossible. Some efforts can be made to increase the linear response range of in situ sensors by pumping buffer solution at various flow rates through the system in order to dilute the analyte in the measuring compartment between membrane and biological component. The accuracy of this dilution method is drastically affected by changes in the permeation rate through the membrane.

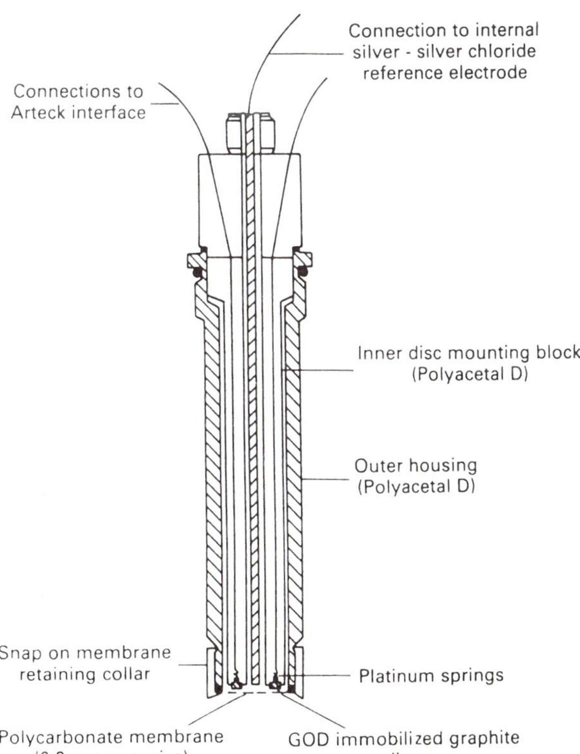

Figure 22-3.
Schematic diagram of an in situ glucose biosensor (reproduced with permission from Reference [21]).

22.3.2 Flow-Injection Analysis

Flow-injection analysis (FIA) has gained tremendous importance for bioprocess control during the last few years [23–27]. Although this is not an in situ or a real on-line analytical technique, it can be automated and operated at very high analysis cycle frequencies (quasi on-line). One of the most important advantages of FIA is the use of very small sample volumes.

When used in biotechnology, the first part of a FIA system (see Figure 22-4) is a filtration probe [28–32]. A cell-free sample is drawn from the bioprocess unit (eg, a fermenter) via this probe. Thus, these probes are sterile barriers between the bioprocess and the analysis system, protecting the bioprocess from infections and the analysis system from clogging by cells. Only a few FIA applications utilizing cell-containing samples are known [33]. In those cases, cells and medium are separated inside the FIA system.

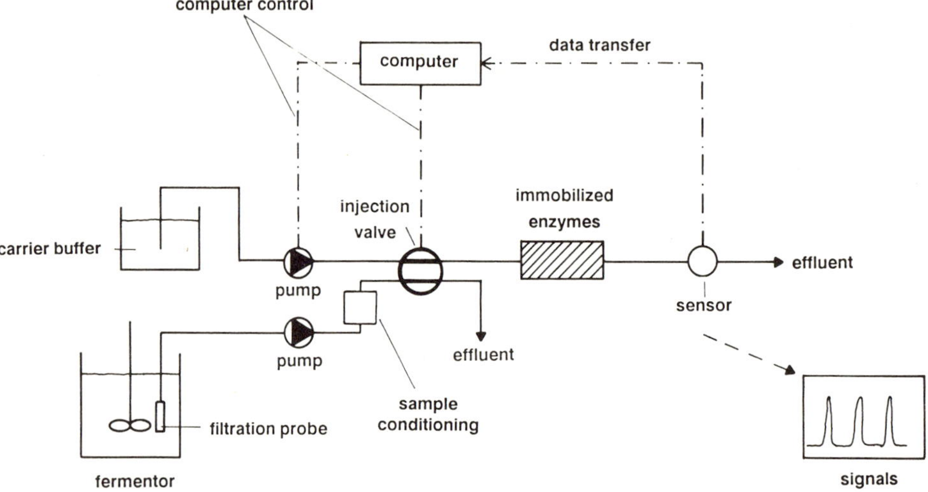

Figure 22-4. Principles of FIA systems for process monitoring in biotechnology. If the assays include enzymatic reactions, the enzymes can be immobilized in front of the detector or can be directly bound to the detector surface.

Different filtration probes can be used. In Figure 22-5, a filtration probe for sampling in a fermenter bypass is shown [28, 34]. Since flat membranes are used in this probe, many types of micro- and ultrafiltration membranes can be used. On the other hand, the increased risk of infection can make bypass sampling problematic. The membranes used should be selected for cut-off size and material (eg, adsorption characteristics) as required by the process conditions. They should provide a sample stream that has the same analyte concentrations as the liquid in the fermenter. An in situ probe employing tubular membranes is shown in Figure 22-6. Tubular membranes are not available in the same variety as flat membranes, but this probe offers the advantages of a very small dead volume and simple connection to a fermenter via an ordinary sensor port.

The principle of FIA is shown in Figure 22-7. Defined sample volumes are injected periodically into a buffer carrier stream (C). When an analytical reagent (R) is combined with the buffer stream, the sample and reagent can react in a reaction coil (RC) before the reaction products are monitored by a detector system (D). Mixing of the sample with the buffer and reagent flow occurs by controlled dispersion. The sample is transported through the system by a rapidly flowing buffer stream. It is not necessary for the analytical reaction to reach its equilibrium. When all of the parameters (eg, buffer flow rate, reaction time) are constant, reliable detection can be made before the reaction equilibrium is attained (dynamic measurement). This results in high frequencies of analysis and short analysis times. A commercially available FIA system built up from single FIA modules (eg, pump, injector, detector) is shown in Figure 22-8.

Another advantage of FIA is the provision for sample preconditioning before reaction and monitoring. Thus, the reaction can be run under its optimal conditions. Sample precondition-

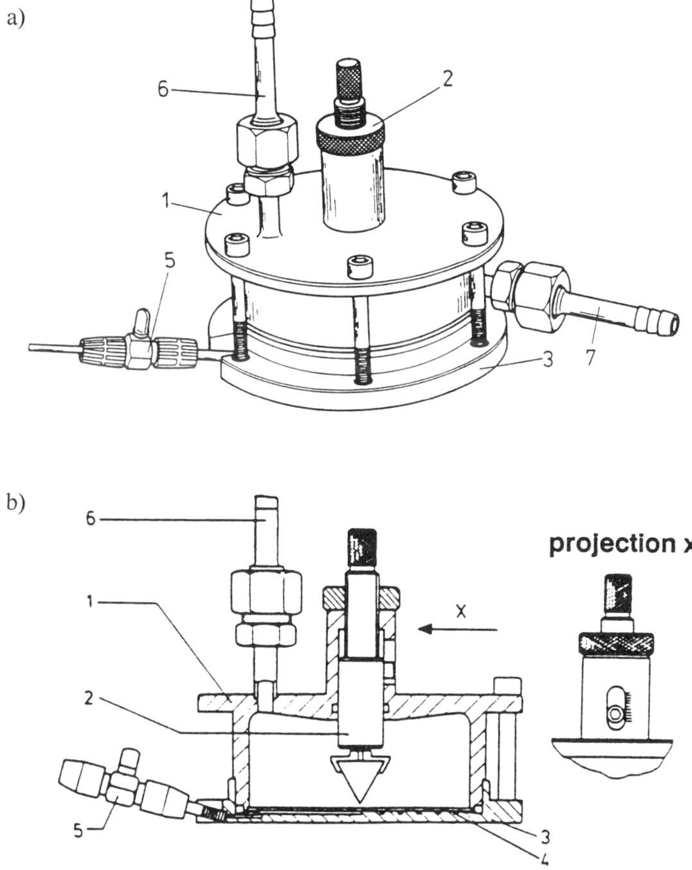

Figure 22-5. Schematics of BIOPEM (B. Braun Melsungen), a magnetically stirred filtration cell for continuous sampling under sterile conditions. a) whole module; b) schematic drawing with adjustable stirrer assembly. (1: top plate; 2: adjustable magnetic stirrer assembly; 3: bottom plate; 4: channels; 5: shut-off valve; 6: outlet line; 7: inlet line) (reproduced by permission of B. Braun Melsungen, Germany).

Figure 22-6.
In situ filtration probe utilizing tubular membranes (membrane not shown). The probe can be connected to the fermenter via standard ports. The tubular membrane fits directly on the inner support. Channels in this support are neccessary to collect the cell-free medium (developed at the Institut für Technische Chemie of the Universität Hannover, Hannover, Germany).

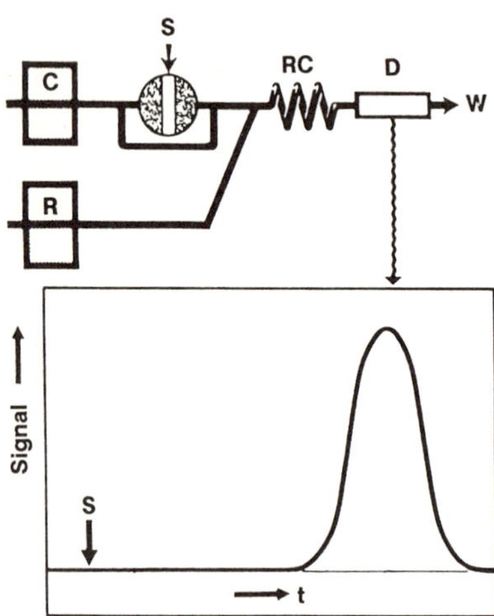

Figure 22-7.
Principle of flow injection analysis. (C: carrier flow, R: reagent flow, S: sample, RC: reaction coil, D: detector, W: waste (adapted from Reference [26]).

ing can involve achievement of an optimal reaction matix and/or an optimal sample dilution [35].

As shown in Figure 22-4, immobilized enzymes can be used in the detection step. The injected sample passes the immobilized enzyme and the enzymatic reaction is measured by a special transducer. It is not necessary to have direct contact between the immobilized biological component and the sensor.

FIA systems can be recalibrated frequently by injecting calibration solutions instead of a sample. The entire analytical system (i. e., enzyme and sensor) is calibrated under real analysis conditions. Deactivated enzyme preparations or sensors can be exchanged without any problems since the fermenter is protected by the sterility barrier of the filtration probe.

In summary, FIA offers tremendous advantages compared to in situ sensors:

— small sample volume
— sample preconditioning

Figure 22-8. Modular FIA system (Eppendorf). The system consists of a sampler, selector, four-channel pump, sample injector, reaction system (manifold), master module (detector, controller for data processing, and master system control) (by permission of Eppendorf, Hamburg, Germany).

- high frequencies of analysis
- short detection times
- sterile barrier to the bioprocess
- recalibration at any time
- use of non-sterilizable sensors is possible
- replacement of deactivated enzymes or transducers

However, there are some disadvantages to the use of FIA systems, including:

- discrete data and the requirement for peak analysis
- time delay due to sampling
- automation of a complex process
- expensive ($ 20.000–$ 40.000 per channel)

22.3.3 Autoanalyzer Systems

The use of autoanalyzer systems has been reported for several biotechnological applications [36]. Like FIA systems, autoanalyzers are also able to perform different analytical procedures automatically. Several aspects of autoanalyzer operation (eg, sampling, use of detectors) are comparable to FIA. However, in contrast to the FIA technique, sample and reagent solutions are pumped continuously through an autoanalyzer system by multichannel peristaltic pumps. No complex sample injection valves are used; instead, relatively simple mixing units and reaction coils are employed. Any detector suitable for flow-through operation can be used. These sensors deliver a continuous signal which can be used for process control without complex signal processing. To avoid backmixing, air is usually injected periodically to segment the flow. In such air-segmented autoanalyzer systems, bubble separators must be installed before the flow reaches the detector unit. Air segmentation causes several other technical problems.

Problems also arise regarding the calibration of autoanalyzers since the whole system is continuously affected by the complex sample solution. Chemical deposits change the flow behaviour, cause clogging, and change the sensor's sensitivity drastically (eg, optical detectors). In addition, microbial growth can occur in the system. Thus, frequent calibration is necessary or cleaning steps have to be included in the analytical procedure.

Some advantages of autoanalyzer systems are:

- continuous signal
- inexpensive (about $ 10.000 per channel)
- simple to automate

Disadvantages of autoanalyzer systems are:

- large sample and reagent volumes needed
- longer response times (to obtain reaction equilibrium)
- frequent cleaning and calibration steps necessary

22.4   Sensors of the Physical Environment

Sensors and sensor systems capable of quantifying various aspects of the physical environment are perhaps the most widely used in biotechnology. The signals from these probes are frequently utilized in process analysis and control schemes (eg, temperature, shaft power, foam detection, and liquid volume/level). Many of the instruments used to make the measurements shown in Figure 22-1 are familiar to workers in several fields and thus they will be discussed only briefly here. Details on these sensors can be found elsewhere [eg, 37–40].

Since biological systems are extremely sensitive to changes in temperature and cells typically have sharp optima for growth and product formation (often different values), temperature sensors are the most common probes in biotechnology. Thermistors, platinum resistance sensors, and thermocouples are all used to provide an on-line signal; thermometer bulbs are also employed but cannot be used for automated monitoring.

In bioprocessing, gas-phase pressure measurements are utilized for evaluation of the partial pressures of various gaseous components, to monitor the performance of outlet gas filters, and to control sterilization procedures. Since they are compatible with the aseptic requirements of biotechnological processes, membrane-type pressure sensors are most commonly used.

Liquid volume/level measurement is important in fermenter operation for control in continuous and fed-batch processes, as well as in the filling phase of any type of process. Since the bioreactor geometry is known, volume and level measurements can be interconverted. Several kinds of sensors can be used: membrane pressure devices (to measure the hydrostatic pressure difference between the bottom of the vessel and the headspace; knowledge of the liquid density is required to calculate the volume) and strain gauges (again requiring the liquid density) give measurements of liquid volume, while the liquid level can be determined with capacitance sensors, conductivity probes, or floating sensors.

Liquids that are highly aerated and/or contain high protein concentration are often susceptible to foaming, which may interfere strongly with the bioprocess in question. The amount of foam can be measured with sensors based on electrical conductivity, heat conductivity, capacitance, or light scattering [41].

The viscosity of a bioprocess liquid is an important factor, affecting mass transfer, heat transfer, and power consumption. It is affected by the concentrations and types of microorganisms, substrates, products, and solids present in the liquid, and can vary during a fermentation or separation process. Unfortunately, on-line viscosity measurements are difficult, especially when significant amounts of air bubbles and suspended solids are present, and thus viscosity probes are not in widespread use. Sensors that have been applied for viscosity determination (usually in laboratory studies) are often various types of rotational viscosimeters [42].

Parameters such as impeller speed and shaft power (in a stirred bioreactor) and fluid velocity are indicators of the degree of mixing and thus play an important role in the control of mass transfer. Impeller speed is easily monitored with a tachometer (electronic or mechanical) [39], but the measurement of shaft power input is not as straightforward. The most common method utilizes a torsion dynamometer attached to the impeller drive; however, this technique includes losses due to friction in the drive shaft. Better data can be obtained from balanced strain gauges mounted on the impeller [37]. On-line measurement of the liquid velocity in a flowing or stirred system can be obtained by a heat-pulse method in which a resistance thermometer is used to measure a brief temperature increase caused by an upstream pair of electrodes [43]. Use of this sensor system has been limited to laboratory applications.

22.5 Sensors of the Chemical Environment

In cultivation processes, the chemical environment (normally called the fermentation broth) is one of the main foci in biotechnological analysis [36]. Substrates, products, and several metabolites can usually be found in the fermentation broth. Substrates are consumed by the microorganisms and products or metabolites are released by the cells into the medium. Thus, detailed on-line monitoring of these components can reflect not only the progress of the bioprocess but can also give an insight into the state and metabolism of the microorganisms.

The most frequently used sensors of the chemical environment in biotechnology are potentiometric pH and amperometric pO_2 electrodes (see Figures 22-9 and 22-10). A huge variety of sterilizable electrodes of these types is offered by several companies. Since these sensors are used in situ, problems can arise because recalibration is not possible when no calibration system (see 22.3.1) is used (see also Chapter 7).

Redox electrodes are often used to measure the global redox potential in fermentation media [17]. This parameter is usually hard to interpret and in most cases the redox potential has the same curve as the pO_2 signal.

Several ion-selective electrodes are of interest for bioprocess control [11–13, 44] (see also Chapter 10.1). Only a few of them can be sterilized, and thus they have to be used in FIA or autoanalyzer systems. A pCO_2 electrode that is sterilizable in situ is offered by Ingold Meßtechnik. This sensor is based on a pH electrode and can be recalibrated in situ.

Analysis of the gas phase is important for the monitoring of the metabolic activity of the organisms during the cultivation process. Oxygen and carbon dioxide are the most important components of the gas phase in many bioprocesses; their concentrations are usually measured with paramagnetic and infrared analyzers, respectively. More complex analyses can be performed with on-line mass spectroscopy [45–47], on-line gas chromatography [48], semiconductor-based gas sensors [49], or on-line laser Raman spectroscopy [50] (see also Chapter 10-2). Volatile compounds such as ethanol and acetic acid can also be quantified with these on-line instruments. In this case, liquid samples are withdrawn from the fermenter by special sampling devices [28–32]. The use of on-line mass spectrometers has become increasingly popular, even in industrial bioprocess monitoring applications.

During the last few years, on-line HPLC (high performance liquid chromatography) has also become an interesting analytical technique for bioprocess monitoring. Cell-free samples are injected into the HPLC system periodically, allowing the determination of several analytes (less volatile medium compounds, metabolic intermediates, and products) at the same time. It is possible to analyze quite different compounds such as sugars, antibiotic metabolites, and amino acids in a single HPLC run [51–54].

The application of autoanalyzers and FIA systems for bioprocess control has gained more and more importance. Several wet chemical or biochemical assays can be run automatically

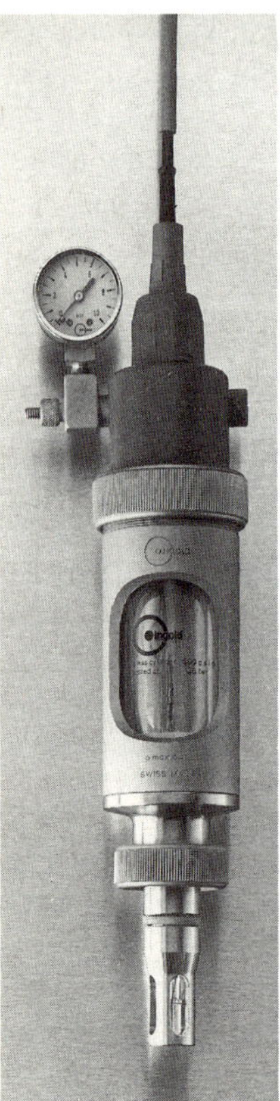

Figure 22-9.
Sterilizable pH electrode (by permission of Ingold Meßtechnik, Urdorf, Switzerland).

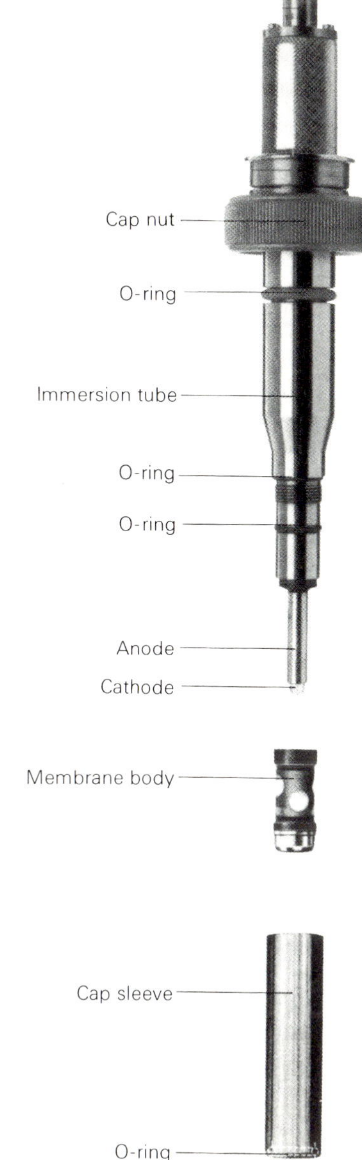

Figure 22-10.
Sterilizable pO_2 electrode (by permission of Ingold
Meßtechnik, Urdorf, Switzerland).

in these systems. When interfaced to bioprocesses via adequate sampling devices, an on-line
(autoanalyzer) or quasi on-line (FIA) analysis can be provided. The time lag between sampling
and detection depends upon the apparatus and various analysis parameters. Since the
literature for these analyses is vast, we refer the reader to the excellent books and reviews of
Ruzicka and Hansen [23–27].

Examples of the use of these sample-flow systems are numerous. Protein concentration
monitoring applications are especially well documented. For instance, total protein monitor-

Table 22-1. Applications of biosensors for process monitoring in biotechnology (CD: calibration data only, no on-line monitoring data).

Compound	Process/Organism	Type of Sensor	Analysis System	Reference
glucose	*E. coli*	amperometric	in situ	[20]
glucose	*S. cerevisiae*	amperometric	automated off-line analysis	[68,69]
glucose	*S. cerevisiae*	amperometric	in situ	[21, 22]
glucose	yeast	spectrophotometric	FIA principle	[70]
glucose	*E. coli*	pO_2 electrode	in situ	[71, 72]
glucose	yeast	amperometric	off-line, CD	[73]
glucose	yeast	microbial sensor pO_2 electrode	FIA principle, CD	[74]
glucose	yeast	enzyme electrode	FIA principle, CD	[75]
glucose	*E. coli, S. cerevisiae*	Gambro glucose analyzer	FIA principle	[76]
glucose	yeast	YSI glucose analyzer	FIA principle, CD	[77]
assimilable sugar	glutamic acid production	pO_2 electrode	FIA principle, CD	[78]
L-leucine, α-keto acid, glucose	enzyme membrane reactor	spectrofluorometric, amperometric	FIA principle	[79]
glucose, sucrose	*C. acremonium,* enzyme reactor, yeast	enzyme thermistor	FIA principle	[80–86]
glucose, lactate	lactic acid production	chemiluminescence	FIA principle	[33, 87]
glucose, glutamate, lactate	animal cell culture	pO_2 electrode	autoanalyzer, CD	[88]
L-lysine, glucose	developed for fermentation control	pO_2 electrode	FIA principle	[89, 90]
cellobiose	enzyme reactor	enzyme thermistor	FIA principle	[91]
ethanol, methanol	yeast	pO_2 electrode	off line	[92]
ethanol	yeast	pO_2 electrode	FIA principle, CD	[93]
ethanol	yeast	pO_2 electrode	in situ, unsterile	[94]
ethanol	yeast	enzyme optrode	FIA principle, CD	[95]
penicillin	*P. chrysogenum*	pH electrode	FIA principle, off-line samples, CD	[96]
penicillin	*P. chrysogenum*	pH electrode	off line	[97]
penicillin	*P. chrysogenum*	pH electrode	in situ	[98]
penicillin G and V	*P. chrysogenum*	enzyme thermistor	FIA principle	[99–101]
formic acid	*A. formicans*	microbial sensor amperometric	FIA principle, CD	[102]
cephalosporin	fermentation media	microbial sensor pH electrode	FIA principle, CD	[102]
glutamic acid	*E. coli*	microbial sensor CO_2 electrode	FIA principle, CD	[103]
acetic acid	acetic acid production	microbial sensor pO_2 electrode	FIA principle, CD	[104]
proteins	animal cell culture	on-line immuno analysis	FIA principle	[60]

ing can be performed by sample-flow systems. A number of different applications have been reported, including the measurement of the total protein concentration with the Biuret method (FIA) [55], the Lowry method (FIA and autoanalyzer) [56, 57], the Bradford method (FIA) [34], and a modified bicinchoninic acid method (FIA) [58]. An assay for the specific determination of protein is also possible in these systems: the selective analysis of formate dehydrogenase and leucine dehydrogenase in cell disintegration processes and the concentration of antithrombin III or pullulanase in fermentation processes have been reported [59, 60].

Whether they are used as in situ probes or as a part of FIA or autoanalyzer system, biosensors provide the ability to analyze, in a highly selective and sensitive manner, single analytes in the very complex fermentation broth (chemical environment) [61–67]. For optimal process monitoring and control, several important analytes (eg, substrates or products) must be analyzed (see also Chapter 14). Biosensors provide the opportunity to analyze these substances in situ or quasi on-line, instead of the complex off-line analysis assays used up to now.

As shown in Table 22-1, biosensor applications in bioprocess monitoring often focus on the determination of the glucose concentration. This can be explained by the fact that glucose is one of the main growth substrates used in biotechnology. Furthermore, it is obvious that most biosensors are integrated into a FIA system; very few in situ sensors are described. Based on experiences with biosensors for glucose and the other "easily detectable" analytes listed in Table 22-1, biosensors for more complicated analytes (eg, biologically active proteins, metabolic intermediates) will gain in importance for bioprocess control.

The problems of adequate sampling are obvious upon examination of Figures 22-11 and 22-12. In both cases, an enzyme thermistor was used for quasi on-line glucose determination of a fed-batch cultivation of *Cephalosporium acremonium* [86]. Sampling was performed by micro- or ultrafiltration. Only the analysis of ultrafiltered samples shows good agreement between enzyme thermistor and off-line data.

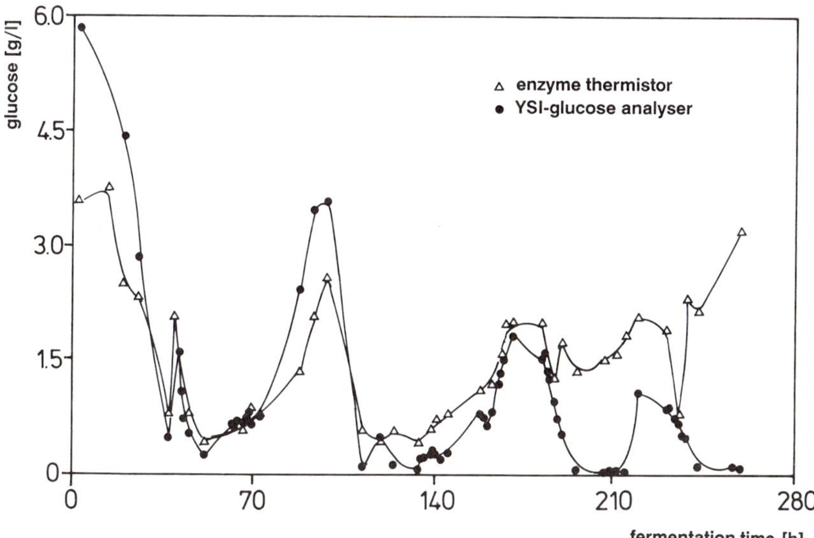

Figure 22-11. Monitoring of glucose with an enzyme thermistor during a fed-batch cultivation of *C. acremonium*. Samples were withdrawn from the fermenter via a microfiltration probe. (● : Yellow Springs glucose analyzer data, △ : enzyme thermistor data).

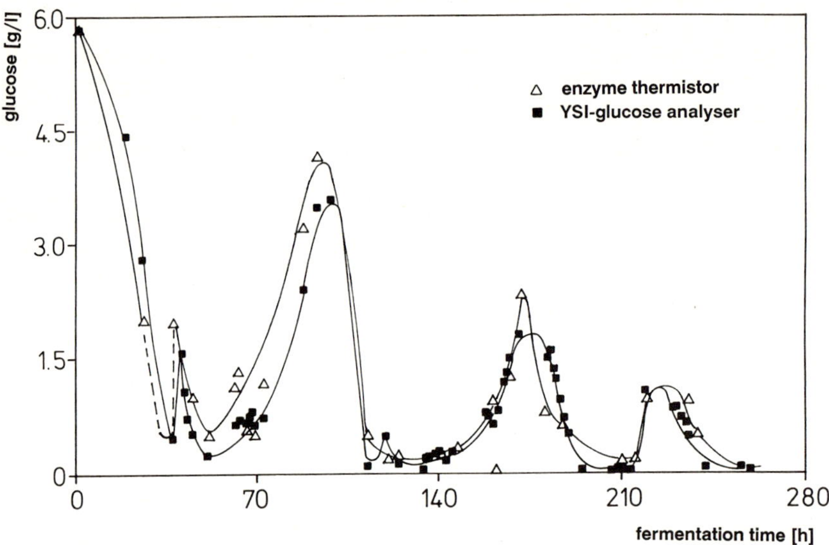

Figure 22-12. Monitoring of glucose with an enzyme thermistor during a fed-batch cultivation of *C. acremonium*. Samples were withdrawn from the fermenter via an ultrafiltration probe. (■ : Yellow Springs glucose analyzer data, △ : enzyme thermistor data).

22.6   Sensors of the Biological System

These sensors are used to provide on-line and in situ information about the biomass concentration and about cell components and cell characteristics. This type of information is urgently needed for a better understanding of bioprocesses and for effective process optimization. Although sensors of this kind are still mostly under development, some types have already shown their tremendous potential [7].

22.6.1   Biomass Concentration Estimation

Biomass concentration is one of the most important parameters in bioprocesses involving active cells. Biomass can be defined either as the amount of viable, productive cells or the total amount of microbial cells. On-line biomass concentration estimation has been performed with several types of sensors:

- optical sensors (nephelometric or spectrofluorometric methods)
- calorimetric devices
- filtration probes
- viscosity sensors
- electrochemical sensors (impedimetric, potentiometric, or amperometric methods)
- acoustic sensors

Nephelometric sensors measure the turbidity of a cell-containing sample. Different in situ versions that quantify transmitted or scattered light are described in the literature for on-line monitoring [105–111]. Turbidity is a function of *all* light scattering particles in the suspension (eg, cells, solid particles, bubbles) and thus only the total biomass can be measured with this type of sensor. In addition, fouling on the surface of the optical components in the fermentation broth can cause problems. Since the linear correlation between turbidity and cell concentration is resticted to small extinction values (about 0.5 [112]), this method can only be applied to relatively low biomass concentrations. Otherwise the lightpath has to be decreased with increasing biomass concentration. Despite these problems, nephelometric sensors are available from several companies (Hatch, BTG, Monitek, Aquasant, Guided Wavelength).

Special in situ fluorometers (BioChem Technology, Ingold Meßtechnik) measure the amount of intracellular NAD(P)H inside living cells (culture fluorescence). These reduced adenine dinucleotides fluoresce at 460 nm when the cells are irradiated with 340–360 nm (UV) light. Figure 22-13 shows a schematic drawing of a similar fluorometer that can be used for the simultaneous detection of two different wavelengths. Several publications have shown that the monitoring of the intracellular NAD(P)H pool can be used for biomass estimation

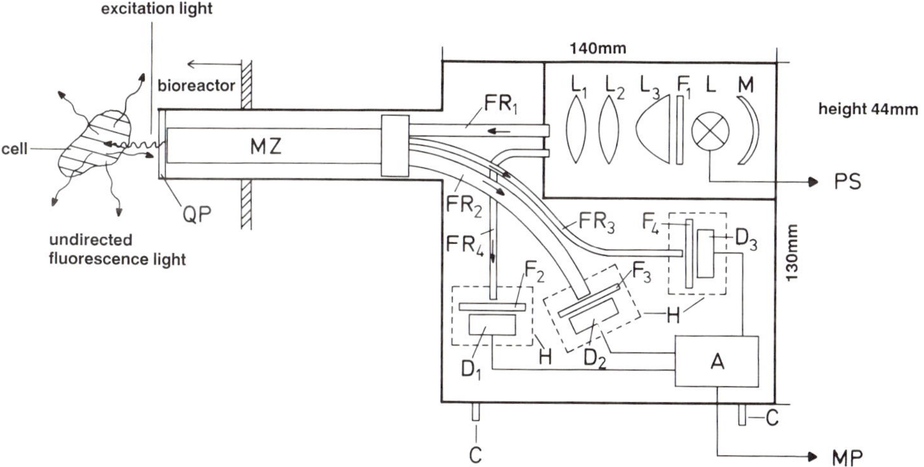

Figure 22-13. Schematic design of a fluorescence probe for the simultaneous detection of two different wavelengths.

M	mirror
L	lamp
L_{1-3}	lenses
F_{1-4}	filters
FR_{1-4}	fiber cables
D_{1-3}	photodetectors
H	housing
MZ	mixing zone of fibers
QP	quartz plate
A	amplifier
C	cooling
MP	microprocessor unit
PS	power supply

[113–122]. However, since the adenine dinucleotides are important coenzymes for several anabolic and catabolic reactions in the cell, the NAD(P)H pool changes when the metabolism of the microorganisms is affected. These changes as well as some abiotic factors (eg, bubbles or fluorescent components in the medium) could cause problems during biomass estimation with culture fluorescence.

The heat produced during the growth of microorganisms can be also be used for biomass concentration estimation. Different calorimetric devices (external-flow, twin-type, and heat-flux calorimeters) and different calorimetric techniques (dynamic and continuous calorimetry) have been used for on-line biomass estimation [8]. In most cases, the experimental setup is complicated and measurements are restricted to relatively small volumes (less than 1 L). Larger devices (continuous calorimeters for volumes up to 14 L) were studied by Luong and Volesky [123–125]. One of the best devices seems to be the "heat-flux calorimeter" developed by Marison and von Stockar. Several applications to bioprocess monitoring are given by the authors [126–129].

Since biomass can be separated from the medium by filtration, automated filtration probes can be used for biomass estimation. During the filtration process, the filtration volume, pressure and the growth of the filtration cake are monitored. Nestaas et al. [130–132] and Reuß et al. [133–134] showed that these probes could be successfully applied for biomass concentration estimation during the cultivation of *Penicillium chrysogenum* and during the Pekilo process, respectively. These devices are relatively complex since sampling, defined filtration, rinsing, sterilization, and monitoring have to be controlled. One measuring cycle lasts approximately thirty minutes.

In situ rotational viscosimeters have also been used for biomass estimation [135]. This technique utilizes the fact that the viscosity of the medium increases with increasing biomass concentration. However, the application of such devices is relatively rare.

Different electrochemical sensors have been developed for cell concentration measurement. The most promising of these sensors are based on impedimetric measurements. A commercial version of a sensor that measures the frequency-dependent permittivity is available from Aber Instruments Ltd [137–139]. Another type of electrochemical probe measures the potential changes in the cell suspension caused by the production of electroactive substances during cell growth [140–143]. To date, no on-line applications of these potentiometric sensors under real cultivation conditions have been reported. Other types of probes, such as amperometric and fuel-cell sensors, measure the current produced during the oxidation of certain compounds in the cell membrane. Mediators are often used to increase the sensitivity of the technique [143–145].

Piezocrystals can be used for biomass concentration estimation since the compressibility of a sample is a function of cell concentration and affects the voltage output of the crystal strongly [146]. A more promising technique is acoustic resonance densitometry (ARD) [147, 148]. In this method, the specific gravity or relative density of the cell-containing sample is measured. Blake-Coleman et al. report the application of this method during on-line biomass concentration estimation for cultivations of *Erwinia chrysanthemi* and *E. coli* [147, 148], while Kilburn et al. monitored mammalian cell cultures by ARD [149]. The measurements are affected by gas bubbles and foam and thus they must be performed in an external flow loop if these factors are significant.

22.6.2 Cell Characterization

The analysis of cell components and cell characteristics gives information on the metabolic and physiological states of the microorganisms that could be valuable for process monitoring and control. Unfortunately, on-line monitoring systems for cell components or cell characteristics are rare.

On-line, in situ fluorometers provide information on the metabolic state of microorganisms by monitoring the intracellular NAD(P)H pool (see also 22.6.1). Some applications of fluorescence measurements are given in Table 22-2. These probes can be used to monitor suspended as well as immobilized cells. In both cases, environmental and metabolic changes can be detected under real cultivation conditions.

Automated wet analytical reactions can be performed in autoanalyzer systems to monitor intracellular compounds [167]. Ahlmann et al. reported the development of an on-line system for analysis of intracellular penicillin G amidase produced by genetically modified bacteria. The time delay between sampling and photometric detection was 30 min. The whole system included automated sampling, cell disruption, and enzyme activity determination [167].

One of the most promising techniques for on-line monitoring of intracellular components is nuclear magnetic resonance spectroscopy (NMR), which offers the possibility of measuring several intracellular components directly. Some applications are listed in Table 22-3. Although this method is still restricted to relatively small bioreactors (a few mL), requires high cell densities, and often entails complicated analytical procedures, it offers tremendous potential for biotechnology [169–173].

Table 22-2. Different phenomena studied with NAD(P)H-dependent fluorescence monitoring.

Phenomenon Studied	Organism	Reference
Aerobic-anaerobic transition	*K. aerogenes*	[150]
	S. cerevisiae	[113, 118, 151, 152]
	C. tropicalis	[114]
	E. coli	[119, 153]
	C. guilliermondii	[154]
Aeration rate	*P. chrysogenum*	[155]
Addition of carbon source to starved cells	*S. cerevisiae*	[113–115, 151, 152]
	C. tropicalis	[156]
	E. coli	[119]
Diauxic growth	*S. cerevisiae*	[152]
Dilution rate changes	*S. cerevisiae*	[115]
	E. coli	[157]
	P. putida	[158]
Culture synchrony	*S. cerevisiae*	[118, 159]
Glycolytic oscillations	*S. cerevisiae*	[160–163]
Metabolic shifts	*Thermoactinomyces* sp.	[113]
	C. acetobutylicum	[116, 164–166]
Addition of metabolic uncoupler	*S. cerevisiae*	[152, 160]

Table 22-3. Cellular parameters measurable by NMR (adapted from Reference [168]).

Nucleus	Natural Abundance	Relative Receptivity	Applications
^{1}H	99.985 %	1.00	metabolic intermediates
^{13}C	1.108 %	$1.76 \cdot 10^{-4}$	metabolic intermediates
^{14}N	99.63 %	0.001	NH_3; amino acids; urea
^{15}N	0.37 %	$3.85 \cdot 10^{-6}$	protein turnover; nitrogen assimilation; intracellular pH
^{19}F	100 %	0.834	intracellular Ca^{2+}, Zn^{2+}, and pH
^{23}Na	100 %	0.0927	intracellular Na^+
^{31}P	100 %	0.0665	ATP; ADP; NAD(P)(H); intracellular pH and Mg^{2+}; phosphorylated intermediates
^{35}Cl	75.53 %	$3.56 \cdot 10^{-3}$	intracellular Cl^-
^{39}K	93.1 %	$4.75 \cdot 10^{-4}$	intracellular K^+

22.7   Conclusions

In this chapter, we have described a wide variety of sensors, ranging from pH probes to biosensor-based FIA systems, that can be used to monitor biotechnological processes. At this time, however, not many of the newer instruments (especially biosensors and probes for biomass characterization) presented here are commercially available, and of those that are, only a few have achieved widespread industrial use. This situation probably exists for a number of reasons: these sensors have only been developed recently (most within the last 5–10 years) and thus potential users may be unaware of their existence; many of these newer probes are expensive (primarily because they are new); and finally, the bioprocess industry has yet to fully accept this new generation of sensors. In addition, many factors are known to affect the accuracy and precision of these more complex instruments, requiring experienced personnel for operation and data interpretation.

Nonetheless, this condition is destined to change rapidly. With the ever-increasing variety of sensors being developed, more will become commercially available. Furthermore, the dramatic increase in the incidence of computer-controlled bioprocesses means that control and monitoring algorithms can be employed to take full advantage of the type of information provided by the newer, on-line sensors. Finally, the production of high-value, sensitive compounds justifies the slightly higher expense of some of these probes.

22.8   References

[1] Bailey, J. E., Ollis, D. F., *Biochemical Engineering Fundamentals,* 2nd edition; New York: McGraw-Hill, 1986, pp. 658–725.

[2] Erickson, L. E., Stephanopoulos, G., in: *Chemical Reaction and Reactor Engineering,* Carberry, J. J., Varma, V. (eds.); New York: Marcel Dekker, 1985.

[3] Wang, N. S., Stephanopoulos, G. N., *CRC Critical Rev. Biotechnol.* **2**, No. 1 (1984) 1103.

[4] Cooney, C. L., in: *Biotechnoloy* Vol. 1, Rehm, H. J., Reed, G., (eds.); Weinheim: VCH, 1981, pp. 73–112.

[5] Merten, O.-W., Palfi, G. E., Steiner, J., *Adv. Biotechnol. Proc.* **6** (1986) 111–178.

[6] Harris, C. M., Kell, D. B., *Biosensors* **1** (1985) 17–84.

[7] Reardon, K. F., Scheper, T., in: *Biotechnology* Vol. 3, 2nd edition, Rehm, H. J., Reed, G. (eds.) Weinheim: VCH, in press.

[8] von Stockar, U., Marison, I. W., *Adv. Biochem. Eng.* **40** (1989) 93–136.

[9] Twork, J. V., Jacynych, A. M., *Biotechnol. Prog.* **2**, No. 2 (1986) 67–72.

[10] Hitchmann, M. L. O., Hill, H. A. O., *Chem. in Britain* **12** (1986) 1117–1124.

[11] Clarke, D. J., Kell, D. B., Morris, J. G., Burns, A., *Ion-Sel. Electrode Rev.* **4** (1982) 75–131.

[12] Arnold, M. A., Solsky, R. L., *Anal. Chem.* **58**, No. 5 (1986) 84R–101R.

[13] Arnold, M. A., Meyerhof, M. E., *Anal. Chem.* **56**, No. 5 (1984) 20R–48R.

[14] Pungor, E., Cserfalvi, T., *Anal. Chim. Acta* **190** (1986) 99–106.

[15] Clarke, J. R. P., *Anal. Chim. Acta* **190** (1986) 1–11.

[16] Reeves, P., *Anal. Chim. Acta* **190** (1986) 45–65.

[17] Kjaergaard, L., in: *Adv. Biochem. Eng.,* Ghose, T. K., Fiechter, A., Blakeborough, N. (eds.); Berlin: Springer, pp. 131–150.

[18] J. Müller, Ingold Meßtechnik AG, personal communication.

[19] Kok, R., Hogan, P., *Biosensors* **3** (1987/1988) 89–100.

[20] Brooks, S. L., Ashby, R. E., Turner, A. P. F., Calder, M. R., Clarke, D. J., *Biosensors* **3** (1987/1988) 45–56.

[21] Bradley, J., Kidd, A. J., Anderson, P. A., Dear, A. M., Ashby, R. E., Turner, A. P. F., *Analyst* **114** (1989) 375–379.

[22] Bradley, J., Turner, A. P. F., Schmidt, R. D., in: *Biosensors — Applications in Medicine, Envrionmental Protection and Process Control,* GBF-Monographs 13, Schmid, R. D., Scheller, F. (eds.); Weinheim: VCH, 1989.

[23] Ruzicka, J., Hansen, E. H., *Chem. Techn.* **9** (1979) 756–764.

[24] Ruzicka, J., Hansen, E. H., *Anal. Chim. Acta* **114** (1980) 165–178.

[25] Ruzicka, J., Hansen, E. H., *Anal. Chim. Acta* **179** (1986) 1–58.

[26] Ruzicka, J., Hansen, E. H., *Flow Injection Analysis,* 2nd edition; New York: Wiley, 1988.

[27] Ruzicka, J., Hansen, E. H., "Homogenous and Heterogeneous Systems, Flow Injection Analysis Today and Tomorrow", *Anal. Chim. Acta* **214** (1988) 1–27.

[28] Kroner, K. H., Kula, M. R., *Anal. Chim. Acta* **163** (1984) 3–15.

[29] Schügerl, K., Lübbert, A., Scheper, T., *Chem.-Eng.-Techn.* **59**, No. 9 (1987) 701–714.

[30] Mandenius, C. F., Danielsson, B., Mattiasson, B., *Anal. Chim. Acta* **163** (1984) 135–141.

[31] Spohn, U., Voß, H., *BTF-Biotech.-Forum* **6**, No. 4 (1989) 274–288.

[32] Zabriskie, D. W., Humphrey, A. E., *Appl. Eur. Microbiol.* **35**, No. 2 (1978) 337–343.

[33] Nielsen, J., Nikolajsen, K., Villadsen, J., *Biotechnol. Bioeng.* **33** (1989) 1127–1134.

[34] Recktenwald, A., Kroner, K.-H., Kula, M.-R., *Enzyme Microb. Technol.* **7** (1985) 146–149.

[35] Garn, M., Gisin, M., Thommen, C., Cevey, P., *Biotechnol. Bioeng.* **34** (1989) 423–428.

[36] Schügerl, K., in: *Biotechnology* Vol. 3, 2nd edition, Rehm, H. J., Reed, G. (eds.); Weinheim: VCH, in press.

[37] Aiba, D., Humphrey, A. E., Mills, N. F., in: *Biochemical Engineering:* New York: Academic Press, 1965, Chapter 10.

[38] Wang, D. I. C., Cooney, C. L., Demain, A. L., Dunnhill, P., Humphrey, A. E., Lilly, D., *Fermentation and Enzyme Technology;* New York: J. Wiley, 1979, pp. 212–237.

[39] Tannen, L. P., Nyiri, L. K., in: *Microbial Technology,* Vol. 2, 2nd Ed., Peppler, H. J., Perlman, D. (eds.), New York: Academic Press, 1979 pp. 331–374.

[40] Solomons, G. L., *Material and Methods in Fermentation;* New York: Academic Press, 1969.

[41] Viesturs, U. E., Kristapsons, M. Z., Levitans, E. S., *Adv. in Biochem. Eng.* **21** (1982) 169–224.

[42] Charles, M., *Adv. in Biochem. Eng.* **8** (1978) 1–62.

[43] Lübbert, A., Larson, B., Korte, T., in: *Bioreactors and Biotransformations,* Moddy, G. W., Baker, P. B. (eds); London: Elsevier Applied Science Publishers, 1987, pp. 76–86.

[44] Schindler, J. G., Schindler, M. M., *Bioelektrische Membran-Elektroden;* Berlin: de Gruyter, 1983.

[45] Lloyd, D., Bohátka, S., Szilágyi, J., *Biosensors* **1** (1985) 179–212.

[46] Heinzle, E., Lafferty, R. M., *Eur. J. Appl. Microbiol. Biotechnol.* **11** (1980) 17–22.

[47] Schmidt, W. J., Meyer, H.-D., Schügerl, K., Kuhlmann, W., Bellgardt, K. H., *Anal. Chim. Acta* **163** (1984) 101–109.

[48] Groboillot, A., Pons, M.-N, Engasser, J.-M., *Appl. Microbiol. Biotechnol.* **32** (1989) 37–44.

[49] Vorlop, K. D., Becke, J. W., Klein, J., *Anal. Chim. Acta* **163** (1984) 287–291.

[50] E. H. Dunlop, personal communication.

[51] Möller, J., Hiddessen, R., Niehoff, J., Schügerl, K., *Anal. Chim. Acta* **190** (1986) 195–203.

[52] Ebel, S., Reyer, B., Werner-Busse, A., *Fresenius Z. Anal. Chem.* **327**, No. 2 (1987) 193–197.

[53] Lenz, R., Bölcke, C., Peckmann, U., Reuss, M., *IFAC Proc. Ser.* (Modell. Control Biotechnol. Proc.) **10** (1986) 85–90.

[54] Mathers, J. J., Dinwoodie, R. C., Talurovich, M., Mehnert, D. W., *Biotechnol. Lett.* **8**, No. 5 (1986) 311–314.

[55] Shideler, C. E., Stewart, K. K., Crump, J., Wills, M.-R., Savory, J., Rence, B. W., *Clin. Chem.* **26**, No. 10 (1980) 1454–1458.

[56] Salerno, R. A., Odell, C., Cyanovich, N., Bubnis, B. P., Morges, W., Gray, A., *Anal. Biochem.* **151** (1985) 309–314.

[57] Kusov, Y. Y., Kalinchuck, N. A., *Anal. Biochem.* **88** (1978) 256–262.

[58] Davis, L. C., Radke, G. A., *Anal. Biochem.* **161** (1987) 152–156.

[59] Recktenwald, A., Kroner, K.-H., Kula, M.-R., *Enzyme Microb. Technol.* **7** (1985) 607–612.

[60] Scheper, T., Anders, K. D., Freitag, R., Hundeck, H. G., Müller, W., Schelp, C., Bückmann, A. F., Reardon, K. F., in: *Biosensors: Applications in Medicine, Environmental Protection and Process Control,* GBF Monographs 13, Schmidt, R. D., Scheller, F. (eds.); Weinheim: VCH, 1989, pp. 253–262.

[61] Scheller, F., *Studia Biophys.* **119** (1987) 221–224.

[62] Clarke, D. J., Calder, M. R., Carr, R. J. G., Blake-Coleman, B. C., Moody, S. C., Collinge, T. A., *Biosensors* **1** (1985) 213–320.

[63] Schmid, R. D., *Biosensors* **3** (1987/1988) 239–249.

[64] Luong, J. H. T., Mulchandani, A., Guilbault, G. G., *TIBTECH* **6** (1988) 310–316.

[65] Guilbault, G. G., Luong, J. H., *Sel. Electrode Rev.* **11** (1989) 3–16.

[66] Guilbault, G. G., Mascini, M., *Sel. Electrode Rev.* **10** (1988) 33–48.

[67] Turner, A. P. F., *Sens. Actuators* **17** (1989) 433–450.

[68] Geppert, G., Asperger, L., *Studia Biophys.* **119** (1987) 159–162.

[69] Geppert, G., Asperger, L., *Bioelectrochem. Bioenerg.* (a section of J. Electroanal. Chem., 231) **17** (1987) 399–407.

[70] Mandenius, C. F., *Biotechnol. Bioeng.* **32** (1988) 123–129.

[71] Cleland, N., Enfors, S. O., *Eur. J. Appl. Microbiol. Biotechnol.* **18** (1983) 141–147.

[72] Cleland, N., Enfors, S. O., *Anal. Chem.* **56** (1984) 1880–1884.

[73] Gründig, B., Krabisch, Ch., *Anal. Chim. Acta* **222** (1989) 75–81.

[74] Karube, I., Mitsuda, S., Suzuki, S., *Eur. J. Appl. Microbiol. Biotechnol.* **7** (1979) 343–351.

[75] Chotani, G., Costantinides, A., *Biotechnol. Bioeng.* **24** (1982) 2743–2745.

[76] Holst, O., Hakanson, H., Miyabayashi, A., Mattiasson, B., *Appl. Microbiol. Biotechnol.* **28** (1988) 32–36.

[77] Parker, C. P., Gardell, M. G., Di Biasio, D., „A complete system for fermentation monitoring", *Internat. Biotechnol. Lab.* June (1986) 33.

[78] Hikuma, M., Obana, H., Yasuda, T., Karube, I., Suzuki, S., *Enzyme Microb. Technol.,* **2** (1980) 234–238.

[79] Kittstein-Eberle, R., Ogbomo, I., Schmidt, H.-L., *Biosensors* **4** (1989) 75–85.

[80] Danielsson, B., Mandenius, C. F., Winquist, F., Mattiasson, B., Mosbach, K., *Proc. 5th Internat. Fermen. Symp., London, Canada, July 20–25,* 1980.

[81] Danielsson, B., Mosbach, K., in: *Biosensors – Fundamentals and Applications,* Turner, A. P. F., Karube, I., Wilson, G. S. (eds.); Oxford: Oxford Science Publications, 1987, 575–597.

[82] Danielsson, B., Mattiasson, B., Karlsson, R., Winquist, F., *Biotechnol. Bioeng.* **21** (1979) 1749–1766.

[83] Mandenius, C. F., Bülow, L., Danielsson, B., Mosbach, K., *Appl. Microbiol. Biotechnol.* **21** (1985) 135–142.

[84] Mandenius, C. F., Danielsson, B., Mattiasson, B., *Acta Chem. Scand.* **B34**, No. 6 (1980) 463–465.

[85] Mattiasson, B., Mandenius, C. F., Axelsson, J. P., Danielsson, B., Hagander, P., *Ann. N. Y. Acad. Sci.* **413** (1983) 193–196.

[86] Wehnert, G., Sauerbrei, A., Bayer, T., Scheper, T., Schügerl, K., Herold, T., *Anal. Chim. Acta* **200** (1987) 73–78.

[87] Nikolajsen, K., Nielsen, J., Villadsen, J., *Anal. Chim. Acta* **214** (1988) 137–145.

[88] Romette, J. L., in: *Biosensors International Workshop 1987,* GBF-Monographs 10, Schmidt, R. D., (ed.); Weinheim VCH, pp. 81–86.

[89] Romette, J. L., Yang, J. S., Kushabe, H., Thomas, D., *Biotechnol. Bioeng.* **25** (1983) 2557–2566.

[90] Kernevez, J. P., Konante, L., Romette, J. L., *Biotechnol. Bioeng* **25** (1983) 845–855.

[91] Danielsson, B., Matthiasson, B., Mosbach, K., *Appl. Biochem. Bioeng.* **3** (1981) 97–143.

[92] Belghith, H., Romette, J. L., Thomas, D., *Biotechnol. Bioeng.* **30** (1987) 1001–1005.

[93] Hikuma, M., Kubo, T., Yasuda, T., Karube, I., Suzuki, S., *Biotechnol. Bioeng.* **21** (1979) 1845–1853.

[94] Verduyn, C., Zomerdijk, T. P. L., van Dijken, J. P., Scheffers, W. A., *Appl. Microbiol. Biotechnol.* **19** (1984) 181–185.

[95] Scheper, T., Bückmann, A. F., *Biosens. Bioelectron.* **5** (1990) 125–135.

[96] Gnanasekaran, R., Mottola, H. A., *Anal. Chem.* **57** (1985) 1005–1009.

[97] Nilsson, H., Mosbach, K., Enfors, S.-O., Molin, N., *Biotechnol. Bioeng.* **20** (1978) 527–539.

[98] Enfors, S. O., Nilsson, H., *Enzyme Microb. Technol.* **1** (1979) 260–264.

[99] Mattiasson, B., Danielsson, B., Winquist, F., Nilson, H., Mosbach, K., *Appl. Env. Microbiol.* **41** (1981) 903–908.

[100] Decristoforo, H. in: *Immobilized Enzymes and Cells. Part D, Methods in Enzymology,* Colowick, S. P., Kaplan, N. O. (eds.); Orlando: Academic Press, 1988, p. 137.

[101] Decristoforo, G., Danielsson, B., *Anal. Chem.* **56** (1984) 263–268.

[102] Karube, I., in: *Biosensors − Fundamentals and Applications,* Turner, A. P. F., Karube, I., Wilson, G. S. (eds.); Oxford: Oxford Science Publication, 1987, 13–29.

[103] Hikuma, M., Obana, H., Yasuda, T., Karube, I., Suzuki, S., *Anal. Chim. Acta* **116** (1980) 61–67.

[104] Hikuma, M., Kubo, T., Yasuda, T., Karube, I., Suzuki, S., *Anal. Chim. Acta* **109** (1979) 33–37.

[105] Hancher, C. W., Thacker, L. H., Phares, E. F., *Biotechnol. Bioeng.* **16** (1974) 475–484.

[106] Lee, C., Lim, H., *Biotechnol. Bioeng.* **22** (1980) 639–642.

[107] Lee, Y. H., *Biotechnol. Bioeng.* **23** (1981) 1903–1906.

[108] Lima Filho, J. L., Ledingham, W. M., *Biotechnol. Techniques* **1**,No. 3 (1987) 145–150.

[109] Metz, H., *Chemie-Technik* **10** (1981) 691–696.

[110] Merten, O.-W., Palfi, G. E., Stäheli, J., Steiner, J., in: *Developments in Biological Standardization* **66** (1987) 357–360.

[111] Okashi, M., Watabe, T., Ishikawa, T., Watanabe, Y., Miwa, K., Shode, M., Ishikawa, Y., Ando, T., Shibata, T., Kitsunai, T., Kamiyama, N., Oikawa, Y., *Biotechnol. Bioeng. Symp.* **9** (1979) 103–116.

[112] Koch, A. L., *Anal. Biochem.* (1970) 252–259.

[113] Zabriskie, D. W., Humphrey, A. E., *Appl. Eur. Microbiol.* **35**, No. 2 (1978) 337–343.

[114] Beyeler, W., Einsele, A., Fiechter, A., *Eur. J. Appl. Microbiol. Biotechnol.* **13** (1981) 10–14.

[115] Scheper, T., Schügerl, K., *Appl. Microbiol. Biotechnol.* **23** (1986) 440–444.

[116] Srinivas, S. P., Mutharasan, R., *Biotechnol. Bioeng.* **30** (1987) 769–774.

[117] Zabriskie, D. W., *Biotechnol. Bioeng. Symp.* **9** (1979) 117–123.

[118] Scheper, T., Lorenz, T., Schmidt, W., Schügerl, K., *Ann. N. Y. Acad. Sci.,* **506** (1987) 431–445.

[119] Meyer, H.-P., Beyeler, W., Fiechter, A., *J. Biotechnol.* **1** (1984) 341–349.

[120] Arminger, W. B., Lee, J. F., Montalvo, L. M., Forro, J. R., *Proceedings: 190th ACS Meeting, Chicago, 1985, MBTD 40.*

[121] Groom, C. A., Luong, J. H. T., Mulchandani, A., *J. Biotechnol.* **8** (1988) 271–278.

[122] MacMichael, G., Arminger, W. B., Lee, J. F., Mutharasan, R., *Biotechnol. Techn.* **1**, No. 4 (1987) 213–218.

[123] Luong, J. H. T., Yerushalmi, L., Volesky, B., *Enzyme Microb. Technol.* **5** (1983) 291–297.

[124] Luong, J. H. T., Volesky, B., *Can. J. Chem. Eng.* **60** (1982) 163–167.

[125] Luong, J. H. T., Volesky, B., *Adv. Biochem. Eng. Biotechnol.* **28** (1983) 1–40.

[126] Birou, B., von Stockar, U., *Enzyme Microb. Technol.* **11** (1989) 12–16.

[127] Birou, B., Marison, I. W., von Stockar, U., *Biotechnol. Bioeng.* **30** (1987) 650–660.

[128] Marison, I. W., von Stockar, U., *Biotechnol. Bioeng.* **28** (1986) 1780–1793.

[129] Marison, I., von Stockar, U., *Enzyme Microbiol. Technol.* **9** (1987) 33–43.

[130] Nestaas, E., Wang, D. I. C., *Biotechnol. Bioeng.* **23** (1981) 2803–2813.

[131] Nestaas, E., Wang, D. I. C., Suzuki, H., Evans, L. B., *Biotechnol. Bioeng.* **23** (1981) 2815–2824.

[132] Nestaas, E., Wang, D. I. C., *Biotechnol. Bioeng.* **25** (1983) 1981–1987.

[133] Reuß, M., Boelcke, C., Lenz, R., Peckmann, U., *BTF-Biotech.-Forum* **4** (1987) 2–12.

[134] Reuß, M., in: *Biochemical Engineering,* Chmiel, Hammes, Bailey (eds.); Stuttgart: Fischer Verlag, 1987, pp. 149–168.

[135] Shimmons, B. W., Svrcek, W. Y., Zajic, J. E., *Biotechnol. Bioeng.* **18** (1976) 1793–1805.

[136] Perley, C. R., Swartz, J. R., Cooney, C. L., *Biotechnol. Bioeng.* **21** (1979) 519–523.

[137] Kell, D. B., *J. Gen. Microbiol.* **133** (1987) 1651–1665.

[138] Harris, C. M., Todd, R. W., Bungard, S. J., Lovitt, R. W., Morris, J. G., Kell, D. B., *Enzyme Microbiol. Technol.* **9** (1987) 181–186.

[139] Harris, C. M., Kell, D. B., *Biosensors* **1** (1985) 17–84.

[140] Wilkins, J. R., Stoner, G. E., Boykin, E. H., *Appl. Microbiol.* **27** (1974) 949–952.

[141] Wilkins, J. R., *Appl. Env. Microbiol.* **36**, No. 5 (1978) 683–687.

[142] Wilkins, J. R., Young, R. N., Boykin, E. H., *Appl. Env. Microbiol.* **35**, No. 1 (1978) 214–215.

[143] Matsunaga, T., Karube, I., Suzuki, S., *Appl. Env. Microbiol.* **37**, No. 1 (1979) 117–121.

[144] Matsunaga, T., Karube, I., Suzuki, S., *Eur. J. Appl. Microbiol. Biotechnol.* **10** (1980) 125–132.

[145] Ramsay, G., Turner, A. P. F., *Anal. Chim. Acta* **215** (1988) 61–69.

[146] Ishimori, Y., Karube, I., Suzuki, S., *Appl. Env. Microbiol.* **42**, No. 4 (1981) 632–637.

[147] Blake-Coleman, B. C., Calder, M. R., Carr, R. J. G., Moody, S. C., Clarke, D. J., *Trends in Anal. Chem.* **3** (1984) 229–235.

[148] Blake-Coleman, B. C., Calder, M. R., Carr, R. J. G., Moody, S. C., *Biotechnol. Bioeng.* **28** (1986) 1241–1249.

[149] Kilburn, D. G., Fitzpatrick, P., Blake-Coleman, B. C., Clarke, D. J., Griffiths, J. B., *Biotechnol. Bioeng.* **33** (1989) 1379–1384.

[150] Harrison, D. E. F., Chance, B., *Appl. Microbiol.* **19**, No. 3 (1970) 446–450.

[151] Arminger, W. B., Forro, J. R., Montalvo, L. M., Lee, J. F., *Chem. Eng. Commun.* **45**, No. 3 (1986) 197–206.

[152] Müller, W., Wehnert, G., Scheper, T., *Anal. Chim. Acta* **213** (1988) 47–53.

[153] Gebauer, A., Scheper, T., Schügerl, K., *Bioprocess Eng.* **2** (1987) 13–23.

[154] Maneshin, S. K., Arevshatyan, A. A., *Appl. Biochem. Microbiol.* **8** (1972) 273–275.

[155] Scheper, T., Lorenz, T., Schmidt, W., Schügerl, K., *J. Biotechnol.* **3** (1986) 231–238.

[156] Einsele, A., Ristroph, D. L., Humphrey, A. E., *Eur. J. Appl. Microbiol. Biotechnol.* **6** (1979) 335–339.

[157] Scheper, T., Gebauer, A., Schügerl, K., *The Chem. Eng. Journal* **34** (1987) B7–B12.

[158] Li, J., Humphrey, A. E., *Biotechnol. Lett.* **11**, No. 3 (1989) 177–182.

[159] Scheper, T., Hitzmann, B., Rinas, U., Schügerl, K., *J. Biotechnol* **5** (1987) 139–148.

[160] Betz, A., Chance, B., *Arch. Biochem. Biophys.* **109** (1965) 759–584.

[161] Chance, B., Estabrook, R. W., Gosh, A., *Proc. Natl. Acad. Sci.* **51** (1964) 1244–1251.

[162] Kuchenbecker, D., Bley, T., Schmidt, A., *Stud. Biophys.* **86**, No. 2 (1981) 92–96.

[163] Doran, P. M., Bailey, J. E., *Biotechnol. Bioeng.* **29** (1987) 892–897.

[164] Reardon, K. F., Scheper, T., Bailey, J. E., *Biotechnol. Lett.* **8**, No. 11 (1986) 817–822.

[165] Reardon, K. F., Scheper, T., Bailey, J. E., *Biotechnol. Prog.* **3**, No. 3 (1987) 153–167.

[166] Rao, G., Mutharasan, R., *Appl. Microbiol. Biotechnol.* **30** (1989) 59–66.

[167] Ahlmann, N., Niehoff, A., Rinas, U., Scheper, T., Schügerl, K., *Anal. Chim. Acta* **190** (1986) 221–226.

[168] Harris, R. K., *Nuclear Magnetic Resonance: A Physiochemical View;* Harlow, UK: Longman House, 1986.

[169] Kanamori, K., Roberts, J. D., *Acc. Chem. Res.* **16** (1983) 35–41.

[170] Roberts, J. K. M., Jardetzky, O., *Biochim. Biophys. Acta* **639** (1981) 53–76.

[171] Gupta, R. K., Gupta, P., Moore, R. D., "NMR Studies of Intracellular Metal Ions in Intact Cells and Tissues," *Ann. Rev. Biophys. Bioeng.* **13** (1984) 221–246.

[172] Gadian, D. G., *Nuclear Magnetic Resonance and Its Applications to Living Systems,* London: Oxford University Press, 1982.

[173] Fernandez, E. J., Clark, D. S., *Enzyme Microb. Technol.* **9** (1987) 259–271.

23 Clinical and Respiration Gas Analysis

Hansjörg Albrecht, Laser-Medizin-Zentrum Berlin, FRG

Contents

23.1   Introduction

23.1.1   Objectives of Clinical Measurements

During diagnostic and therapeutic procedures, quantitative values are used as information on the actual health state of a patient. These measured values complete the clinical signs gathered by visual observation of doctors and nurses. More or less sufficient information and a mental model of disease and therapy based on the experience of the physician result in a specific therapy which aims to change or improve the actual state of the patient. Monitoring the progress of therapy is based essentially on the observation of selected parameters and the recording of their changes.

In particular, the therapy of severely ill patients with deficiencies of one or several vital systems is based on the surveillance of the important body functions. In addition, the correct function of the medical equipment used and the connections to the patient need to be monitored because a malfunction of devices assisting vital body functions may be fatal for the patient. During intensive care, for continuously monitored parameters automatic alarm systems are used which signal a deviation from preset alarm limits by visual and/or audible alarms and which automatically start a recorder to record selected parameters during the alarm situation.

These devices for the measurement of functions and parameters of patients and medical devices and for alerting staff in case of incidents are called "monitors". The transducer or sensor is the essential part which collects or picks up the information for the monitor by use of various physical or chemical methods and transforms it to an electrical signal. This signal will be processed by an electronic signal interface and fed to a data processor which presents the data to the user, stores the data, processes derived values, and originates alarms in the case of signal deviations from preset limits.

In modern clinical practice, competing methods for measuring the same or similar parameters are used and parameters are preferred that have been proved to be easy to apply and reliable in everyday use. Nevertheless, several methods are usually used simultaneously as the resulting parameters yield similar data on the same organ. Apart from changing opinions on approved parameters and the development of new procedures, the selection to methods depends on the intended application and the information needed. In addition to global parameters, supplementary parameters exist which are used jointly to compose a detailed picture.

Regarding the totality of parameters that are measured, eg, during intensive care of a ventilated patient, one can obtain an impression of a "measuring or parameter chain". Figure 23-1 presents an example of the monitoring of gas exchange with global parameters, such as inspiratory oxygen concentration, FiO_2, or oxygen saturation, SaO_2, or specific parameters, such as oxygen uptake or oxygen partial pressure in target organs.

"Measuring Chain" (Ventilation Therapy)

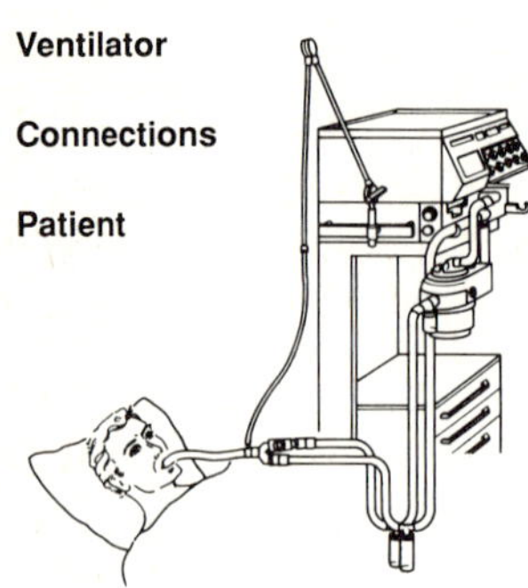

Figure 23-1.
Measuring or parameter "chain" in the case of a ventilated patient.

23.1.2   Gas Analysis

Gas exchange and metabolism are two of the human vital functions. Several organ systems contribute to these functions, failure of which cause life-threatening states of the patient. Gas exchange represents a central function occurring at the cellular and organ levels. Gas exchange includes the lung for the external part of respiration, the cardiovascular system for the transportation of oxygen to the organs, and diffusion in the tissue to provide the cells with oxygen as the internal part of respiration. The most important task is the supply of oxygen to all organs, as oxygen is the fundamental prerequisite for metabolic processes. Organs such as the brain and myocardium, with a high activity in energy metabolism, have an especially high demand for oxygen.

Owing to the lack of larger oxygen reservoirs in the body, in the event of an insufficient oxygen supply the organism reacts with a reduction in certain vital functions, such as unconsciousness with a reduction in heart and brain activity.

A second task is the removal of metabolites which are produced in the tissues, such as water and carbon dioxide, which would cause intoxication otherwise.

Intensive care aims at the replacement or support of deficient central functions. Important functions are respiration and a sufficient cardiac and circulation function. Monitoring the gas exchange in intensive care especially of ventilated patients requires, in addition to gas analysis at several measuring sites in the body, information on the heart function, eg, heart rate, on the pressure of the circulation system, eg, arterial pressure, and on the actual utilization of the available amount of oxygen. With regard to the methods required to achieve information that is as complete as possible, the clinical task consists in the determination of gas concentrations or partial pressures in the gas phase of the lung, airways, and equipment hoses, in the transportation system of the blood, and in the tissue. In recent years there has been an increasing importance of non-invasive methods, as invasive procedures are combined with risk and additional stress for the patient.

23.1.3 Sensors

Depending on the measuring site in the body, various methods are used. The transducer should be understood as a sensor that transforms a physical or chemical parameter into an electrical signal which can be processed further. Modern sensors, being partly still under development, usually function with self-calibration or store their specific characteristics by the use of their own processors and memory to produce normalized signals which can be digitally processed. Another continuing trend is the miniaturization to develop sensors that are smaller, of higher intelligence, and suitable for diverse applications. Depending on the time of application, sufficient tissue compatibility and stability of calibration and of sensitivity are required.

In the following sections, clinical experience with various types of sensor will mainly be discussed. Sensor fundamentals and typical specifications will be discussed here only to the extent that clinical results can be explained. Sensor fundamentals are the topics of other chapters.

Partial pressures or "concentrations" of gases are still measured by medical personnel in units such as mmHg or Torr instead of the SI unit the pascal (Pa). Fractions are measured in volume-% or %. Partial pressure values depend on atmospheric pressure and humidity and require calibration to ambient air in cases such as O_2 concentrations measured in %O_2.

Typical monitor data, such as response time, accuracy, or sensitivity, do not depend only on sensor characteristics. For example, the overall response time is composed of the sensor response time, of delays due to ambient influences such as water, of sample flow in the case of sidestream sensors, or of gas exchange time of the cell or detector volume.

23.2   Fields of Clinical Application

23.2.1   Applications

Typical clinical situations for the monitoring of gas exchange are found in all cases of possible or actual restrictions of respiration, eg,

— ventilated patients during intensive care (adult intensive care);
— spontaneously breathing or ventilated patients during anesthesia;
— treatment of premature babies in incubators (neonatal intensive care);
— respiratory therapy of acute or chronically ill patients.

The applications are distinct with respect to the quality and kind of information needed. Thus, respiratory therapy requires discrete values with larger time intervals to review the progress of a normally slow recovery (curing) process. During intensive care and anesthesia, the prevention of sudden changes in gas exchange is needed additionally, and therefore continuous measurement with an automatic alarm system will be preferred.

Ventilation means forced gas exchange with several risks that are caused by the replacement of spontaneous breathing with suppression of the natural respiratory drive, the application of increased pressures (injury to lung tissue), dehydration and possible harm from toxic concentrations of gas blends (O_2 in the case of neonates, anesthetic gases generally).

Gas analytical measurements alone are not sufficient for monitoring the gas exchange. The surveillance has to include "accompanying" parameters which reflect the physical parameters of the gas blends analyzed.

Important parameters for monitoring ventilation therapy are listed below. Some of them are mandatory in national regulations and international or national standards. Figure 23-2 shows an example.

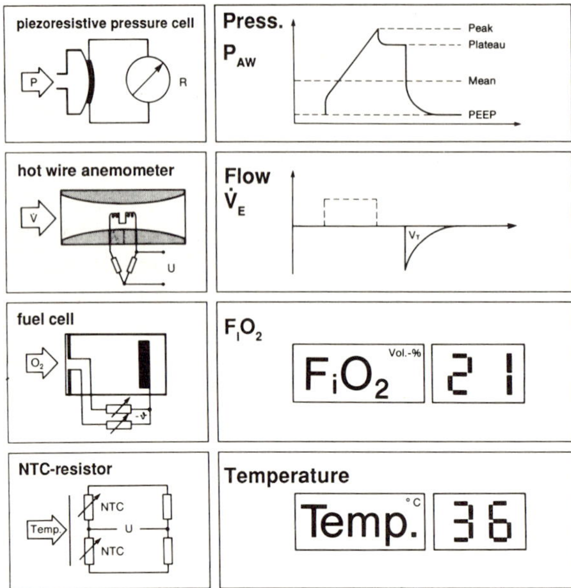

Figure 23-2.
Ventilation monitoring, eg, according to standard DIN 13254.

− airway pressure:
as a function of time, and maximum or minimum pressure values measured at distinct sites of the system consisting of ventilator, hoses, tubus, and lung tissue (alveoli);

− volume:
eg, flow, tidal volume and minute volume (tidal volume means the inhaled and exhaled breath volume; minute volume sums all breaths during 1 min);

− compliance and resistance:
of ventilator, hoses, airways and lung (compliance measures the dilatability or flexibility of lungs or hoses);

− climate:
of respiration gases, such as temperature and humidity;

− gas concentrations:
especially oxygen concentration in the gas phase, in blood, and in target organs.

During expiration, mainly carbon dioxide is an additional important parameter for monitoring the exhaled gases. During anesthesia and partly during the subsequent stay in the recovery room, the following parameters are determined: inspiratory and expiratory concentrations of O_2, N_2O, and volatile anesthetic agents, expiratory CO_2 and possible inspiratory residual concentrations of CO_2. The gas blends used depend on the kind of anesthesia administered. Gas anesthesia means typically the application of blends of O_2 and N_2O with volatile halogenated hydrocarbons. The additional monitoring of volatile anesthetic agents is aimed at detecting over- or underdosing on the one hand to guarantee a sufficient depth of anesthesia and on the other to limit the influence on blood pressure, which could affect perfusion and gas exchange in myocardium and brain.

Monitoring the humidity of respiration gases and the supervision of dehydration needs more technical expenditure. Therefore, during adult ventilation the climatization of respiratory gases is solved indirectly by use of a humidifier or by use of an artificial nose, or the water loss during short gas anesthesias is accepted. The application of a rebreathing system with a CO_2 absorber during long-term anesthesias also provides a limitation of dehydration.

The loss of water is more critical during intensive care of premature babies, as these babies show a greater extent of skin respiration with a greater loss of water. The convection heating in the incubator causes additional dehydration. Therefore, a state-of-the-art incubator for neonatal intensive care is equipped with sensors to monitor oxygen concentration, humidity, and temperature and with an automatic control to adjust these parameters by controlling the gas blends and adding water vapor.

23.2.2   Measurement of Physical and Chemical Gas Properties

Depending on the information content and the need for further processing, different measuring sites are used. Figure 23-3 presents two important sites for the location of sensors. Basically, measurements in the alveoli or in the target organ are to be preferred. However, the methods for obtaining this information are either invasive procedures or could be not applied continously. Therefore, measuring sites are selected which offer determinations with reduced risks for the patient but with accepted restrictions in accuracy and information content in relation to the severity of the disease. This follows the general trend of using more non-invasive procedures with fewer risks.

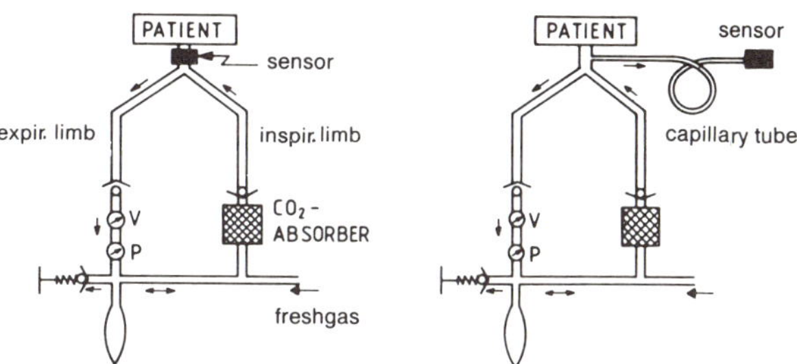

Figure 23-3. Measuring sites with a circle system for mainstream or sidestream sensors.

23.2.3   Measuring Sites

1. in the ventilator for metering, to monitor malfunctions of the device, to monitor the connections to the patient;
2. near the patient at the Y-piece in the mainstream of the air flow;
3. distant from the patient in the sidestream with a connector at the Y-piece;
4. inside the cover of an incubator in the air space above the baby.

In addition to the habits of the physician, the selection of the measuring site depends also on the special conditions of the surgical procedure, as some problems may occur, eg, during operations in a sitting position with a heavy-weight mainstream sensor tearing at the tubus, or when monitoring restless intensive care patients, or due to the small size of monitored neonates.

The analysis of the gas mixture can be carried out in the mainstream or in the sidestream. Figure 23-3 presents two examples of a circle system of an anesthesia machine with the location of the respiratory gas sensors at the Y-piece for direct measurements in the mainstream of the breathing system, or connected by a thin capillary tube or hose in the sidestream distant from the patient. (A circle system is part of the ventilation system in anesthesia machines dedicated to so-called semi-closed or closed anesthesia, which reuses the exhaled respiration gases after removing CO_2 by an absorber.)

Both procedures provide different information with different significance. Depending on the application and the clinical situation, the advantages or disadvantages of one or other procedure may prevail. In anesthesia there is a trend to prefer multiparameter monitors for concentrations of O_2, N_2O, CO_2, and agents. Therefore, a sidestream monitor may combine several separate sensors for different gases which utilize the same gas sample.

The advantage of sidestream measurements distant from the patient is the negligible obstacle of the only light-weight and non-bulky small connector for the sampling line.

During intensive care, normally only a small number of gas concentrations such as O_2 and, less important, CO_2, are monitored so that in the case of a small and light-weight sensor an inspiratory and expiratory measurement at the Y-piece could be possible. A breath-resolved recording requires a short rise time in addition. Normally the applied O_2 concentration is measured in the inspiration hose. The determination of gas exchange in the lung, of blood concentrations, and of tissue concentrations requires different sites and methods. Different sites for the same parameter yield different information. The combination of the different single values depict a detailed and useful picture. The most important measurements, their advantages, and their drawbacks are discussed below in detail.

23.3   Blood Gas Analysis

The already mentioned transportation function of blood determines different values with different meanings depending on the selected site.

Information on oxygen uptake is obtained by measuring the arterial partial pressure, paO_2, information on oxygen consumption by measuring the mixed venous partial pressure, information on the effect of ventilation by measuring the arterial partial pressure, $paCO_2$, and information on the acid-base excess by measuring the pH value. Clinical measurements of blood gases are carried out in large numbers as in vitro analyses of blood samples. The analysis is carried out by automatic instruments by use of electrochemical sensors such as Clark electrodes for pO_2 or the method according to Stow-Severinghaus using glass electrodes for pCO_2. Measured values are pO_2, pCO_2 partial pressures and pH.

Depending on the sample site in a capillary, or arterial, or venous blood vessel, the values give different information and are repeated at different intervals. Based on the assumption of standard conditions, oxygen saturation, SaO_{2calc}, can be calculated by use of partial pressure values whereas the accuracy deteriorates with shifts of the oxygen-hemoglobin affinity curve. Such effects are caused by changes in pCO_2, pH, or temperature, or the presence of dysfunctional hemoglobin types or metabolites such as 2,3-DPG.

In recent years, the optical CO-Oximeter has been increasingly used in clinical laboratories to measure directly oxygen saturation and hemoglobin types. Instruments with 5–7 wavelengths present not only the fractional saturation value but also the amounts of all types of hemoglobin such as functional and dysfunctional types. Today, blood gas analysis of samples to determine partial pressures is the so-called "gold standard". Nevertheless, considerable errors occur which depend on mistakes during sampling or transportation, especially storage of the samples, use of plastic syringes, etc. In addition, the procedures of arterial puncturing is time consuming and risky.

To reduce the invasivity, numerous supplementary methods have been evaluated to determine their usefulness in replacing some invasive methods or obtaining additional data. Some of these tested methods involve transcutaneous sensors for pO_2 and pCO_2 partial pressures in the tissue ($tcpO_2$ and $tcpCO_2$) and transcutaneous measurements of oxygen saturation in peripheral vessels or invasive intravasal measurements of oxygen saturation.

23.4. Transcutaneous Sensors

23.4.1 Transcutaneous Gas Sensors for pO_2 and pCO_2 Measurements

The interesting target organs the brain and myocardium are not directly accessible to non-invasive methods. Therefore, auxiliary parameters are chosen whose measured values provide indirect information on gas exchange in these organs if certain conditions are considered. Assuming physiologically stable circulation conditions, muscle and skin tissue are provided with sufficient oxygen in such a way that the partial pressure values reflect the general health state.

23.4.1.1 *Fundamentals of $tcpO_2$ and $tcpCO_2$ Sensors*

Transcutaneous measurement of partial pressures is based on the gas permeability of human skin. An electrochemical sensor is placed on the skin which is heated to increase arterial blood in superficial blood vessels [1].

The electrochemical measurement of pO_2 by use of a polarographic Clark cell offers the advantage of designing small and compact sensors which show a linear response to oxygen partial pressure. Figure 23-4 presents a sectional view of a combined transcutaneous sensor for $tcpO_2$ and $tcpCO_2$.

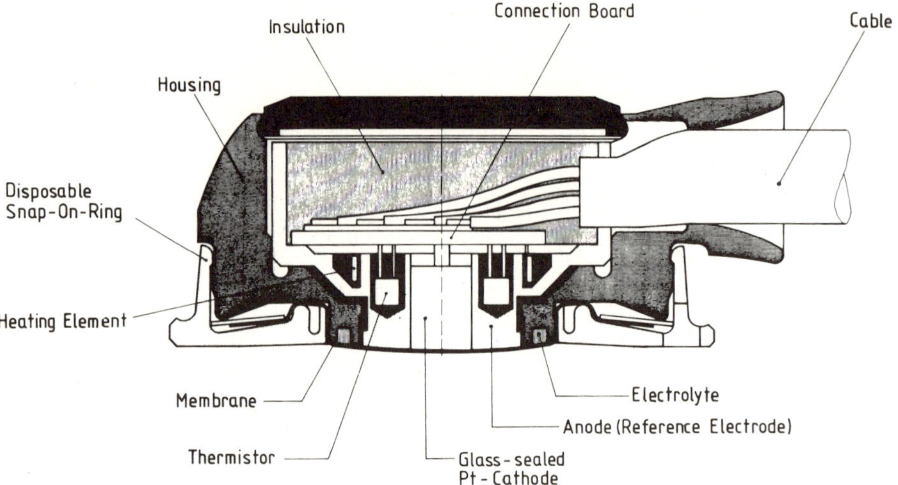

Figure 23-4. Sectional view of transcutaneous pO_2/pCO_2 sensor (Dräger).

Oxygen diffuses through the synthetic membrane to a negatively polarized platinum electrode to be electrochemically and quantitatively reduced. The membrane acts as a diffusion barrier to provide a linear response of the current with respect to the pO_2 partial pressure. The electrical signal is measured relative to an Ag/AgCl electrode as anode [2].

The measurements of CO_2 according to Stow-Severinghaus utilizes a separate glass electrode based on the reaction of diffused CO_2 with water and determines the resulting pH shift. Figure 23-5 shows the sectional view of a combisensor for $tcpO_2$ and $tcpCO_2$, which consists of two separate compact sensors for O_2 and CO_2 and which includes an integrated

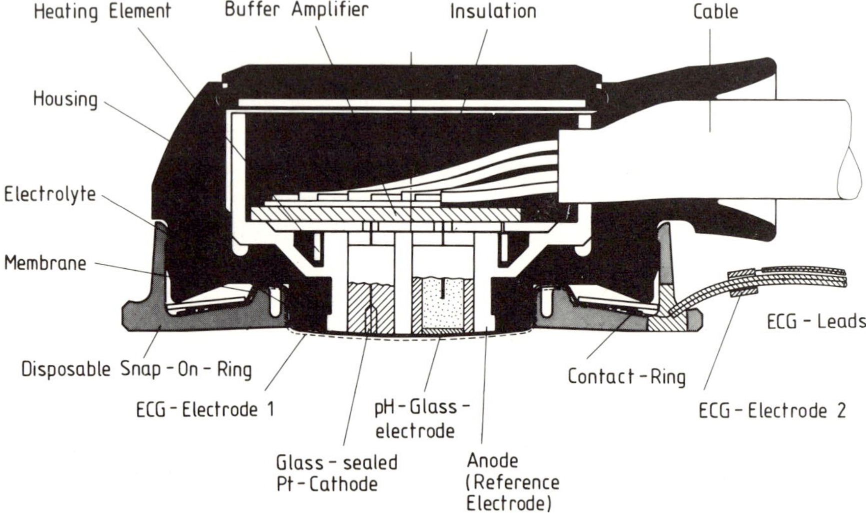

Figure 23-5. Combisensor for pO_2, pCO_2 and heart rate (Dräger).

miniaturized preamplifier and an additional ECG electrode. The high resistance of miniature glass electrodes requires the integrated preamplifier to reduce signal interferences [3].

The ECG electrode is deposited as a microporous gas-permeable gold layer, which can be easily penetrated by the blood gases or tissue gases. Figure 23-6 shows the application of this combisensor to a neonate.

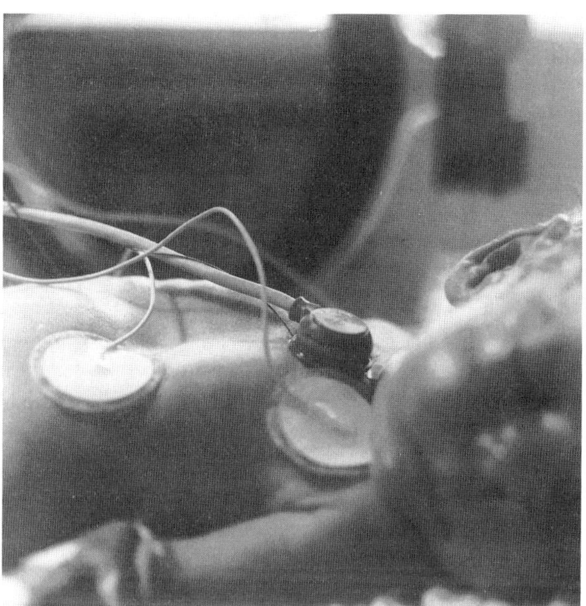

Figure 23-6.
Dräger combisensor applied to a neonate with additional ECG electrodes to monitor heart rate.

Table 23-1. Typical specifications of transcutaneous combisensor [5].

Weight	5 g
Lifetime	> 1 year
Warm-up time	30 min after membrane changes 10 min hyperemization time for neonates
O_2	
Response time ($T_{0..90}$)	10 s
Drift	< 0.5 %/h
Cross sensitivity	
70 vol% N_2O	10 Torr
0.8 vol% halothane	20 Torr
CO_2	
Response time ($T_{0..90}$)	40 s
Drift	< 1%/h
Cross sensitivity	
70 vol% N_2O,	
100 vol% O_2,	
4 vol% halothane	< 3 Torr

This type of combisensor offers the change of reducing the number of electrodes needed to monitor a neonate with its small body surface and very sensitive skin by integrating a $tcpO_2$, a $tcpCO_2$ and one of the ECG electrodes needed to monitor heart rate [4].

This combisensor developed by Dräger is under clinical evaluation. Table 23-1 shows typical specifications.

23.4.1.2   Clinical Behavior and Interpretation

The value of the partial pressure measured at the skin surface depends in a complex way on blood partial pressure, constitution of the skin, local perfusion, metabolism in the associated tissue, cardiac output, and application temperature. An increased temperature of 43 °C raises the gas permeability and expands the capillary vessels of skin which are filled with more artial blood. The local hyperemia has the disadvantage of limiting the application time at a certain site. Assuming stable circulation conditions, transcutaneously measured values correlate with arterial partial pressure by a factor of 1.2 (neonates) to 1.0 (small children) [1]. The measured value for adults proved to be very unreliable. In the case of unstable conditions or shock with a reduction of peripheral blood flow, the transcutaneous value drops very early. Inconvenience in routine use is caused by long preparation times of the sensor, the need for periodic membrane changes, the long run-in time of freshly prepared sensors, the necessity for periodic calibrations and the slow response time to changes in partial pressure.

Additional risks are caused by the heat, which may burn the very sensitive skin of babies in case of defects or too long application times, and the requirement for potentially abrasive adherence to the skin [6].

For use during anesthesia with O_2-N_2O gas blends, these sensors show a considerable cross sensitivity to N_2O and agents which prevents their reliable clinical use.

23.4.2   Transcutaneous Sensors for Oxygen Saturation

Oxygen is transported in the blood in two physical forms, freely dissolved in the plasma water and bound reversibly to hemoglobin within the red blood cells. Under normal conditions of breathing ambient air only 2% of the total blood oxygen content is dissolved. The relationship between blood partial pressure of oxygen and the amount of oxygen bound to hemoglobin is commonly expressed as "oxygen-hemoglobin affinity". Shifts of this S-shaped curve control the uptake and release of oxygen at different sites in the body. A knowledge of the amounts of functional and dysfunctional kinds of hemoglobin simplifies the determination of saturation and oxygen content of blood. The measurement of arterial saturation gives information on oxygen content or the amount of oxygen available to the organs. Under shock conditions, the oxygen supply of less important organs is reduced to the benefit of the brain and myocardium. An additional but invasive measurement of venous saturation provides data for determining oxygen utilization.

23.4.2.1 Fundamentals of Pulse Oximetry

In vivo, non-invasive O_2 saturation relies on the transcutaneous measurement of the absorption of specific wavelengths of red and infrared light by hemoglobin and oxyhemoglobin as the light passes through skin, tissue, and blood. Figure 23-7 shows the absorption spectra of hemoglobin and the pulsatile absorption in tissue to provide the signal for the evaluation of saturation [7].

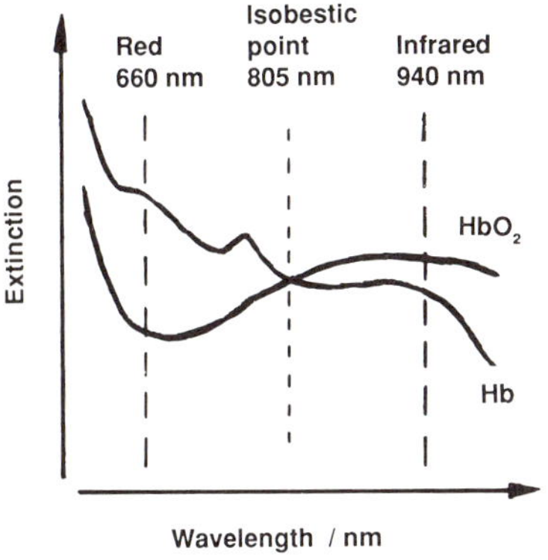

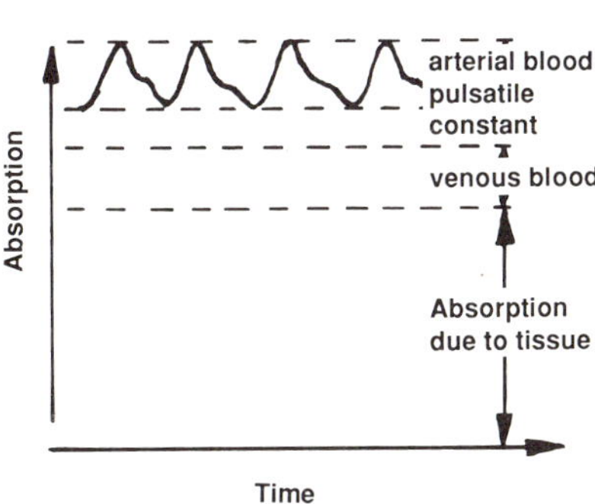

Figure 23-7. Pulsatile absorption in tissue and blood.

To measure arterial O_2 saturation and to separate the different parts of the signal, pulse oximeters record the transmitted light with reference to the arterial pulse of blood. A constant absorption signal is generated by pigmented skin, tissue layers, bones, and venous blood. Pulsating arterial blood flow causes alternating amounts of blood to fill the vessels, which leads to an absorption signal that is variable with time. This alternating signal is evaluated to calculate saturation. By alternating the transmission of red and infrared light, the measured signal can be normalized. The calibration is based on a generalized characteristic which take into account the difference between absorption in blood solutions and in tissue [8]. Table 23-2 shows typical specifications.

Table 23-2. Typical specifications of pulse oximeters [9].

Measurement	pulsatile absorption at about 660 and 940 nm
Calibration	functional or fractional saturation
Measuring sites	finger, toe, ear nose, forehead (reflectance) foot, hand (pediatric, infant probes)
Probe types	finger-clip, flex-probes
Useful range	80–100% saturation
Interference	low perfusion motion, ambient light

23.4.2.2 Clinical Experience

Normally, sensors of pulse oximeters are applied at peripheral vessels in the finger, toe, or ear lobe. Their signal depends also but to a minor extent on the circulation situation. In the case of shock or centralization, one manufacturer recommends the use of sensors applied at the bridge of the nose to gain access to a side branch of a brain-supplying vessel.

Pulse oximetry offers fabourable advantages over tcpO$_2$ monitors by use of unheated sensors, by simple handling, by possible applications to dermal edema, and by the combination with the blood pulse for monitoring also in cases of bad perfusion.

Figure 23-8 presents a variety of typical sensors for different application sites.

Limitations of the method are caused by misreading of dysfunctional hemoglobins which do not contribute to oxygen transport. These differences can be expressed by the use of different terms. The reading of pulse oximeters based on the functional part of hemoglobin is called "functional saturation", $SaO_{2\,func}$, and the reading of a CO-Oximeter is called "fractional saturation", $SaO_{2\,frac}$, relying on the total amount of hemoglobin. To distinguish between saturation measured by a pulse oximeter or by determinations of arterial blood, the term SpO_2 has been proposed in addition.

In clinical practice, this distinction offers less value. The importance of pulse oximetry lies in the possibility of a simple to use method for monitoring hypoxemia, which can be used with

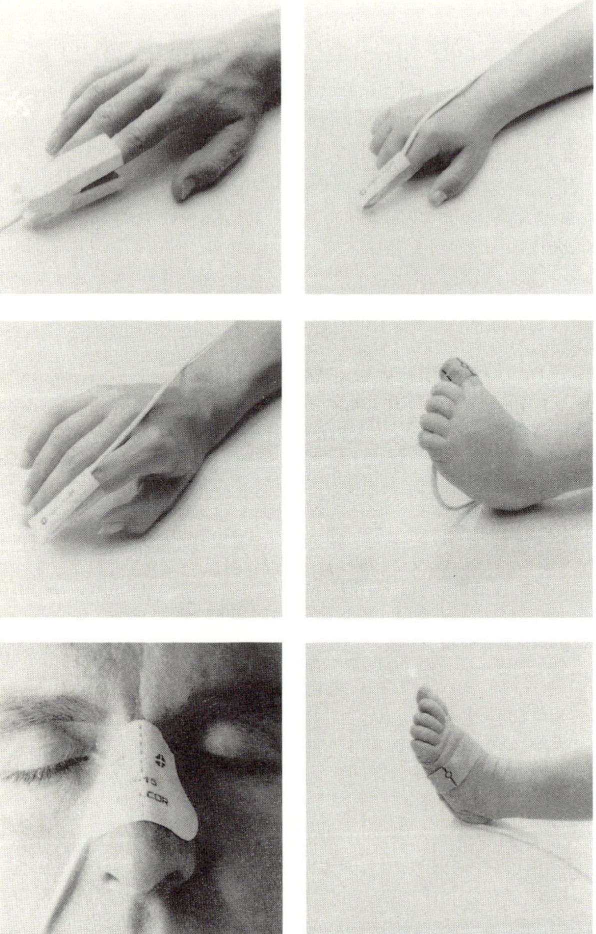

Figure 23-8.
Sensors for pulse oximeter (Nellcor).

patients normally not monitored. By introducing pulse oximetry into clinical use, it has been possible to identify how often and unexpectedly hypoxic or hypoxemic states occur during normal procedures of intensive care and anesthesia. Possible interference from dyes applied to determine cardiac output are of only minor clinical importance.

23.5   Physical Properties of Respiration Gases

Basically, the measured value and the parameter of clinical interest have to be distinguished, whereas the clinical parameter may be calculated by use of one or more measured values. The physical variables pressure, flow, and temperature yield information of different kinds: curves as a function of time, distinct values measured at certain moments of respiration phases, and derived values as presented in Figure 23-2.

Pressure:
pressure curve, peak, mean, plateau and endexpiratory
pressure;

Temperature:
breathing gas temperature, humidity;

Flow:
flow curve, breath volume, minute volume, compliance, resistance.

23.5.1   Pressure Sensors

Piezoresistive sensors automatically normalized to ambient pressure are used as clinical pressure sensors today. They have to be protected against water by use of a filter in the sensing line. The installation is set up in the sidestream by means of a thin hose as the sensing line in which only a negligible flow occurs. Electronic ventilators utilize additional pressure sensors for device control.

23.5.2   Temperature Sensors

In general, thermistors with a linearized characteristic are used to measure gas temperature.

23.5.3   Flow and Volume Sensors

23.5.3.1   Methods and Clinical Applications

Several physical methods are used to measure flow and volume. Some of them offer the advantage of a mechanical and sturdily built design and can also be used with flammable gas blends. The accuracy of these devices is limited and usually they are lacking an electrical output signal. Table 23-3 shows a selection of methods and devices arranged in three groups according to the basic procedure depending on measurement of volume, flow vector, or transportation properties.

Table 23-3. Methods and devices for flow and volume.

Method	Device
1. Volume-dependent methods (mainly mechanical or pneumatic)	
Bellows with extension measurement	metering for ventilators
Volumeter	flow meter for circle systems
Gas meter	gas flow in general
Piston	passive spirometer
	active ventilator
2. Flow-dependent methods (mainly pneumatic)	
Rotameter	gauge tubes for anesthesia machines
Turbine	spirometer
Pressure drop (Fleisch, Pitot)	spirometer
Deflection of a vane (electrical measurement of deflection)	ventilator
3. Dependent on transport properties (electronic)	
Vortex (counting of vortices)	ventilator
Hot wire anemometer	ventilator
Doppler ultrasound	spirometer, ventilator

23.5.3.2   *Fundamentals of Hot Wire Anemometer*

All clinically used methods to measure flow and volume show adequate accuracy. A higher performance is required in the case of processing with other parameters, eg, gas concentrations. For this purpose, only methods with an electrical output and a short response time are suitable, such as the hot wire anemometer or the use of Doppler ultrasound. Figure 23-9 shows a special kind of temperature difference hot wire anemometer. Two platinum wires as parts of an electronic bridge circuit are used to measure the heat loss due to gas flow and the temperature of the flowing gas. In practice, the current to maintain a constant temperature difference is measured. The heat loss is approximately proportional to the square root of the flow. Depending on the low heat capacity of a thin wire of about 10 µm diameter, a fast time response to resolve changes up to 100 kHz and a broad dynamic range can be obtained [10].

This type of sensor has been used clinically for about 10 years as disposable sensor. The response time of 10 ms depends on the selected design. By modification of the geometrical arrangement of the two platinum wires and by adding a larger "shadow wire" in the proximity of one wire, the electronic interface can determine the flow direction and distinguish inspiration from expiration. This sensor type is applied to mainstream measurements during neonatal ventilation and is located between the patient tube connector and Y-piece. Figure 23-10 shows a sectional view and a photograph of the application. Two fine-meshed sieves on both sides of the sensor element prevent mechanical damage to the wires and deposition of mucus or secretions on the hot wires. Tables 23-4 and 23-5 show typical specifications of both sensor types.

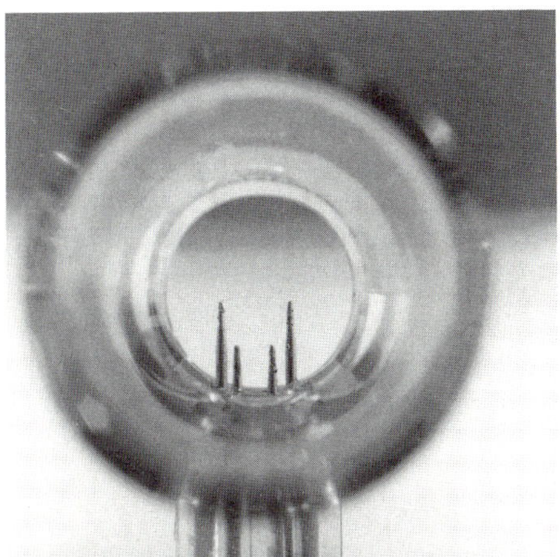

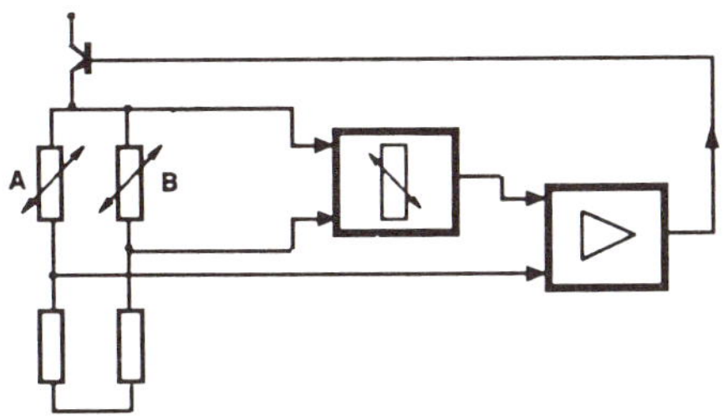

Figure 23-9. Measuring principle and view into a Spirolog sensor (Dräger); two hot wires (A, B) to determine flow, temperature, and gas blend components.

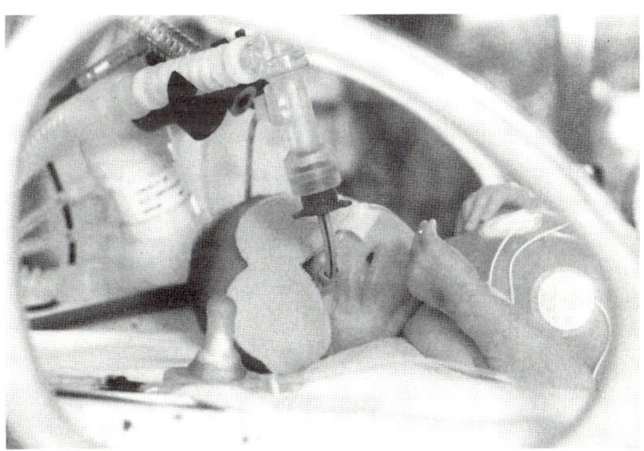

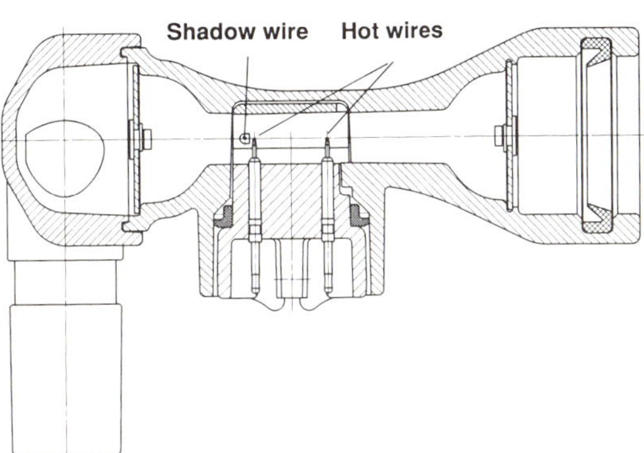

Figure 23-10.
Neonatal flow sensor (Dräger): cross section and application; two hot wires and a shadow wire to determine flow and flow direction.

Table 23-4. Typical specifications of Spirolog flow sensor (Dräger) [10].

Measurement	flow
Type	calibrated disposable sensor
Flow range	0.2–200 L/min
Accuracy	$<5\%$
Reproducibility	1%
Time resolution	10 ms
Gas mixture	compensated
Modification	flow direction sensitive, for neonatal monitoring

Table 23-5. Typical specifications of neonatal babylog 8000 flow sensor (Dräger).

Measurement	flow, flow direction sensitive
Type	disposable sensor
Flow range	0.017–56 L/min
Inaccuracy	$<10\%$
Time resolution	10 ms
Dead space	1.6–3.4 mL depending on tube connector
Temperatur drift	$0.8\%/°C$
Resistance of Y-piece incl. sensor	approx. 5 mbar/(L/s) at 10 L/min

23.5.4   Processing of Display Values of Pressure, Volume, Compliance, and Resistance

Depending on the physical method used, continuous values of flow and/or pressure are recorded at the measuring site. However, the waveform of the pressure curve displayed on a screen offers only part of the information needed. In addition, the pressure values during certain respiration phases are required, calculation of which is based on an analysis of the waveform with additional control signals of the ventilator or by use of the flow signal.

Typical examples are the following:
— peak pressure as the maximum pressure value during inspiration which should be limited to prevent damage of the lung tissue and which is used to generate stenosis alarm;
— plateau pressure, a positive pressure with no flow at the end of inspiration which is used to obtain an improved distribution of fresh gas in poorly ventilated parts of the lung;
— endexpiratory pressure with no flow at the end of expiration, which is used with positive values as PEEP to increase the active surface in the lung for gas exchange during certain phases of intensive care;
— minute volume, tidal volume, which are flow-derived values to control the ventilation of the patient Cand to set the ventilator.

More extended calculations are needed to derive the values of airway resistance and compliance of the lung. Processing of compliance of the ventilator and the hoses to the patient are used to compensate for their influence on applied volumes especially in the case of small tidal volumes. Compliance and resistance of airways and lung are processed by use of pressure and flow as functions of time. Compliance gives an optimization parameter to adjust PEEP as the best compromise between lung distension to increase the number of ventilated compartments in the lung and the generated obstruction of pulmonary blood circulation.

23.6 Oxygen Sensors

Maintaining the oxygen supply of the patient and prevention of hypoxic or hypoxemic states are fundamental tasks during intensive care and anesthesia. Monitoring of selected vital parameters by means of reliable methods combined with automatic alarm systems is used to prevent underdosing and, in the case of neonatal intensive care, also overdosing with oxygen. Whereas parameters such as saturation provide data on the amount of oxygen transported by the blood and in the case of an incident give global information on an oxygen supply that has been interfered with in some way, measurements in the gas phase allow the monitoring of gas compositions and in the case of an incident an early warning before the oxygen reservoirs of the gas volumes in the ventilator and hoses are exhausted.

Oxygen sensors for measurements in the gas phase are required for incubators, ventilators, and anesthesia machines where monitoring of the gases administered to the patient is essential. These applicational conditions determine the required properties. The response time of possible concentration changes in a circle system depends for example, on the degree of rebreathing, on the fresh gas flow, and on the gas volumes of the ventilator and hoses and amounts to several minutes. Therefore, a reasonable response time of oxygen sensors should be about 10–30 s. In addition, oxygen sensors have to operate also in the presence of water (liquid or vapor), of carbon dioxide, and of anesthetic agents. Accuracy and stability of the method have to comply with minimum requirements, eg, standards such as ISO 7767 [11], DIN 13252, or DIN 13254. Depending on the fields of application, a low cross sensitivity and high stability vs. the above-mentioned gases are required. The sensor lifetime is important with regard to two points:

1. Application time and storage time are important for the organization of the replacement of exhausted sensors to handle the supply.

2. Cost calculation is based on the purchase and maintenance costs of monitors, replacement costs of sensors, and material consumed during the period of application.

23.6.1 Review of Methods of Measuring O_2 Concentration

Mainly three types of oxygen sensors are used in clinical applications. Whereas the two electrochemical types have the largest market shares [12], paramagnetic sensors as the third type are important in combination with IR sensors for sidestream multi-gas analysis in anesthesia monitoring. Real multi-gas sensors such as mass spectrometers or Raman spectrometers are of minor importance.

23.6.2   Electrochemical Sensors

Electrochemical sensors include amperometric cells, especially galvanic fuel cells, and polarographic cells. Both types are available from several manufacturers world-wide in large numbers for clinical use. Higher stability with fewer technical problems makes the galvanic cell currently the most popular oxygen sensor in the operating room environment [12].

23.6.2.1   Fundamentals of Polarographic Cells

The working principle of the polarographic cell according to Clark has been already mentioned in Sections 3 and 4. Figure 23-11 shows the simplified set-up of a cell with reference electrode.

Oxygen diffuses through the membrane into the sensor and is reduced at the cathode. The current is porportional to the O_2 concentration and is temperature dependent. Table 23-6 gives typical specifications of this type of sensor.

Polarographic sensors contain no anode which is consumed during application, but the water loss from the electrolyte and the poor stability vs. anesthetic gases limit the lifetime. Response times are typically 10 s or more since an increased membrane thickness to reduce water loss increases the response time. A wide variety of types exist to fit various applications.

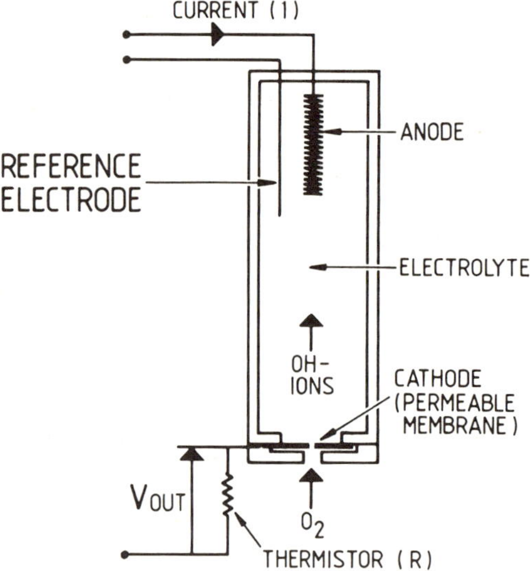

Figure 23-11. Polarographic oxygen cell.

Table 23-6.   Typical specifications of polarographic cells [13].

Response time	3–15 s (depending on membrane thickness)
Lifetime	6 months (electrolyte and membrane can be replaced by skilled staff)
Cross sensitivity	
N_2O and halothane	significant

The characteristic is linear but during application sudden shifts may occur. Therefore, periodic calibrations are essential. Long exposure to N_2O causes absorption and swelling, which lead to destruction of the sensor.

23.6.2.2 Fundamentals of Fuel Cells

Fuel cells have a guaranteed application time of about 12–15 months, the thickness of the membrane reflects a compromise between response time and reduction of water loss. Various brands show differences in quality, eg, application time, accuracy, and reliability, and in price.

The Oxycom cell (Dräger) has a special feature, as can be seen from the sectional view shown in Figure 23-12. Oxygen molecules diffuse through the membrane into the electrochemical cell and are reduced at noble metal electrodes. At the same time, a lead electrode is oxidized. The output signal is temperature dependent, which is compensated for by means of thermistors in the sensor housing.

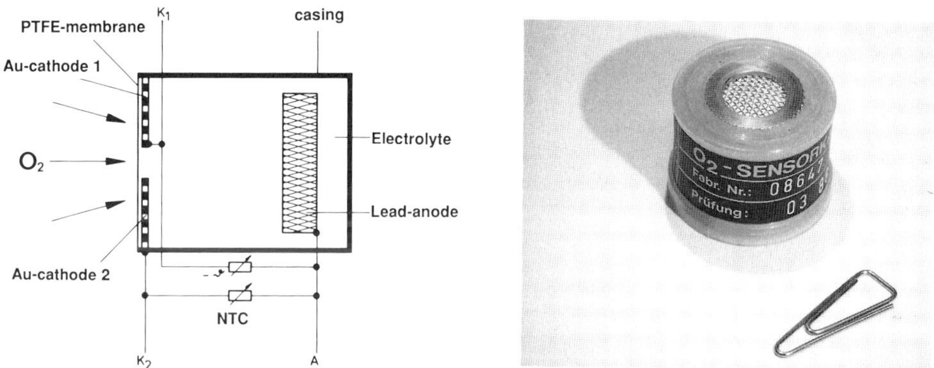

Figure 23-12. Oxygen fuel cell with two cathodes, Oxycom sensor (Dräger).

The sensor consists of two half-cells with separate cathodes which offer a redundancy to cover the most frequent failures by comparing the output signals continously in the monitor. In the case of signal differences larger than preset values, the monitor initiates an alarm. In addition, careful and tight process control during manufacturing results in a precisely defined characteristic which offers an accuracy of better than 1% in the case of a two-point calibration. The metal net prevents lifting of the membrane which would increase the diffusion path for oxygen. This feature guarantees almost constant response times of less than 20 s.

A tight sensor casing reduces additional water loss and increases the application time [14]. Table 23-7 gives typical specifications of a fuel cell.

A storage time of more than 18 months without a reduction in lifetime is obtained by a special packing that stores the sensor in N_2. The quaranteed lifetime is characterized by the product of the exposure time and concentration of O_2 and Co_2 — the so-called O_2-%-hours or CO_2-%-hours, respectively.

Table 23-7. Typical specifications of fuel cell Oxycom sensor (Dräger).

Accuracy	less than 1 vol% O_2
(two-point calibration)	(range 5–100 vol% O_2)
Linearity error	$<2\%\,O_2$
Temperature error	$<3\%$ of measured value (range 15–40 °C)
Response time	<15 s
Lifetime	1.5 years
Cross sensitivity 70% N_2O, 2% CO_2, 4% halothane	$<1\%$ full scale

23.6.2.3 Clinical Experience

The main application fields, ventilation and anesthesia, differ in the kind of gas blends and procedures used. Rebreathing or circle systems with long response times to concentration changes require a wash-out time of several minutes to establish defined gas concentrations for sensor calibration. A practical approach is to detach the oxygen sensor from the anesthesia machine by use of plug-in connectors and calibrate it in ambient air after a few minutes delay time. This procedure of a one-point calibration results in an inaccurancy of less than 5% in the range of 100% oxygen.

23.6.3 Paramagnetic O_2 Sensors

23.6.3.1 Fundamentals of Paramagnetic Sensors

During the development of multiparameter monitors for anesthetic gases, paramagnetic sensors have been designed which have the features of good linearity and fast response times. A well-known sensor operating as a dynamic difference sensor is produced by Datex. Figure 23-13 shows the principle of the dynamic paramagnetic cell based on the magneto-acoustic effect.

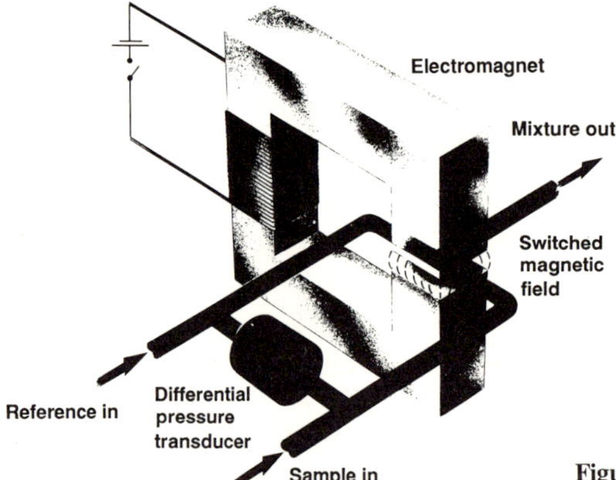

Figure 23-13. Paramagnetic oxygen sensor (Datex).

The sample and reference gases are mixed within a homogeneous magnetic field between the poles of an electromagnet. The switching frequency of about 100 Hz generates a pressure signal that is proportional to the oxygen content difference, which can be picked up by a differential microphone. To avoid magneto-mechanical effects, the whole sensor has to be carefully designed. Second-order pneumatic filters reduce pressure effects. Water and water vapor have to be prevented from entering the measuring cell, as accidental pumping of water into the sensor will cause irreversible damage [15]. Table 23-8 gives typical specifications of a paramagnetic O_2 sensor.

Table 23-8. Typical specifications of paramagnetic O_2 sensors.

Dimensions	65 × 75 × 130 mm
Weight	0.8 kg
Response time (sensor) T_{10-90}	150 ms at 100 mL/min
Response time (monitor)	450 ms
Cross sensitivity	<2% for 100% N_2O
Drift	<2%/day

23.6.3.2 *Clinical Experience*

The sensor is applied in the sidestream, with a requirement to separate water to prevent unnoticed shifts of the measured value due to aspirated water. The main advantage is the possibility of an easy in-line combination with IR optical sensors which allow the essential anesthetic gases to be measured with one monitor in the sidestream. The fast response of this arrangement cannot be used to calculate derived parameters as the application in the sidestream generates undefined and changing delay times.

23.6.4   Other Oxygen Sensors

Multi-gas sensors are based on mass and Raman spectrometric procedures (see also Section 10), of which up to now only mass spectrometers have been applied in clinical practice with certain acceptance. They are used mainly in North American hospitals with central instruments and long tubes to up to twenty patients or operating rooms. In recent years less expensive stand-alone systems have become available for use with a single patient at a time.

23.6.4.1 *Zirconium Oxide Sensor*

ZrO_2 sensors, which are used in process control and automotive monitoring, are of very low significance in the medical field. The sensor must work at temperatures of about 800 °C owing to the requirement for high conductivity of the solid electrolyte together with a rapid attainment of equilibrium at the electrodes. The sensor offers a high dynamic range and a response time of 100 ms. The application is limited to metabolic monitoring only since, eg, monitoring of anesthetic agents does not allow such high temperatures. The reason is not the potential flammability but the probable decomposition of agents by hot surfaces and the intoxication of patients from rebreathing systems.

23.6.5   Clinical Interpretation

The most important parameter in monitoring oxygen administration is the inspiratory concentration, FiO_2, to prevent mishaps and incidents. A secondary task is to prevent over- and underdosage of oxygen especially in neonatal care. Additional information on lung function can be derived by using time-resolved wash-in and wash-out curves. Some users state that similarly to CO_2, the difference in FiO_2 and FeO_2 gives information on oxygen uptake during anesthesia. However, there are completely different relationships between gas content in blood and partial pressure of O_2 or CO_2. CO_2 is characterized more or less by a linear relationship. Oxygen content is characterized by the typical S-shape behaviour of hemoglobin affinity, which when breathing normal air causes only 2% to be physically dissolved. Applying increased partial pressures of oxygen adds only small amounts of oxygen to the content of blood. Therefore, difference values between inspiratory and expiratory partial pressures or fractions of oxygen are strongly dependent on the actual partial pressure and show no practical significance with regard to oxygen uptake. However, oxygen sensors with fast response times offer the possibility of recording the O_2 waveform breath by breath with sufficient time resolution. The interpretation of these curves is more difficult than that of CO_2 curves, particularly in the case of application of an increased oxygen concentration during anesthesia. Oxygen uptake requires the processing of amounts by means of a fast and undelayed flow signal and a direct and fast O_2 measurement.

New developments in small and light-weight O_2 fuel cells may allow future modifications for mainstream sensors with fast response times. Figure 23-14 shows a prototype of such cells, which are intended in the first instance to replace today's heavy mainstream sensors with full redundancy for use in monitoring FiO_2 or for monitoring inspiratory and expiratory oxygen concentrations simultaneously.

Figure 23-14. New type of a small oxygen fuel cell (Dräger).

23.7   Carbon Dioxide

In clinical applications, three different types of sensors are used to determine CO_2 partial pressures, depending on the measuring site. Figure 23-15 shows the carbon dioxide transport from the tissue to the lungs [16]. Partial pressure values at all three sites are used for clinical monitoring.

For measurements of partial pressures in tissues and in liquids such as serum or blood, sensors according to Severinghaus are used. A membrane separates the external liquid or gas phase from the electrolyte surrounding a glass electrode, the potential of which is measured against an Ag/AgCl electrode (see Sections 23.3 and 23.4).

For measurements in the gas phase, sensors based on absorption of infrared radiation by CO_2 molecules are preferred.

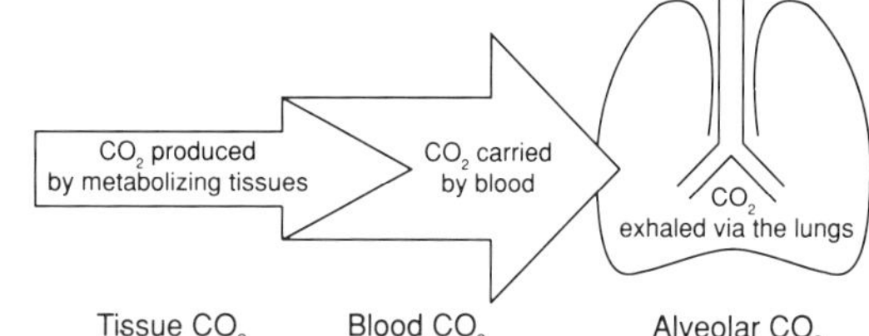

Figure 23-15.
Carbon dioxide transport (courtesy of Nellcor).

23.7.1   Fundamentals of IR Optical Sensors

The fundamental principle of CO_2 IR sensors is based on a "light barrier". Figure 23-16 shows the fundamental method of CO_2 and N_2O measurement.

A light source irradiates the gas to be analyzed and a detector records the transmitted light. The absorbing molecules diminish the light intensity passing through the gas cell [17]. By selecting appropriate wavelengths to irradiate the gas sample or by spectral dispersion of transmitted light and by recording the modified intensities, a CO_2-specific signal is obtained. According to the Lambert-Beer law, the reduction in light intensity is an exponential function of the number of CO_2 molecules per unit volume, of the absorption or extinction coefficient, and of the absorption length. The advantage of this method is that all triatomic and polyatomic molecules in the gas phase show distinct absorption bands due to rotational transitions of the molecules. By use of this method, gases such as CO_2, N_2O, volatile anesthetic vapors, water vapor, ethanol, and acetone are detectable, although the last two are of minor clinical significance.

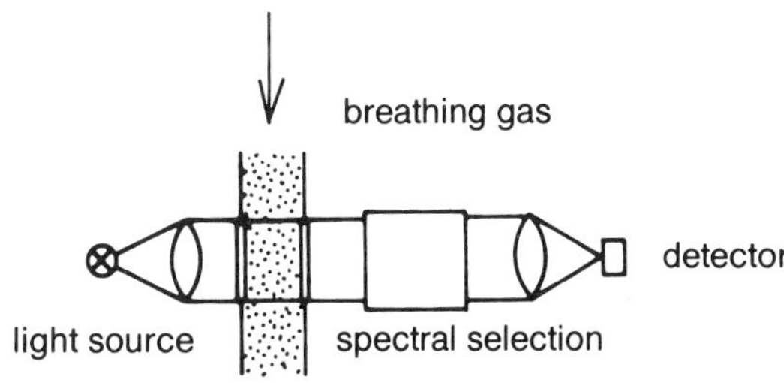

Figure 23-16.
Principle of IR sensor with black-body source.

Historically, the first sensor to be used clinically, more than 10 years ago, was based on a nonoptical detector involving two sensor absorption chambers with a membrane in between (see Figure 23-17). The irradiating light is split into two beams, one passing through a reference gas cell and the other through the measuring cell. The transmitted radiation from each channel is absorbed in the appropriate sensor chamber. Depending on the transmitted energy, the pressure increases and moves the membrane. The deflection of the membrane is recorded electrically [18]. Today's IR sensors use electro-optical or pyro-electrical photo-detectors.

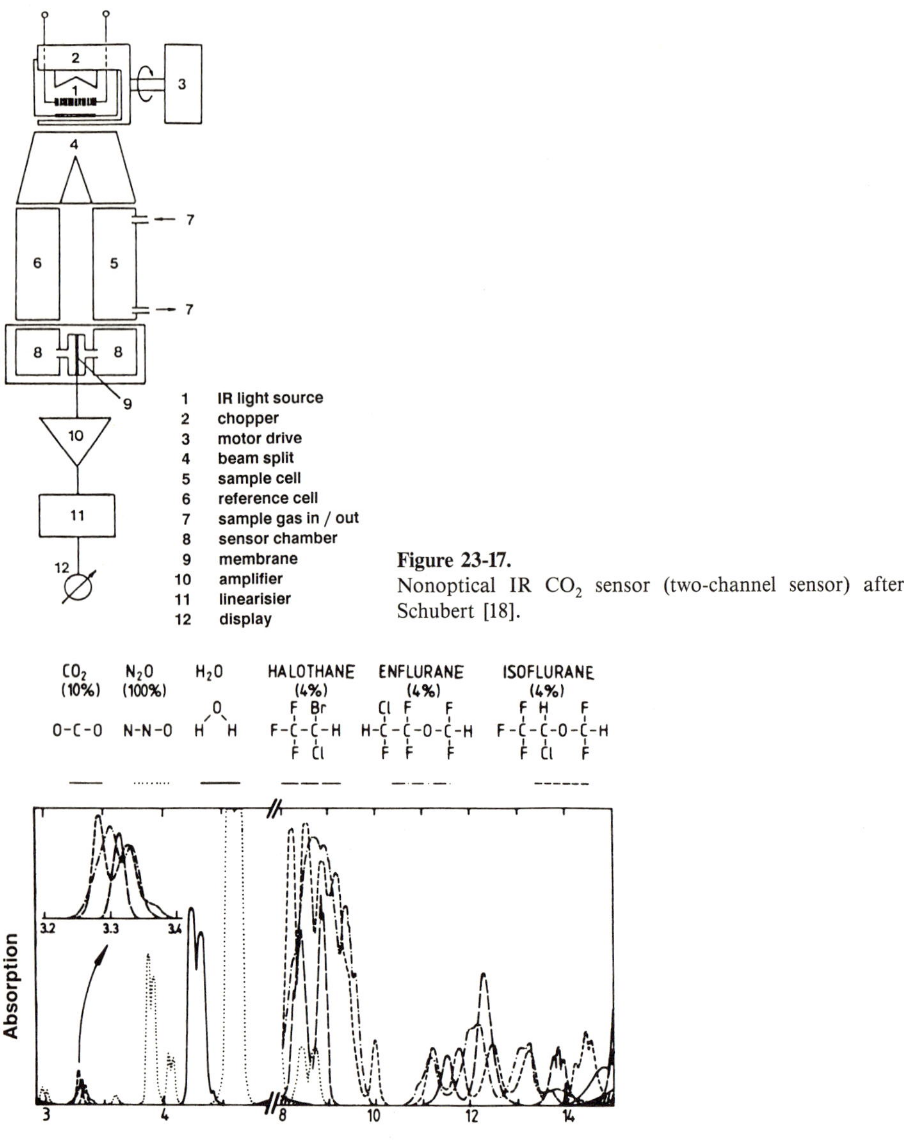

Figure 23-17.
Nonoptical IR CO_2 sensor (two-channel sensor) after Schubert [18].

Figure 23-18. Infrared spectra of respiratory gases.

Figure 23-18 presents typical absorption spectra with the essential bands of CO_2, N_2O, anesthetic agents, and water vapor. Halogenated hydrocarbons such as Halothane, Ethrane, and Isoflurane are used as anesthetic agents [19].

These gases and vapours are detectable separately in all cases when at least one absorption band is available in a spectral region where no interference from bands of other components of the sample occur. In addition, dispersive elements such as interference filters with sufficiently steep cut-off regions and fast detectors have to be available. As an optical method, several effects may interfere with the components of the sensor and influence the signal, such as deposits on optical surfaces, drift of electro-optical parts, and aging of the light source. Therefore, technical and application-related modifications are required in order to achieve a reliable measurement. In addition, the simultaneous detection of several gases requires the feature of modifying the molecule-specific dispersive elements.

23.7.1.1 *Design of IR Sensors*

Figure 23-19 shows the configurations of a one-channel detector and a two-channel detector with a filter-wheel which offers the possibility of determining additional components. The special features caused by the application-related location in the mainstream or sidestream have to be added, including the effects on clinical interpretation.

Figure 23-19a shows the fundamental structure of the light barrier with a pulsed light source to be used in the case of simple interface electronics such as a lock-in-amplifier. The filter for the spectral dispersion is placed in a parallel light beam. The band pass filter for the 4.26 µm region can consist of two components. The one-channel method either works without periodic automatic calibration or calibrates on CO_2-free inspiration gas as, eg, during normal ventilation in intensive care. Zero-point calibration uses gas blends without CO_2. Full-scale or span calibration is carried out by use of specific test gas blends.

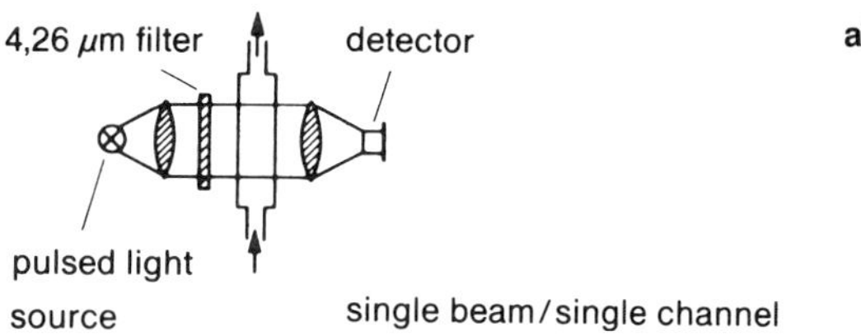

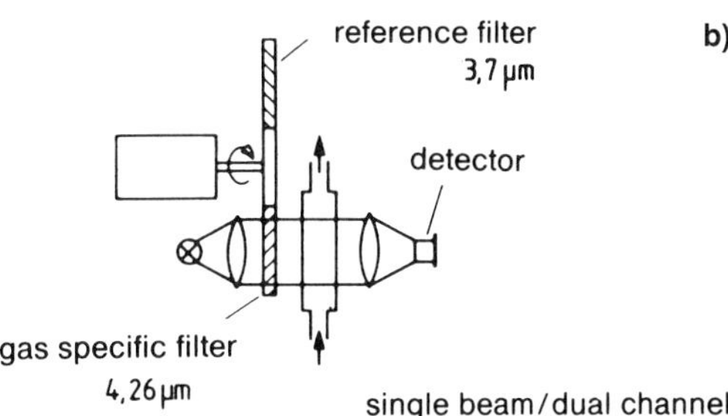

Figure 23-19.
Sectional view of different types of IR sensors: (a) one-channel sensor; (b) sensor with filter-wheel.

An absolute calibration carried out periodically requires a second optical channel, which can be performed by an optical parallel channel as displayed in Figure 23-21, or by use of one filter in a filter-wheel (see Figure 23-19b or Figure 23-20) [20].

The arrangement in Figure 23-21 (see below) has no moving parts and is suitable for compact sensors which are less shock sensitive than arrangements with a filter-wheel. The filter-wheel offers the analysis of additional gases with only one optical detection channel to avoid additional detectors with differing drifts. A filter placed in a spectral region free from absorption, typically at 3.7 µm, calibrates the total sensor for all gases. This solution, like the sensors from Nellcor or Datex, offers benefits for multi-gas sensors with additional gas-specific filters.

The increased volume of a filter-wheel sensor and the higher technical expenditure are of minor importance in the case of arrangements in the sidestream, which are preferred for today's clinical use [21]. A defined or negligible delay between CO_2 and flow signal is required only in special applications such as the determination of the total amount of exhaled carbon dioxide or the calculation of the dead-space volume, which is of interest during ventilation. Current trends in clinical application are to use, during anesthesia monitoring, multi-gas sensors in sidestream and, just beginning, during intensive care, CO_2 mainstream sensors of the "second" generation.

Mainstream sensors in clinical use are offered by companies such as Hewlett-Packard, Dräger, Siemens, Novametrix and Nihon Kohden. Sensors from Dräger and Siemens are one-channel sensors which are automatically calibrated in a relative manner by use of inspiration air. Sensors from Hewlett-Packard and Nihon Kohden utilize filter-wheels to calibrate absolutely the measured CO_2 and N_2O concentrations. Filter-wheel sensors have proved to be fragile and shock-sensitive, which causes additional repair costs in the usually rough clinical use. Normal conditions involve stress and shock caused by movement of patients, by changing beds, by exchange of breathing hoses, and aspiration of secretions.

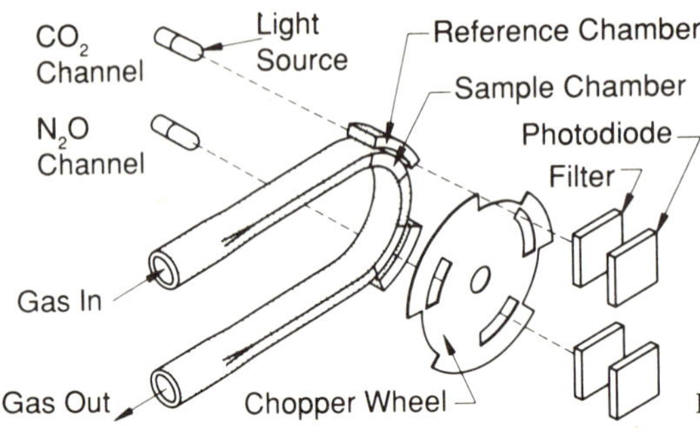

Figure 23-20. Filter-wheel IR sensor (Nellcor).

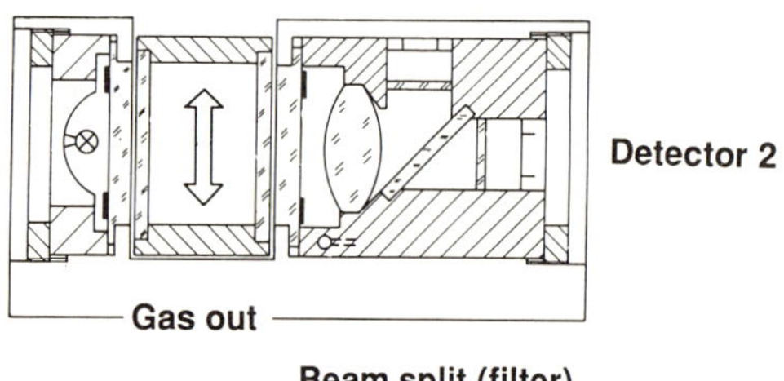

Figure 23-21.
Sectional view of new compact CO_2 sensor (Dräger).

The main problems with mainstream sensors are sensor weight and volume, which act as obstacles or additional load charging the Y-piece and the tubus. Mechanical supports for the Y-piece, hoses, or sensor have the disadvantage of an inflexible fixation of the patient. To utilize the benefits of a signal being free from delay and interference, a sensor as small and light as possible and without moving parts is required. The new mainstream sensors from Novametrix and Dräger come close to these requirements. The differences between the two types are caused by different optical arrangements of the calibration channel, which offers different performances in the case of secretion in the gas cell.

Figure 23-21 presents a sectional view of this new sensor type without a third channel to determine N_2O concentrations. This omission copes with the trend towards multi-gas sensors in anesthesia monitoring to measure additional gases than N_2O alone. However, this sensor fits the demands of ventilation monitoring.

A common problem with CO_2 sensors with dispersive elements is the need to use high-quality interference filters for accurate measurements. An alternative to avoid this problem is a sensor based on an electrical discharge source instead of on a black-body radiator. This technology is most suitable for the measurement of one gas component such as CO_2. The electric discharge source consists of a glass container with electrodes filled with CO_2 gas. By applying a high voltage and a high-frequency electromagnetic field, the gas is excited and emits its discrete IR excitation spectrum. By the optimum match of radiation and absorption spectrum, the sensor is highly sensitive to CO_2 but almost insensitive to other gases present in the gas sample.

This kind of plasma source avoids the need for expensive optical filters and allows a simple and straightforward design. Figure 23-22 shows the principle of this sensor, which is under clinical evaluation.

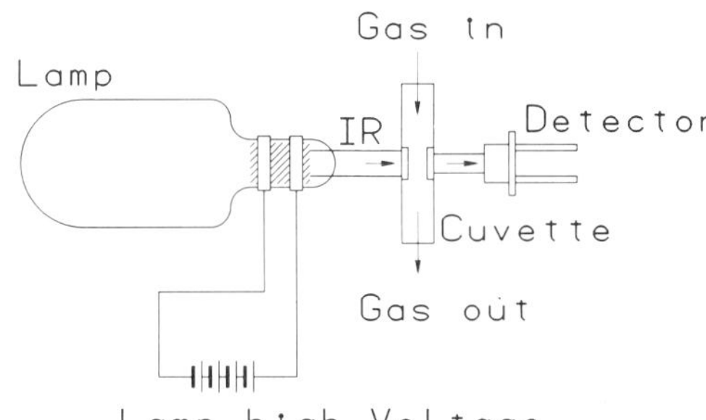

Figure 23-22.
Optimized CO_2 sensor with discharge light source.

23.7.1.2 Correction of Measured Values

When applying one-channel sensors such as that in Figure 23-21 during gas anesthesia with N_2O-O_2 blends, the N_2O concentration should be determined by the use of other methods, eg, reading the settings of rotameters or of output signals of an electronic anesthesia machine. This input value with reduced accuracy can be used to correct the measured CO_2 value, which is influenced by the gas composition. Physical processes in gas blends generate the so-called broadening of absorption bands by collisions of molecules. Therefore, the gas-specific filter picks up slightly different absorption signals which generate an increased reading. The correction of measured CO_2 partial pressures to compensate for this influence is for N_2O

10%, for O_2 3%, and for N_2 1% of the measured value. In addition, the "trivial" influences of ambient pressure and temperature of breathing gases have to be considered.

23.7.1.3   Problems in Clinical Use

In addition to possible miscalibrations caused by CO_2 in inspiration air, practical problems with mainstream sensors occur also by filling the gas cell with water, secretions, or mucus. However, this sensor type has proved to be less sensitive than sidestream sensors. Depending on the design of the optical path for calibration, this effect can be more or less compensated for. In the case of abnormal intensity reduction by filling more than half of the cell an alarm will be generated.

Sidestream sensors such as the Datex, Engström and Andros types show completely different problems. Sensor volume and weight are of minor significance, although compact and small multi-parameter monitors are preferred by hospital staff to reduce the total area in operation rooms covered by anesthesia equipment. The main problem with sidestream sensors is to maintain a constant flow to aspirate continuously small volumes of sample gas. Today, gas returns to the breathing circle are avoided because of potential hygiene hazards. Long sidestream lines generate signal delays of several seconds which have to be considered in the alarm system. Mixing with inspiration gas and turbulences combined with a low sample flow distort the rise time and shape of the CO_2 waveform. Liquid water and mucus in the sample line may obstruct the aspiration. Modern sidestream sensors monitor the pressure drop of the sample line to detect obstructions. Typical flow rates are 250 and 150 mL/min with response times of 150 or 250 ms, respectively. Newly developed monitors use sample flows of 50 mL/min with reduced rise times.

The humidification of respiration gases during long-term ventilation or long operations cause larger amounts of condensed water which have to be removed before entering the measuring cell of sidestream sensors by applying water separators. Water separators consisting of tubes, eg, Nafion tubes, which are permeable to water vapor are effective up to 1 mL/h.

To remove amounts of water larger than 10 mL/h, mechanical water traps (see Figure 23-23) use sharp turns in the gas flow direction to separate the part with water droplets [22]. The gas flow is split to collect the liquid water in the jar of the water trap.

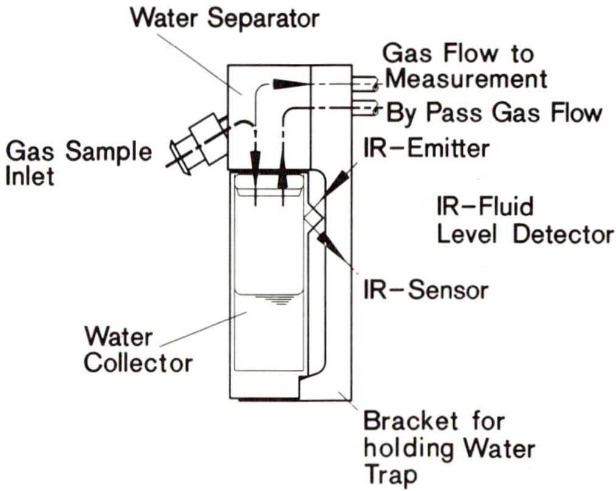

Figure 23-23.
Water separator (Dräger Capnodig).

Undefined delay times and distortion of the rise time reduce the derivable information of waveform analysis but the end-tidal value and the average value can be used clinically. Figure 23-24 presents an example of a flow mismatch with distortion of the CO_2 waveform. The inspiratory value may be used to detect system failures. During the warm-up time and the subsequent first hour, the IR sensor typically shows increased drift and reduced accuracy until temperature balance is reached. Zero-point calibration can be effected periodically by automatically aspirating ambient air. Drifts of the measuring system are calibrated manually or automatically over longer periods by use of test gases.

To combine the advantages of both types, Nellcor designed a sensor (see also Figure 23-20) which uses only 1 m of sample line and an extremely small sample cell and offers the advantage of short delay times of about 1 s, low flows of 50 mL/min and short rise times of about 85 ms [23]. These features also offer acceptable monitoring of neonates. In practical use, the features are slightly influenced by the normal problems in sidestream sensors with condensed water also which are suposed to be overcome by the "proximal-diverting" monitor.

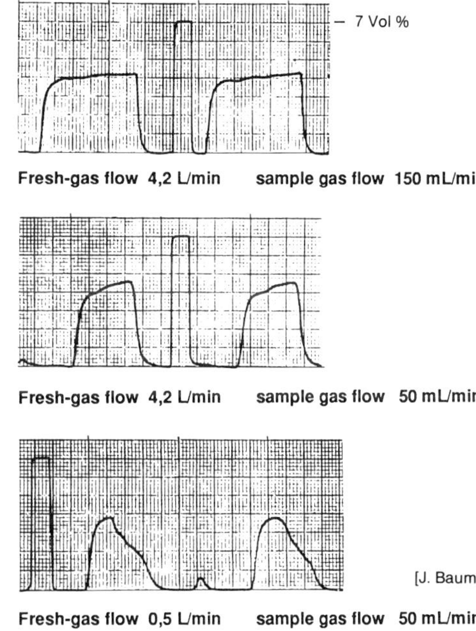

Figure 23-24.
Mismatch of sample flow and fresh-gas of a circle system.

23.7.2 Interpretation of Measured Values

The discussion of interferences, advantages, and disadvantages of both methods indicates the clinical relevance of several parameters. Figure 23-25 displays the essentials of the normal capnographic waveform, where point D is the maximum partial pressure at the end of exhalation.

This end-tidal partial pressure ($EtCO_2$) in the gas phase corresponds to the mixed alveolar value ($pACO_2$) which is correlated with the arterial partial pressure ($paCO_2$), depending on perfusion, ventilation, and distribution disorder. The difference between arterial (a) and

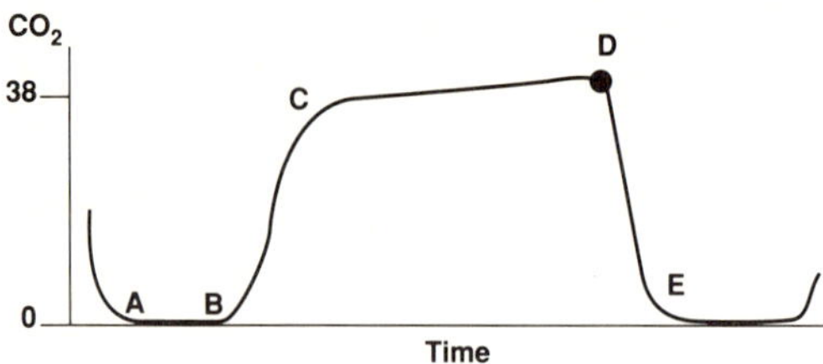

- Zero baseline (A-B)
- Rapid, sharp rise (B-C)
- Alveolar plateau (C-D)

- End-tidal value (D)
- Rapid, sharp downstroke (D-E)

Normal a-ADCO$_2$ = 2-3 mmHg

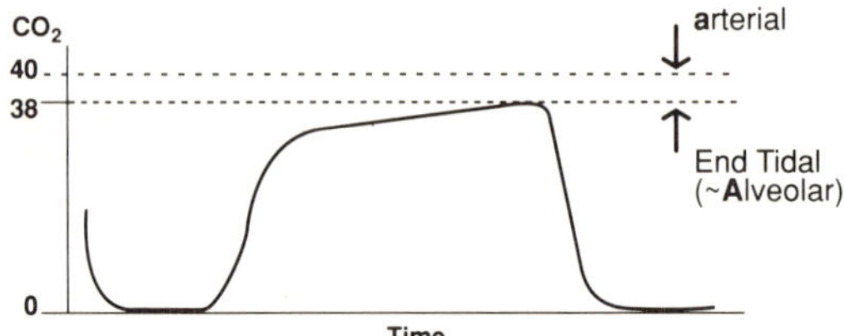

Figure 23-25.
Essentials of normal capnographic waveform (courtesy of Nellcor).

alveolar (A) partial pressure is the parameter a-ADCO$_2$. Higher values indicate a poor sampling technique (see Figure 23-26) or a ventilation/perfusion mismatch.

During anesthesia with mainly patients with healthy lungs, a direct correlation and a low a-ADCO$_2$ are assumed. In this case slow changes in EtCO$_2$ indicate hypo- or hyperventilation. Sudden changes indicate changes in perfusion or air emboli. The last interpretation may be equivocal and misleading in some situations as similar effects may occur caused by blood pressure drops. An additional N$_2$ measurement (see multi-gas sensors, Section 9) would allow the two cases to be distinguished.

During intensive care, the probability of distribution and perfusion disorders is increased, which leads to an increased value of a-ADCO$_2$, and therefore only relative changes can be used for therapy control depending on the individual state.

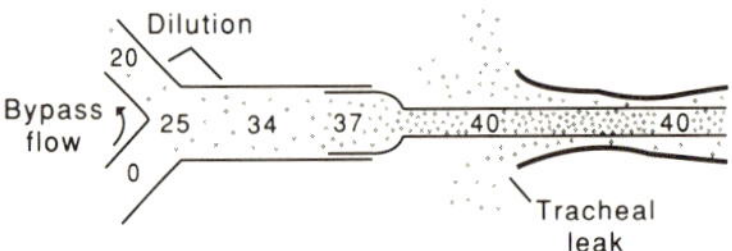

Inaccurate readings may result when:

- Airway adapter is placed too close to bypass (or fresh gas) flow

- Large leak around the endotracheal tube leads to loss of tidal volume

- Monitor with slow rise time is used in a patient with rapid respiratory rates

- Monitor with high sample flow volume is used in a patient with small tidal volumes

Figure 23-26.
Inaccurate readings by poor sampling technique (courtesy of Nellcor).

By comparing the continuously measured CO_2 partial pressure by IR sensors with determinations based on blood samples, a correlation factor is achieved which is typical for the monitored patient in his or her specific health state. By use of this "calibration" value, the frequency of blood samples can be reduced significantly.

In combination with additional parameters such as saturation, derived parameters such as oxygen uptake, and circulation parameters, the state of the patient can be characterized sufficiently.

Table 23-9 presents typical specifications for three types of sensor.

Table 23-9. Typical specifications of IR CO_2 sensors.

Mainstream

Weight	< 100 g
Rise time	100–200 ms
Accuracy	2% full scale
N_2O interference	direct compensation in case of filter-wheel sensor or by data processing

Sidestream

Sample flow	150 mL/min
Rise time	150–250 ms
Accuracy	3% full scale
Drift	5% full scale/day
N_2O interference	direct compensation in case of filter-wheel sensor or data processing
Calibration	ambient air, test gases

Midstream

Sample flow	50 mL/min
Rise time	85 ms
Accuracy	2% full scale
Drift	u. a.
N_2O interference	direct compensation
Calibration	ambient air, test gases

23.8 Sensor for Volatile Anesthetic Agents

Measurements of the fraction or partial pressure of N_2O in anesthetic gas blends is usually carried out as "waste product" by means of an infrared CO_2 multi-channel sensor. The measured value has only minor clinical significance. It can be used in addition to the O_2 signal to characterize N_2 wash-out and in addition to agent concentration to determine the dynamics of agent saturation and steady state, whereas the times required to reach the steady state are different for N_2O and the agent. Another application of the N_2O value has been discussed with regard to monitoring of closed circle anesthesia to determine the N_2 concentration indirectly as a complement to 100%.

The measurement of halogenated hydrocarbons such as Halothane, Enflurane, and Isoflurane, which are used as vaporized anesthetic agents, is of moderate clinical relevance. The main applications are monitoring of over- and underdosing, to control anesthesia depth, brain perfusion, or blood pressure during specific phases of the surgical procedure [19, 24]. In addition, this type of measurement is used to determine malfunctions of anesthetic agent vaporizers.

23.8.1   Fundamentals of Agent Sensors

Only two methods, absorption on the surface of an oscillator crystal and infrared absorption, show practical significance in comparison with various other methods.

Sensors such as the rubber-band sensor are historical examples of early non-electronical sensors which used changes in thickness and length of a rubber band by absorbed agents to move the hand of a mechanical instrument. The accuracy was limited.

The measuring procedure of the oscillator crystal method expresses its limitations by its basic function. The detection principle is based on detuning of a crystal oscillator by absorbed substances in a silicone layer: all substances which can be absorbed generate a signal by detuning. The first sensor of this kind applied in the mainstream was discredited by erroneous measurements caused by humid agent vapors, by absorption of other hydrocarbons, ethanol, and acetone, and by temperature drift, which caused deviations of the reading of the same order of magnitude. The new generation of sensors offered by Engström and Siemens is placed in the sidestream and the sample gases are dried prior to measurement. Mixtures of gases are neither detectable nor composites separately measurable and thus generate erroneous readings. A possible application is the location in the fresh gas limb of an anesthesia machine to monitor vaporizer output. In applications to monitor respiration gases of semi-closed and closed anesthesia machines with rebreathing systems, inaccuracies are caused by incompletely removed humidity.

The basic principle of infrared absorption explained in Section 7 for CO_2 sensors holds for these sensors also. Normally filter-wheel sensors are applied, which are placed in the sidestream owing to their larger size and weight.

The only exception is the Irina sensor from Dräger, which can be placed in the mainstream. Owing to its bulky sensor, location in the inspiration or expiration limb is preferred to locations between the Y-piece and tubus. Figure 23-27 presents a sectional view with the filter-wheel and the multi-pass cell.

To distinguish and identify the various anesthetic agents, a high spectral resolution is required. The 3 µm spectral region offers separate but weak absorption bands. For this spectral range, appropriate optical filters and fast detectors are available. The weak absorption bands require a multi-pass of the transmitting light through the gas sample to increase the absorption signal. The accuracy obtained is 1–2% [19]. This sensor offers the possibility of quantitive determinations of agent partial pressures which will offer new online applications in the future. Today no clinical significance exists. With regard to future requirements to monitor closed anesthesia systems, the sensor should be modified by reducing the size, which can be achieved by separating the optical part of the radiation source and multi-pass cell from the optical detection part with the filter-wheel, detector, and sensor electronics. The objective should be to design a sensor of the size of a CO_2 sensor of the second generation. This sensor, combined with a precise flow sensor, would open up the application of the determination of amounts of agent instead of only partial pressure values.

In recent years, sample amounts for laboratory testing of a new compact and integrated double sensor have become available which can be stacked in-line to measure several gases simultaneously. The first prototype showed a poor signal-to-noise ratio, long-term drift, and some inaccuracies. However, this kind of sensor may have additional development potential.

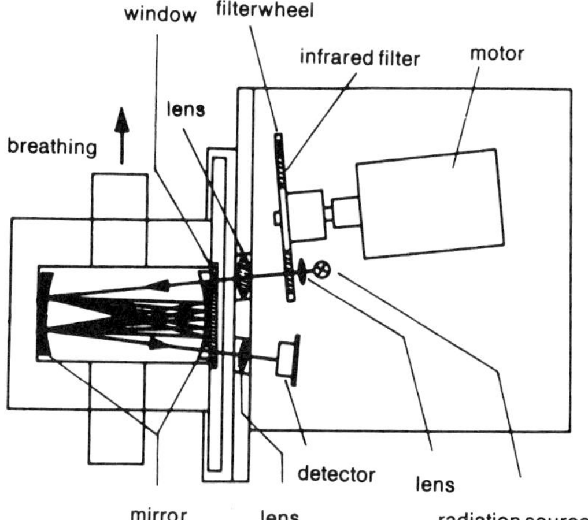

Figure 23-27.
Sectional view of mainstream agent sensor Irina (Dräger).

23.8.2 Clinical Experience

Basically all remarks on problems with sidestream sensors hold also for agent monitoring. The main difference is the inverted direction of deviations of the correct reading. In both cases the difference between inspiratory and expiratory partial pressures is reduced. In CO_2 monitoring this mean a smoothing of the shape of the capnogram and a reduction in end-tidal value. In agent monitoring during anesthesia, the inspiratory value is higher than the expiratory value, and therefore a disturbance will increase the end-tidal value.

In general, separate measuring cells are used for CO_2 with a small volume and a fast detector and larger cells for agents in a line. The rise time for CO_2 is 150–200 ms, depending on flow rate and sample volume. Agent sensors show rise times of about 400–600 ms. This causes an additional reduction in waveform resolution and in the accuracy of inspiratory and end-tidal values. This reduced performance correlates with the reduced clinical significance of agent monitoring compared with CO_2 monitoring. A new developed sensor, Nellcor N-2500, shows a rise time of 175 ms a flow rate of 50 mL/min.

Based on the availability of multi-gas sensors for sidestream applications, the use of monitors for inspiratory and expiratory concentrations has increased in recent years, despite the interferences mentioned. This trend seems surprising as the measured value offers only indirect determinations of, eg, anesthesia depth. In addition, essential information can also be obtained without measurement in respiration gases, eg, by the kind of anesthesia management, fresh-gas measurements or the use of calibrated and reliable vaporizers. During induction and in cases of high degrees of rebreathing, significant differences between fresh-gas and circle system concentrations occur. The statement concerning reliable and precise vaporizers has to be qualified, as vaporizers applied at the limits of the specifications or in the case of mishaps and failures will give concentrations different from the settings.

Figure 23-28 shows the flow dependence of two brands of Enflurane vaporizers which exceed the limits set by certain national standards in several countries when the device is operated at the limits of the specifications [25].

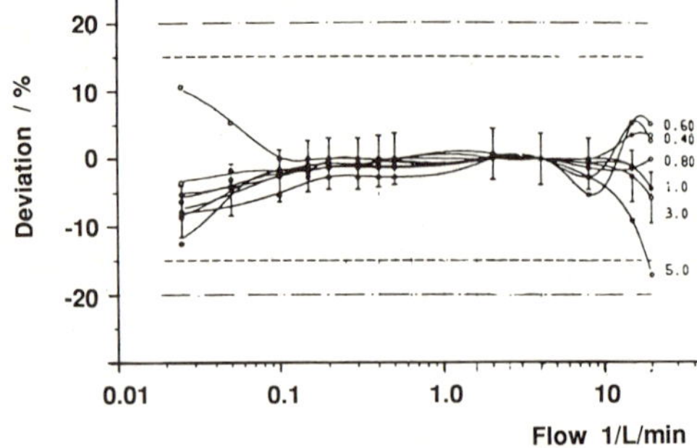

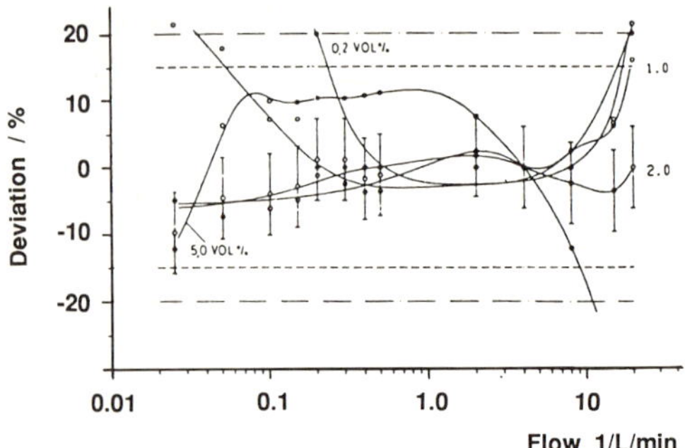

Figure 23-28.
Flow dependence of two different enflurane vaporizers (according to Gilly, and Dvoracek [25].

23.8.3   Clinical Applications

Owing to the simple handling of sidestream sensors in clinical application, only sidestream agent sensors have a substantial market share. With an increasing tendency these sensors are being offered as multi-parameter monitors as part of complete anesthesia work-place system. However, in addition a large market exists for the so far only partly equipped anesthesia machines, completion of which requires additional monitors with new parameter combinations. These additional monitors and complete monitoring systems are competing in the market. The total monitoring market has increased as the value of the monitoring part of an anesthesia workplace system has increased continuously in recent years compared with the value of the "gas machine". This trend to simpler anesthesia machines and to more sophisticated monitoring reflects the increase in combination anesthesias in clinical practice.

Mandatory monitoring of new anesthesia machines for the detection of device-related maldosage to comply with the Medical Device Regulations (MedGV) is an exception for the FRG only. For this purpose, all agent monitors can be used, but to comply only with the regulations or monitoring the vaporizer output only, inexpensive solutions are available. For example, the Iris sensor from Dräger is specially designed to be placed in the fresh-gas line without the need to protect the one-channel sensor against water condensation or contamination with secretions. Figure 23-18 shows the absorption spectra with strong absorption bands between 10 and 14 µm. This spectral region lies in a so-called water window. This sensor could be designed without a compensation beam and without moving parts. Indentification of agents is not possible, the inaccuracy is about 5–8% absolute.

With regard to possible additional substances in expired gases which may cause absorption in the spectral region used for various IR agent sensors, such as ethanol in the case of surgical treatment after car accidents or acetone in the case of diabetic patients, the agent sensors should able to detect unusual gas blends to avoid misreading, with potentially fatal effects. So far only two sensors show this feature: Irina from Dräger and the N2500 from Nellcor. Table 23-10 presents typical specifications of the different sensor types.

Table 23-10. Typical specifications of IR agent sensors.

Fresh-gas sensor (Iris)

Inaccuracy	7% absolute
Drift	<5% full scale/day
Cross sensitivity	mixture of agents
Rise time	1 s

Sidestream sensor

Inaccuracy	5% full scale
Drift	4% full scale/day
Cross sensitivity	mixture of agents
Rise time	750 ms at 150 mL/min

Mainstream sensor (Irina)

Inaccuracy	2% absolute
Drift	<1%
Cross sensitivity	detection of mixtures
Rise time	120 ms

Midstream sensor (N-2500)

Inaccuracy	<5% full scale (0–2%)
Drift	n. a.
Cross sensitivity	detection of mixtures
Rise time	175 ms at 50 mL/min

23.9   Tracer Gases

23.9.1   Clinical Applications

The term "tracer gases" means gases which are either no direct part of metabolism but which are administered for diagnostic purpose as tracers, or are end products of metabolism and indicate special diseases. At present this application field of gas analysis is still of minor significance in intensive care and anesthesia monitoring. One reason is the lack of appropriate and economically priced sensors which do not need the large expenditure associated with multi-gas sensors such as mass spectrometry (see Section 10). Typical examples are measurements of rare gases, O_2, CO, and SF_6 for lung function analysis to determine parameter such as diffusion index based on O_2 or CO wash-out curves, determination of the functional residual capacity (FRC) by means of rare gases or SF_6, or monitoring of urological surgical procedures by adding ethanol to the flushing liquid to detect unclosed venous vessels. In general, part of these measurements to determine lung parameters are carried out during diagnostic lung function testing by means of expensive laboratory equipment. The application of selected procedures may offer the chance of further improving the effectiveness of ventilation therapy during intensive care. There are intentions to use existing sensors with minor modifications or to open up the market for new sensors incorporated in ventilators. The combination of therapeutic devices and sensors completed with data processing power offers novel opportunities to carry out diagnostic procedures by means of the therapeutic device in-line or in separate procedures with the effect of avoiding or reducing the risks of transporting the patient frequently through the hospital to the distant lung-function laboratory.

In addition to multi-gas sensors, which are discussed later (see Section 10), direct-measuring O_2 and IR sensors can be used for this kind of measurement. By use of fast O_2 sensors, O_2 and N_2 wash-out curves are measurable whereas N_2 is determined as a complement to 100%. Ethanol, acetone and SF_6 can be measured by means of modified IR sensors. For the separation of interfering composites such as additional gases in sample blends like anesthetic gases in the case of ethanol measurements, an appropriate expenditure is required.

23.9.2   Future Applications

23.9.2.1   *Lung Function Testing during Intensive Care*

In general, the use of a paramagnetic sidestream sensor is also possible for recording of time-resolved O_2 curves after breathing during a short period pure oxygen for O_2 wash-out or pure nitrogen for N_2 wash-out. However, a mainstream sensor offers the possibility of using control signals of the ventilator additionally as no delay affects the subsequent procedures. When determining amounts of inhaled and exhaled substances, a direct measuring system combined with a fast flow sensor is a prerequisite. CO and SF_6 measurements will generate precise results by means of a modified Irina sensor which may be placed in the expiration limb with or without a flow sensor. Sidestream sensors would require larger expenditures.

The potential success of this kind of combination of therapeutic device and on-line procedures will be based on improvements in mainstream sensors to produce reliable results of clinical significance.

23.9.2.2 Monitoring during Urological Operations

During various endoscopic surgical procedures in urology, such as tumour resection in the bladder, a flushing system with pressurized saline solution continuously clears the operation site. In the case of not completely closed small venous vessels, the flushing liquid infiltrates the blood circulation system without being observed. All clinical signs and standard monitoring are of low sensitivity until an amount of about 1 L of additional liquid has been infiltrated. In cases of suspected infiltration, the surgical procedure has to be discontinued for several days until recovery of the patient. By adding ethanol as a tracer substance to the flushing solution to obtain a 2% concentration, this tracer cannot pass the bladder wall. However, in the case of retrograde infiltration of the venous system the tracer is transported to the lung, passes the alveolar membrane and ethanol appears in the expired air. This method is very sensitive and shows the infiltration of amounts of less than 50 mL of liquid.

Basically, ethanol sensors designed for use in traffic control can be used for this purpose. However, their function is based on a cooperative proband [26]. Sedated or anesthetized patients are not able to carry out the necessary procedures. In addition, the electrochemical sensors used suffer interference from most anesthetic gases. For future application, IR sensors which distinguish between ethanol and anesthetic gases have to be designed on the basis of today's IR sensors.

23.9.2.3 Gas Analysis in Emergency Care

Gas analysis in emergency care can be applied in several situations:

— pulse oximetry to monitor ventilated patients;
— acetone in exhaled gases to detect metabolic deviations of diabetic patients;
— organic solvents in exhaled gases in cases of intoxication.

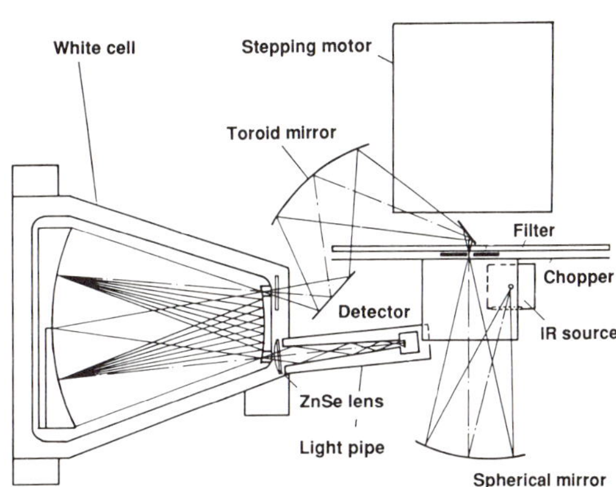

Figure 23-29.
Isolde adjustable IR sensor (Dräger).

For the last application an indication of the kind of solvent used can speed up the treatment and save valuable time. However, this demand requires another type of sensor, eg, an IR sensor which consists of an adjustable dispersive optical element to select specific absorption bands and compare the intensities with spectra stored in a built-in memory [27]. Figure 23-29 presents a prototype of a "tunable IR sensor", Isolde, which can be used for the detection of solvents in expired gases. This sensor has been clinically evaluated as a prototype but has not yet been offered commercially.

23.10   Multi-Gas Sensors

As already mentioned several times, the joint processing of signals generated by different types of sensors can cause various problems as different and in some cases undefined delay times occur. In the case of monitoring slow changes in parameters, the interferences can be neglected. However, the need to maintain different sensor systems with different technologies remains. Multi-gas sensors offer an improvement as one system is able to measure all gases of interest.

Only mass spectrometry has certain clinical significance in different arrangements. Mass spectrometry is a standard procedure in scientific research. Its introduction into routine clinical application has been restricted by the technical and organizational expenditure needed. Attempts have been made to reduce the costs by using cost-sharing multiplex procedures such as simultaneous use for several operating rooms or intensive care beds. In North American hospitals, installations have been established with central devices and sample lines with up to twenty measuring sites. All respiration gases including tracer gases such as rare gases are detectable as time-resolved concentration curves and derived display values. Normally a patient is monitored for several minutes, then the sampling system is switched to the next patient. After the complete cycle, the first patient is monitored again. In the meantime, the patient is lacking sufficient monitoring of these parameters and critical situations may occur undetected for a long period. Therefore, in the USA additional monitors such as pulse oximeters or IR CO_2 monitors now have to be used for continuous monitoring and providing alarms in critical situations to cover the time gaps.

In Europe, mass spectrometers are used clinically only for research purpose. Dedicated systems for a single patient have become available with reduced prices and improved features with regard to simplified operation. The prices of this new type are comparable to those of combination monitors offered for anesthesia monitoring, for example. However, combination monitors offer the possibility of integrating further parameters such as saturation or circulation monitoring.

Mass spectrometry offers the advantage of detecting all parameters of interest with the same delay. Influences caused by water or secretions are similar to those with other sidestream monitors. New generation devices avoid previous problems of vacuum technology, complicated handling, and calibration. The market potential remains uncertain.

New in the market are multi-gas monitors based on the effect of Raman scattering. In the mid-1970s prototypes were established for evaluating the feasibility, which demonstrated the possibility of designing an instrument for clinical gas analysis [28]. Figure 23-30 shows the principle of a external multi-pass arrangement for increasing the efficiency of Raman scattering in respiration gases and to offering the possibility of simultaneous recording of all gases of interest.

Raman scattering is based on inelastic scattering of incident laser light which shows shifts in colour or wavelength depending on the gases transmitted. All diatomic and multiatomic gases show Raman shifts which reflect as energy differences vibrational or rotational transitions of the molecules affected. In the evaluation of vibrational lines, all gases of clinical significance are detectable with the exception of rare gases. Compared with IR absorption sensors, additional gases such as O_2, CO, and N_2 can be measured with the same system. As mentioned before, CO and N_2 measurements are also of certain clinical significance. CO can be used to determine the diffusion index of the lung tissue. Anesthesia monitoring with additional N_2 recording identifies air emboli during surgical procedures. In recent years a new monitor, Rascal, based on this effect was introduced for clinical monitoring by Albion [29]. This device uses an intracavity scattering arrangement instead of an external multi-pass cell to increase the scattering efficiency. The main cost factor of Raman spectrometers is the laser, and special procedures are used to increase the yield to reduce the technical expenditure. The clinical applications and the experience with this new system still have to prove that the identically generated signals for all gas concentration can be used for processing derived values.

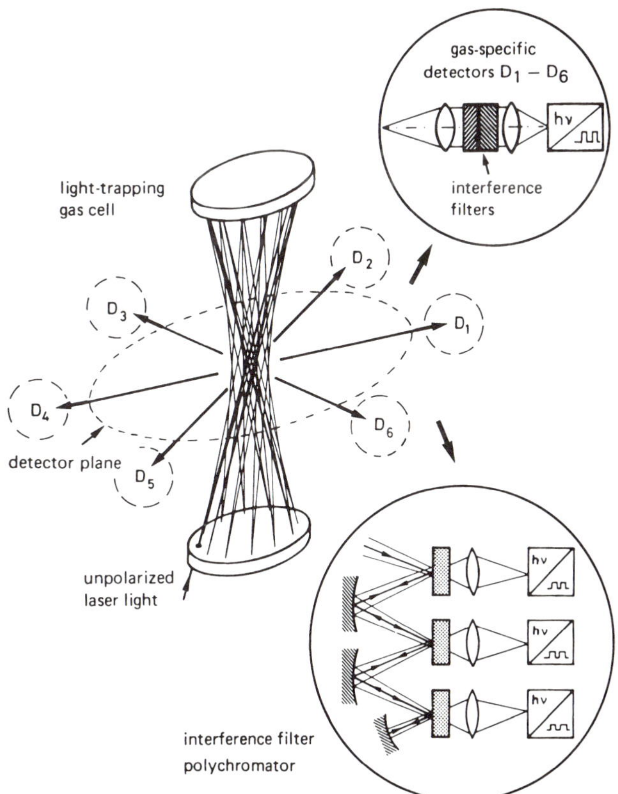

Figure 23-30. Multi-gas sensor based on Raman scattering, external multi-pass cell.

23.11   Derived Parameters − Metabolic Monitoring

The example of compliance calculation (Section 5) demonstrated that with the availability of the flow signal as a function of time, several parameters could be obtained which show diagnostic and therapeutic relevance. Similar processing is possible by use of flow and concentration signals of in- and exhaled gases. Processing of amounts of O_2, CO_2, and anesthetic gases is of clinical interest. Oxygen uptake, CO_2 production, uptake and release of N_2O and agents, in addition to partial pressure values, are parameters which reflect the dynamics of gas exchange and metabolism during intensive care and anesthesia [30].

Processing of oxygen uptake and CO_2 elimination can be based on two different methods which require different performances of the sensors used.

23.11.1   Determination by Means of Concentration and Flow Measurements

Prerequisites are fast and precise concentration and flow measurements with constant and defined delays and response times, whereas the inaccuracies of both methods result in additional deviations in worst case. The required performances are as follows:

− flow measurement with less than 10 ms resolution;
− O_2 and CO_2 measurements with less than a 100 ms response time;
− inaccuracy in each case better than 2%.

These specifications can only be reached by reconvolution of a determined transfer function of the mainstream sensors. In the case of CO_2 these requirements are met approximately by the new mainstream sensors from Dräger and Novametrics. In the case of O_2 sensors the performance of currently available sensors does not meet the specification needed. Fast O_2 sensors under development may meet these requirements in the future, but are not yet available.

Diagnostic lung function test equipment is based on other more precise and more expensive sensors, but can be used only off-line without intergration in therapeutic equipment.

23.11.2   Compensation Method

The use of a compensation method avoids the problems with fast and highly precise concentration and flow measurements. Monitors based on compensation methods have been developed in recent years, such as the Oxiconsumeter by Dräger for oxygen consumption and the Deltatrac by Datex for oxygen consumption and CO_2 production.

The measuring principle of the Oxiconsumeter is based on precise difference measurements and a highly precise dosing of consumed oxygen, the so-called replenishment method. Figure 23-31 shows a block diagram of the replenishment method. Dried and temperature-compensated inspiratory gas is measured in the first of two parallel polarographic oxygen cells, and in the second cell similar treated samples from the mixing chamber are measured continuously [31].

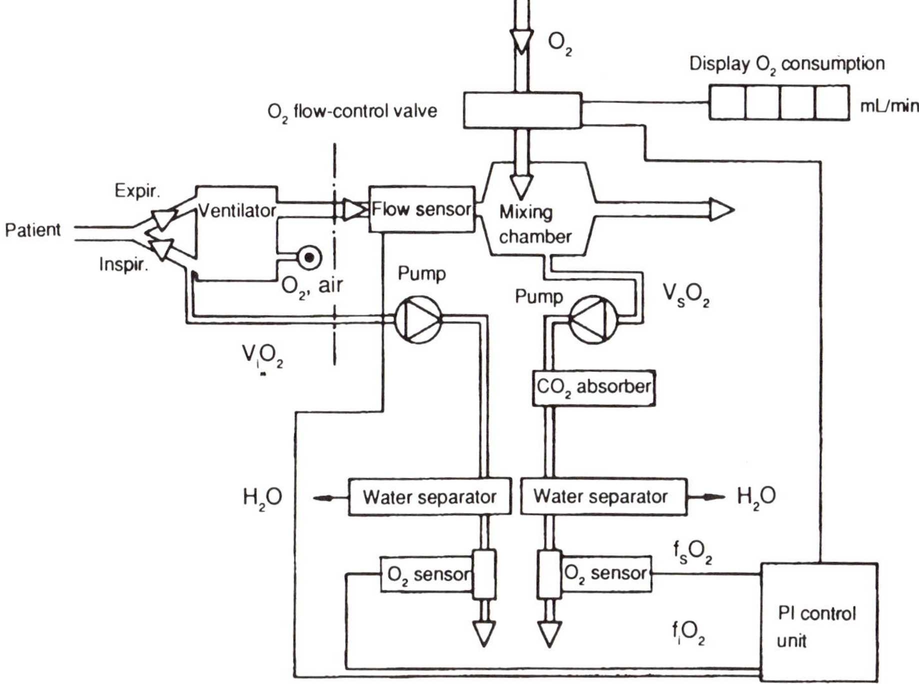

Figure 23-31. Oxiconsumeter replenishment method (Dräger).

The difference is used to control the on-line addition of oxygen by means of a precise electronic high-pressure valve. The inaccuracy of dosage is less than 1%. Differences in sensitivity are compensated for by frequent calibrations with the same gas for both sensors. The flow sensor controls the response time of the system in case of changes in inspiratory oxygen concentration, oxygen consumption, and respiration rate. Depending on the oxygen concentration, the total inaccuracy is about 10%.

The Deltatrac (Figure 23-32) uses for the determination of oxygen consumption and CO_2 production a dilution method with a precise constant-flow source which transforms the measurement of absolute concentrations into measurement of concentration differences and constancy of flow [32].

In both cases the accuracy requirement is shifted from the sensors to the oxygen dosage or constant-flow generator. Processing the respiratory quotient, RQ, requires inaccuracies of oxygen uptake and CO_2 production of less than 3% to result in RQ values with deviations of less than 6–8%, which can be used to control the nutrition of patients in the post-operation phase. To complete the parameters needed, the nitrogen excretion has to be determined by clinical laboratory methods.

The continuous measurement of oxygen uptake with ventilated or spontaneously breathing patients has proved to be a special and not routine monitoring procedure. Future applications may involve additional special procedure such as lung water determinations or non-invasive monitoring of cardiac output changes.

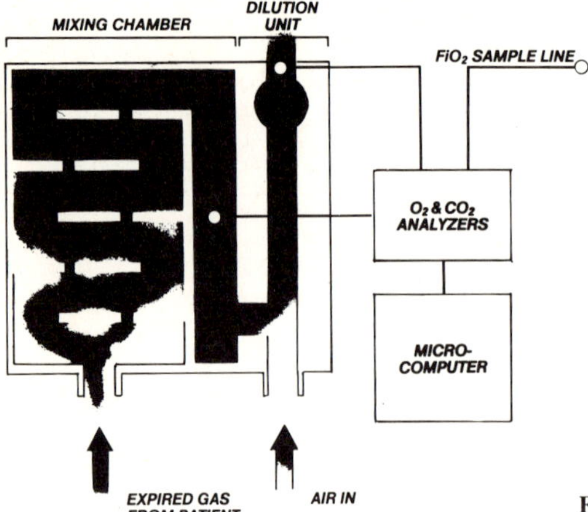

Figure 23-32. Deltatrac dilution method (Datex).

23.12   Outlook

Monitoring of gas exchange by means of various but partly competitive sensors has proved to be a valuable tool in clinical monitoring during intensive care or anesthesia. In addition to prevention of mishaps or the early detection of potential life-threatening situations with the chance of timely intervention, the on-line surveillance of the state of patients offers the required information on therapy results. The significance of distinct parameters depends strongly on the availability of reliable sensors. The developments in recent years indicate that by means of technical improvements well-known parameters such as oxygen saturation suddenly increase in clinical significance when sensors and monitors become reliable in routine clinical practice and are easy to operate. The rapid spread of the pulse oximeter for anesthesia monitoring and the subsequent collapse of the transcutaneous sensor market in pediatric care are typical examples of how technical availability changes the habits of medical personnel. These changes are accompanied by intense discussions which are sometimes very theoretical and unbalanced as no parameter alone can present a complete picture of such a complex system as the human body. The clinical problem should determine the selection of the required parameters.

"Disaster prevention" has acquired increased importance in recent years, especially in the USA [33]. Current opinion is reflected by the state of discussion on "essential monitoring" or "Primärmonitoring", which has influenced national and international standards, recommendations of professional associations and in some countries governmental regulations. The discussions in North American hospitals and professional associations have yielded a summary of potential mishaps and a rating of typical monitors [34] according to their ability to detect these mishaps. Table 23-11 presents a matrix of mishaps and monitors.

Discussions in Europe started some years later, including the clinical questions, and distinguished several levels of monitoring. These levels concern, in order of decreasing importance, equipment monitoring, monitoring of vital transportation functions, monitoring of metabolism, and monitoring of special organ functions. For the most important levels of equipment and vital transportation function, monitoring the function as an early warning system is the most important feature. Early warning depends on the time between the detection of the deviation of a parameter and the remaining reaction time for correction until an irreversible change in patient state occurs. Sensors to monitor gas exchange are regarded as some of the most important devices for this purpose.

Table 23-11. Monitors to prevent mishaps and as early warning system [34].

Monitors/Parameters:	Disconnection	Hypoventilation, Obstruction	Esophageal Intubation	Bronchial Intubation	Hypovolemia	Circuit Hypoxia	Halocarbon (Agent) Overdose	Pneumothorax	Air Embolism	Arrhythmia	IV Drug Overdose	Hyperthermia
Pulse Oximeter	2	2	2	2	0	2	0	2	2	1	0	1
Capnograph	3	2	3	0	1	0	0	1	2	0	1	3
Flow/Tidal Volume	3	3	1	0	0	0	0	1	0	0	1	1
Blood Pressure	0	0	0	0	3	0	2	1	2	0	2	0
Stethoscope	2	1	2	2	0	0	1	1	0	1	0	0
Halometer/Agent	2	0	0	0	0	0	3	0	0	0	0	0
Circuit O_2 Analyzer	0	0	0	0	0	3	0	0	0	0	0	0
ECG	0	0	0	0	0	0	1	0	0	3	0	0
Thermometer	1	0	0	0	0	0	0	0	0	0	0	0

Scale of values as early warning system:

3   high
2   moderate
1   low
0   none

23.12.1 Discussion of Current Trends

In contrast to forecasts of 10 years ago, biosensors with biochemical reagents fixed to a solid-state chip placed on an intravasal catheter have acquired no clinical significance so far. One reason in addition to the lack of availability is the trend towards less invasive methods. This opinion is based on the increased clinical use of more precise sensors to monitor gas exchange and metabolism. The future availability of faster and more precise mainstream sensors and the extended processing of "smart" systems will offer new non-invasive global parameters. This will include non-invasive monitoring of hemodynamic parameters such as cardiac output by means of saturation and oxygen consumption, or arteriovenous shunts.

The competition between dedicated sensors and multi-gas sensors will continue. The example of the fast oxygen fuel cell demonstrates that electrochemical sensors can be expected to have additional reserves for development and application.

23.13   References

[1] Tremper, K. K., "Transcutaneous PO_2 measurement" in *Can. Anaesth. Soc.* **31** (1984) 664–677.

[2] Busack, H. J., et al., "Anmerkungen zur Technik der transkutanen Sauerstoff-Partialdruckmessung" in *Medizintechnik* **4** (1979) 135–139.

[3] Gambert, R., "Transkutane Gas-Sensoren", Fachbeilage Mikroperipherik; in *Hard and Soft*, Düsseldorf: VDI-Verlag, Juli/August 1987, 4.

[4] Hölscher, U., "A Novel Approach for an ECG Electrode Integrated into a Transcutaneous Sensor" in *Continous Transcutaneous Monitoring*, Huch, A., Huch, R., Rooth, G., (eds.); New York: Plenum Press, 1988, pp. 291–293.

[5] *Transcutaneaous Sensors for O_2 and CO_2, Instructions for use*, Dräger, Lübeck, September 1987.

[6] Barker, S. J., Tremper, K. K., "Transcutaneous Oxygen Tension: A Physiological Variable for Monitoring Oxygenation" in *J. Clin. Mon.* **1** (1985) 130–134.

[7] Yelderman, M., New, W., "Evaluation of Pulse Oximetry" in *Anesthesiology* **59** (1983) 349–352.

[8] New, W., "Pulse Oximetry" in *J. Clin. Mon.* **1** (1985) 126–129.

[9] ECRI (eds.), "Pulse Oximeters" in *Health Devices* **18** (1989) 183–230.

[10] Schwanbom, E., "Ein Hitzdrahtanemometer zur Strömungsmessung in der Medizintechnik"; Jahrestagung DGAI, Lübeck-Travemünde, 1976 *Dräger informiert ... aus der Medizintechnik,* **1** (1977).

[11] "Oxygen analyzers for monitoring patient breathing mixtures – Safety requirements"; ISO Standard ISO 7767 first edition, 1988.

[12] Meriläinen, P. T., "Sensors for Oxygen analysis: Paramagnetic, electrochemical, polarographic, and zirconium oxide technologies" in *Biomedical Instrumentation & Technology,* Nov./Dec. 1989, 462–466.

[13] *Oxywarn 100/100R, Operating Manual*, Dräger, Lübeck, January 1985.

[14] *Dräger Sensor O_2, Technical Data*, Dräger, Lübeck, March 1989.

[15] Meriläinen, P. T., "A fast differential paramagnetic O_2-sensor" in *Int. J. Clin. Mon. Comp.* **5** (1985) 187–195.

[16] Feaster, W. W., Jost, K. A., Swedlow, D. B., *Capnography: A Quick Reference;* Nellcor Inc., Hayward CA., USA, 1988.

[17] Pruss, D., "Optische Gasanalyse", Fachbeilage Mikroperipherik; in *Hard and Soft,* Düsseldorf: VDI-Verlag, Juli/August 1987, 9–10.

[18] Schubert, H., *Sensorik in der medizinischen Diagnostik, Praxiswissen Medizin und Technik;* Köln: Verlag TÜV Rheinland, 1989.

[19] Hattendorff, H.-D., Leiss, M., *Measurement of Anaesthetic Agents and Nitrous Oxide 8th World Congress of Anaesthesiologists,* Manila, Philippines, Jan. 26, 1984.

[20] "Automatically Self-Calibrating Capnography in the Nellcor N-1000 Multi Function Monitor", *Capnography Note number 2*, Nellcor Inc., Hayward CA, USA, 1989.

[21] *Airway Gas Monitoring, an emerging standard of anesthesia care, leaflet,* Datex Instrumentation Corp. Helsinki, Finland.

[22] Mogue, L. R., Rantala, B., "Capnometers" in *J. Clin. Monit.* **4** (1988) 115–121.

[23] "Proximal-Diverting Capnography in the Nellcor N-1000 Multi-Function Monitor" in *Capnography Note Number 3* Nellcor Inc., Hayward CA, USA, 1989.

[24] Whitcher, C., "Monitoring of Anesthetic Halocarbons: Self-Contained ("Stand-Alone") Equipment" in *Seminars in Anesthesia,* Vol. V, **3** (1986) 213–224.

[25] Gilly, H., Dvoracek, G., "Low and high flow characteristics of vaporizers: Evaluation of TEC and VAPOR systems" in *Abstracts II, VII. European Congress of Anaesthesiology, Vienna Sept., 7–13, 1986;* Vienna: Maudrich, 1986, pp. 53–54.

[26] Rancke, R., "Überwachung der Einschwemmung bei transurethralen Prostataeingriffen durch Zusatz von Äthylalkohol zur Spülflüssigkeit", PhD Thesis, Medizinische Universität, Lübeck, 1989.

[27] Stark, H., "IR-Absorptionsspektroskopisches Analysegerät zur Bestimmung von Typ und Konzentration Organischer Lösungsmittel in der Atemluft vergifteter Personen" in *Abschlußbericht BMFT, FKZ 07033609,* Bundesministerium für Forschung und Technologie, Bonn, FRG, 1987.

[28] Albrecht, H., Müller, G., Schaldach, M., "Application of Laser Raman Spectroscops to Medical Diagnosis II" in *Proceedings of the Sixth International Conference on Raman Spectroscopy,* Bangalore, India, September 4–9, 1978, Vol. 2, pp. 526–527.

[29] Rascal, *Principle of Operation,* Leaflet, Albion Instruments, Salt Lake City UT, USA , 1988.

[30] Dantzker, D. R., "Oxygen Transport and Utilization" *Respir. Care,* 33 (1988) 874–880.

[31] Albrecht, H., "Meßprinzip des Oxyconsumeters" in *Methodische Fragen zur indirekten Kalorimetrie,* Kleinberger, G., Eckart, J., (eds.), Workshop Salzburg 1986; Munich: Zuckschwerdt, 1988.

[32] Deltatrac Metabolic Monitor, leaflet Datex Instrumentation Corp., Helsinki, Finland.

[33] Whitcher, C. et al., *Anesthetic Safety and Cost Effective Monitoring,* Scientific Exhibit Presented at 1986 Annual Meeting ASA, Las Vegas, Oct. 19–21, 1986.

[34] Whitcher, C. et al., *Mishap Prevention by Improved Monitoring: Benefits, Costs, and Funding Strategies,* Scientific Exhibit Presented at 1987 Annual Meeting ASA, Las Vegas, Oct. 19–21, 1987.

24 Chemical Sensors in Clinical Diagnostics

Martin Gerber, Boehringer Mannheim Corporation, Indianapolis, USA
Karl Wulff, Boehringer Mannheim GmbH, Mannheim, FRG

Contents

24.1 Introduction

The field of chemical sensors and biosensors is very fast developing. Hence, as in other areas of special scientific focus the boundary between science and speculation is ill-defined and moving.

Because of the overwhelming number of publications the purpose of this particular article is not to completely cover published results and ideas in this field but rather to focus on those close to reality or already in the market. Biosensors in general are well-covered by excellent reviews (see Chapter 14 and references given there, and [1–5]).

When we think of applications of sensor technology in clinical diagnosis we specifically have in mind the application in routine testing. Here the quality of performance of the existing photometric methods provides a great challenge for any new technological approach. Any new method, if sensor-based or else, must show clear advantages over existing technology at least in one of the features: reliability, safety, sensitivity, analytical range and linearity, speed, and – last, not least, – it must be massproducible at a low price.

Some of the technical problems of the application of sensor technology in clinical chemistry have been outlined by Russell and Rawson [6].

Gas sensors will be dealt with in Chapter 23 of this volume, pH-electrodes, since decades a well-established technology, also will not be a topic of this article.

Chemical sensors are composed of a specifier, which will specifically recognize the analyte, and in spatial intimate contact with the transducer will generate a chemical signal to be converted by the transducer into an optical or electrical signal, which is further processed. Most authors, eg, [3, 5] define a "biosensor" as a chemical sensor with a biologically derived specifier. For practical reasons and to include the ion selective electrodes, where the specifier quite frequently is a synthetic organic molecule, we define a biosensor as a chemical sensor being specific for an analyte of biological significance.

The entity composed of the specifier and transducer is the "biosensor". As we will see below, the interface between specifier and transducer is the most critical one in this technology.

Table 24-1. Concentration ranges of clinical diagnostic analytes.

Analyte	Concentration range in serum (mol/L)	Current technology
Glucose	10^{-4}–10^{-1}	Photometry Amperometry
Cholesterol	$5 \cdot 10^{-3}$–10^{-2}	Photometry
Creatinin	$5 \cdot 10^{-5}$–10^{-4}	Photometry
Bilirubin	0–$5 \cdot 10^{-4}$	Photometry
Urea	10^{-4}–10^{-1}	Photometry
IgG	10^{-4}	EIA
Therapeutic drugs	10^{-4}–10^{-6}	Photometry, EIA
Digoxin	10^{-9}	EIA
Hormones	10^{-8}–10^{-12}	EIA, RIA

The range of concentrations relevant for clinical diagnosis is spread over more than 8 decades from high concentrated metabolites, like glucose, serum proteins and therapeutic drugs down to the concentration of hormones.

Table 24-1 gives a list of selected analytes, relevant either as targets for sensor technology or as examples for our further discussion.

Out of the manifold of transducers only a few are relevant for chemical sensors in clinical diagnosis, either in already existing devices or through their potential in future developments.

24.2 Electrolytes

Since ion selective electrodes are dealt with quite extensively in this volume they will only briefly be mentioned here. They represent the most advanced field of sensor application in clinical routine diagnosis. A huge number of publications [7] and a variety of very successfully performing devices in the market reflect the mature state of this technology.

Ion selective electrodes measure an interface potential at the surface of the electrode. The specifier is either generated by a specific property of the surface (glass electrode) or by a selective carrier transporting the ion across a membrane. The main problem here as with potentiometry in general is the stability of the output signal. Consequently the production of a stable reference electrode junction potential is very critical.

Table 24-2 shows the narrow concentration ranges of the major blood electrolytes. Therefore the demands for reliability and precision are extreme. The most frequent needs in the routine laboratory to assay for Na^+ and K^+ in the same sample and to measure Ca^{2+} are well-covered by the various devices on the market.

The activity of sodium ions is commonly measured with a Na^+ selective glass electrode.

Table 24-2. Main blood electrolytes.

Ion	Concentration range in mmol/L
Na^+	135.0 –150.0
K^+	3.5 – 5.8
Ca^{2+}	0.45– 0.8
Cl^-	21.3 – 26.5

Potassium ions are assayed with a selective membrane electrode employing a poly(vinylchloride)-membrane doped with a specific ionophore. The classical K^+ carrier is the antibiotic valinomycin. Active research is going on at many places to develop synthetic ionophores for both cations, sodium and potassium.

Chloride ions are assayed with polymeric membranes based on selective classical ion exchangers.

Calcium ions are determined by a selective membrane electrode based on organophosphate ion exchangers or natural carriers [10].

24.3   High Concentration Metabolites

For the assay of low molecular weight metabolites usually enzymes are employed as specifiers which turn over the metabolites to form a reaction product. Its concentration can be measured potentiometrically or amperometrically [11, 12].

Since most of the enzymatic reactions used are redox reactions and since the electrode processes are redox reactions as well, special attention has to be given to substances present in the sample interfering with redox reactions, especially ascorbate, urate, and bilirubin.

24.3.1   Glucose

One of the most attractive analyte for amperometric enzyme electrodes is blood glucose. The specifier of choice is the enzyme glucose oxidase (EC 1.1.3.4, Enzyme Classification acc. to Enzyme Nomenclature of the IUB) catalyzing the oxidation of glucose to gluconolactone. The oxidation takes place in two steps:

$$\text{Glucose} + \text{Glucose Oxidase}_{ox} \;\rightarrow\; \text{Gluconolactone} + \text{Glucose Oxidase}_{red} \qquad (24\text{-}1\,a)$$

$$\text{Glucose Oxidase}_{red} + O_2 \;\rightarrow\; H_2O_2 + \text{Glucose Oxidase}_{ox} \qquad (24\text{-}1\,b)$$

Since the active center of redox enzymes is shielded by the protein structure, reduced glucose oxidase cannot be directly reoxidized at a metal or carbon electrode. Mediator molecules must be used for the electron transfer. The natural mediator is oxygen yielding hydrogen peroxide. This mediator, however, has several disadvantages:

a) The Michaelis constant of glucose oxidase for oxygen is unfavorably high, thus limiting the dynamic range of glucose measurement.

b) Oxygen or hydrogen peroxide are causing problems in direct electrochemical measurement because of the interferences of reductive serum constituents.

Since the amperometric oxidation of hydrogen peroxide requires a rather high voltage, the reducing serum constituents have to be kept away from the electrode by one or several layers of selective semi-permeable membranes [13]. Such membranes may have a higher permeability for oxygen than for glucose. A relative increase in local oxygen concentration will be the result extending the upper limit of the dynamic range for glucose.

Two systems on the market employing this membrane based approach to assay for blood glucose concentrations are the "30/30 Direct" from Elco Diagnostics Company for home monitoring use and the benchtop instrument "23A" from Yellow Springs Instruments for use in larger laboratories. Both systems use membranes with immobilized glucose oxidase. These membranes can be reused at least for a time period of 4 weeks.

A more ingenious approach is to employ a chemical redox pair, either water soluble or matrix bound, which needs a lower potential for reoxidation at which the interfering substances are not yet oxidized. Also the oxidized form of this redox pair is present in a steady state concentration well above the saturation concentration for glucose oxidase. Hence, the upper limit of the dynamic range of glucose assay is in the range of the highest pathological glucose value.

The classical approach to a water soluble mediator is the redox pair hexacyanoferrate(II)/hexacyanoferrate(III). A more convenient approach is the use of the redox pair ferrocene/ferricinium immobilized to the electrode surface [14, 15]. The ferrocene is oxidized at the electrode to regenerate ferricinium ion. The sensor is insensitive to oxygen and operates at relatively low potentials. This technological approach is the basis for the "ExacTech" glucose sensor being marketed by Baxter Travenol in the United States.

The electron transfer rate can be improved by using an electrically conductive polymer like polypyrrole [16] as an interface between the electrode and the non-conductive polymer matrix bearing the glucose oxidase and the mediator. Even better results can be achieved in a copolymerization of enzyme and mediator into a polypyrrole film on top of a platinum electrode [17, 18]. Here, the mediator ferrocene is copolymerized with the pyrrole.

Another approach uses a cobalt tetrakis(o-aminophenyl)porphyrin polymer film, prepared by electrooxidative polymerization of the monomer on top of the electrode as conductive film-mediator-couple [19].

A direct electron exchange between glucose oxidase and the electrode can be achieved by covalently binding mediator molecules to the enzyme protein as electron relays. Using this method, Degani and Heller [20] by attaching on the average one electron-transfer relay to each 12000–75000 Dalton of enzyme succeeded to reoxidize reduced glucose oxidaze directly at gold, platinum or carbon electrodes.

Another very interesting microfabrication approach to glucose electrode technology is used by Ikarijama et al. [21–23] who directly incorporated glucose oxidase into the micropores of a platinized microelectrode by platinizing in the presence of the enzyme. This way they constructed a microelectrode of 10 to 200 μm diameter with a linear range from 5×10^{-7} to 10^{-2} mol glucose/L.

By using pulse voltammetry with this electrode, glucose could be assayed in volumes as small as 2 μL. The signal was volume-independent and the response time less than 1 s [24].

Watanabe et al. [25] constructed a glucose electrode by immobilizing glucose oxidase through covalent linkage to the surface of a SnO_2-electrode. The dynamic range at normal oxygen pressure extended to 30–50 mmol/L.

A universally useful parameter for a chemical reaction is the reaction enthalpy produced by this reaction. If heat is the input signal for the transducer, the interface between specifier and transducer in a biosensor is easily defined and independent from the mechanism of the chemical reaction occuring at the specifier. This theoretically makes calorimetry almost universally applicable to biosensor development. In practice, however, the usefulness of this approach is severely limited by heat generated from other sources than the specific chemical reaction and by thermal eddies in the sample solution. Three types of transducers have been employed in experimental devices, thermistors [26, 27], thermopiles [28, 29], and temperature sensitive integrated circuits [30].

Potentiometric methods are less useful in metabolite assays because of the instability of the signals and the need of calibration. A special form of a potentiometric enzyme electrode, the enzymatically coupled field effect transistor is reviewed by Caras and Janata [31].

24.3.2 Cholesterol

By the action of cholesterol esterase (EC 3.1.1.13) free cholesterol is liberated from ester linkage which is then oxidized by oxygen catalyzed by cholesterol oxidase (EC 1.1.3.6).

$$\text{Cholesterol ester} + H_2O \rightarrow \text{Cholesterol} + \text{fatty acid} \tag{24-2}$$

$$\text{Cholesterol} + O_2 \rightarrow \text{Cholest-4-en-3-one} + H_2O_2 \tag{24-3}$$

Clark et al. [32] have published in 1981 a method employing both enzymes in solution and potentiometrically measuring the concentration of the hydrogen peroxide formed. For an amperometric cholesterol assay Yao et al. [33, 34] used immobilized cholesterol esterase and cholesterol oxidase in a reactor assaying the hydrogen peroxide formed at a peroxidase electrode with the redox pair hexacyanoferrate(II)/hexacyanoferrate(III) as a mediator.

At first hydrogen peroxide is used by the peroxidase to oxidize the mediator:

$$H_2O_2 + 2\,Fe(CN)_6^{4-} + 2\,H^+ \rightarrow 2\,Fe(CN)_6^{3-} + 2\,H_2O$$

which is then electrochemically reduced at -50 mV at a platinum vs. Ag/AgCl electrode. At this low voltage reductive serum components do not interfere with the electrode reaction.

24.3.3 Other Metabolites

Electrochemical assays for therapeutic drugs, like paracetamol or acetyl-salicylate have been described [35]. Urea can be assayed potentiometrically by a H^+ sensitive ISFET with a SiO_2 gate insulator covered by a Si_3N_4 film to which urease (EC 3.5.1.5) is covalently attached [36].

24.4 Low Concentration Analytes

The direct measurement of analyte concentrations below approx. 10^{-4} mol/L with electrochemical sensors is so far only possible with the advanced microfabrication process of Ikarijama et al. [21–23]. Usually a chemical amplifier system has to be used for increased sensitivity. The only analytes which already provide this amplification system themselves are enzymes catalytically turning over their substrates. If the product of the enzymatic reaction can be assayed with a sensor, the activity of the respective enzyme can also be determined. As an example, an amperometric assay for glucose oxidase using benzoquinone as an oxidant has been published [37]:

$$\text{Glucose} + \text{Benzoquinone} \rightarrow \text{Gluconolactone} + \text{Hydroquinone} \tag{24-4}$$

The rate of hydroquinone formation is followed amperometrically.

For sensing low concentrated analytes without an enzymatic activity, these can be "labeled" with an enzyme via an immunologic reaction using an enzyme labeled antibody specific for the respective analyte. This results in an enzyme immunoassay with electrochemical indication.

Also, systems monitoring the heat generated by the enzymatic reaction have been employed for enzyme immunoassays using a thermistor as a transducer [38].

Other methods of monitoring immunoassays with electrochemical sensors use the change of interface potential at an antibody or antigen loaded membrane upon reaction with the respective analyte [39].

A ChemFET with human serum albumin immobilized to the gate surface was described in an assay for anti-human serum albumin antibodies [40].

The minor changes occurring at an antibody or antigen loaded surface resulting from an immune reaction with the respective counterpart (antigen or antibody, resp.) can be monitored as a mass change using a piezo-electric transducer [41] or optically, eg, by surface plasmon resonance [42].

The general problem with all these methods following minor changes at an active surface is the reproducibility of signal and background as well as specificity within complex biological samples. So far none of these approaches have reached a stage of development allowing the launch of a commercially available product. However, there is no doubt that finally further progress in microfabrication techniques, especially in the integration of biomolecules into electrical or optical transducers, will result in the availability of reliable devices for clinical diagnostics.

24.5 Discussion

By obvious reasons the most advanced sensor technology is represented in the assays for inorganic ions.

Next is the challenging field of glucose assays. The first glucose sensors for routine use are already on the market. Here, however, especially in the application to patient selfmonitoring the new sensor technologies face the well-established test strip based technology which provides reliable results, is convenient and, rather inexpensive. The confidence of doctors and patients in this mature technology, based on valid arguments, is quite high. Why are researchers investing time, effort and, money into glucose sensors? The reason for many of them is the final goal of implantable glucose sensor and later a closed loop, sensor driven insulin pump.

To be accepted by the patient, an implantable sensor must have a needle-like configuration. In vivo this sensor will be in contact with interstitial fluid rather than with blood. The glucose concentration in blood and in the interstitial compartment are usually in equilibrium. However, fast changes in blood glucose concentration are only slowly followed by the concentration changes in the interstitial fluid [43]. Another important problem to consider are the mechanisms in the human body to reject a foreign implant. The implantable sensor must, eg, be protected against the immune system, against phagocytosis and against encapsulation. The main issue is then the reliability of the sensor from the viewpoint of technology.

The stability of the most critical component, the electrode bound glucose oxidase, has been demonstrated for at least 6 months under certain conditions in an animal model [44]. To prevent a potential drift over the long time of in vivo use besides working and auxiliary electrode a reference electrode seems to be mandatory [45]. The chemical problems of the in vivo environment, especially the one of oxygen pressure, have been discussed by Pickup et al. [46]. A further problem of all implantable devices will be toxicology studies to be performed with the device itself and its components, eg, mediator compounds.

A needle-type glucose sensor covered by a biocompatible membrane has been described recently [47]. However, no in vivo results are reported.

We can conclude that many problems along the way towards an implantable glucose sensor have been addressed and partly solved. However, many open questions are still unsolved and further problems will arise as work progresses.

Sensors for other analytes are under active research at many places. Their way towards commercialization will be the longer the lower the concentration range of the respective analyte. We expect here a scientifically exciting development to take place.

24.6   References

[1] Guilbault, G. G., Luong, J. H., *Selective Electrode Rev.* **11** (1989) 3–16.
[2] Campanella, L., Tomassetti, M., *Selective Electrode Rev.* **11** (1989) 69–110.
[3] Higgins, I. J., Lowe, C. R., *Philos. Trans. R. Soc. London* **B316** (1987) 3–11.
[4] Regnault, W. F., Picciolo, G. L., *J. Biomed. Mater. Res.: Applied Biomaterials* **21**, No. A2 (1987) 163–180.
[5] Owen, V. M., Turner, A. P. F., *Endeavour, New Series* **11** (1987) 100–104.
[6] Russell, L. J., Rawson, K. M., *Biosensors* **2** (1986) 301–318.
[7] Solsky, R. L., *Anal. Chem.* **60** (1988) 106R–113R.
[8] Byrne, T. P., *Selective Electrode Rev.* **10** (1988) 107–124.
[9] Nabet, P., *Analusis* **15** (1987) 379–385.
[10] Sena, S. F., Bowers, jr., G. N., *Methods in Enzymology* Vol. 152, Berger, S. K., Kinnen, A. R. (eds.): New York: Academic Press, 1988, pp. 320–334.
[11] Scheller, F., Kirstein, D., Kirstein, L., Schubert, F., Wollenberger, U., Olsson, B., Gorton, L., Johansson, G., *Philos. Trans. R. Soc. London* **B316** (1987) 85–94.
[12] Guilbault G. G., Kauffmann, J.-M., *Biotechnol. Applied Biochem.* **9** (1987) 95–113.
[13] Harrison, D. J., Turner, R. F. B., Baltes, H. P., *Anal. Chem.* **60** (1988) 2002–2007.
[14] Cass, A. E. G., Davis, G., Francis, G. D., Hill, H. A. O., Aston, W. J., Higgins, I. J., Plotkin, E. V., Scott, L. D. L., Turner, A. P. F., *Anal. Chem.* **56** (1984) 667.
[15] Green, M., Hill, H. A. O., *J. Chem. Soc., Faraday Trans. 1* No. 82 (1986) 1237–1243.
[16] Mizutani, F., Asai, M., *Bull. Chem. Soc. Jpn.* **61** (1988) 4458–4460.
[17] Foulds, N. C., Lowe, C. R., *Anal. Chem.* **60** (1988) 2473–2478.
[18] Iwakura, C., Kajiya, Y., Yoneyama, H., *J. Chem. Soc., Chem. Commun.* (1988) 1019–1020.
[19] Oyama, N., Ohsaka, T., Mizunuma, M., Kobayashi, M., *Anal. Chem.* **60** (1988) 2534–2536.
[20] Degani, Y., Heller, A., *J. Am. Chem. Soc.* **110** (1988) 2615–2620.
[21] Ikariyama, Y., Yamauchi, S., Yokiashi, T., Ushioda, M., *Anal. Lett.* **20** (1987) 1791–1801.
[22] Ikariyama, Y., Yamauchi, S., Aizawa, M., Yukiashi, T., Ushioda, M., *Bull. Chem. Soc. Jpn.* **61** (1988) 3525–3530.
[23] Ikariyama, Y., Yamauchi, S., Yukiashi, T., Ushioda, M., *J. Electrochem. Soc.* **136** (1989) 702–706.
[24] Ikariyama, Y., Shimada, N., Yamauchi, S., Yokiashi, T., Ushioda, H., *Anal. Lett.* **21** (1988) 953–964.
[25] Watanabe, T., Okawa, Y., Buzuki, H., Yoshida, S., Nihei, Y., *Chem. Lett.* (1988) 1183–1186.
[26] Mosbach, K., Danielsson, B., *Anal. Chem.* **53** (1981) 83A.
[27] Danielsson, B., *Appl. Biochem. Biotech.* **7** (1982) 127.

[28] Guilbeau, E. J., Towe, B. C., Muehlbauer, M. J., *Trans. Am. Soc. Artif. Intern. Organs* **33** (1987) 329–335.

[29] Muehlbauer, M. J., Guilbeau, E. J., Towe, B. C., *Anal. Chem.* **61** (1989) 77–83.

[30] Muramatsu, H., Dicks, J. M., Karube, I., *Anal. Chimica Acta* **197** (1987) 347–352.

[31] Caras, S., Janata, J., *Methods in Enzymology* Vol. 137, Mosbach, K. (ed.); New York: Academic Press, 1988, pp. 247–255.

[32] Clark, jr., L. C., Duggan, C. A., Grooms, T. A., Hart, L. M., Moore, M. E., *Clin. Chem.* **27** (1981) 1978–1982.

[33] Yao, T., Sato, M., Kobayashi, Y., Wasa, T., *Analyt. Biochem.* **149** (1985) 387–391.

[34] Yao, T., Wasa, T., *Anal. Chim. Acta* **207** (1988) 319–323.

[35] Frew, J. E., Green, M. J., *Anal. Proc.* **25** (1988) 276–277.

[36] Karube, I., Moriizumi, T., *Methods in Enzymology* Vol. 137, Mosbach, K. (ed.); New York: Academic Press, 1988, pp. 255–260.

[37] Aubrée-Lecat, A., et al., *Anal. Biochem.* **178** (1989) 427–430.

[38] Danielsson, B., Mattiasson, B., Mosbach, K., *Pure Appl. Chem.* **51** (1979) 1443.

[39] Aizawa, M., *Philos. Trans. R. Soc. London* **B316** (1987) 121–134.

[40] Karube, I., Tamiya, E., Dicks, J. M., Gotoh, M., *Anal. Chim. Acta* **185** (1986) 195.

[41] Davis, K. A., Leary, T. R., *Anal. Chem.* **61** (1989) 1227–1230.

[42] Kooyman, R. P. H., Kolkman, H., Van Gent, J., Greve, J., *Anal. Chim. Acta* **213** (1988) 35–45.

[43] Fischer, U., et al., *Diabetologica* **30** (1987) 940–945.

[44] Clark jr., L. C., Spokane, R. B., Sudan, R., Stroup, T. L., *Trans. Am. Soc. Artif. Intern. Organs* **33** (1987) 323–328.

[45] Velho, G., Froguel, P., Sternberg, R., Thevenot, D. R., Reach, G., *Diabetes* **38** (1989) 164–171.

[46] Pickup, J. C., Shaw, G. W., Claremont, D. J., *Biosensors* **3** (1987/88) 335–346.

[47] Yamasaki, Y., et al., *Clin. Chim. Acta* **180** (1989) 93–98.

25 Solid-State Electrochemical Potentiometric Sensors for Gas Analysis

HANS-HEINRICH MÖBIUS, Ernst-Moritz-Arndt-Universität, Greifswald, FRG

Contents

25.1 Introduction

Sensors with oxide-ion-conducting solid electrolytes are usually only considered useful in oxygen determination. Actually, with this kind of sensor other gas components such as H_2O, H_2, CO, NH_3 or the concentration ratios of these gases can be measured under certain circumstances. In Figure 25-1 the electrode potentials of zirconium dioxide solid-electrolyte sensors are plotted against temperature [1, 2]. The regions demarcated show how the electrode potentials vary in some of the currently known fields of application. The numerous applications for electrochemical potentiometric gas sensors, as indicated in the diagram, are summarized in this chapter. First, the electrode potential equations necessary for practical applications are described. Then the actual physical chemical conditions for the measurement of sensor signals yielded by these equations are discussed. Various practical examples show what has been achieved.

The possibility of determining other gas constituents besides oxygen using sensors with oxide-ion-conducting solid electrolytes arises from the thermodynamic equilibria that are set up with oxygen in the gas phase or between particles in the gas phase and the solid electrolyte.

Investigations of galvanic solid-electrolyte gas cells with chemical equilibria played a significant role in the development of gas potentiometry. Haber et al. in 1905/6 published extensive results of measurements with oxyhydrogen gas cells, cells with CO, CO_2, C and hydrogen- and oxygen-concentration cells. For these, glass was used as a solid electrolyte between 350 and 570 °C whereas between 800 and 1000 °C, porcelain served [3, 4]. Obviously, the measurements were not reliable enough to think in terms of gas analytical applications for such cells. Only the precise results of electrochemical-thermodynamic investigations of the equilibria $CO + 1/2\ O_2 \rightleftharpoons CO_2$ and $C + CO_2 \rightleftharpoons 2\ CO$ [5] using oxide-ion-conducting solid electrolytes, suggested the idea of gas analytical applications. They lead to the registration of the first patent for potentiometric gas analyzers in 1958 [6], more than 100 years after Gaugain had discovered the galvanic solid-state electrolyte gas cell in 1853 [7].

Thermodynamic equilibria are also of principal importance, if one wishes to determine carbon dioxide or sulfur dioxide with carbonates or sulfates which show conductance for sodium or potassium ions. It is not the ion migrating through the solid, but the electrochemical equilibrium between molecules in the gas phase, particles in the solid electrolyte and electrons in the electrical conductor that determine the electrode potential utilizable in sensors.

Systems with cell potential differences corresponding to thermodynamic equilibria have important advantages over many other types of sensors. Measurements of potential differences, without current flow, allow the direct calculation of concentrations or concentration ratios using well-known laws. Calibration with, for example, test gases is often not necessary. Over large temperature and gas concentration ranges it is posssible to obtain sensor signals independent of geometric quantities, material conductivities and impurities. This explains why sensors of this type have been reported that determined the correct values of oxygen concentration in industry gases, for more than three years, without correction. Such a reliability is necessary when signals from different types of sensors are to be mathematically combined, to continually determine several components of a gas mixture using ratios.

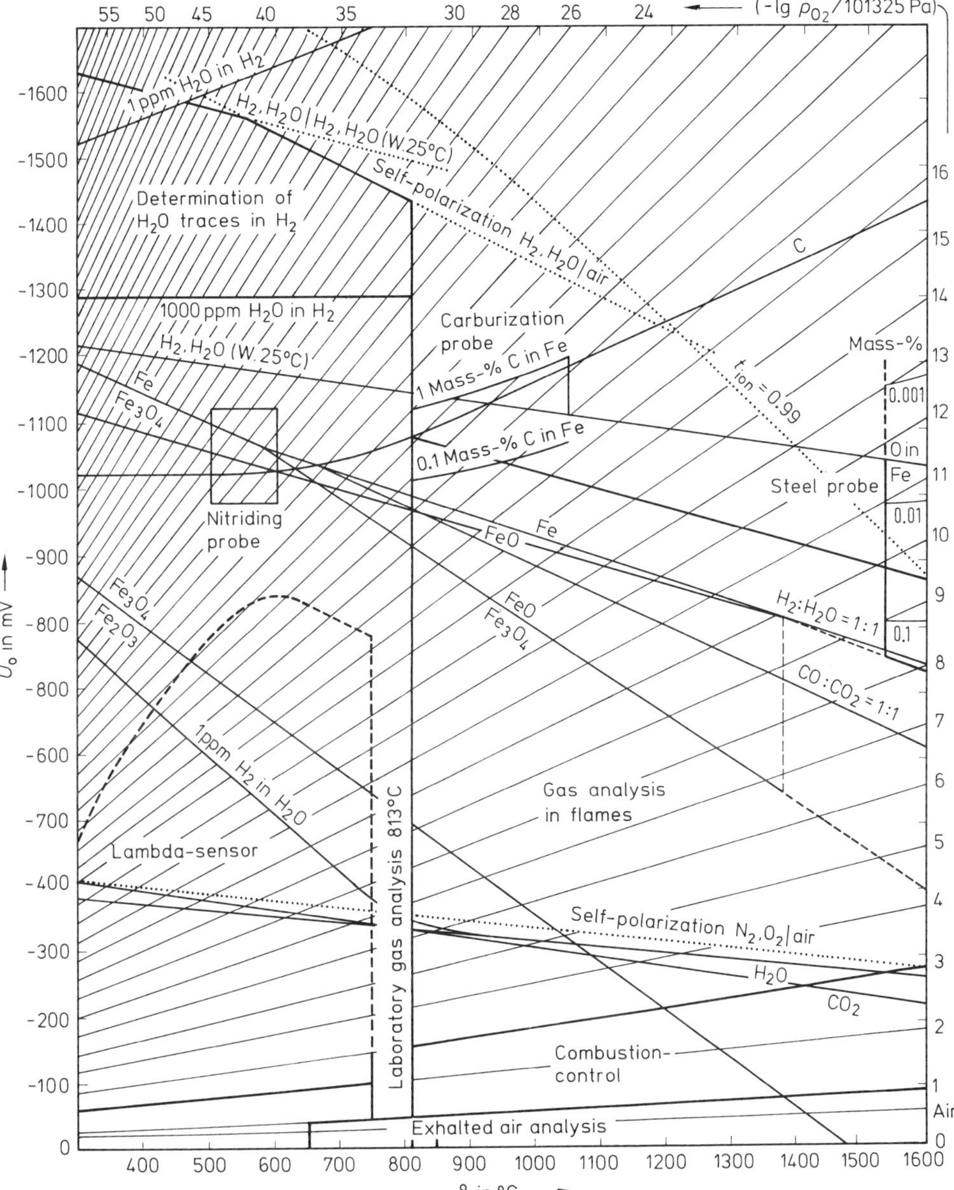

Figure 25-1. Summary of the calculated electrode potentials of electrodes with oxide-ion-conducting solid electrolytes showing the limits of application of ZrO$_2$ solid electrolytes and the areas of application of devices and probes with solid electrolyte cells [1, 2].

25.2 Signal Functions of Potentiometric Sensors with Oxide-Ion-Conducting Solid Electrolytes

The signal from an electrochemical potentiometric sensor is an electrical potential difference U, made up of the emf supplied by the galvanic cell and of interfering potential differences which are more or less avoidable. These disturbances will be ignored in this chapter and only the emf or in another terminology, the equilibrium potential difference U_{eq} will be considered. (It should be noted that both potential differences and emf will be used here with the same sign, unlike earlier definitions [8, 9].) U_{eq} is the cell potential difference given by an isothermal galvanic cell when no current flows and equilibrium is completely reached [10].

In practice, the dependence of the sensor signal on the concentration φ of certain components B in one of the electrode compartments, is of interest. It is useful to divide U_{eq} between the electrodes of the cell in order to ascertain how the signal function $U(\varphi_B)$ arises. One defines electrode potentials and considers electrode potential equations which describe the dependence on temperature and composition quantities of the potential-forming substances.

In classical electrochemistry the electrode potential in aqueous electrolyte solutions is defined as the potential difference of the cell measured between the particular electrode and the Nernstian hydrogen reference electrode — symbol U_H with specifying parameters in parentheses [8–10]. For technical reasons it is not possible to measure the potentials of galvanic solid electrolyte cells at high temperatures against the Nernstian hydrogen reference electrode. Because a great number of thermodynamically defined electrodes with ion-conducting solid electrodes exist, it would be advantageous if these could be treated in a uniform way. Thus, electrodes in contact with pure oxygen at atmospheric pressure (standard pressure 101325 Pa, tabular pressure p_t) were chosen as a convenient reference base in 1964 [11]. The solid electrolyte of the standard oxygen electrode and of the counterelectrode should be the same pure ion conductor. But one can choose any electronic conducting material inert against oyxgen to form the electrode. Any arbitrary temperature can be chosen. (It must be stressed that "standard" in thermodynamics does not necessarily mean 25 °C [12]).

Thus, for galvanic cells with oxide-ion-conducting solid electrolytes, electrode potentials are the potential differences between the electrode in question and the standard oxygen electrode. The symbol is U_O with specifying parameters in parentheses. Its sign is negative when the oxygen partial pressure is smaller than atmospheric pressure, in accordance with the rules used for potential series in electrochemistry. The potential differences between different electrodes are calculated by taking the difference of the two electrode potentials. The potential differences of cells have no definite sign. The difference between the electrode potentials can be formed to always yield a positive sign. For this the electrode with the smaller oxygen partial pressure should always stand on the left of the cell diagram to follow the international convention [13]. An example is given below:

$$\text{Pt, } O_2\,(p'_{O_2}) \;|\text{stabilized } ZrO_2|\; O_2\,(p''_{O_2}),\text{ Pt} \tag{25-1}$$

$$U = U''_0 - U'_0 > 0\,, \quad \text{if} \quad p'_{O_2} < p''_{O_2}\,. \tag{25-2}$$

Under the aforementioned conditions of equilibrium the standard electrode potential of the oxygen electrode $U_O^{\ominus}(O_2)$ is always zero. If a considered electrode is at equilibrium likewise, the condensed substances that take part in the electrode reaction are present as defined equilibrium phases and only ideal gases at standard pressure p_t are present, then a standard electrode potential is obtained, eg, $U_O^\ominus(\text{Ni}, \text{NiO})$, $U_O^\ominus(\text{H}_2, p_t, \text{H}_2\text{O}, p_t)$. If the condensed phases contain impurities and if the partial pressures of the reactive components in the gas phase differ from standard pressure, then only electrode potentials can be measured, eg, $U_O(\text{NiO}, \text{Ni}_x\text{Mg}_{1-x}\text{O})$, $U_O(\text{H}_2, p_1, \text{H}_2\text{O}, p_2)$.

The standard electrode potentials can be calculated from thermodynamic data of the substances participating in the equilibrium reaction. Electrode potentials contain additional contributions which depend on the composition of the mixed phases. These additional contributions are the origin for possible analytical applications.

Isothermal galvanic cells with only one known cell reaction can be calculated from the molar reaction Gibbs energy, $\Delta_r G$, and the number of electrons exchanged in the reaction, z_r, ie,

$$U_{eq} = \Delta_r G/(z_r F) \ . \tag{25-3}$$

If the cell reaction is only the change of molecular oxygen from a higher to a lower chemical potential then:

$$\Delta_r G = \mu(\text{O}_2)' - \mu(\text{O}_2)'' = RT \cdot \ln(p'_{\text{O}_2}/p''_{\text{O}_2}) \quad \text{and} \tag{25-4}$$

$$U_{eq} = (RT/4F) \cdot \ln(p'_{\text{O}_2}/p''_{\text{O}_2}) \ . \tag{25-5}$$

In this case we have a *concentration cell.*

Equation (25-5) can also be used if oxygen reacts in one of the electrode compartments with other substances (eg, H_2 or Ni). However, from this reaction together with the transfer of oxygen an overall reaction results that can be used as cell reaction to calculate U_{eq} from Equation (25-3).

Combining

$$\Delta_r G = \Delta_r G^\ominus + RT \cdot \ln \prod \alpha_i^{\nu_i} \tag{25-6}$$

with Equation (25-3) and setting $U^\ominus = \Delta_r G^\ominus/(z_r F)$, we obtain the general Nernst equation

$$U_{eq} = U^\ominus + [RT/(z_r F)] \cdot \ln \prod \alpha_i^{\nu_i} \ . \tag{25-7}$$

$U^\ominus$ is the standard cell potential difference, which is determined only by the reactants in definited standard states. This quantity $U^\ominus$ results as the difference of standard electrode potentials. The power term $\prod \alpha_i^{\nu_i}$ contains the corrected composition quantities α_i (fugacities and activities) with the stoichiometric coefficients ν_i of the gases and condensed substances taking part in the cell reaction [10, 12]. If a sensor at equilibrium delivers signals in agreement with Equation (25-7) then we have a *reaction cell.* In this case at solid electrolytes with oxide ion vacancies V_O, two reactions can be found:
besides

$$1/2 \ \text{O}_2(\text{g}) + V_O(\text{s}') + 2\,\text{e}^-(\text{s}'') \rightleftharpoons \text{O}^{2-}(\text{s}') \tag{25-8}$$

also reactions without molecular oxygen, eg,

$$H_2O\,(g) + V_O\,(s') + 2\,e^-\,(s'') \rightleftharpoons O^{2-}\,(s') + H_2\,(g)\ . \tag{25-9}$$

The concept of electrode potentials, described here, has great advantages over considerations based on thermodynamic data calculated with measured potential differences of cells: for application of solid-electrolyte potentiometric sensors it is simple to understand, results follow immediately and thus it is very helpful in practical cases.

25.2.1 Oxygen Gas Electrodes

The partial pressures used in the thermodynamic equations are proportional to the volume concentrations $\varphi_B = v_B/v$:

$$p_B = p \cdot \varphi_B \tag{25-10}$$

where p is the total pressure of the gas phase. The differences between the pressures p' and p'' in both electrode compartments are often so small, that the *ratio* of the oxygen concentrations can be used in Equation (25-5). Deviations from ideal behavior (which could be taken into account by fugacity coefficients) can usually be ignored in gas potentiometry.

If one electrode of an oxygen concentration cell is the standard oxygen electrode, then U_{eq} is the electrode potential of the second electrode. With the values of the fundamental constants, the electrode potential equation of the oxygen electrode is given by

$$\begin{aligned}
U_O\,(O_2)/mV &= 0.021543\,(T/K) \cdot \ln\,(p_{O_2}/p_t) \\
&= 0.049606\,(T/K) \cdot \lg\,(p_{O_2}/p_t)
\end{aligned} \tag{25-11}$$

$$U_O\,(O_2; p_t)/mV = 0.049606\,(T/K)\,\lg\,\varphi_{O_2}\ . \tag{25-12}$$

In Figure 25-1 a net of lines with the origin at 0K shows the variation of such electrode potentials with temperature.

The large range of the variation of partial pressure or oxygen concentration means that the use of decadic steps and, for some purposes, of pO similar to pH is sensible [11, 14]:

$$pO = -\lg\,(p_{O_2}/p_t)\ . \tag{25-13}$$

The majority of sensors with ZrO_2 solid electrolytes use air as a reference gas. For dry air [15] at standard pressure p_t the theoretical electrode potential equation is given by:

$$U_O\,(air,\,dry;\,p_t) = 0.049606\,(T/K) \cdot \lg\,0.2093 = -0.03369\,T/K\ . \tag{25-14}$$

A diagram of the air electrode potential is shown in Figure 25-1. The absolute value of the air electrode potential rises with increasing moisture content and falls with increasing total air pressure. Eg, one can find an increase from 36.16 to 36.43 mV at 800 °C caused by 50%

relative humidity (at 20 °C) and a descrease to 35.93 mV due to 100 mm water gauge pressure over p_t. In practice, these sources of error can often be compensated considerably.

For a potentiometric oxygen sensor with an air reference electrode and the sensor signal

$$U(\text{Sensor}) = U_O(\text{air, dry}; p_t) - U_O(\text{gas}; p_t) \tag{25-15}$$

the oxygen concentration in the test gas in vol. % is given by

$$\varphi_{O_2}/\text{vol.-\%} = 20.93 \exp\{-46.42\,(U/\text{mV})/(T/\text{K})\}\ . \tag{25-16}$$

Apart from the small sources of errors given above, other disturbances can affect the use of this equation (see Section 25.4).

25.2.2 Water Vapor and Carbon Dioxide Electrodes

The following dissociation equilibria occur at electrodes under water vapor or carbon dioxide:

$$H_2O \rightleftharpoons H_2 + 1/2\,O_2 \tag{25-17}$$

or

$$CO_2 \rightleftharpoons CO + 1/2\,O_2\ . \tag{25-18}$$

In pure water vapor or carbon dioxide the following equivalence relationships are valid:

$$p_{H_2} = 2\,p_{O_2} \quad \text{and} \quad p_{CO} = 2\,p_{O_2}\ . \tag{25-19}$$

The equilibrium constants K_1 and K_2 from the law of mass action, where:

$$K_1 = \frac{p_{O_2}^{1/2}\,p_{H_2}}{p_{H_2O}\,p_t^{1/2}} \tag{25-20}$$

and

$$K_2 = \frac{p_{O_2}^{1/2}\,p_{CO}}{p_{CO_2}\,p_t^{1/2}} \tag{25-21}$$

are given approximately between 700 and 1400 °C [16] by:

$$\lg K_1 = 2.947 - 13008\ \text{K}/T \tag{25-22}$$

and

$$\lg K_2 = 4.505 - 14700\ \text{K}/T\ . \tag{25-23}$$

Combining Equations (25-11) and (25-19) to (25-23) we obtain the electrode potential equations:

$$U_O(H_2O)/mV = 0.049606\,(T/K)\cdot(2/3)\cdot\lg(K_1\,p_{H_2O}/2\,p_t)$$

$$= -430.1 + [0.0875 + 0.0331\,\lg(p_{H_2O}/p_t)]\,T/K \tag{25-24}$$

$$U_O(CO_2)/mV = -486.1 + [0.1390 + 0.0331\,\lg(p_{CO_2}/p_t)]\,T/K . \tag{25-25}$$

The variation with temperature of the electrode potentials for pure water vapor and pure carbon dioxide at standard pressure is shown in Figure 25-1.

If deviations from the equivalence relationships (25-19) are given, ie, if an excess of oxygen, hydrogen or carbon monoxide is present, then the temperature curve of the electrode potential deviates from the curve calculated using Equations (25-24) or (25-25) as soon as the excess concentrations are larger than the concentration of the dissociation products in the equilibria given in Equations (25-17) and (25-18). Graphs with large curvature are obtained then. Figure 25-2 [11] shows such curves for inert gases that contain saturated water vapor (25 °C) and an excess (stepped in decades) of hydrogen or oxygen. In the case of hydrogen excess the curves lie above the water vapor curve and in the case of oxygen excess below it.

The electrode potential equations in the transition regions [17] (generalized in comparison to [11]) are given by:

$$U_O(H_2O \text{ with } H_2 \text{ or } O_2; \text{ or } CO_2 \text{ with } CO \text{ or } O_2)/mV =$$

$$= 0.0992\,(T/K)\cdot\lg\left[\sqrt[3]{\frac{aC}{4} + \sqrt{\left(\frac{aC}{4}\right)^2 + \left(\frac{b}{3}\right)^3}} + \right.$$

$$\left. + \sqrt[3]{\frac{aC}{4} - \sqrt{\left(\frac{aC}{4}\right)^2 + \left(\frac{b}{6}\right)^3}}\right] \tag{25-26}$$

or when $b > 6\,(aC/4)^{2/3}$ (casus irreducibilis)   with $\cos\varphi = aC/[4\,(b/6)^{3/2}]$:

$$U_O(H_2O \text{ with } H_2 \text{ or } O_2, \text{ or } CO_2 \text{ with } CO \text{ or } O_2)/mV =$$
$$= 0.0496\,(T/K)\cdot\lg[(2b/3)\cos^2(\varphi/3)] \tag{25-27}$$

where C is either K_1 from Equation (25-22) or K_2 from Equation (25-23) depending on the dissociating gas; a is the partial pressure of H_2O or CO_2, and b is the partial pressure of H_2 or CO or the double oxygen pressure in the state without dissociation of H_2O or CO_2, ie, at room temperature.

Figure 25-2 shows not only the transition plots of the electrode potentials in the electrode potential curve of the aforementioned inert gas,water vapor mixture, but also the electrode potentials $U_O(H_2O)$ from Equation (25-24) for water vapor partial pressures in decadic steps. This family of curves gives the dependence on temperature and pressure of the equivalence points at the potentiometric titration of hydrogen with oxygen [18]. Figure 25-2 uses the example of the water,hydrogen,oxygen system to give the electrode potentials that can be theoretically expected in a "neutral" or "almost neutral" gas.

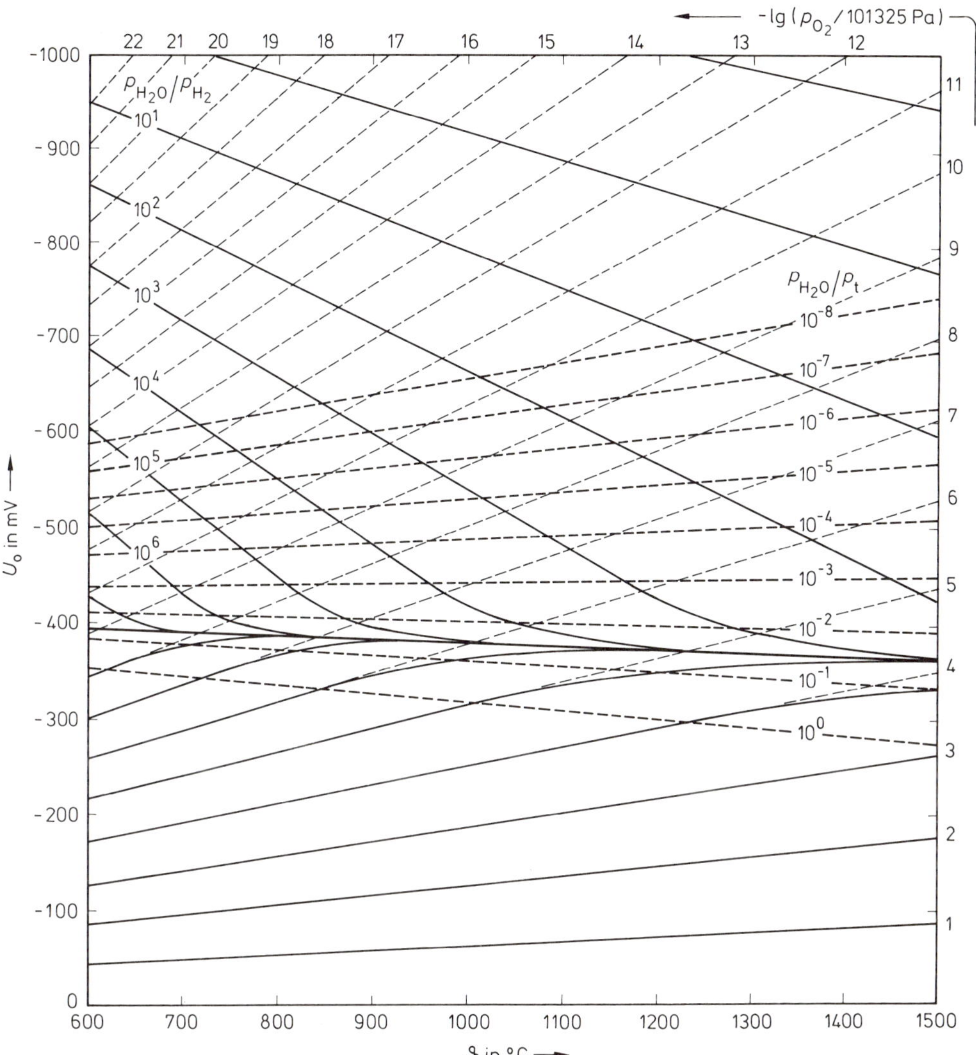

Figure 25-2. U/T diagrams for oxygen and hydrogen, water vapor electrodes for gases saturated with water vapor at 25 °C and for water vapor electrodes at various pressures [11].

25.2.3 H_2,H_2O and CO,CO_2 Electrodes

Figure 25-2 shows that, on cooling, the hydrogen or oxygen excess curves yield the appropriate linear electrode potential. How soon this happens depends on the oxygen concentration or H_2,H_2O ratio. The oxygen curves follow Equation (25-11). The H_2,H_2O curves are linear, as soon as the partial pressure of oxygen in the equilibrium given by Equation (25-17) is small enough to be ignored compared to the hydrogen partial pressure. Combining Equations (25-11), (25-20) and (25-22) one obtains the electrode potential equation in this region:

$$U_O\,(H_2, H_2O)/mV \;=\; -1290.6 + [0.2924 - 0.0992 \cdot lg\,(p_{H_2}/p_{H_2O})]\,T/K \qquad (25\text{-}28)$$

and similarly with Equations (25-11), (25-21) and (25-23)

$$U_O\,(CO, CO_2)/mV \;=\; -1458.4 + [0.4470 - 0.0992 \cdot lg\,(p_{CO}/p_{CO_2})]\,T/K\;. \qquad (25\text{-}29)$$

At about 813 °C, the equilibrium constants K_1 and K_2 given by Equations (25-22) and (25-23) are equal. The electrode potential curves for pure water vapor and carbon dioxide (at the same pressure) as well as the curves for the same ratios of H_2,H_2O and CO,CO_2, intersect at this temperature (see Figure 25-1). H_2 and CO with H_2O and CO_2 lead to the same pO and have the same effectiveness in redox reactions in this region of temperature. For this reason 813 °C is recommended as the fixed temperature for the sensor in devices that are to be used to measure combustible gas mixtures [11, 79].

Away from the transition region with its crooked curves (see Figure 25-2), the electrode potentials $U_O\,(H_2,H_2O)$ and $U_O\,(CO,CO_2)$ are independent of the degree of dilution with inert gas of the H_2,H_2O or CO,CO_2 mixtures. Dilution, however, is of importance for the pO

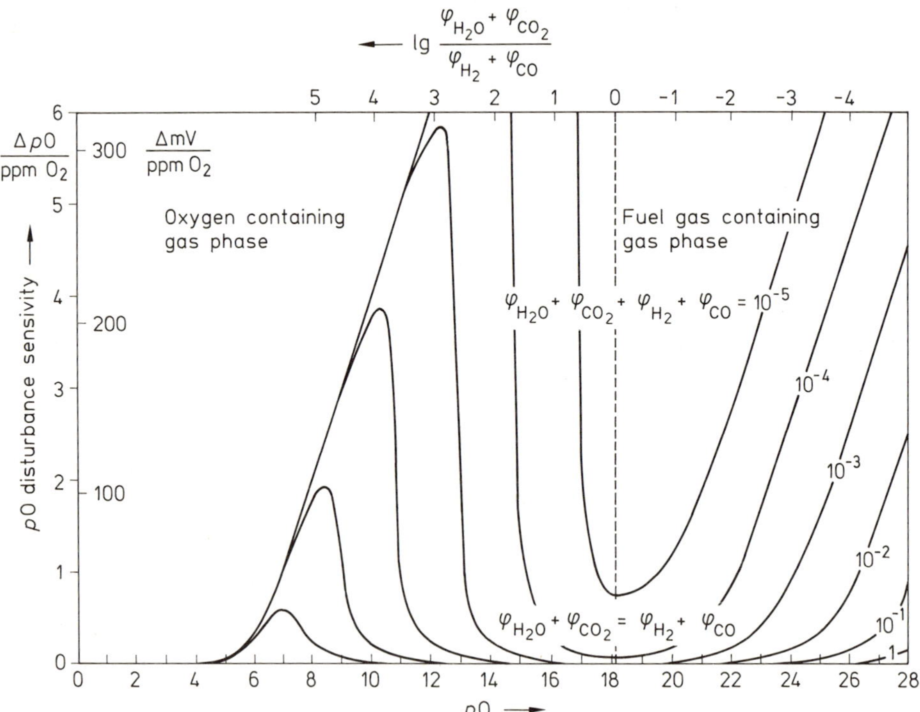

Figure 25-3. Change of pO on the addition of 1 vol.-ppm oxygen to different gas mixtures at 813 °C as a function of pO [11].

disturbance sensitivity [11] of a gas, which corresponds to the sensitivity of the electrode potential to oxygen inleakage in the sensor arrangement. Figure 25-3 shows the pO disturbance sensitivity vs. pO for various combinations of H_2O, CO_2, H_2 and CO concentrations. The largest buffer capacity of gaseous mixtures containing combustible gases is given in the range where the concentration of fuel gas molecules is equal to that of the burnt gas molecules. The smallest decrease of the pO-buffer capacity occurs in this region on dilution of the equilibrium gas mixture.

It can be seen from the family of electrode potential plots of the oxygen electrodes (Figure 25-1), that the oxygen concentration for a H_2, H_2O ratio of $1:1$ at $1000\,°C$ is under 10^{-14} and at $600\,°C$ is about 10^{-24}. The number of molecules in gases at standard pressure is very large, eg, $5.76 \cdot 10^{18}$ per cm^3 at $1000\,°C$. Thus, there are $16\,600$ oxygen molecules per cubic centimeter in the above example at $1000\,°C$. Potentiometric sensors with oxide-ion-conducting solid electrolytes give correct results even when only a fraction of one molecule per cubic centimeter is present, eg, at $600\,°C$ ca. 10^{-5} molecules/cm^3. This becomes clear when examining the reaction given by Equation (25-9). We can therefore assume, that in oxygen sensors with oxide-ion-conducting solid electrolytes, on changing from a higher to a lower oxygen partial pressure at the electrode, besides the electrode reaction shown as Equation (25-8) electrode reactions of type (25-9) completely without molecular oxygen take place increasingly.

25.2.4   Water Gas Electrodes

Equilibria (25-17) and (25-18) often occur in flames or hot reducing gas mixtures where there can only be one and the same oxygen partial pressure. The gas components come to equilibrium when the gas is hot enough, due to the water gas shift reaction:

$$CO_2 + H_2 \rightleftharpoons CO + H_2O . \tag{25-30}$$

In a water gas mixture the redox ratio

$$Q = (p_{CO_2} + p_{H_2O})/(p_{CO} + p_{H_2}) \tag{25-31}$$

and the carbon/hydrogen ratio

$$V = (p_{CO} + p_{CO_2})/(p_{H_2} + p_{H_2O}) \tag{25-32}$$

are constant parameters.

Combining these parameters and Equations (25-20) to (25-23) we obtain the oxygen partial pressure in water gas (WG), a simplified version of the equation given in reference [11]:

$$p_{O_2}(\mathrm{WG}) = \frac{1}{4}\left\{ \frac{Q+1}{V+1}(VK_1 + K_2) - K_1 - K_2 + \right.$$

$$\left. + \sqrt{\left[\frac{Q+1}{V+1}(VK_1 + K_2) - K_1 - K_2\right]^2 + 4K_1 K_2 Q} \right\}^2 . \tag{25-33}$$

Substitution into Equation (25-11) yields the water gas electrode potential equation:

$$U_O\,(H_2O, CO, H_2, CO_2)/mV = 0.049606\,(T/K)\,\lg\,(p_{O_2}\,(WG)/p_t) \qquad (25\text{-}34)$$

The electrode potential curves lie between those for pure H_2, H_2O and CO, CO_2 mixtures (Figure 25-1). They are steeper if the proportion of CO, CO_2 is larger than H_2, H_2O.

At 813 °C, K_1 equals K_2 and V is no longer important. Combining a reduced version of Equation (25-33) with Equations (25-34) and (25-22) results in

$$U_O\,(H_2O, CO, H_2, CO_2;\ 813\,°C)/mV = -973.0 + 107.8\,\lg\,Q\,. \qquad (25\text{-}35)$$

If U_O is measured at a temperature different from 813 °C, we can calculate V applying the determined Q. It is not possible to measure all four components in water gas directly using gas potentiometry. But it should be possible theoretically on the basis of the Equations (25-33) to (25-35) to determine the two important parameters V and Q by simple measurements at two temperatures. In order to determine V, T has to be as far as possible from 1086 K (813 °C) and both measurements must be very exact.

Other equilibria possible in water gas mixtures are the following:

$$CO + 2\,H_2 \rightleftharpoons CH_3OH \qquad (25\text{-}36)$$

and

$$CO + 3\,H_2 \rightleftharpoons CH_4 + H_2O\,. \qquad (25\text{-}37)$$

In most cases, however, the kinetic conditions of these reactions are unfavorable and the concentrations achieved at equilibrium are negligibly small.

25.2.5 Carbon Electrodes

Not only the partial pressure of oxygen has a distinct value in the water gas equilibrium (25-30), but also the partial pressure of carbon. This is due to the following equilibria which exist simultaneously:

$$C + CO_2 \rightleftharpoons 2\,CO \qquad (25\text{-}38)$$

$$C + H_2O \rightleftharpoons CO + H_2\,. \qquad (25\text{-}39)$$

For the equilibria (25-30), (25-38) and (25-39) and ignoring correction coefficients, we obtain:

$$p_{CO}\,p_{H_2O}/(p_{CO_2}\,p_{H_2}) = K_3 \qquad (25\text{-}40)$$

$$p_{CO}^2/(p_{CO_2}\,p_t\,x_C)\quad = K_4 \qquad (25\text{-}41)$$

$$p_{CO}\,p_{H_2}/(p_{H_2O}\,p_t\,x_C) = K_5 \qquad (25\text{-}42)$$

where p_t is the standard pressure (101.325 kPa) and $x_C = n_C/\Sigma n_i$ is the mole fraction of carbon in a solid phase. The equilibrium constants and their temperature dependence are accessible by the equations:

$$K_3 = K_2/K_1 \tag{25-43}$$

$$\lg K_4 = 9.913 - 8920\,\mathrm{K}/T \tag{25-44}$$

$$K_5 = K_4 K_1/K_2 \,. \tag{25-45}$$

The mass action relationships for (25-38) and (25-39) can also be described using equilibrium partial pressures of carbon, p_C. These relationships show that p_C increases with decreasing temperature. The value of saturation, $p_{C,sa}$, can be reached at higher or lower temperatures depending on composition. If there are no retardations then reactions (25-38) and (25-39) run reversely with decreasing temperature. Because of carbon precipitation, p_C is always equal to $p_{C,sa}$; p_{CO_2} and p_{H_2O} become larger and p_{CO} and p_{H_2} smaller, ie, Q and the oxygen partial pressure increase and hence the absolute value of U_O (water gas) decreases.

The electrode potential curve for the equilibrum of pure carbon and a pure CO,CO_2 mixture (Boudouard-Equilibrium) at a total pressure p_t is shown in Figure 25-1. This is the border line of all electrode potential curves of pure CO,CO_2 mixtures. The plot is linear at low and high temperatures, because the concentration of either CO in comparison to CO_2 or vice-versa can be ignored. In the limiting cases there are the equilibria

$$\text{below } 500\,^\circ\mathrm{C}: \quad C + O_2 \rightleftharpoons CO_2 \tag{25-46}$$

$$\text{above } 900\,^\circ\mathrm{C}: \quad C + 1/2\,O_2 \rightleftharpoons CO \,. \tag{25-47}$$

The equilibrium constants for Equations (25-46) and (25-47) are

$$\lg K_6 = 0.127 + 20562\ \mathrm{K}/T \text{ for } 100 \ \ldots \ 500\,^\circ\mathrm{C} \tag{25-48}$$

$$\lg K_7 = 4.568 + 5916\ \mathrm{K}/T \text{ for } 900 \ \ldots \ 1300\,^\circ\mathrm{C} \,. \tag{25-49}$$

These equilibrium constants can be used to calculate the electrode potentials:

$$U_O\,(C,CO_2;100 \ \ldots \ 500\,^\circ\mathrm{C})/\mathrm{mV} = -1020 - (0.00635 - 0.0496\,\lg p_{CO_2})\,T/\mathrm{K} \tag{25-50}$$

$$U_O\,(C,CO;900 \ \ldots \ 1300\,^\circ\mathrm{C})/\mathrm{mV} = -587 - (0.453 - 0.0992\,\lg p_{CO})\,T/\mathrm{K}\,. \tag{25-51}$$

Between 500 and 900 °C the electrode potential plots are curved and depend mainly on K_4. With the pressure parameter

$$P = (p_{CO} + p_{CO_2})/p_t \quad \text{one obtains}$$

$$U_O\,(C,CO,CO_2)/\mathrm{mV} = -1458 + [0.417 + 0.0992\,\lg\,(\sqrt{1 + 4P/K_4} - 1)]\,T/\mathrm{K}\,. \tag{25-52}$$

If the gas at the carbon electrode is diluted with an inert gas or with H_2 or H_2O, then at standard pressure we have $P < 1$ and the carbon plot is displaced upwards. Therefore the carbon precipitation from a water gas or a water gas, inert gas mixture starts at higher absolute values of the electrode potential.

In contact with a solid material, eg, steel, exceeding the carbon electrode potential for a gas means that carbon can precipitate as pure graphite on the solid. The solid saturates with carbon, until the excess settles out on the surface.

However, before this stage of carbon precipitation is reached, carbon exchange between solids and water gas can occur. At equilibrium between condensed and gaseous phases the chemical potentials of carbon in both phases are equal following Henry's law. By taking into account solubility constants, the corresponding electrode potential equations can be derived. These serve as a basis for the control of carburizing and decarburizing processes at the surface of steels with solid-electrolyte sensors.

25.2.6   Metal,Metal Oxide Electrodes

If two not completely miscible condensed phases can achieve a chemical equilibrium via gas phase oxygen, eg,

$$NiO \rightleftharpoons Ni + 1/2\,O_2 \tag{25-53}$$

then a distinct oxygen partial pressure arises (due to the law of mass action), that is only dependent on temperature (Gibbs' phase rule). If the metal and metal oxide are *pure,* we can obtain a standard electrode potential with the appropriate thermodynamic data. With experimental electrochemical data determined with an air reference electrode [19] it was found that

$$U_O^{\ominus}\,(Ni, NiO)/mV = -1202 + 0.431\,T/K \quad \text{for } 650 \ldots 1100\,°C\,. \tag{25-54}$$

If the oxygen partial pressure in the surrounding of an Ni,NiO mixture is greater than that calculated from Equations (25-54) and (25-11) then nickel is oxidized. In the reverse case, (the oxygen partial pressure is smaller) it is reduced. If a gas flows over a Ni,NiO mixture, then by comparing the electrode potential measured in the gas with that calculated from Equation (25-54) and taking temperature into account, we can decide whether oxidation of Ni or reduction of NiO occurs.

On the other hand, Ni,NiO mixtures that are contained in a limited volume or in an inert gas, deliver at constant temperature a definite constant oxygen partial pressure, which can be used in reference electrodes. The electrode reaction can run without molecular oxygen (cf. Equation (25-9)):

$$NiO\,(s') + V_O\,(s'') + 2\,e^-\,(s''') \rightleftharpoons O^{2-}\,(s'') + Ni\,(s''')\,. \tag{25-55}$$

The law of mass action for equilibrium (25-53),

$$p_{O_2}^{1/2}\,a_{xNi}/(a_{xNiO}\,p_t^{1/2}) = K_8, \tag{25-56}$$

leads to deviations from Equation (25-54), if the solid phases are impure, ie, the mole fractional activities a_x are not equal to unity, eg, in nickel base alloys or in NiO-MgO mixed crystals.

The electrode potential curves for equilibria with the three pure iron oxides FeO, Fe_3O_4 and Fe_2O_3 are shown in Figure 25-1. Electrochemical measurements yield for the three plots at high temperatures [20]:

$$U_O^{\ominus} (Fe, Fe_xO)/mV \quad = -1387.3 + [0.45396 - 0.0140 \cdot \ln(T/K)] \, T/K \quad (25-57)$$

$$U_O^{\ominus} (Fe_yO, Fe_3O_4)/mV = -1407.1 - [0.89187 - 0.19116 \cdot \ln(T/K)] \, T/K \quad (25-58)$$

$$U_O^{\ominus} (Fe_3O_4, Fe_2O_3)/mV = -1049.4 - [0.71879 - 0.17787 \cdot \ln(T/K)] \, T/K \,. \,(25-59)$$

The regions of stability of the various iron oxides are found between the electrode potential curves. The electrode potential resulting with a gas mixture must lie between these curves, if one wishes to convert iron oxides into this particular state by influence of the gas mixture.

It is made clear in Equations (25-57) and (25-58) by the stoichiometric coefficients x and y, that FeO in equilibrium with iron has a stoichiometry different from FeO in equilibrium with Fe_3O_4. In the region of stability of the FeO phase the stoichiometry varies from x to y with change in p_{O_2}. Thus, conditions can be adjusted by oxygen sensors to produce various defect concentrations.

If Fe_2O_3 is mixed homogeneously with Al_2O_3 in ceramic components, then the reduction of these mixed phases cannot be exactly controlled using Equation (25-59). Special investigations must be carried out in order to measure the reduction potentials. These studies are important for the porcelain industry. The color of ceramics and glazes is influenced to a large extent by the oxygen partial pressure in the kiln gas during firing [21].

Equations (25-54) and (25-57) to (25-59) are valid only for solid phases. For systems containing liquid phases, eg, for molten FeO over liquid iron, analogue relationships are achieved using other parameter values. The multiplicity of metal,metal oxide electrodes has been shown here only with a few characteristic examples.

25.2.7 Oxygen Solution Electrodes

We can determine the oxygen partial pressure above oxygen containing solutions with an oxygen gas sensor. If the coefficient of oxygen solubility in the condensed phase is known and equilibrium is achieved, it is hence possible to determine the oxygen concentration in the solution. It is, however, more advantageous to determine the oxygen activity in solutions directly with contact of oxygen sensors. These can contain gaseous or solid reference electrodes.

The cell reaction is often that oxygen react as molecular species at the reference electrode and is dissolved as atoms at the measurement electrode (eg, in silver or iron):

$$1/2 \, O_2 \, (g) \; \rightleftharpoons \; O \, (solution) \,. \tag{25-60}$$

Another variation is that an oxide is formed and dissolved (eg, in sodium). Henry's law about the distribution of the same species between two ideal mixtures holds only rarely. As a rule, the process of dissolution is a chemical reaction and can be expressed using the law of mass action. Thus oxygen sensors for solubility determinations are reaction cells. The equilibrium constants depend on the nature of the reacting substances, the solvent and the temperature, but are almost independent of the total pressure.

Mostly in the case of solutions of oxygen and always in the case of solutions of oxides in metals, the upper solubility limit is given by the appearance of an oxide phase. The electrode potential equation in this case is derived in two steps.

Initially we need the electrode potential at equilibrium between the oxide-saturated metal phase and the oxide phase. An example is given by the cell reaction:

$$2\,Na\,(\ell, sa.Na_2O) + 1/2\,O_2\,(g) \rightleftharpoons Na_2O\,(\ell)\,. \tag{25-61}$$

The temperature function is of the type given in Section 25.2.6. It can be experimentally determined using solid-electrolyte cells with solid-state reference electrodes [22].

Diluted solutions of Na_2O in Na correspond to more negative potentials. The change in potential on dilution can be described thermodynamically by the term $(RT/2F) \cdot \ln{(c/c_{sa})}$, if activity coefficients correcting non-ideal behavior can be ignored. Here c is the concentration of Na_2O or O in the diluted solution; c_{sa} is the concentration of Na_2O or O in the saturated solution. To obtain the sensor function $U(c, T)$ it is necessary to determine c_{sa} as a function of temperature in the second step and substitute it. In the example given, experimental data [22] together with Equation (25-54) yield the electrode potential equation

$$U_O\,(Na_2O \text{ dissolved in } Na,\,\ell)/mV = -2093 + [0.660 + 0.0992\lg{(c/c_{sa})}]\,T/K$$

$$= -1809 - (0.042 - 0.0992\lg{(c/\text{mass ppm O})})\,T/K \tag{25-62}$$

For measurements of oxygen dissolved in pure iron the following equation is determined from the data in [23, 24]:

$$U_O\,(O \text{ dissolved in } Fe,\,\ell)/mV = -1404 + [0.355 + 0.0992 \cdot \lg{(c/c_{sa})}]\,T/K$$

$$= -818 - [0.105 - 0.0992 \cdot \lg{(c/\text{mass}\%O)}]\,T/K \tag{25-63}$$

Figure 25-1 shows the calculated results, giving the range in which the steel industry uses dipping probes extensively.

Of course, other relationships hold for iron alloys, where the interaction of oxygen with alloy components can be taken into account by characteristic coefficients [25]. In particular, the addition of aluminium reduces the concentration of oxygen in the molten steel. This method is applied everyday in the steel industry. It is controlled by solid-state electrolyte sensors which use Cr,Cr_2O_3 mixtures as reference system and are stable in molten steel for only 15–20 s.

25.3 Signal Functions of Potentiometric Oxygen Sensors with Oxoanionic Solid Electrolytes

For a long time the principle was assumed to be valid that solid electrolytes for gas analysis have to contain such ions as the migrating species which could be formed from particles of the gas phase [6]. According to that, oxide-ion-conducting solid electrolytes are useful for electrode reactions as quoted in Equations (25-8) and (25-9), and in connection with halogenide ion-conducting solid electrolytes electrode reactions such as

$$\frac{1}{2}\,Cl_2\,(g) + V_{Cl}\,(s'') + e^-\,(s') \rightleftharpoons Cl^-\,(s'') \,. \tag{25-64}$$

For a potentiometric determination of H_2S, sulfide-ion-conducting solid electrolytes were searched for [26].

However, the experiments of Gaugain [7] with galvanic cells such as

$$Pt,\ C_2H_5OH\,(g)\ |glass|\ air,\ Pt \tag{25-65}$$

and the special measurements of Haber et al. [3, 4] with similar cells had already shown that redox reactions of gases with normal glass and porcelain as solid electrolytes could be potentiometrically examined. Ever since Warburg's measurements on amalgam cells [27], it has been known that in simple glasses the electric conductivity depends on the migration of sodium ions. In the glass electrodes for measurements of the activity of H^+ ions in aqueous solutions no protonic conductivity could be found [28]. Glass and silicate ceramics were the first alkali-ion-conducting oxoanionic solid electrolytes to be used in gas potentiometry.

In 1977 Gauthier et al. [29, 30] inspired a new development using sulfates, carbonates and nitrates. With galvanic cells containing pure K_2SO_4, K_2CO_3, and $Ba\,(NO_3)_2$ as solid electrolytes the concentration of SO_x, CO_2, or NO_x was measured in gases. Furthermore it was suggested to measure AsH_3 with $Ca_3\,(AsO_4)_2$, PH_3 with $Ca_2P_2O_7$, and SeH_2 with K_2SeO_4 [31].

The galvanic potential differences in the electrodes of galvanic cells with oxoanionic solid electrolytes are determined by gas components which are not in reaction equilibrium with the cations, whose mobility determines the conductivity. Galvanic potential differences result according to the law of mass action for electrode reactions as follows:

$$CO_2\,(g) + \frac{1}{2}\,O_2\,(g) + 2e^-\,(s') \rightleftharpoons CO_3^{2-}\,(s'') \quad \text{and} \tag{25-66}$$

$$SO_3\,(g) + \frac{1}{2}\,O_2\,(g) + 2e^-\,(s') \rightleftharpoons SO_4^{2-}\,(s'') \,. \tag{25-67}$$

Pure ionic conductivity of the solid electrolytes caused by anyone type of ions is only necessary in order to be able to measure the sum of all galvanic potential differences in the electric circuit of the galvanic cell without losses.

In practice, sensors with oxoanionic solid electrolytes are less successful till now, especially in tests of long-term stability. There are many reasons for this, a fundamental reason is given by the electrode processes taking place during unevitable current flows. Every direct current causes on one side a loss of solid electrolyte material in consequence of alkali ion migration and gas delivery. On the other side the discharge of alkali ions causes chemical reactions with gas components forming compounds like oxides, hydroxides, basic salts or hydrates which do not correspond to the solid electrolyte material. Every flow of direct current produces an asymmetry in the body of the oxoanionic solid electrolyte. At the cathode, besides the reactions (25-66) and (25-67), simultaneously electrode reactions are possible, for example,

$$H_2O\,(g) + \frac{1}{2}\,O_2\,(g) + 2\,e^-\,(s') \rightleftharpoons 2\,OH^-\,(s'') \tag{25-68}$$

and, therefore, mixed potentials and cross sensitivities result. Even when introducing the same gas into both electrode compartments, asymmetric potential differences will occur whose constance as a supposition for electronic compensation is not guaranteed.

First, we will disregard such disturbances and consider the equilibrium potential differences of the most important sensors with pure homogeneous oxoanionic solid electrolytes. Concepts for a solid-state electro-chemistry with electrode potentials of the kind used in Section 25.2 do not appear to be useful here.

25.3.1 Sensors with Carbonate Solid Electrolytes

The simplest type of sensors with oxoanionic solid electrolytes is the one with alkali carbonates. The electrode reaction (25-66) on both sides produces the following cell reaction:

$$CO_2\,(g') + \frac{1}{2}\,O_2\,(g') \rightleftharpoons CO_2\,(g'') + \frac{1}{2}\,O_2\,(g'') \tag{25-69}$$

and the equilibrium potential difference of a concentration cell:

$$U\,(CO_2, O_2) = \frac{RT}{F}\,\ln\,\frac{p''_{CO_2}\,p''^{1/2}_{O_2}}{p'_{CO_2}\,p'^{1/2}_{O_2}}\,. \tag{25-70}$$

If one feeds one electrode with air of a constant composition (for example, from a compressed-air store) and the other electrode with air from the environment, in which the variation of the CO_2 concentration due to meteorological and biological processes is small, then the signal functions simplify themselves due to practically constant O_2 partial pressures. For equal total pressures at the electrodes the following equation for determination of CO_2 is valid:

$$\varphi_{CO_2} = \varphi_{CO_2,ref}\,\exp\left[\frac{U\,(CO_2)/mV}{0.0992\;T/K}\right]\,. \tag{25-71}$$

If, however, both the O_2 concentration and the CO_2 concentration vary heavily at the measuring electrode, for example, in breathing gas, then one only achieves an applicable analytical signal function with the help of O_2 concentration cells. In the case of a measuring electrode, which is connected over a carbonate solid electrolyte oppositely to a CO_2,O_2 reference electrode and over an oxide-ion-conducting solid electrolyte oppositely to an O_2 reference electrode, one obtains an arrangement which is in principle simple [32, 33]:

$$\text{Au, } O_2\text{(air)} \quad\Big|\quad \text{stab. } ZrO_2 \quad\Big|\quad \text{Au, measuring gas} \quad\Big|\quad \text{carbonate} \quad\Big|\quad \text{Au, } CO_2, O_2 \qquad (25\text{-}72)$$
$$\text{pole 1} \qquad\qquad\qquad\qquad\qquad \text{pole 2} \qquad\qquad\qquad\qquad\qquad \text{pole 3}$$

In the case of equal total pressures of the gases in the electrode compartments, between the three poles the following equilibrium potential differences result:

$$U_{1,2} = \frac{RT}{4F} \ln \frac{\varphi_{O_2,2}}{\varphi_{O_2,1}} \tag{25-73}$$

$$U_{2,3} = \frac{RT}{2F} \ln \frac{\varphi_{O_2,3}^{1/2}\, \varphi_{CO_2,3}}{\varphi_{O_2,2}^{1/2}\, \varphi_{CO_2,2}} \tag{25-74}$$

$$U_{1,3} = U_{1,2} + U_{2,3}$$
$$\qquad = -\frac{RT}{2F} \ln \varphi_{CO_2,2} + \frac{RT}{4F} \frac{\varphi_{O_2,3}\, \varphi_{CO_2,3}^{2}}{\varphi_{O_2,1}} \tag{25-75}$$

$U_{1,3}$ varies only with φ_{CO_2} of the measuring gas, when $\varphi_{O_2,3}$, $\varphi_{CO_2,3}$ and $\varphi_{O_2,1}$ in the reference gases remain constant. Simultaneously to φ_{CO_2} $U_{1,2}$ yields the O_2 concentration in the measuring gas, which without influence on $U_{1,3}$ is allowed to vary.

Equation (25-75) is valid for $U_{1,3}$, independent from a metal electrode being fixed or left between the oxide-ion-conducting and the carbonate solid electrolyte. At the direct contact between the solid electrolytes the reaction

$$CO_3^{2-}\,(s') \rightleftharpoons CO_2\,(g) + O^{2-}\,(s'') \tag{25-76}$$

is possible. This reaction is the basis of the sensitivity of the electrode contacts for CO_2 [34].

Instead of the uncomfortable gas reference electrodes with certain CO_2 and O_2 concentrations several other solid reference systems have been suggested and tested [29]. Metallic silver in contact with silver-containing carbonates delivers no long-term stable reference basis because the silver-ion activities in the carbonate phase cannot be kept constant [32]. Systems like $CaO,CaCO_3$ provide a thermodynamic defined CO_2 partial pressure only dependent on temperature and can be examined exactly with carbonate cells [35]. Such systems, however, can be confined hermetically and temperature-resistantly only under great difficulties, so that in the long-term running disturbances occur due to leaks [32].

25.3.2   Sensors with Sulfate Solid Electrolytes

Before the electrode reaction on sulfates (Equation (25-67)) runs, a chemical reaction takes place:

$$SO_2\,(g) + 1/2\,O_2\,(g) \rightleftharpoons SO_3\,(g)\,. \tag{25-77}$$

Only when this chemical reaction is brought into equilibrium, a clear dependency of the galvanic potential difference at sulfate electrodes on the SO_2 concentration can be expected. For the equilibrium constant of the reverse reaction of Equation (25-77),

$$\frac{p_{SO_2}\,p_{O_2}^{1/2}}{p_{SO_3}\,p_{t}^{1/2}} = K_9 \tag{25-78}$$

with tabulated dates [36] one obtains as a first approximation

$$\lg K_9 = 4.846 - 5107\,\mathrm{K}/T\,. \tag{25-79}$$

In the chemical equilibrium between SO_2 and SO_3, depending on temperature, various different mixture ratios occur, which are also dependent on the O_2 concentration (Figure 25-4 [33]). The part of SO_2 of the total SO_x content rises from nearly 0 at 300 °C to 1 above 1000 °C. In air a considerable portion of SO_x is present as SO_3 up to high temperatures. Both other curves in Figure 25-4 for 0.2 vol.% SO_x in gases with 0.5 and 5 vol.% O_2 describe typical cases for the analysis of flue gases.

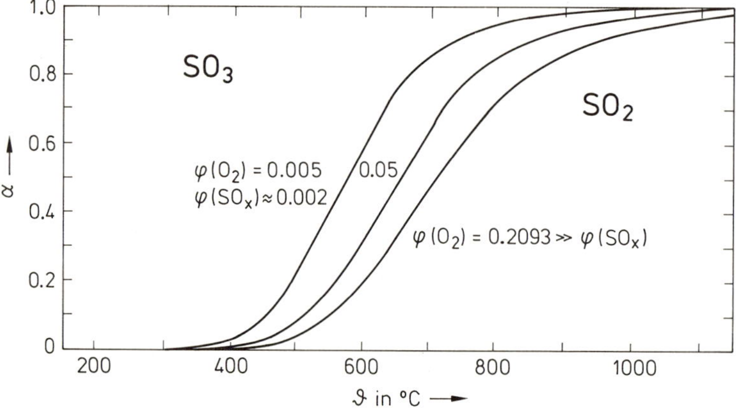

Figure 25-4. SO_3 and SO_2 as fraction of SO_x at various oxygen partial pressures at normal pressure in chemical equilibrium as function of temperature [33].

For a concentration cell with the cell reaction

$$SO_3\,(g') + \frac{1}{2}\,O_2\,(g') \rightleftharpoons SO_3\,(g'') + \frac{1}{2}\,O_2\,(g'') \tag{25-80}$$

with equal total pressures at both electrodes the potential-difference equation is as follows:

$$U(SO_3, O_2) = \frac{RT}{2F} \ln \frac{\varphi''_{SO_3}\, \varphi''^{1/2}_{O_2}}{\varphi'_{SO_3}\, \varphi'^{1/2}_{O_2}} \, . \tag{25-81}$$

Feeding an electrode with gas, whose SO_3 and O_2 concentration is known, and detecting the O_2 concentration in the measuring gas via $U(O_2)$ of a O_2 sensor, one can determine the SO_3 concentration in the examined gas (independently of the establishment of the equilibrium (25-77)) according to the equation

$$\varphi(SO_3) = \exp\left[\frac{2F}{RT}\left[U(SO_3, O_2) - U(O_2)\right]\right] + \ln \frac{(\varphi_{SO_3}\, \varphi^{1/2}_{O_2})_{RS}}{(\varphi^{1/2}_{O_2})_{RO}} \tag{25-82}$$

Here $(\varphi_{SO_3}\varphi_{O_2})_{RS}$ indicate concentrations at the reference electrode of the sulfate cell and $\varphi_{O_2, RO}$ the O_2 concentration at the reference electrode of the O_2 sensor.

Taking care with the help of catalysts [37, 34] that in both electrode compartments the reaction (25-77) remains in the equilibrium belonging to the cell temperature, then the following equation is valid also:

$$U(SO_2, O_2) = \frac{RT}{2F} \ln \frac{\varphi''_{SO_2}\, \varphi''_{O_2}}{\varphi'_{SO_2}\, \varphi'_{O_2}} \, . \tag{25-83}$$

For measurements in air against a reference gas consisting of air and a very small part of SO_2, one obtains in consequence of $\varphi'_{O_2} \cong \varphi''_{O_2}$ and with the mentioned supposition the simple equation

$$\varphi_{SO_2} = \varphi_{SO_2, ref} \exp \frac{U(SO_2)/mV}{0.0992\, T/K} \, . \tag{25-84}$$

If at room temperature the reference gas contains no SO_3, but only a known concentration of SO_2 then for the presupposed case the sensor signal at, eg, 800 °C delivers the SO_2 concentration, according to Equation (25-84). Actually, the SO_2 concentration is stable only above 1000 °C (see Figure 25-4) and metastable at room temperature. In spite of this fact, the correct SO_2 concentration is obtained because the degree of conversion from SO_2 into SO_3 (α in Figure 25-4) on both electrodes at measuring temperature is nearly equal by the help of catalysts and reduced in Equation (25-84).

For measurements in flue gas, variations of O_2 concentration and also great differences between α in the measuring and reference gas are to take into consideration. The amount looked for is the sum of the concentrations of SO_3 and SO_2, and in equilibrium this is

$$\varphi_{SO_x} = \varphi_{SO_3}\left(1 + K_9/\varphi^{1/2}_{O_2}\right) . \tag{25-85}$$

With that, Equation (25-81) changes to

$$U(SO_x, O_2) = \frac{RT}{2F} \ln \frac{\varphi''_{SO_x}\, \varphi''_{O_2}}{\varphi'_{SO_x}\, \varphi'_{O_2}} + \frac{RT}{2F} \ln \frac{\varphi'^{1/2}_{O_2} + K_9}{\varphi''^{1/2}_{O_2} + K_9} \, . \tag{25-86}$$

In the case of parallel determination of the O_2 concentration in the measuring gas with an O_2 sensor, this equation can be analytically used with the help of a computer program. In rough, the second term of Equation (25-86) can be omitted at high temperature because the value of K_9 exceeds that of the O_2 concentration. With the above used index symbols the equation for the determination of SO_x is in this case as follows:

$$\varphi_{SO_x} = \exp\left[\frac{2F}{RT}\left[U(SO_x, O_2) - 2U(O_2)\right]\right] + \ln\frac{(\varphi_{SO_2}\,\varphi_{O_2})_{RS}}{(\varphi_{O_2})_{RO}}. \tag{25-87}$$

Also for sensors with sulfate electrolytes in the region of application above 800 °C no stable Ag/Ag^+ reference electrodes could be found [37]. Better results were achieved with the solid state reference systems $MgO,MgSO_4$ and $MnO,MnSO_4$ [37]. Again, in the case of sulfate solid electrolyte sensors the most stable and relieable reference electrodes were those with flowing reference gas.

For direct contacts between oxide-ion-conducting and sulfate solid electrolytes equations analogue to Equations (25-75) and (25-76) are valid. The sensitivity of such contacts were examined in detail with and without catalysts [34] because without a metallic electrode phase a better long-term stability of the sensors with sulfate solid electrolytes is expected.

25.4 Conditions for Determination of Equilibrium Cell Potential Differences of Solid Electrolyte Gas Sensors

Once the functional relation between the electric quantity and the concentration of a gas constituent is well known for a solid/gas system, numerous development steps remain until a sensor can be put to practical use. The determining factor is the goal of application, for example the gas analysis in the laboratory, industrial gas analysis, its use in medical devices or in motor vehicles. In any case, long-term stable physico-chemical systems have to be established which have to fulfill certain conditions. These concern for example the sample extraction and treatment, the temperature, the stability of total pressure and of material qualities, and the construction in a miniature or mechanically and thermally robust form. With solid electrolytes of the same kind one has come across to completely different sensor designs, depending on the application (Figure 25-1).

The various problems which have to be considered in complete gas sensor systems are mainly dealt with in patent papers. A few general aspects shall be mentioned here and explained by examples.

A widespread method of sensor usage today consists of two steps: first regularly automatic ascertaining the sensor signal function with different calibration gases and applying signal processing techniques and second analyzing on a basis of such a calibration the sensor signals continually up until the next calibration. By usage of the calibration method the demands on the sensor are relatively small, the equipment, however, is expensive and requires much service and the uncertainties of the signal analysis continually remain between the calibrations.

Ultimately, those sensors are cheaper whose signals follow thermodynamic relations and whose reliability is guaranteed to be long-term stable by their construction and by the materials selections (calibration-free-method). Devices with such sensors are far-reachingly free of maintenance and are checked only temporarily or in the case of striking disturbances.

Evident examples for the variety of the construction of measuring-systems with solid electrolyte sensors are found in devices for potentiometrical recording of the oxygen concentration in flue gases. Two main ways with various advantages and disadvantages are walked on:

a) Measurements on industrial plants taken directly in the gas medium or in bypasses with probes whose heads contain sensors with or without electric heating. At surrounding temperatures up to 1600 °C one can receive signals practically without delay (which are therefore well-suited for control purposes) and without special sampling and maintenance.

b) Measurements in devices at the outside-wall of the measuring gas room or further away in measuring-watch-towers, whereby an intermediate-treated (cleaned, dried) gas stream is fed to the constantly temperated sensor. The resulting timeloss can be so large that the signals are useless for automatic regulations. However, in the prepared gas sample more than one gas component can be determined comfortably with sensors of different kinds and temperature.

Solid electrolyte cells with solid and gaseous reference systems were already provided for in the first proposal (Figure 25-5; [6]). Solid reference systems (cf. Section 25.2.6) seem to be very advantageous because based on them, small simple sensors can be produced [38]. The reference electrode potential of such sensors is however neither under oxidizing nor reducing conditions long-term stable [39]. In consequence of the oxygen permeability of oxide-ion-conducting solid electrolytes, which was examined because of this question [40], totally oxidized or reduced layers with increasing thickness in the reference system arise which are the cause of a growing concentration polarization of the cell. Only in measurements of traces in inert gases (cf. Section 25.5) and in short-term measurements in steel meltings (cf. Section 25.2.7), solid reference systems like Cu,Cu_2O and Ni,NiO or Cr,Cr_2O_3 stand the test. The system Pd,PdO [41] possesses advantages for the temperatures from 400 to 800 °C because of the relatively high oxygen partial pressure (1 bar at 875 °C), and systems like Sn,SnO_2, Ga,Ga_2O_3 and In,In_2O_3 [42] possess advantages because of the mobility of the melted metal phase; these systems can be regenerated by electrolysis.

Also when applying the air reference electrode, measurements of gases with large pO-disturbance-sensitivity (eg, of inert gases with small oxygen concentration or of hydrogenium with small water vapor concentration, cf. Figure 25-3) are disturbed in consequence of the oxygen permeation through the solid electrolyte (cf. borderlines of selfpolarization in Figure 25-1). Otherwise one receives long-term stable reference electrode potentials safest with streaming reference gases, eg, in steel meltings at 1600 °C over several minutes [43] and in industrial gases over weeks and years [44].

The conditions for the accomplishment of correct results with gas concentration cells correspond widely to those which have to be met in measurements of the transport number of the ions of solids in air/oxygen-cells [45]. At temperatures below 700 °C traces of oxydizable materials, the thickness of the electrode layers and pretreatment of the cell can influence the results of oxygen concentration measurements [46].

The use of potentiometric gas sensors without calibration by realizing equilibrium cell potential differences demands that above all the following is guaranteed:

a) thermically and chemically stable solid electrolyte material without open pores and with an ionic transport number $>99\%$;

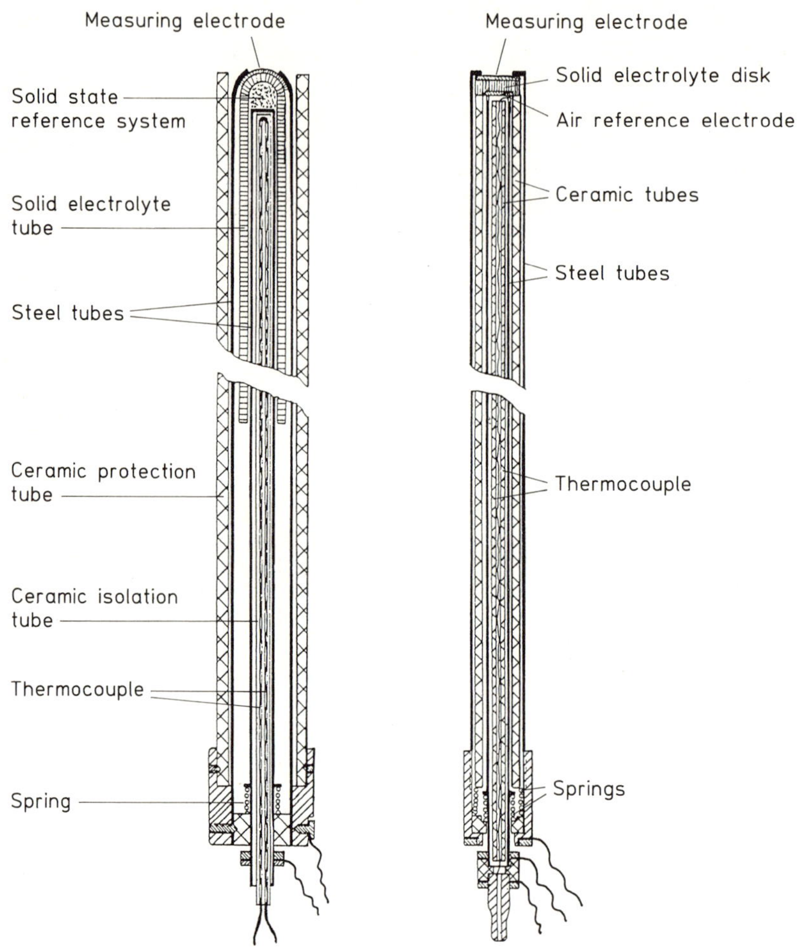

Figure 25-5. 1958 suggested probes for the measurement of the oxygen partial pressure in industrial gases with solid and with gaseous reference systems [6].

b) isothermic-isobaric electrode compartments with uniformly mixed gas phases at the contacts between electronically and ionically conducting phases;

c) electric conduction wires made out of the same material without contacts with the solid electrolyte material outside the electrode compartments;

d) gas streams which do not get contaminated by leaks and do not get changed by thermodiffusion in the area of temperature gradients;

e) sensor temperatures in the range where the desired chemical and electrochemical equilibria are reached to the full extend under the catalytical effects of the different parts of the sensor.

In some cases, eg, in measurements in nitriding gases of the hardening technology (NH_3, H_2, H_2O, N_2), proper results are already received above 500 °C without electric heating of the sensor [47–49]. At surrounding temperatures above 800 °C, in general the necessity to heat sensors electrically with ZrO_2 solid electrolytes is left off. Up to 1000 °C one can still work with metallic probe tubes (made of CrAl- or CrNiSi-steels, eg, in probes for carburizing or carbo-nitriding in hardening technology). Above 1000 °C, for applications in the glass and ceramic industry, only the ceramic constructed probes show long-term stability (tubes of sintered alumina; above 1200 °C mounted vertically, Figure 25-6, [50]). Short-term measurements are possible in gas compartments with temperatures up to about 1600 °C. Continuous long-term measurements are possible when moving the sensor from the hottest zone back to the wall of the kiln (Figure 25-7, [51, 52]). However, leaks and gas plaits can falsify the measurement then.

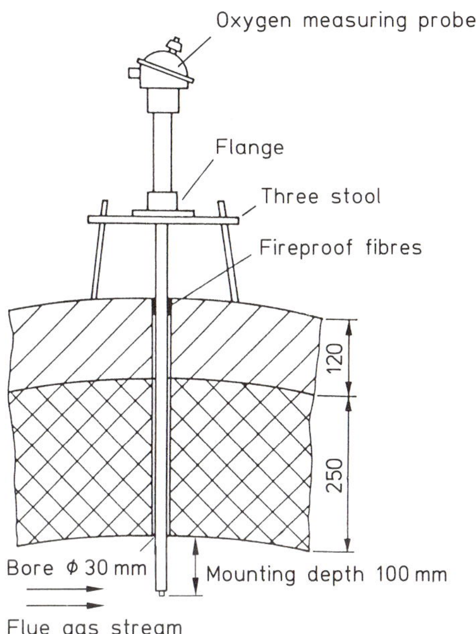

Figure 25-6.
Vertical mounting of a probe in the flue gas channel of a glass melting tank with temperatures >1200 °C [50].

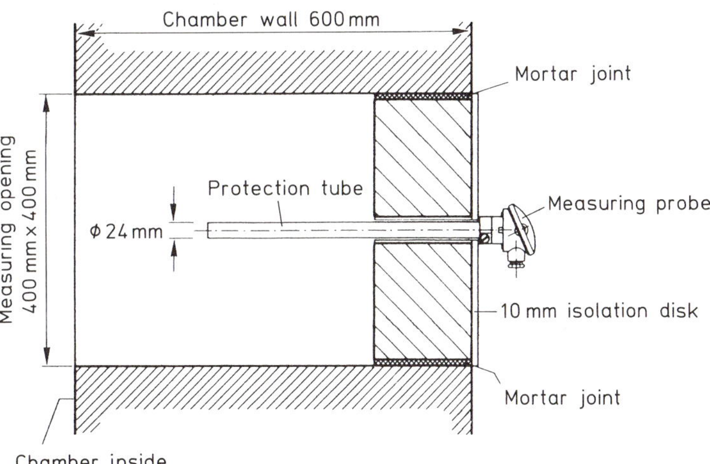

Figure 25-7. Horizontal mounting of a probe in a measuring opening in the chamber wall of a glass melting plant [51].

Probes without electric heating (high-temperature, HT probes) can be constructed easily with one-sided closed solid electrolyte tubes (Figure 25-8, [51]). As the ceramic industry produces tubes made of suitable solid electrolyte material only with lengths up to about 60 cm, the probe length is limited accordingly, if the seal between solid electrolyte tube and protection tube shall be attached outside the kiln plant. Longer probes can be produced with only 6 up to 10 cm long, both-sides open solid electrolyte tubes by mounting with ceramic mortar in alumina tubes [48, 50, 53, 54]. In this case, disturbances of the measuring electrode due to leaks of the ceramic mortar can be prevented by suitable guiding of the gas streams.

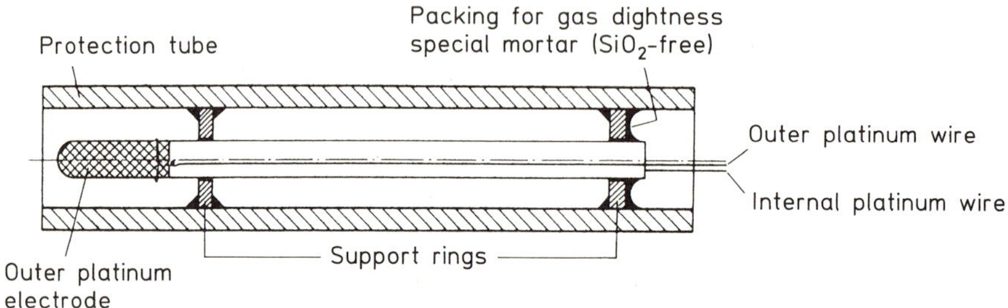

Figure 25-8. High temperature probe with one-end closed solid-electrolyte tube on supporting rings in a protecting tube [51].

The working time of HT-probes decreases quickly over glass meltings with increasing temperature [55]. Vapors of the oxides, above all of sodium, potassium, silicium, lead, antimon and arsen condense on the probe material in the temperature gradient from the inside to the outside of the tank wall. The condensed substances penetrate into the ceramic material over grain boundaries and lead to tube bursts. Thereby the stabilized zirconium dioxid is more sensitive than alumina. Components which contain SiO_2 must not be used in HT-probes [56].

The part of the solid electrolyte tube with Pt or PtRh electrodes being located outside the condensation zone in the measuring gas supplies largely correctly the thermodynamically expected signals, because equal temperatures and reaction equilibria adjust themselves on both electrodes in the big hot rooms of industrial plants. HT-probes are usually therefore applied without calibration and without in-situ checking. Comparing examinations of probes before and after insertion into the flue gas stream of glass melting tanks showed that the correctness of the signals practically does not change [57].

More often, probes in exhaust gases with temperatures below 800 °C are needed. In Figures 25-9 to 25-12 four technically carried out and tested examples are presented in order to make clear the different realization possibilities of probes with electric heating.

The solid electrolyte is soldered into a steel mounting as a round disk (Figure 25-9, [58]), or fastened into a ceramic binding substance as a one-sided closed tube (Figure 25-10, [59]), or held in a steel mounting as a both-sided open tube with special glasses (Figure 25-11, [44]), or totally embedded into ceramic mortars (Figure 25-12, [60]). The electric heating is arranged in air on one side of the solid electrolyte disk (Figure 25-9), whereby great differences between the temperatures of the electrodes are inevitable, or is laid axial-symmetrically around the sensor in the flue gas (Figure 25-10 to 25-12). In the arrangements shown in Figures 25-11 and 25-12, in addition the gas filter is heated, by which a sooting is prevented and a great working time of the filters is achieved.

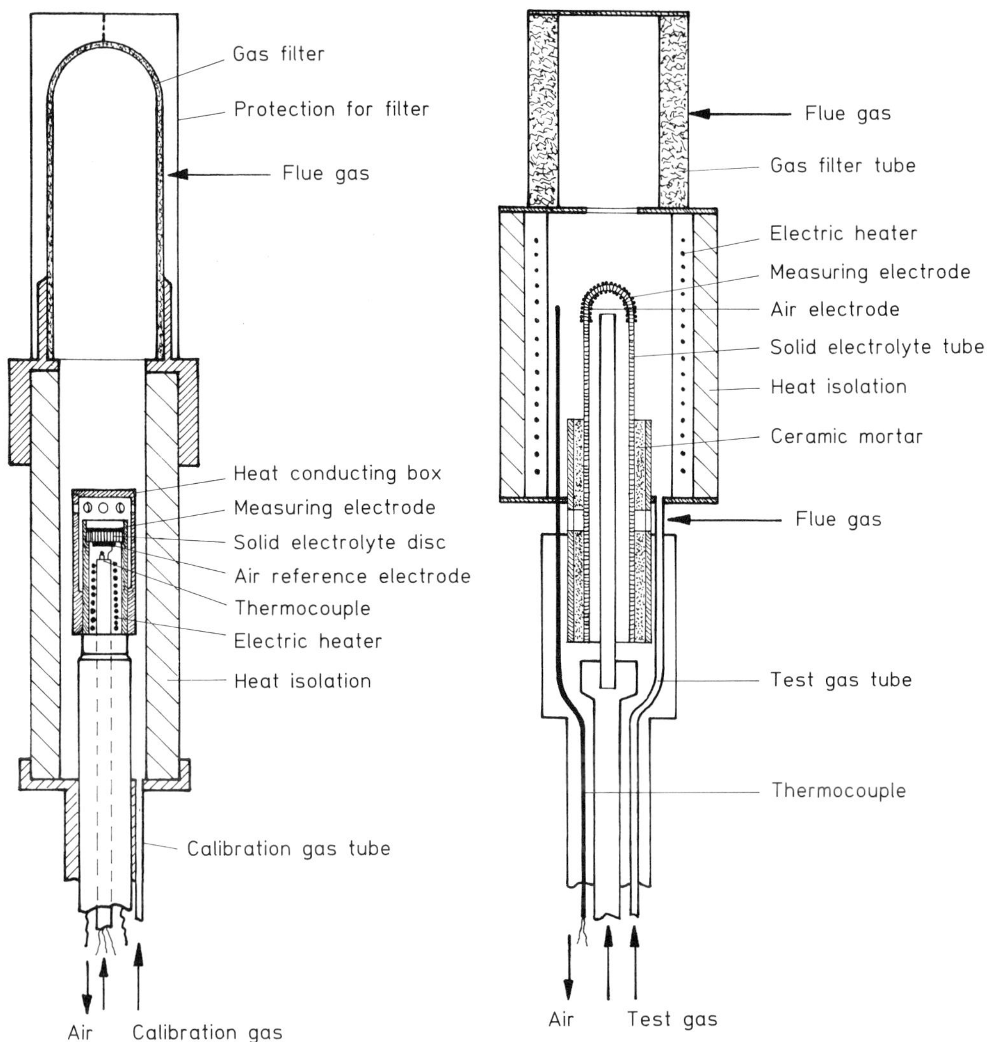

Figure 25-9.
Electrically heated head of a probe with a solid-electrolyte disk [58].

Figure 25-10.
Electrically heated head of a probe with a one-end closed solid-electrolyte tube [59].

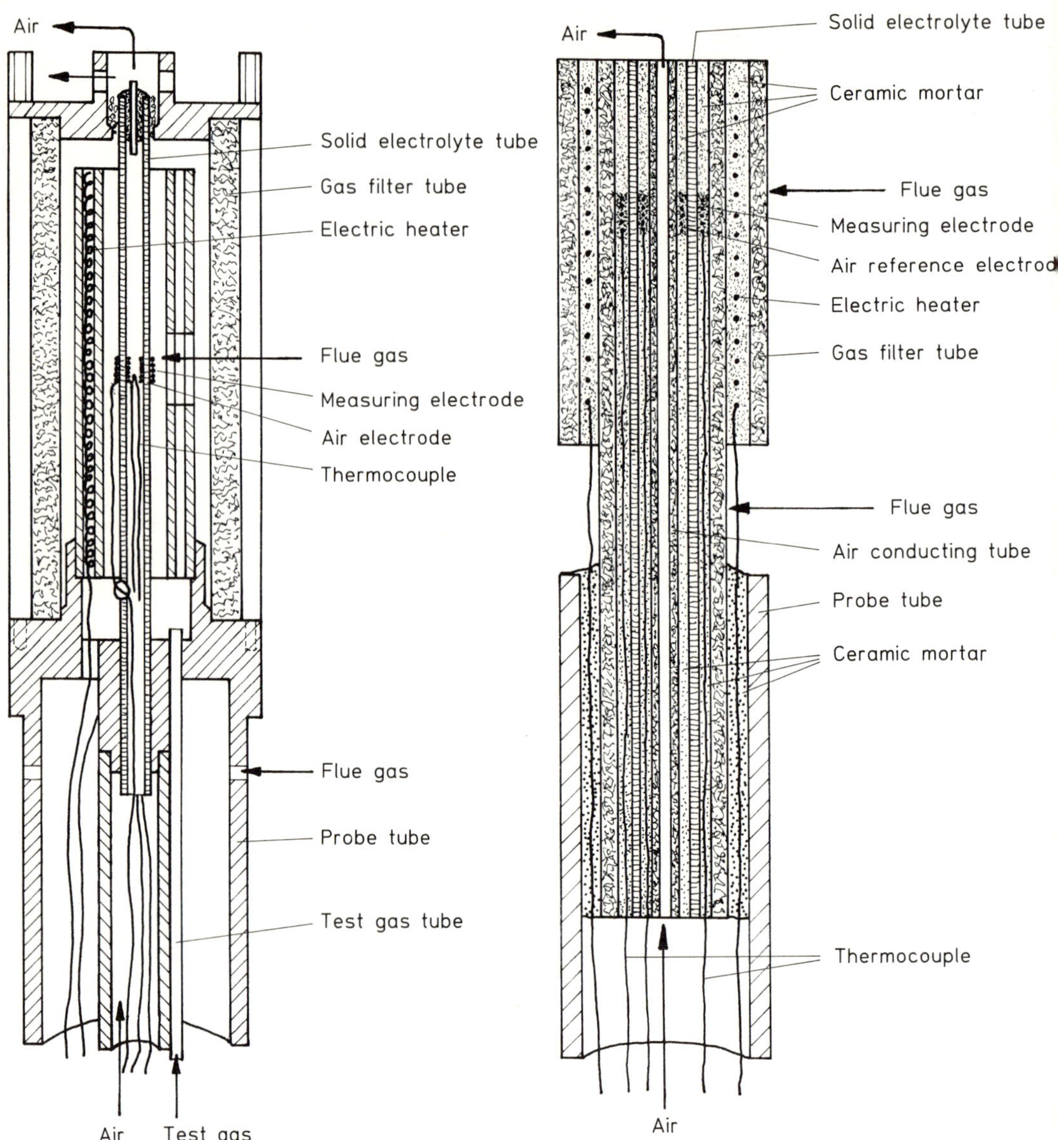

Figure 25-11.
Electrically heated head of a probe with a
solid-electrolyte tube open on both sides [44].

Figure 25-12.
Electrically heated head of a ceramic probe
with a solid-electrolyte tube open on both
sides and with powder electrodes [60].

Disturbances of the measuring electrode by air from leaks existing in the means to install the solid electrolyte tubes, are prevented in the arrangements shown in Figures 25-10 to 25-12 by rinsing the areas between the measuring electrode and the air leading tube with flue gas.

The air for the reference electrode is conducted back to the outside in Figures 25-9 and 25-10, whereas it is delivered into the passing flue gas in Figures 25-11 and 25-12. In the latter case it is possible to have the air transported with the help of the chimney suction. The arrangement shown in Figure 25-9 has to be calibrated regularly with gases of different known compositions because of the large temperature gradient over the solid electrolyte, in the other arrangements the correctness of the signals is only occasionally controlled by a test gas or by other means.

The arrangement shown in Figure 25-12 stands out for containing embedded powder electrodes between the solid electrolyte tube and the filter tubes instead of the usual adhering layer electrodes. In powder electrodes potential jumps arise at uncountable contact points between loose ionic or electronic conducting powder particles and compact electric conductors. The thermal spalling resistance does not depend on the expansion coefficients of the used materials. Noble metal wires can be embedded into solid electrolyte powder, but also arrangements totally free of noble metals for example with NiCr-wires in powders made of lanthan-strontium-manganite or -chromite can be produced [54, 61, 62]. For technical purposes, the temperature dependence of the signals is easily to compensate by counter-connecting ther-

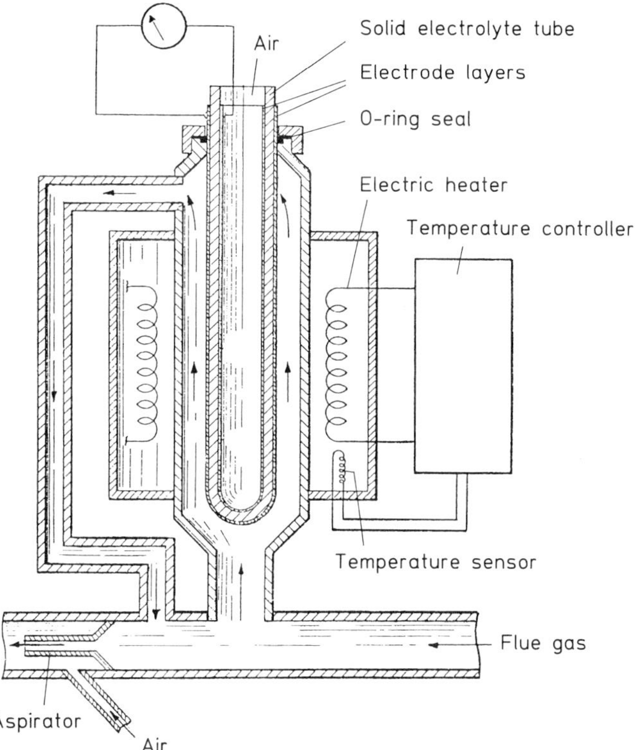

Figure 25-13. Oxygen measuring device with a solid-electrolyte cell in flue gas transported by an aspirator and by a convective flow in consequence of temperature differences [64].

mocouples [63]. The catalytical qualities of the electrode material can be varied easily in the case of powder electrodes.

In technical practice different oxygen-measuring devices are also used which work outside of the flue gas drain. Figure 25-13 shows an example [64] with flue gas being sucked up by an aspirator. The flue gas reaches the sensor in a slow vertical flow that arises by heating of the sensor, and in which the dust particles stay back by themselves on account of their weight. Additional sensors can be easily mounted in the stream of measuring gas, eg, pellistors to prove burning-gas that did not react completely with oxygen.

The technical usage of carbonates, sulfates and nitrates as solid electrolytes brings various difficulties, above all because these materials are hard to manufacture into mechanically stable gas-tight subjects. That is why vessels made of ceramic gas-tight ion conductors, eg, alkali-ion containing mullit, have been produced and their bottom has been contacted on both sides with pellets made of solid electrolyte material (Figure 25-14). Installed in the sensor shown in Figure 25-9 and applying potassium sulfate/platinum electrodes, a probe for SO_2 is created that looks similar to the probe for oxygen. The O_2- and SO_2-probes are worked side by side in flue gas (sensor temperature 843 °C) and deliver signals from which the SO_x concentration is calculated and recorded on the basis of the calibrations according to Equation (25-87). The ceramic ion conductor causes a very large electric resistance of the sensor, which is one of several reasons for frequent disturbances of such measuring systems.

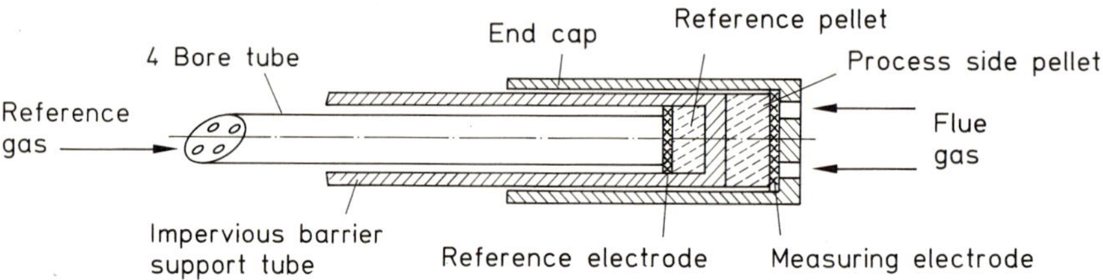

Figure 25-14. Part of the head of a SO_2 measuring probe with potassium sulfate solid electrolytes in pellet shape for mounting inside the arrangement shown in Figure 25-9 [65].

25.5   Examples of the Application of Potentiometric Gas Analysis with Solid-Electrolyte Gas Sensors

The first commercial equipment with zirconia solid-electrolyte cells, which has been produced in the Westinghouse Scientific Equipment Department in Pittsburgh since 1962, served only as an oxygen gauge. Measurements of the oxygen concentration in breathing gas (Figure 25-15) illustrated the fast response of the sensor. The achieved speed of response is the one that is connected with the gas transfer, and not the one of the sensor itself [66].

A survey of the possibilities and limits of the potentiometric gas analysis was obtained by measurements of several different gases vs. air in a cell-temperature range of 650–1600 °C (Figure 25-16). Ideal thermodynamic values (solid lines) have been measured (circles) in some

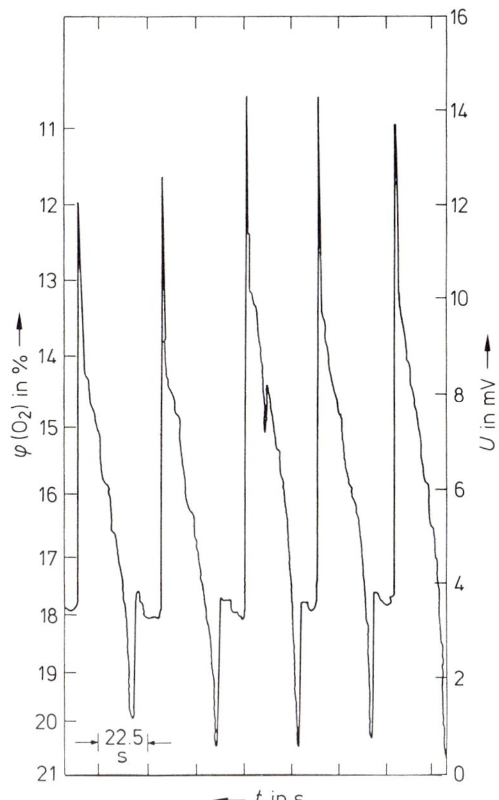

Figure 25-15.
Oxygen measurements with a solid-electrolyte cell on breath during several exhalation cycles (flow rate 2 L/min) [66].

cases for a temperature range up to 800 K. Deviation from such values (dashed lines) occurred in the range of low and high temperatures, the latter particularly in cases where the pO-buffering capacity of the gas was small (Section 25.2.3). Curves 2 and 4 illustrate the transition of oxygen electrode potentials to CO_2 electrode potentials according to Equations (25-12), (25-25), and (25-26). Above 1400 °C the effect of oxygen permeability of the solid electrolytes shows up in all cases, except for pure oxygen. With test gas mixture 3 it becomes noticeable at 1200 °C, with H_2-H_2O mixtures already above 1000 or 800 °C, depending on the degree of dryness and on the flow rate of the gas. In the high-temperature range, the hydrogen that was dried with concentrated sulfuric acid lead to a reduction and probably to a gas permeability of the MgO-stabilized zirconia (black dots).

The oxygen permeability of ceramic tubes, variing widely from material to material, can be observed with a simple set-up as a function of temperature (Figure 25-17). In a gas of 0.6 vol.-ppm O_2, flowing with a rate of 12 L/h through a tube of stabilized zirconia, the oxygen concentration already begins to rise noticeably below 800 °C (Figure 25-18).

The potentiometric oxygen determination is suitable in an analogue way for the examination of the oxygen permeability of rubber and synthetic tubes [68] or membranes [69], in which the use of an additional solid-electrolyte cell is very helpful for the oxygen-dosage ("oxygen pump"). The method can also be modified in such a way that, eg, the permeability for water vapor is measurable.

vapor partial pressure in hydrogen. Here, the oxygen partial pressure between the solid electrolyte and the platinum layer of the measuring electrode rises definitely compared to that in
the measuring-gas compartment. The dotted curves drawn in Figure 25-1 mark the limits of
applicability set by self-polarization, for some important cases. Separate studies of the limits
due to the self-polarization in hydrogen were reported for cells with air-reference electrodes
[14] and with hydrogen/water vapor reference electrodes [70].

At temperatures below 800 °C, potential differences in solid electrolyte cells were often
observed, which exceed the values of Nernst's equation [71]. This puzzling effect could not
be explained for a long time in spite of a number of investigations [72]. They are caused by
contaminations of oxygen-containing gases, which burn up at high temperatures but compete
with the oxygen in electrode reactions at low temperatures [73]. Stable high potential differences can be effected by hydrogen or organic molecules in air and oxygen (compressor oil
vapors), which disappear only by purification (equilibration) of the gases with hot combustion
catalysts added to the system (Figure 25-19). Gas contaminations are also the reason for
changes in the deviation of the potential differences from the ideal values at low temperatures,
when the electrode layers are modified (Figure 25-20), because these deviations depend on the
variable catalytic activitiy of the electrode metal layers.

It has often been suggested to use the sensitivity of the air electrode for combustible gases
as a test for the presence of burning substances in air. Large effects can be observed, eg, in
cells suffused by the same gas, if the electrodes consist of metals of different catalytic activity
such as platinum and gold (Figure 25-21). In order to get reproducible results, electrode
preparation and measuring conditions must be kept strictly constant because of the kinetic
origin of the effects (which can also change due to ageing of the electrodes). In the application

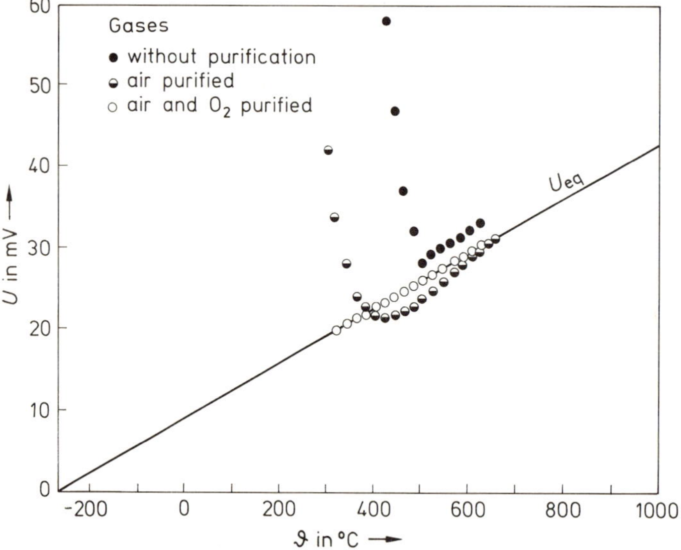

Figure 25-19. Potential differences measured on an air/oxygen cell without purification of the gases, with
purified air and with gases both treated by a combustion catalyst containing MnO_2 and
CuO [73].

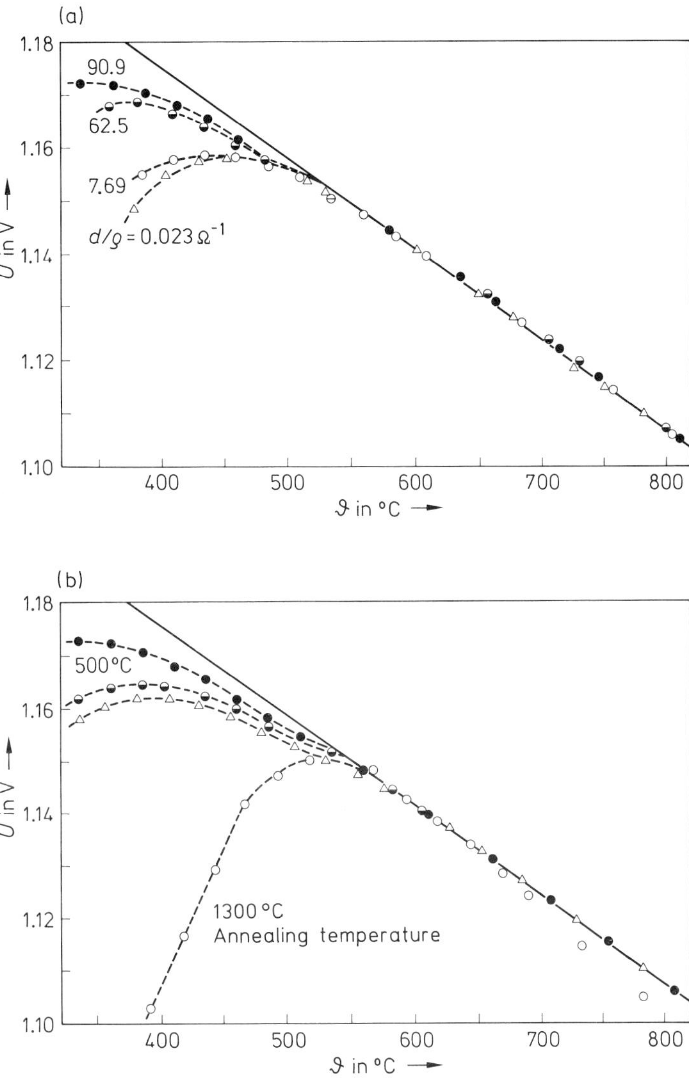

Figure 25-20. Potential differences measured on air/hydrogen, water vapor cells with platinum electrodes prepared with different layer thickness:resistivity ratios d/ρ (a), or after annealing at different temperatures between 500 and 1300 °C (b) [73].

of solid-electrolyte probes with air reference electrode in the heat treatment ovens of the hardening technology (which usually work with ammonia and hydrogen containing gases), one tries to exclude disturbances of the determination of the $H_2O:H_2$ ratio by NH_3 [75]. On the other hand, a commercially available sensor makes use of the NH_3 sensitivity of a solid-electrolyte gas electrode as an NH_3 indicator in the clean-up of exhaust gases [76]. In hydrogen a sensitivity could be found for molecular oxygen, which makes possible a potentiometric indication of metastable oxygen [77].

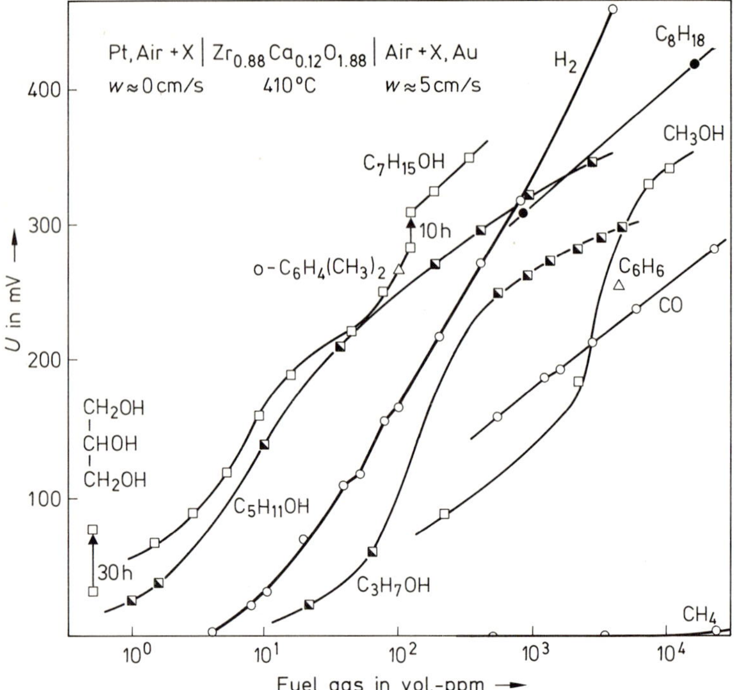

Figure 25-21. Potential differences measured between the electrodes of the cell   Pt | stabilized ZrO$_2$ | Au   at 410 °C in the same gas consisting of air with variable concentrations of different combustible compounds [74].

Disturbances of determinations in inert gas by the oxygen permeability of the solid electrolyte could be more or less prevented, if solid reference systems were used that have decomposition pressures in the range of the minimum of the p and n conduction of the solid electrolyte. As an example, Figure 25-22 shows the results of investigations on the operation of a catalyst for the removal of oxygen from inert gas. It was shown that the catalyst works mainly by means of the hydrogen which it picked up during reduction and which reacts gradually with the oxygen, but which is also partly desorbed into the inert-gas stream. An oxygen partial pressure due to "active copper", ie, equal to the decomposition pressure of Cu$_2$O, was not observed [78].

The partial pressure of oxygen, hydrogen and water vapor in inert gases can be determined continuously, if the inert gas is sent through two solid-electrolyte cells in series and if a well-defined water vapor pressure is established before passage of the second cell [11, 79]. The state of gases before and after purification steps can be recorded by means of bundles of solid-electrolyte cells (Figure 25-23). By applying the fundamentals given in Section 25.2 on the potential differences between the different electrodes of such arrangements, one obtains the necessary equations for calculating the partial pressures. As an example, Figure 25-24 illustrates the efficiency and the capacity of two Leuna catalysts for the purification of inert gases.

In Figure 25-24 rapid changes of the oxygen partial pressure are noticeable after the consumption of hydrogen adsorbed on the catalyst. Such curves are well-known in the poten-

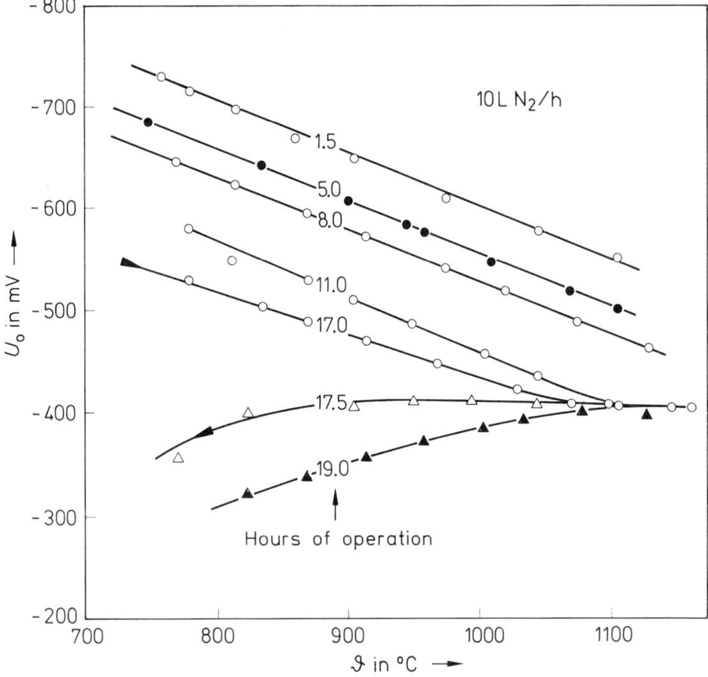

Figure 25-22. Electrode potentials during measurements with a cell Pt,Ni,NiO | stabilized ZrO$_2$ | Pt, measuring gas at temperatures changing between 750 and 1150 °C. Before entering the cell the measuring gas nitrogen passed over 35 g of a Leuna contact Cu (Mg Silicate) which had previously been reduced with hydrogen [78].

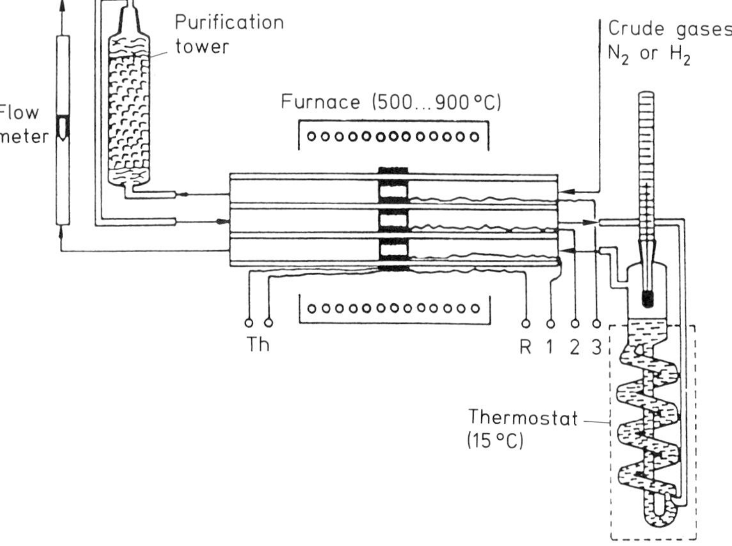

Figure 25-23. Apparatus with a bundle of three solid-electrolyte tube cells, a gas purification tower, a bottle for water vapor saturation and a flow meter (Th thermocouple, R reference system, 1, 2, 3 terminals of inner electrodes) [80].

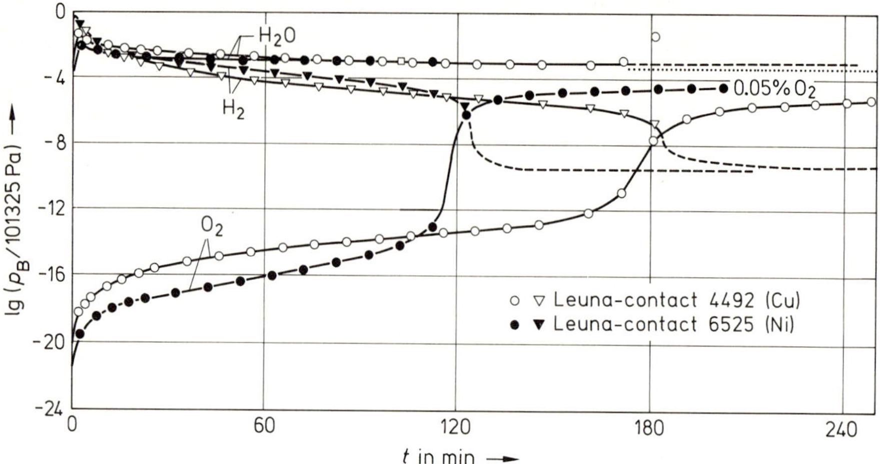

Figure 25-24. The partial pressures of O_2, H_2 and H_2O at 850 °C in nitrogen after passing (5 L/h) through 2 g of contacts at 150 °C, previously reduced with hydrogen [81].

tiometric titration of aequeous and gaseous systems. Figure 25-26 shows just how good the results of gas titrations in a simpe arrangement (Fig. 25-25) agree with the calculations based

The fast response of Pt,O_2, solid-electrolyte electrodes upon changes of the oxygen partial pressure in the range of the equivalent point of air and burning materials is the basis of the lambda sensors for the control of the air/fuel mixture in cars with 3-way catalysts. Lambda is defined as the ratio of the volume of the available air to the volume of air necessary for stoichiometric combustion of the fuel. Between lambda and the oxygen concentration different relations are valid in the range of air surplus and in the range of fuel surplus and these relations depend to some extend on the type of fuel.

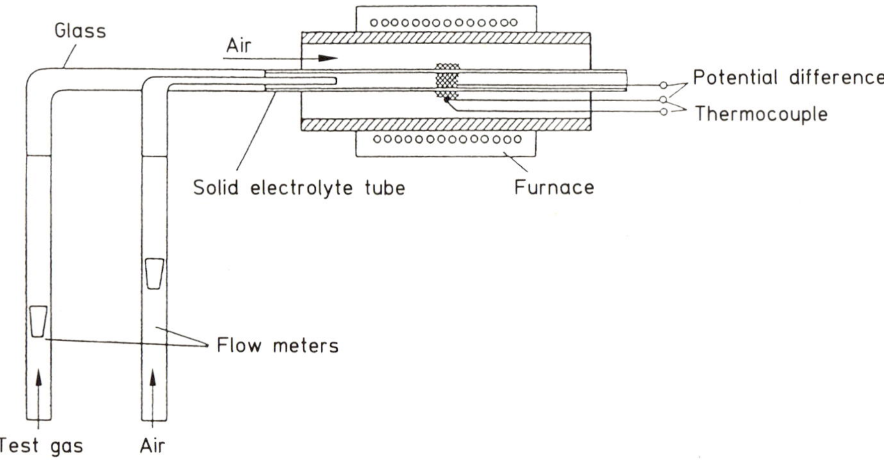

Figure 25-25. Apparatus for titration of gases containing combustible components with air [18].

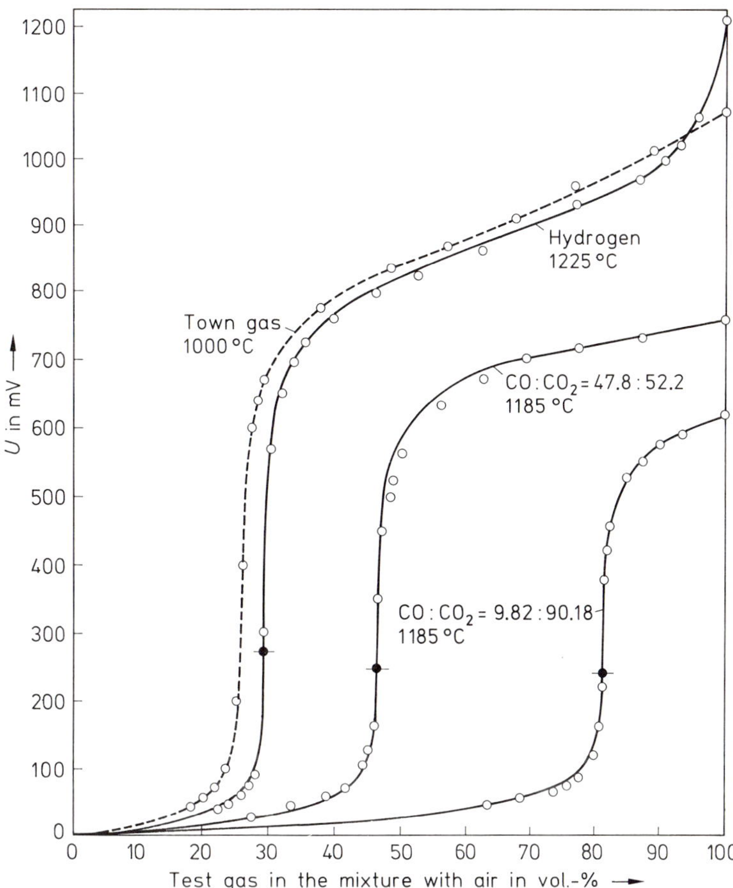

Figure 25-26. Results of measurements (circles) and calculations (curves) during the titration of gases containing combustible components with air [18].

In the "lean" range of burning of $C_n H_m$ the following equation is valid:

$$\lambda = \frac{1}{4\,n/m + 1} \cdot \frac{1 + \varphi\,(O_2 \text{ in air}) \cdot \exp\,[-4\,U_{eq}F/(R\,T)]}{1 - \exp\,[-4F\,U_{eq}/(R\,T)]} \,. \tag{25-88}$$

The oxygen concentration in the exhaust gas, lean or fat, is obtained from the sensor signal according to Equation (26-16), provided that chemical equilibrium is established ($U = U_{eq}$).

In the reaction of "fat" mixtures an exhaust gas containing mainly CO, CO_2, H_2, H_2O and N_2 ("water gas"; see Section 25.2.4) with extremely low oxygen concentrations is produced. Here, it is useful to consider the relation of lambda to the redox quotient Q defined by Equation (25-31) and the sensor signal based on Equation (25-34). The basis is the stoichiometric equation:

$$C_n H_m + \left[\left(1 - \frac{a}{2} \right) n + (1 - b) \frac{m}{4} \right] O_2 \rightarrow$$

$$\rightarrow (1 - a)\, n\, CO_2 + an\, CO + (1 - b) \frac{m}{2} H_2O + b \frac{m}{2} H_2 \qquad (25\text{-}89)$$

and upon it the equation between the stoichiometric coefficients and lambda:

$$\lambda = 1 - \frac{na/2 + mb/4}{n + m/4}. \qquad (25\text{-}90)$$

The two parameters Q (25-31) and V (25-32) turn into

$$Q = \frac{(1 - a)\, n + (1 - b)\, m/2}{an + bm/2} \qquad (25\text{-}91)$$

$$V = 2\, n/m \qquad (25\text{-}92)$$

and give with the Equations (25-33) and (25-34) one of the two equations for the calculation of a and b using the sensor signal at equilibrium U_{eq}. Equation (25-40) gives the second equation to calculate a and b:

$$K_3 = \frac{a(1 - b)}{b(1 - a)}. \qquad (25\text{-}93)$$

On this way, it is possible to get a and b as well as λ, if one knows n and m of the fuel. In the case that chemical equilibrium at the measuring electrode of a lambda sensor is completely reached, the signals for octane follow the curves, given in Figure 25-27 versus lambda and additionally in the lean range versus the oxygen concentration, and in the fat range versus the redox quotient.

In consequence of the high flow rate and rapidly dropping temperature of the exhaust gas, there remain combustible components as well as oxygen. For this reason one achieves potential differences between electrodes which are placed on a solid electrolyte side by side in the exhaust gas if their catalytic properties (for example in the case of Au and Pt) differ. Their use for the control of the air/fuel mixture, however, causes problems because the equivalence point cannot be exactly indicated [82] and the signals become vanishing small with increasing temperature.

The catalytic properties of the electrodes from lambda sensors are of general importance. In commercial sensors with platinum electrodes, protected under porous layers, eg, consisting of $MgAl_2O_4$, oxygen and fuel molecules existing in the exhaust gas react together. In the low-temperature range, in consequence of the fuel gas sensitivity the sensor can show even higher signal values than in the chemical equilibrium are to be expected. Compared to this, lambda sensors with electrodes of specially prepared electronic conducting oxide powders provide signals of the oxygen concentration in the exhaust gas in presence of fuel molecules. Figure 25-28 illustrates with signal curves which have been taken parallel from two sensors in equal

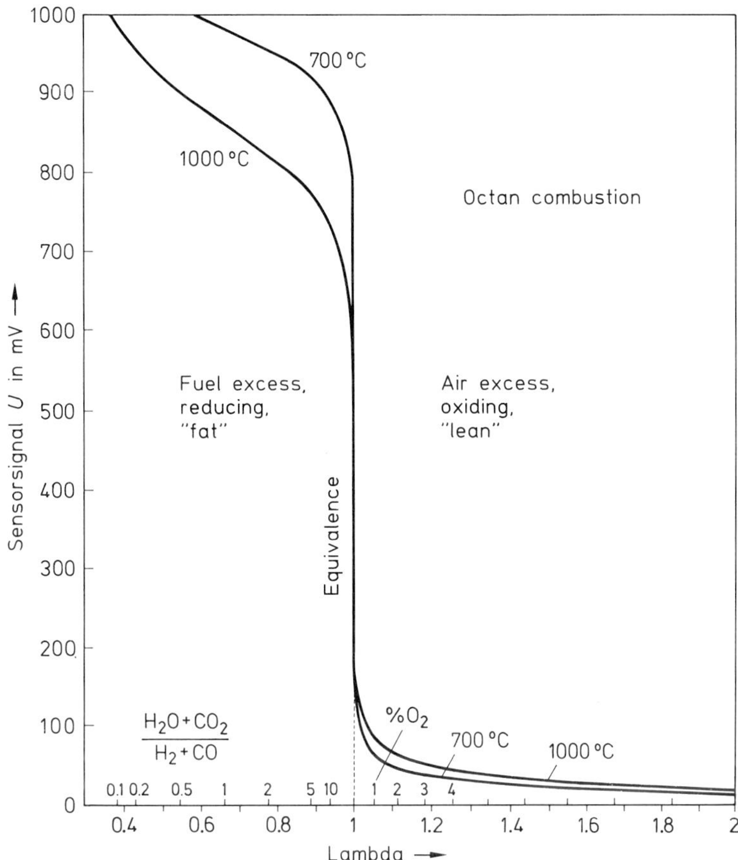

Figure 25-27. Calculated signals of a lambda sensor versus lambda (additionally the graduation for the oxygen concentration in vol.-% in the lean region and for the redox quotient Q in the fat region).

position that the sensor with the platinum electrode indicates fat exhaust gas and the sensor with the powder electrode show around $\lambda = 1$ fluctuating concentrations (Figure 25-28, upper part). In a later situation, the platinum electrode gives signals of heavily fluctuating oxygen concentrations and the powder electrode continuously such in the lean range (Figure 25-28, lower part).

Galvanic solid-electrolyte gas cells can, of course, also be used as detectors in gas chromatography [84]. If the carrier gas is an inert gas with a low oxygen concentration then it is possible to achieve a very sensitive indication of organic substances as a consequence of combustion processes on a catalyst passed in the gas train or on the measuring electrode itself. The signal heights vary (in spite of equal volume concentration) according to the number of C, H, and O atoms in the molecules and the used carrier gas (N_2, CO_2, H_2), so that in some cases a helpful selectivity is achieved. Figure 25-29 illustrates the simple analysis of carbon monoxide from a steel bottle in which the small amount of hydrogen contamination is indicated by a sharp peak, and the admixture of air is indicated by an oxygen peak with reversed sign. In

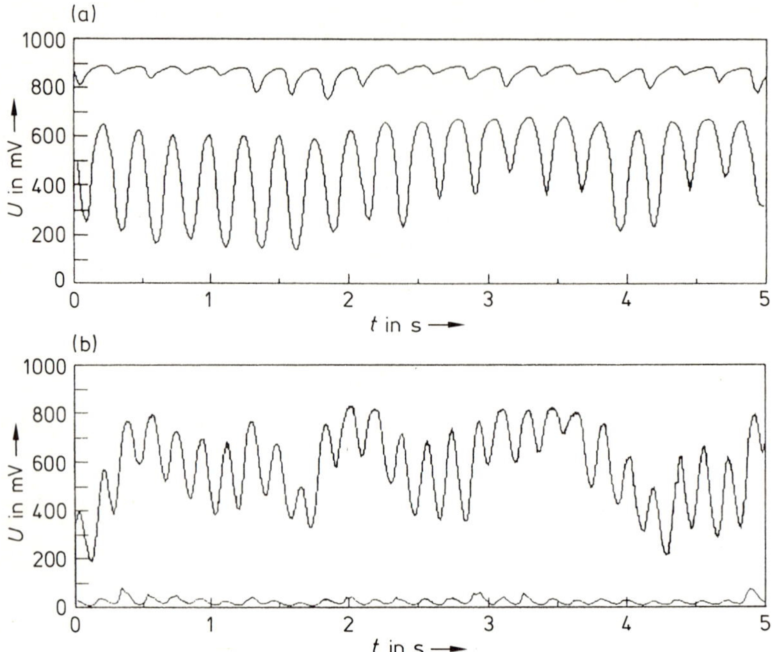

Figures 25-28. Two comparisons of plots of the signals from a Bosch lean lambda sensor (upper plots) and from a lambda sensor with specially prepared electronically conducting oxidic powders. Both sensors were placed together in automobile exhaust gas during periodic variations of the oxygen concentration at about the air/fuel equivalence point [83].

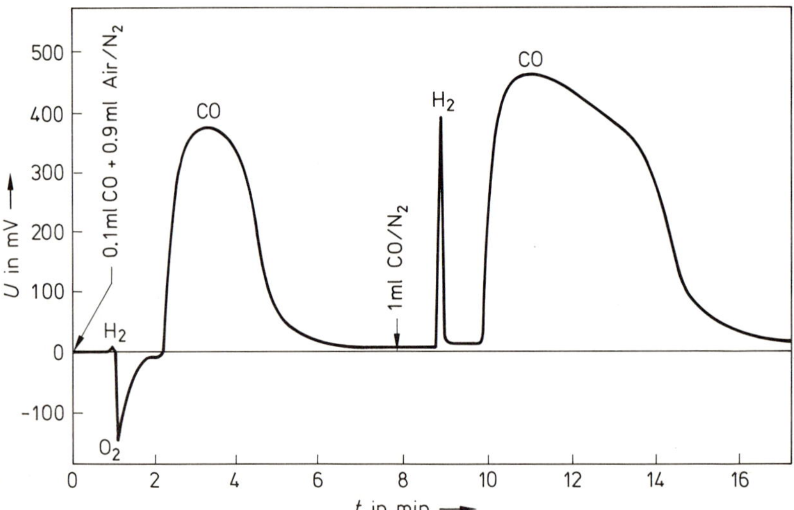

Figure 25-29. Signals from a hot solid-electrolyte gas sensor as a detector behind a gas-chromatographic column analyzing two portions of technical carbon monoxide. The carrier gas was nitrogen with a very small oxygen concentration [85]. In the first run the sample was mixed with air.

practice, this detection method involving many details and logarithmic concentration relations has obviously not been accepted up until now.

Using a probe with a grip for handling directly, there is an impressive possibility to measure the oxygen concentration in the flame of a Bunsen burner, and the oxidation and reduction range of the flame to touch (Figure 25-30). Flame probes are used, for example, to study the burn-out profile of industrial flames (Figure 25-31) [88].

Flame probes are suitable for demonstrating the large rate of response of solid-electrolyte gas electrodes upon sudden changes of the oxygen concentration over several orders of magnitude. That is the reason why the turbulences of the gases in industrial flames can be recorded directly. Figure 25-32 shows examples for the design of turbulence sensors for such purposes.

Measurements on gases with high CO-concentrations [5] are influenced by reactions according to the Boudouard equilibrium (25-30) which cause the precipitation of carbon (border line C in Figure 25-1). In contrast, Equation (25-28) rules widely also in pure hydrogen and allows the determination of low water vapor concentrations. In measurements with the air reference electrode, changes in the water vapor concentration only lead to fluctuations of relatively high signals. Therefore it is more favorable to measure against an electrode with hydrogen, which contains water vapor with a known concentration, or between the inner electrodes of an (inert-gas surrounded) double-tube cell, which contains the reference gas in one tube and the measuring gas in the other one (realized in the device ursalyt G20 of Junkalor GmbH, Dessau). The effect of drying agents was examined and compared with a double-tube cell (Figure 25-33). Thereby phosphorpentoxide proved itself as the most productive agent whose effect only decreases if the whole mass has gone over into a liquid.

A complete potentiometric analysis of "water gas" (cf. Section 25.2.4) is very complicated [11]. In some cases, eg, during the firing of porcelain, in which iron (existing in the ceramic

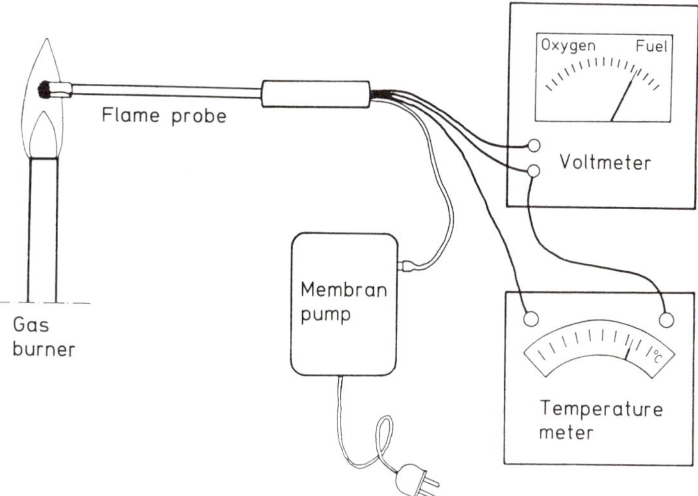

Figure 25-30. Flame probe for the demonstration of the possibility of measurements of the oxygen concentration in oxidizing and reducing hot gases with galvanic solid-electrolyte gas cells [79, 86].

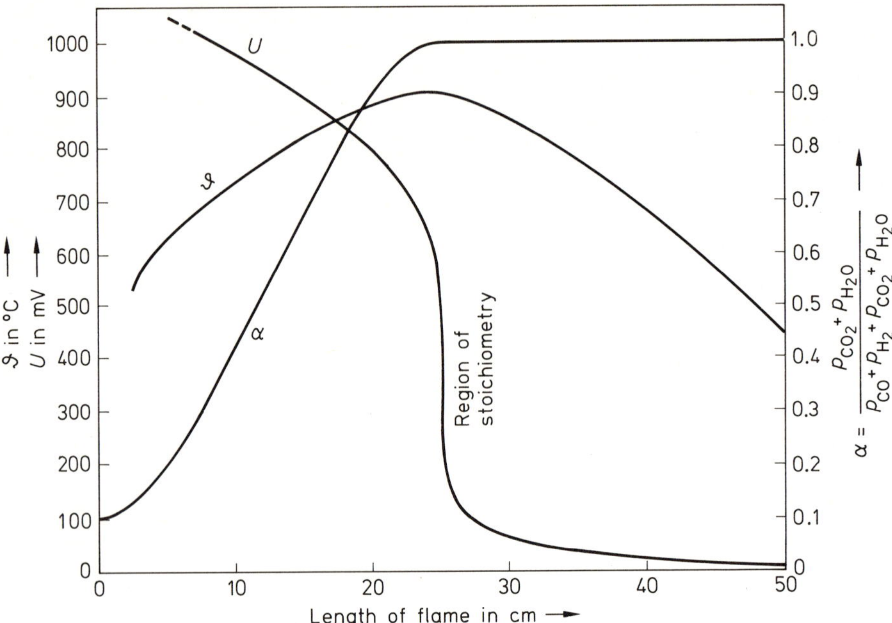

Figure 25-31. Diagram of the Celsius temperature, the signal of a solid-electrolyte flame probe and the degree of combustion versus distance in a long flame [87].

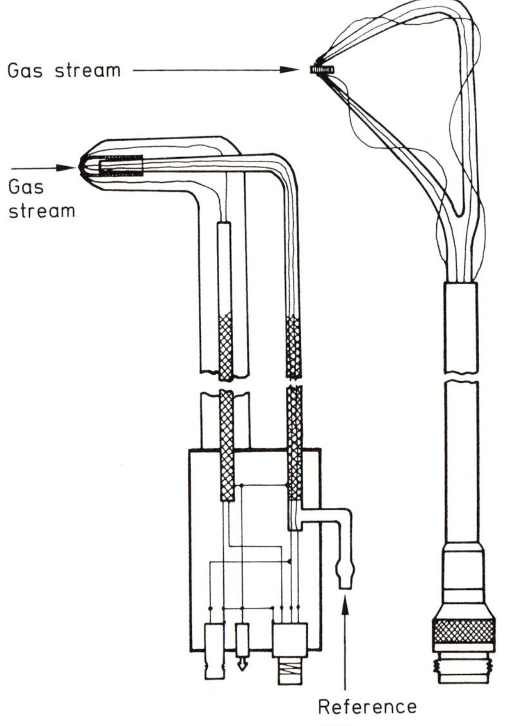

Figure 25-32.
Galvanic solid-electrolyte sensors for the study of turbulences in gases, particularly qualified for flames [89]. The sensor shown at left side is able to record the oxygen concentration at one point and the sensor shown at right is able to record the difference of the oxygen concentration between two points in a hot gas phase.

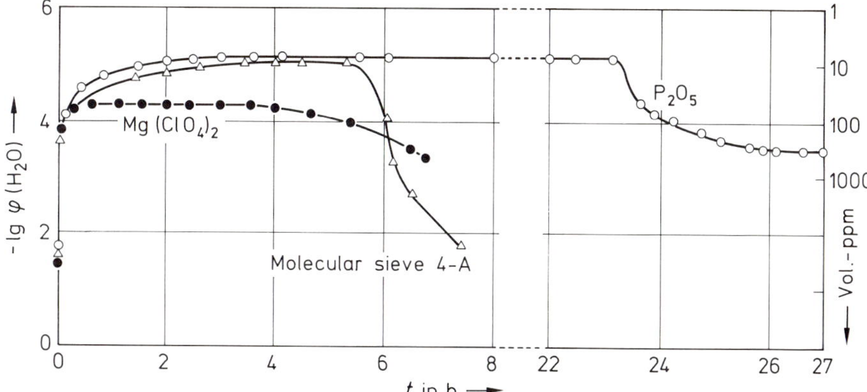

Figure 25-33. The water vapor concentration in a stream of hydrogen (7.5 L/h) with an initial $H_2O:H_2$ ratio of 0.029 after desiccation with the equal mass (4 g) of different substances versus time. The water vapor ratio in the dried gas was calculated from the potential differences measured at 550 °C between the inner electrodes of two parallel galvanic solid-electrolyte tube cells, one of which contained a stream of hydrogen saturated with water vapor at 15 °C [90].

as a contamination) has to be brought into the divalent form, the oxygen concentration of the reducing kiln gas is important alone [21]. During the reduction phase the kiln is containing water gas, in which the oxygen concentration is measurable on the basis of Equation (25-12). Up until now in the reduction phase the CO-concentration is controlled in a sample gas stream outside the kiln. As Figure 25-34 shows, the CO concentration and the potential difference U of an oxygen probe correspond with each other. The signals of the probe reflect the reduc-

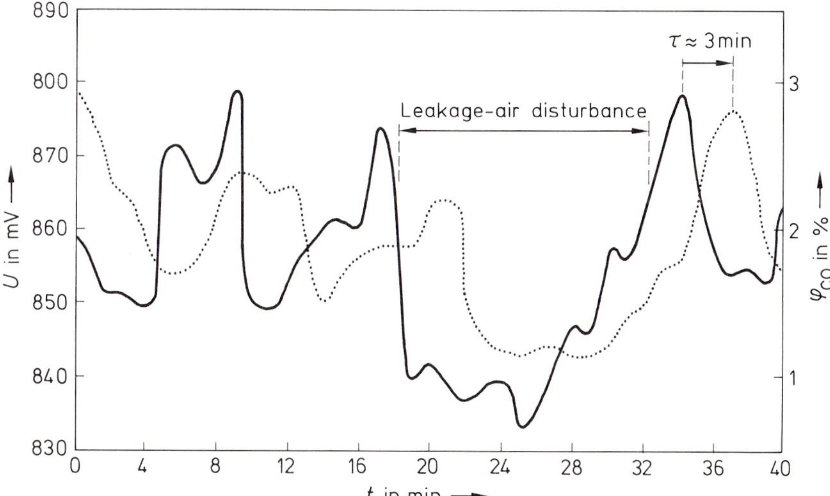

Figure 25-34. Comparison of the plots of the signal from a galvanic solid-electrolyte probe with an air reference electrode and the signal of an infrared device for CO measurements versus time during the reducing phase of burning porcelain [52].

tion power of the gas phase exactly and are available about 3 minutes earlier than the CO signals. Therefore solid-electrolyte probes are suited perfectly for the automation of firing technologies in the ceramics industry [52].

Already in the 70s the successful introduction of solid-electrolyte probes into the heat-treatment technique of metallic materials started. Among others, carburizing (900 to 1000 °C), carbonitriding (900 °C) and nitriding (500 to 600 °C) gas phases are used. Measurements of the oxygen partial pressure serve to keep constant conditions in the heat treatment. Only some new results of the development of solid-electrolyte probes for the nitriding technique shall be discussed here.

As a base for the development of these probes, water vapor partial pressure was firstly determined exactly with the help of a galvanic solid-electrolyte cell in the l, g-equilibrium system H_2O-NH_3 in the temperature range from 0 to 30 °C [91]. The results were used in the investigation of the decomposition of metastable NH_3 at the platinum electrode of a solid-electrolyte cell. It was proved that the NH_3 decomposition in the temperature range from 500 to

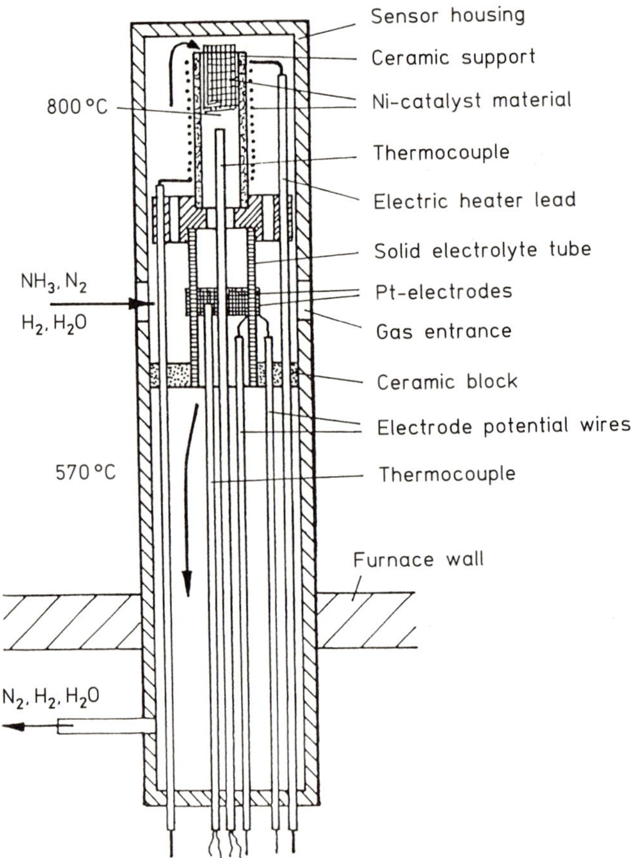

Figure 25-35. Design of a probe for measurements of the quotient $\varphi\,(NH_3)/\varphi\,(H_2)$ in gases for heat treatments of the surface of steel work pieces (*E*-probe, schematic, [48]).

600 °C and in the range of the composition of technically applied gases is neglegibly small [47]. That is why the quotient $Q = \varphi(H_2O)/\varphi(H_2)$ can be measured continuously directly in the heat-treatment ovens during nitriding processes (by a "*Q*-probe"). Furthermore it was found that the quotient $\varphi(NH_3)/\varphi(H_2)$ can be measured if a gas stream is first conducted over an electrode of a solid-electrolyte cell, then the gas is brought into the chemical equilibrium, ie, NH_3 is widely decomposed and after that this gas stream is lead over the second electrode of the same cell [92]. Figure 25-35 shows schematically the construction of a probe, in which the equilibrium achievement (equilibration) is reached at 800 °C on a catalyst ("*E*-probe"). The solid-electrolyte cell of this probe works as a hydrogen concentration cell. For the difference of its electrode potentials the following equation is valid:

$$U(E\text{-probe}) = \frac{RT}{2F} \lg \left[1 + \frac{3}{2} \cdot \frac{\varphi(NH_3)}{\varphi(H_2)} \right]^{-1} . \tag{25-95}$$

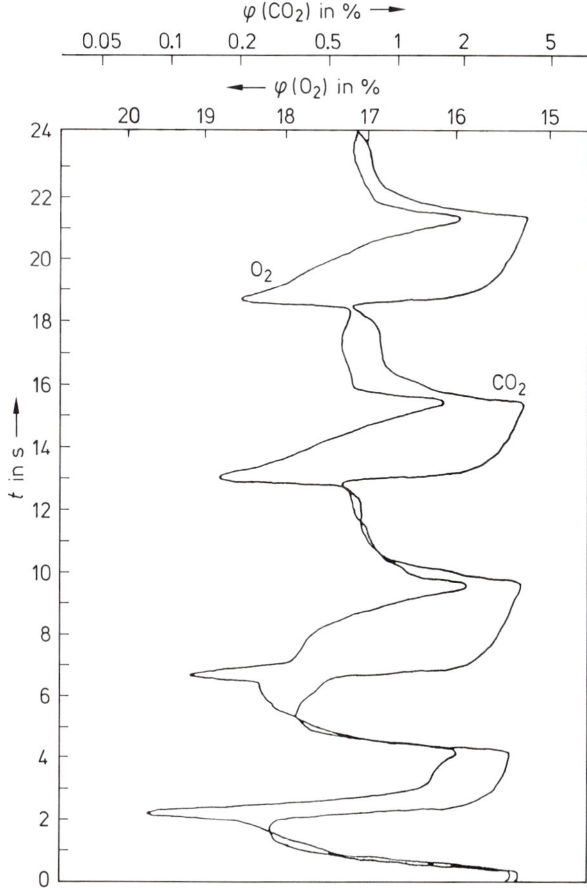

Figure 25-36.
Plots of the signals for the concentration of oxygen and carbon dioxide versus time during exhalation cycles of breath obtained from an arrangement of zirconia and carbonate solid electrolytes with two reference electrodes on the outer sides and the measuring gas electrode between the two electrolytes at 650 °C [32].

For the determination of the four-component system NH_3, H_2, H_2O, N_2 the signals of the Q-and E-probe and the total pressure are only three analysis pieces. Data of the introduced gas mixture is included into the calculations (eg, the relation of NH_3 and air or the O_2 concentration in the gas mixture at normal temperature). One reaches a complete gas-potentiometric determination if a well-known water vapor partial pressure is maintained in the gas by saturation at a constant temperature after the NH_3 decomposition and Q is determined in a third solid-electrolyte cell [93].

Results of the determination of CO_2, SO_2, NO_2 and other gases with carbonates, sulfates and nitrates have been already reported by the inventors of the method [29]. The lower limit for measurements of traces with solid oxoanionic electrolytes in galvanic cells is given by the decomposition pressures of the solid electrolytes at the cell temperature. Trace measurements however are also influenced largely by adsorption equilibria on the surface of the used materials. Installations made of unused annealed materials have to be used for such measurements. The lowest SO_2 concentrations measured at 840 °C run to $4 \cdot 10^{-8}$ [94].

Carbonate and zirconia cells can be united to the three-pole arrangement (25-72), with which (on the base of Equations (25-73) and (25-75)) O_2 and CO_2 can be measured simultaneously and continuously at the same electrode. Such measurements (Figure 25-36) are of importance if the concentration of two gases with practically the same dynamics shall be recorded.

25.6   References

[1] Hartung, R., Guth, U., Zastrow, W., Möbius, H.-H., *V. Wiss. techn. Arbeitstagung: Fortschritte in der metallurg. analyt. Chemie. Freiberg, Bergakademie, 23.–27.10.1983.*
[2] Möbius, H.-H., *Elektrochemische Analysenmethoden, Wiss. Beiträge*; Leipzig: Univ., p. 105–122.
[3] Haber, F., Moser, A., *Z. Elektrochem.* **11** (1905) 593–609.
[4] Haber, F., Fleischmann, F., *Z. anorg. Chem.* **51** (1906) 245–288.
[5] Peters, H., Möbius, H.-H., *Z. physik. Chem., Leipzig* **209** (1958) 298–309.
[6] Peters, H., Möbius, H.-H., *DDR Patent 21673,* 20.5.1958.
[7] Gaugain, M. J.-M., C. R. *Seances Acad. Sc.* **37** (1853) 584–588.
[8] Lange, E., Göhr, H., *Thermodynamische Elektrochemie;* Heidelberg: Hüthig Verlag, 1962.
[9] Van Rysselberghe, P. et al., *Electrochimica Acta* **12** (1967) 748–749.
[10] Ackermann, G., Jugelt, W., Möbius, H.-H., Suschke, H. D., Werner, G., *Elektrolytgleichgewichte und Elektrochemie,* 5th Ed.; Leipzig: Dt. Verlag für Grundstoffindustrie, 1988.
[11] Möbius, H.-H., *Z. physik. Chem., Leipzig* **230** (1965) 396–416.
[12] Möbius, H.-H., Dürselen, W., *Chemische Thermodynamik,* 5th ed.; Leipzig: Dt. Verlag für Grundstoffindustrie, 1988.
[13] Whiffen, D. H., *Manual of Symbols and Terminology for Physicochemical Quantities and Units;* Oxford: Pergamon Press, 1979, p. 28–29.
[14] Hartung, R., Möbius, H.-H., *Chemie-Ing.-Techn.* **40** (1968) 592–600.
[15] Remy, H., *Lehrbuch der anorg. Chemie* Vol. 1; Leipzig: Akad. Verlagsges. Geest & Portig, 1965, p. 806.
[16] Zeise, H., *Thermodynamik auf den Grundlagen der Quantentheorie, Quantenstatistik und Spektroskopie* Vol. 3, part 1; Leipzig: Hirzel-Verlag, 1954.
[17] Möbius, H.-H., *5th Allunionsconf. Electrochemistry, Moscow 29.1.1975.*
[18] Möbius, H.-H., *Z. physik. Chem., Leipzig* **231** (1966) 209–214.
[19] Tews, W., Diplomthesis, Ernst-Moritz-Arndt-Universität, Greifswald, 1968.
[20] Jacobson, E., *Scand. J. Met.* **14** (1985) 252–256.

[21] Möbius, H.-H., Sandow, H., Kämpfer K., Prescher, E., *Silikattechnik* **41** (1990) 169–173.

[22] Alcock, C. B., Stavropoulos, G. P., *Can. Metall.* (quarterly) **10** (1971) 257–265.

[23] Fischer, W. A., Janke, D., *Arch. Eisenhüttenw.* **41** (1970) 1027–1033.

[24] Tankins, E. S., Gokcen, N. A., Belton, G. R., *Trans. Metall. Soc. AIME* **230** (1964) 820–827.

[25] Fischer, W. A., Janke, D., *Metallurgische Elektrochemie;* Düsseldorf: Verlag Stahleisen, 1975.

[26] Schmalzried, H., *Arch. Eisenhüttenw.* **48** (1977) 319–322.

[27] Warburg, E., *Ann. Physik Chemie N.F.* **21** (1884) 622–646.

[28] Schwabe, K., *pH-Meßtechnik;* Dresden: Verlag Theodor Steinkopf, 1963, p. 134.

[29] Gauthier, M., Chamberland, A., *J. Electrochem. Soc.* **124** (1977) 1579–1583.

[30] Gauthier, M., Chamberland, A., Belanger, A., Poirier, M., *J. Electrochem. Soc.* **124** (1977) 1584–1587.

[31] Chamberland, A. M., Gauthier, M., *US Patent 4282078,* 30. 8. 1976, and *US Patent 4388155,* 31. 7. 1980.

[32] Barwisch, F., Thesis, Ernst-Moritz-Arndt-Universität, Greifswald, 1984.

[33] Möbius, H.-H., *Int. Symp. on Systems with Fast Ionic Transport;* Bratislava: Dom Technicky CSVTS, 1985, p. 26–30.

[34] Belanger, A., Gauthier, M., Fauteux, D., *J. Electrochem. Soc.* **131** (1984) 579–586.

[35] Cote, R., Bale, C. W., Gauthier, M., *J. Electrochem. Soc.* **131** (1984) 63–67.

[36] Stull, D. R. et al., *JANAF Thermochemical Tables,* NSRDS-NBS 37; Washington: U.S. Department of Commerce, 1971.

[37] Gauthier, M., Bellemare, R., Belanger, A., *J. Electrochem. Soc.* **128** (1981) 371–378.

[38] Möbius, H.-H., Lang, S., Wilms, K., *DDR Patent 43242,* 22. 5. 1964.

[39] Möbius, H.-H., *Z. physik. Chem., Leipzig* **230** (1965) 414, Fig. 4.

[40] Möbius, H.-H., Hartung, R., *Silikattechnik, Berlin* **16** (1965) 276–280.

[41] Deportes, C. H., Henault, M. P. S., Tasset, F., Vitter, G. R. R., *DE-OS 24 43 037,* 9. 9. 1974 (FR 11. 9. 1973).

[42] Roy, P., Licina, G. J., *DE-OS 26 42 740,* 23. 9. 1976.

[43] Janke, D., *Clean Steel;* London: The Metals Society, 1983, p. 202–231.

[44] Möbius, H.-H., Hartung, R., Guth, U., *messen-steuern-regeln, Berlin* **22** (1979) 269–272.

[45] Shuk, P., Möbius, H.-H., *Z. physik. Chem., Leipzig* **266** (1985) 9–16.

[46] Hartung, R., Maaß, R., *Z. Chem., Leipzig* **21** (1981) 337–338.

[47] Hartung, R., et al., *Z. Chem., Leipzig* **24** (1984) 447–448.

[48] Möbius, H.-H. et al., *DE-OS 36 32 480 A 1,* 24. 9. 1986.

[49] Berg, H. J., Böhmer, S., *Freiberger Forschungshefte B 263;* Leipzig: Dt. Verlag für Grundstoffindustrie, 1988.

[50] Heeleman, H., Guth, U., Zastrow, W., Möbius, H.-H., *Silikattechnik, Berlin* **38** (1987) 224–227.

[51] Meister, R., *Glastech. Ber.* **57** (1984) 147–156.

[52] Kämpfer, K., Prescher, E., Möbius, H.-H., *Cfi/Ber. DKG* **68** (1991) 126–131.

[53] Möbius, H.-H. et al., *DDR Patent 260 419 A 3,* 10. 3. 1981.

[54] Möbius, H.-H. et al., *DDR Patent 261 071 A 3,* 2. 2. 1983.

[55] Oehlschlegel, G., Abraham, K., Kubbilun, H., Werding, G., *Glastech. Ber.* **57** (1984) 157–171.

[56] Weichert, J., *Glastech. Ber.* **55** (1982) 37–40.

[57] Lohmar, J., *Glastech. Ber.* **57** (1984) 141–146.

[58] McIntyre, W. H., Wallace, R. W., Troha, M. J., *US Patent 3 928 161,* 22. 3. 1974, Westinghouse Electric Corp., Pittsburgh.

[59] Wilson, H., Slough, D. A., Rudd, D. A., *DE-OS 25 32 279,* 18. 7. 1975 (GB 18. 7. 1974), George Kent Ltd. Luton.

[60] Möbius, H.-H. et al., *Festkörperchemie und Keramik, Tagungsband, ZfK Rossendorf 1988,* p. 260–262.

[61] Möbius, H.-H. et al., *DDR Patent 260 420 A 3,* 2. 2. 1983.

[62] Möbius, H.-H. et al., *Elektrokhimiya, Moscow* **26** (1990) 1388–1397.

[63] Reckmann, D., Möbius, H.-H., *Energietechnik, Leipzig* **4** (1990) 215–217.

[64] Sayles, D. A., *US Patent 3 869 370,* 7. 5. 1973.

[65] Westinghouse Electric Corp., Pittsburgh, *Instruction Bulletin 106-201,* 1985, p. 7.

[66] Hickam, W. M., *Vacuum Microbalance Techniques* Vol. 4, *Proceed. Pittsburgh Conference 7.–8. 5. 1964;* New York: Plenum Press, 1965.

[67] Möbius, H.-H., *Z. physik. Chem., Leipzig* **233** (1966) 425–429.

[68] Petanides, K., Heimke, G., *CZ-Chemie-Technik* **2** (1973) 166–168.

[69] Bulnheim, J., Guth, U., Möbius, H.-H., Beyrich, Th., *Pharmazie, Berlin* **31** (1976) 859–863.

[70] Hartung, R., Lüders, P., Möbius, H.-H., *Z. physik. Chem., Leipzig* **266** (1985) 1135–1144.

[71] Möbius, H.-H., Thesis, Universität Rostock, 1958.

[72] Hartung, R., Möbius, H.-H., *Z. Chem., Leipzig* **9** (1969) 197–198.

[73] Hartung, R., Maaß, R., *Z. Chem., Leipzig* **21** (1981) 337–338.

[74] Hartung, R., *Z. Chem., Leipzig* **23** (1983) 154–155.

[75] Hartung, R., Hannemann, K., Möbius, H.-H., Berg, H. J., *Z. Chem., Leipzig* **30** (1990) 37–38.

[76] Häfele, E., *System SOLIDOX-NH₃; Karlsruhe: Umweltverfahrenstechnik GmbH.*

[77] Hartung, R., *Z. Chem., Leipzig* **26** (1986) 266–267.

[78] Wilms, H., Diploma Thesis, Ernst-Moritz-Arndt-Universität, Greifswald, 1964.

[79] Möbius, H.-H., *DDR Patent 48722*, 23. 7. 1964.

[80] Müller-Uri, R., Diploma Thesis, Ernst-Moritz-Arndt-Universität, Greifswald, 1969.

[81] Möbius, H.-H., Müller-Uri, R., *Z. Chem., Leipzig* **9** (1969) 158–159.

[82] Hartung, R., *Z. Chem., Leipzig* **22** (1982) 153–154.

[83] Sandow, H., Thesis, Ernst-Moritz-Arndt-Universität, Greifswald (in preparation).

[84] Möbius, H.-H., *DDR Patent 37801*, 28. 6. 1962.

[85] Bertermann, K., Diploma Thesis, Ernst-Moritz-Arndt-Universität, Greifswald, 1962.

[86] Möbius, H.-H., *Naturwissenschaften* **52** (1965) 529–536.

[87] Schwartz, W. et al., *DDR Patent 142390*, 20. 3. 1979.

[88] Harbeck, W., Guth, U., *Gaswärme int.* **39** (1990) 10–24.

[89] Schwartz, W. et al., *DDR Patent 210759*, 4. 10. 1982.

[90] Möbius, H.-H., Müller-Uri, R., *Z. Chem., Leipzig* **9** (1969) 238–239.

[91] Hartung, R., Möbius, H.-H., *Z. physik. Chem., Leipzig* **267** (1986) 422–426.

[92] Teske, K. et al., *Z. Chem., Leipzig* **25** (1985) 96–97.

[93] Möbius, H.-H., Hartung, R., Zastrow, W., Prescher, E., *DDR Patent 284286 A5*, 23. 5. 1989.

[94] Guth, U., Schmidt, P., Möbius, H.-H., *Z. Chem., Leipzig* **28** (1988) 343–344.

26 High-Temperature Sensors for Oxidic Glass-Forming Melts

Friedrich Baucke, Schott Glaswerke Mainz, FRG

Contents

26.1   Introduction

The sensors treated in this chapter are principally simple, electrochemical, oxygen concentration cells without transference consisting of a measuring and a reference platinum electrode and employing a solid electrolyte, eg, oxide-doped zirconias [1–7]. The extreme conditions of their application, ie, temperatures between 700 and 1700 °C and strongly corrosive oxidic glass-forming melts with high basicity, however, have extended their development into a wide field of research.

The problem of corrosion limiting the lifetime of zirconia cell components contacted by the melts could only be solved by special constructions of the zirconia reference electrode [8–11], which, in turn, made the sensors nonisothermal cells and required the measurement of thermoelectric emfs of the solid electrolytes [5, 7]. The physicochemically open systems, represented by the glass-forming melts, and the lack of sufficiently noncorrosive insulating materials necessitated special constructions also of the platinum measuring electrode [5, 7]. The technical and economical demands of employing measuring and reference electrodes at different temperatures requires the knowledge and observation of true thermoelectric emfs also of the glass-forming melts, which could be measured only after zirconia electrodes had been developed [12] and which resulted unexpectedly in additional, fundamental information on the behavior of short-circuited metal electrodes in nonisothermal melts [12]. Owing to the lack of melts with defined oxygen contents, the correct, reversible, response of the sensors had to be verified by thermodynamically based experiments [5].

It is evident that a chapter on these high-demand application electrochemical sensors must treat the fundamental basis and the work done for their special application in some detail and may well provide an example for the development of similar cells in the future. To demonstrate their practical and scientific value, it includes some examples of their applications in basic and applied research.

26.2   Intrinsic Redox System of Oxidic Glass-Forming Melts

Oxidic glass-forming melts are mainly molten silicates, borates, phosphates, and their combinations, eg, borosilicates and aluminosilicates [13]. They are not simply molten oxides or oxide mixtures, as could be concluded from their overall composition, but consist of cations and polyoxyanions with several terminal oxide ions. These anions can be visualized as pieces of the broken network that forms solid glasses. A schematic two-dimensional sketch of the glass structure is given in Figure 26-1. Oxidic glass-forming melts thus belong to molten salts but are distinguished by a special structure which causes their characteristic properties, among which the wide temperature range of high viscosities (eg, see Figure 26-2), the strong tendency for glass formation, and the lack of a defined melting point are the most important. Temperatures as high as 700–1700 °C are required in order to obtain viscosities that are sufficiently small to allow reasonable melting processes. The number of glass-forming melts is extremely large as many elements can function as cations over wide concentration ranges and in almost unlimited combinations.

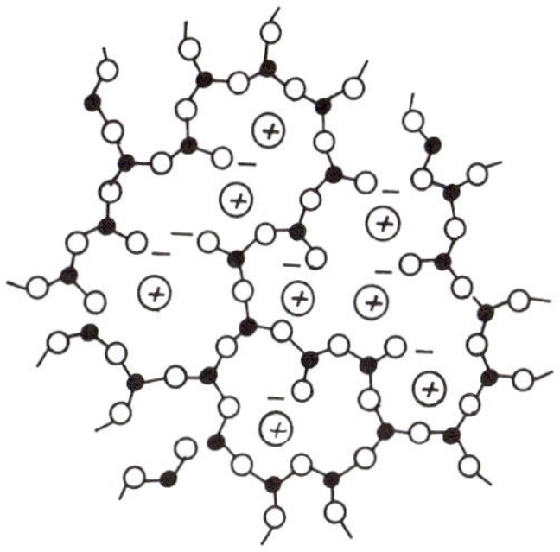

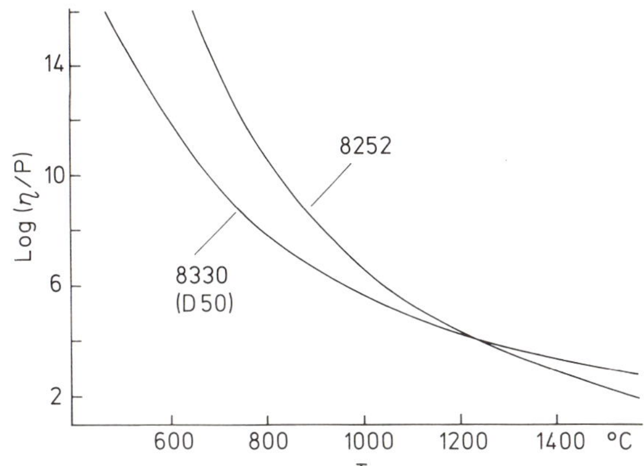

Figure 26-1.
Two-dimensional schematic structure of an alkali metal silicate glass;
● Silicon, ○ Oxygen, ⊕ Alkali metal ions.

Figure 26-2.
Temperature-dependent viscosities of two Schott glass melts showing a wide temperature range of high viscosities and the lack of a defined melting point.

The polyoxyanions of oxidic glass-forming melts formally contain singly bonded terminal oxide ions, which are in equilibrium with a negligible concentration of doubly negative oxide ions. In reality, the negative charges continuously change place because of thermal motion [14] and will be smeared out to a certain extent. Independent of the actual arrangement, the sum of the negatively charged oxidic entities, which are called "oxide ions" for simplicity, forms the reduced component of a redox system whose oxidized component is oxygen. Oxidic glass-forming melts are thus distinguished by the intrinsic oxygen-"oxide" redox system:

$$2\,O^{2-} \rightleftharpoons O_2 + 4\,e^- \tag{26-1}$$

which is characterized by the temperature-dependent equilibrium constant

$$K_{O_2/O^{2-}}(T) = \frac{p^*_{O_2}\,a^4_{e^-}}{a^2_{O^{2-}}}\,. \tag{26-2}$$

The activity of the "oxide", $a_{O^{2-}}$, is basically given by the melt composition at each temperature and, to a good approximation, is constant in most production glasses. The oxygen fugacity, $p^*_{O_2}$, thus determines the oxidizing or reducing power of the intrinsic redox system and the melt, which is symbolized by the fourth power of the electron activity in Equation 26-2.

The following examples of homogeneous and heterogeneous redox equilibria, which are of great importance for practical glass melting, may illustrate the central role of the intrinsic redox system and emphasize the necessity for measuring and controlling the oxygen fugacity during the entire melting process.

(1) Combination of the intrinsic redox system (Equation (26-1)) with the platinum/platinum ion redox system formulated with doubly positive platinum for simplicity,

$$Pt^{2+} + 2\,e^- \rightleftharpoons Pt \tag{26-3}$$

yields the redox equilibrium

$$2\,O^{2-} + 2\,Pt^{2+} \rightleftharpoons O_2 + 2\,Pt\ . \tag{26-4}$$

Its equilibrium constant,

$$K_{Pt/Pt^{2+}}(T) = \frac{p^*_{O_2}\,a^2_{Pt}}{a^2_{O^{2-}}\,a^2_{Pt^{2+}}}\ , \tag{26-5}$$

is the thermodynamic basis of platinum metal corrosion and reprecipitation [15]. High oxygen fugacities of melts in contact with platinum melters or platinum components of melting tanks lead to high concentrations of platinum ions and thus to undesired coloration of melt and glass. If the melt, however, contains platinum ions, the oxygen fugacity must be kept above the limit which excludes precipitation of platinum crystallites, since the solubility of platinum metal is extremely low in most oxidic melts. Unfavorable changes in the redox state of the glass melt can thus result in transport of platinum metal from melter walls into the bulk of the melt via the ions of the metal. In addition to coloration by ionic platinum and precipitated metallic platinum, which render the produced glass useless, the corrosion of the noble metal is a serious economic factor, especially in large-scale glass production. There are glass plants for which a 1 ppm platinum content in the produced glass would mean a loss of several hundred kilograms of platinum per year.

(2) Two redox equilibria also involving metal/metal ion systems are equally important for practical glass melting. The combination of the intrinsic redox system with the lead/lead ion system,

$$K_{Pb/Pb^{2+}}(T) = \frac{p^*_{O_2}\,a^2_{Pb}}{a^2_{O^{2-}}\,a^2_{Pb^{2+}}} \tag{26-6}$$

concerns many optical glasses, since most highly refracting glasses contain large amounts of lead ions, and that with the silver/silver ion system,

$$K_{Ag/Ag^+}(T) = \frac{p^*_{O_2}\,a^4_{Ag}}{a^2_{O^{2-}}\,a^4_{Ag^+}} \tag{26-7}$$

is characteristic of photochromic opthalmic glasses, which turn dark in sunlight and become highly transparent in the dark [16]. Both the lead and the silver ions are readily reducible, and it is important for producing stable melts to keep the oxygen fugacity above values which exclude reduction of these ions and connected precipitation of the liquid metals [17]. In fact, both metals can even be precipitated from certain melts when the oxygen partial pressure is sufficiently reduced by pumping.

(3) Combinations of the intrinsic redox system of glass melts with several polyvalent ionic redox systems, which are often present as trace impurities, are of importance for the transparency of optical glasses [18]. An example of these homogeneous equilibria is the combination with the iron(III)/iron(II) ion system,

$$K_{Fe^{2+}/Fe^{3+}}(T) = \frac{p^*_{O_2}\, a^4_{Fe^{2+}}}{a^2_{O^{2-}}\, a^4_{Fe^{3+}}} \tag{26-8}$$

As iron(III) ions absorb at high and iron(II) ions at low energies of the spectrum [19], the envisaged application of the produced optical glass determines the necessary oxygen fugacitiy of the glass melt at the end of the melting process.

(4) Perhaps the two most important redox equilibria are the combinations of the intrinsic redox system with the arsenic(III)/arsenic(V) or the antimony(III)/antimony(V) polyvalent systems [20], eg,

$$K_{Sb^{3+}/Sb^{5+}}(T) = \frac{p^*_{O_2}\, a^2_{Sb^{3+}}}{a^2_{O^{2-}}\, a^2_{Sb^{5+}}}\;. \tag{26-9}$$

Small amounts, eg, less than 0.1%, of one of these colorless materials are added to glass melts as so-called redox fining agents to remove gaseous impurities, eg, nitrogen, carbon dioxide, and water, present as residues from the raw materials. Redox fining is based on the high oxygen fugacities caused by the change in the equilibrium constant (Equation 26-9) to large values by increasing the melt temperature. This leads to the formation of oxygen nuclei and bubbles, into which the dissolved, undesired, gases diffuse, or to diffusion of oxygen into small bubbles or blisters of the impurities if present and, in both cases, to an increase in the volume and buoyancy of the bubbles. After their removal, the oxygen fugacity of the melt is decreased by lowering the temperature, and remaining pure oxygen bubbles are redissolved.

Relatively large amounts of oxygen can be dissolved by polyvalent ions. This so-called chemical solubility can exceed the physical solubility, which is represented by the oxygen fugacity in Equations (26-2) and (26-5)–(26-9), by many orders of magnitude and is strongly dependent on temperature. Polyvalent ions are thus oxygen buffers with concentration- and temperature-dependent capacity. The buffer action involves the oxide activity (Equation 26-9), which is thus not exactly constant when the total oxygen content of a buffered melt is changed. Since production glasses, however, in contrast to slags, contain only small amounts of polyvalent ions, the oxide activity of glasses as treated in this chapter can be assumed to be constant to a good approximation also when their total oxygen content changes.

26.3 Choice of the Sensor System

As shown by the examples in Equations 26-5 to 26-9, all redox systems of practical interest are connected with, and interconnected by, the intrinsic redox system oxygen/oxide of the oxidic glass-forming melt, which, consequently, characterizes its overal redox state. In addition, provided that the oxide ion activity is approximately constant, ie, independent of the content of chemically dissolved oxygen, as with production glasses containing only negligible concentrations of polyvalent ions, the redox state is characterized by the oxygen fugacity only, which also allows the control of the quality of the melt produced. The necessary relationships between oxygen fugacity and the state of the redox systems concerned must be determined in laboratory experiments, for which an example is given in Section 26.5.2. All measurements must be conducted within the bulk of the melts, since the high viscosities (see Figure 26-2) exclude the sufficiently fast establishment of the distribution equilibrium of oxygen between melt and atmosphere, which, in addition, is often subject to continuous changes due to combustion heating.

An electrochemical sensor was selected because of its possible on-line application yielding continuous, instantaneous, data on the state of the melt at the particular location in the melter or in the crucible, and since highly accurate measurements could be expected over the entire concentration range. Since, however, cells with transference, eg, cells employing a platinum reference electrode in the same melt but with a defined oxygen fugacity, are not suitable because of technical, economic, and reliability reasons, cells without transport were chosen, whose functioning in laboratory applications had been reported in the literature [1, 2]. Zirconia-based solid electrolytes were selected because of their relatively low cost, eg, compared with thoria-based materials, which introduce additional complications due to radiation. In addition, preliminary measurements had shown that the doping ion had to be triply positive, eg, yttrium, as doubly positive ions, eg, magnesium, calcium, and strontium, are subject to ion exchange with contacting melts, which causes structure changes of the ceramic material.

26.4 Electrochemical Cell for Measuring Oxygen Fugacities in Oxidic Glass-Forming Melts

26.4.1 Principle and Characterization

The electrochemical cell for measuring oxygen fugacities in oxidic glass-forming melts consists of a measuring platinum electrode dipping into the melt and a platinum/oxygen reference electrode with defined oxygen partial pressure, which is separated from the melt by a wall of Y_2O_3-doped ZrO_2 with unit transport number of oxide ions [1, 2, 5, 7]. In the first of the two practically possible arrangements, the platinum measuring electrode and the ZrO_2 ceramic have no contact. It is shown in Figure 26-3a and represented by cell system (I):

$$\text{Pt (r), } O_2 \text{ (r)}/ZrO_2/O^{2-} \text{ (m, r), melt, } O_2 \text{ (m, Pt (m)), } O^{2-} \text{ (m, Pt (m))/Pt (m) ,} \quad \text{(I)}$$

where the letters in parentheses indicate the location of the respective electrode or compound, eg, O^{2-} (m, r) = oxide ion in the melt (m) near the reference electrode (r). The electrode reaction of the measuring electrode is given by

$$O_2 \,(m, \text{Pt (m)}) + 4\,e^- \,(\text{Pt (m)}) \rightleftharpoons 2\,O^{2-} \,(m, \text{Pt (m)}) \,, \tag{26-10}$$

and that of the reference electrode by

$$2\,O^{2-} \,(m, r) \rightleftharpoons 2\,O^{2-} \,(ZrO_2, m) \tag{26-11 a}$$

and

$$2\,O^{2-} \,(ZrO_2, r) \rightleftharpoons O_2 \,(r) + 4\,e^- \,(\text{Pt (r)}) \,. \tag{26-11 b}$$

Comparison of Equation (26-10) with Equations (26-11 a) and (26-11 b) shows that the reference electrode (hereafter often "zirconia electrode") functions as a platinum electrode in a melt with defined oxygen partial pressure, since $a_{O^{2-},\, ZrO_2\,(m)} = a_{O^{2-},\, ZrO_2\,(r)}$ because of the uniform solid electrolyte [1]. This is also seen from Equation (26-12) for its galvanic potential:

$$\varepsilon_{ZrO_2} = \varepsilon^0_{O_2/O^{2-}} + \frac{RT}{4F} \ln \frac{p_{O_2,\,r}}{a^2_{O^{2-},\,m,\,r}} \,, \tag{26-12}$$

compared with that of a platinum electrode in the melt:

$$\varepsilon_{\text{Pt (m)}} = \varepsilon^0_{O_2/O^{2-}} + \frac{RT}{4F} \ln \frac{p^*_{O_2,\,m,\,\text{Pt (m)}}}{a^2_{O^{2-},\,m,\,r}} \tag{26-13}$$

where p_{O_2} is the partial pressure, $p^*_{O_2}$ is the fugacity, of oxygen, and $a_{O^{2-}}$ is the activity of oxide ions. The cell reaction of cell (I) is the sum of Equations (26-10), (26-11 a), and (26-11 b):

$$O_2 \,(m, \text{Pt (m)}) + 2\,O^{2-} \,(m, r) \rightleftharpoons 2\,O^{2-} \,(m, \text{Pt (m)}) + O_2 \,(r) \tag{26-14}$$

and, for the isothermal cell in a homogeneous melt, the emf is given by

$$E = \frac{RT}{4F} \ln \frac{p^*_{O_2,\,\text{Pt (m)}}}{p_{O_2,\,r}} \,, \tag{26-15}$$

if an identical standard free enthalpy is assumed for the electrode reactions, Equations (26-10), (26-11 a) and (26-11 b). Owing to the unit transport number of oxide ions in the ZrO_2 ceramic, which is given at all temperatures and oxygen partial pressures of practical interest [21], the cell is without transference, and the emf contains no diffusion potential. The oxygen fugacity of the melt to be measured is finally given by [5, 7]

$$p^*_{O_2,\,m,\,\text{Pt (m)}} = \exp\left(\frac{4\,FE}{RT} + \ln p_{O_2,\,r}\right) . \tag{26-16}$$

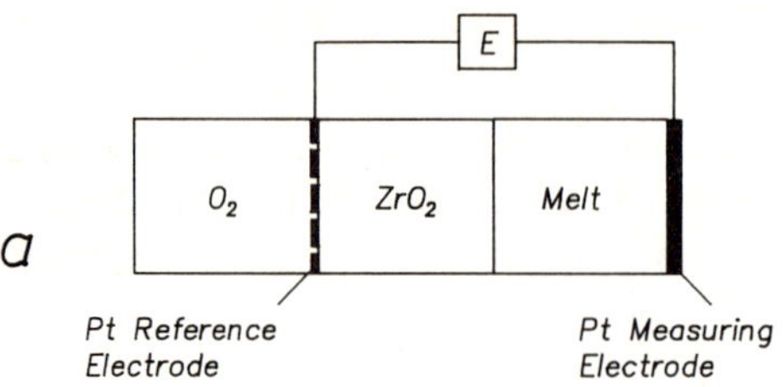

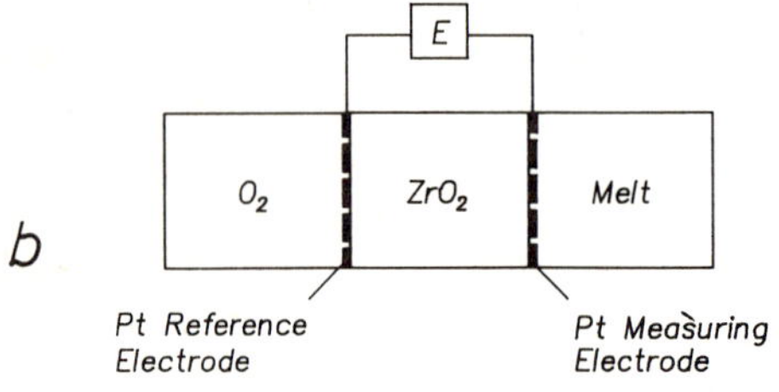

Figure 26-3.
Main electrochemical cell arrangements for measuring oxygen fugacities of oxidic glass-forming melts: (a) separated reference and measuring electrodes; (b) "single-rod electrode".

In the second practical arrangement of the cell, which is shown in Figure 26-3b, the platinum measuring electrode and ZrO_2 ceramic are in intimate contact, and the set-up could be called a "single-rod electrode". The cell scheme is represented by

$$\text{Pt (r), } O_2 \text{ (r)}/ZrO_2/\text{Pt (m), melt, } O^{2-} \text{ (m, r), } O_2 \text{ (m, r)} \qquad \text{(II)}$$

and the electrode reactions:

$$O_2 \text{ (m, r)} + 4\,e^- \text{ (Pt (m))} \rightleftharpoons 2\,O^{2-} \text{ (m, r)} \qquad \text{(26-17)}$$

and Equations (26-11a) and (26-11b) yield the cell reaction

$$O_2 \text{ (m, r)} \rightleftharpoons O_2 \text{ (r)} \qquad \text{(26-18)}$$

and the emf

$$E = \frac{RT}{4F} \ln \frac{p^*_{O_2,\,m,\,r}}{p_{O_2,\,r}} \quad . \qquad \text{(26-19)}$$

Comparison of Equations (26-15) and (26-19) indicates that inhomogeneities of the melt can be of importance only with separated reference and measuring electrodes, since they contact different locations of the melt.

Equations (26-15) and (26-19) are also the basis for testing reference electrodes for proper functioning without removal from the melt, since defined alterations of the reference oxygen partial pressure result in corresponding changes of the emf if the electrodes function correctly [5]. Simultaneous, slow, potential changes of the measuring electrode, which serves as the reference electrode in this case, do not interfere because of the fast electrode response (<20 s for 99.5% of the total potential change).

26.4.2   Electrode Constructions

26.4.2.1   Zirconia Tube Electrode

Various electrode constructions have been developed for laboratory and for process
measurements. Figure 26-4 shows two types of the original or "zirconia tube" electrode
[5, 7, 11]. They consist of a closed zirconia tube containing a four-bore alumina tube which
insulates the Pt reference electrode and the wires of a thermocouple and through which the
reference gas is introduced. An electrode head tightens the arrangement. This construction is
very accurate because of the truly isothermal arrangement of the vital electrode parts, but is
sensitive to thermal shock and has a limited lifetime owing to corrosion of the thin zirconia
wall. Zirconia tube electrodes are applied almost exclusively for laboratory measurements.

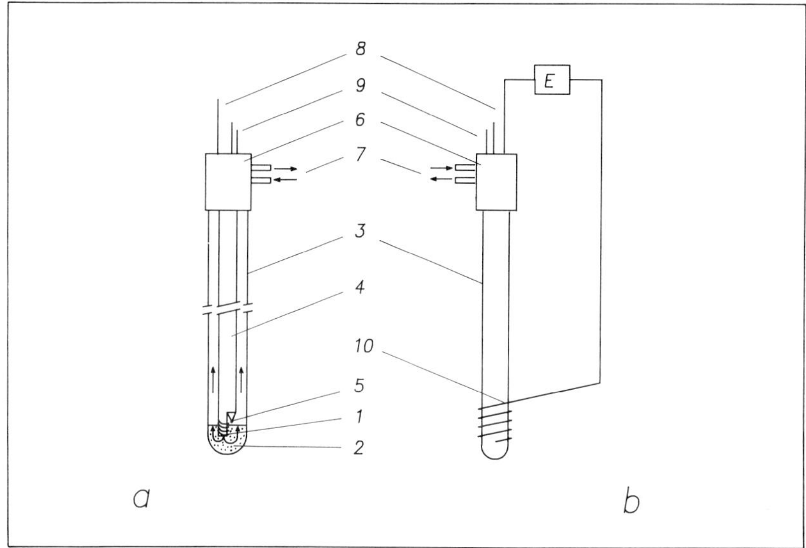

Figure 26-4. Zirconia tube electrodes. (a) Reference electrode; (b) "single-rod electrode" consisting of
reference and measuring electrode. (1) Pt,O_2 reference electrode   (2) zirconia grit ensuring
good contact between Pt and zirconia tube [11]; (3) zirconia tube; (4) four-bore alumina
tube; (5) thermocouple; (6) electrode head; (7) reference gas inlet and outlet; (8) reference
electrode contact; (9) thermocouple wires; (10) Pt measuring electrode; (E) instrument
measuring the emf.

26.4.2.2   Zirconia Disk Electrode

Figure 26-5 presents a construction which is fairly insensitive to thermal shock [22]. A zirconia membrane is tightly sintered to a Pt tube which forms the electrode shaft. The lifetime
also of this arrangement is limited by corrosion of the membrane, which, in addition, is subject to so-called bubble boring, especially in a horizontal position [4].

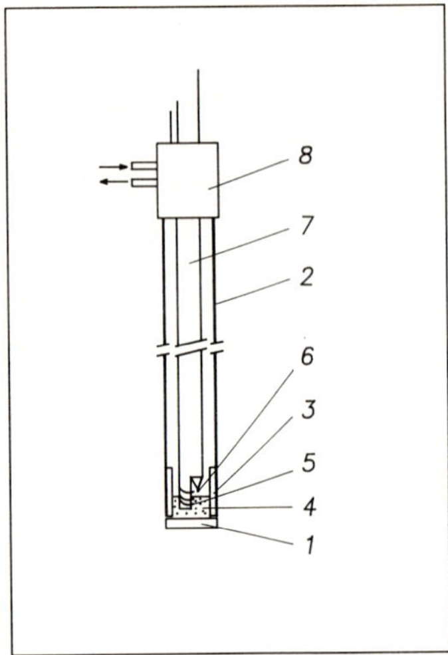

Figure 26-5.
Zirconia disk electrode. (1) Zirconia disk sintered to
(2) platinum tube; (3) alumina tube for insulating zirconia
grit and platinum tube; (4) zirconia grit; (5) Pt,O_2 reference
electrode; (6) thermocouple; (7) four-bore alumina tube;
(8) electrode head.

26.4.2.3 Dissolving Zirconia Electrode

A zirconia electrode with a particularly long lifetime is shown in Figure 26-6 [5, 7–9]. A
zirconia electrolyte bridge, which is inserted into an alumina electrode shaft and whose position is fixed by an alumina bolt, connects the Pt,O_2 reference electrode and the melt. The

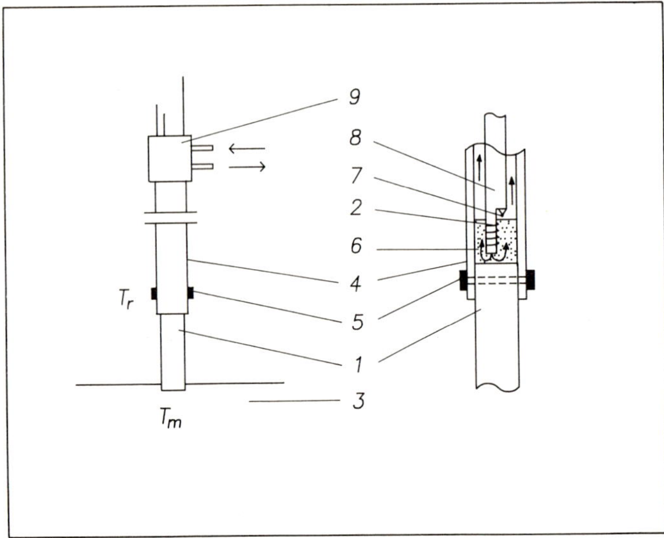

Figure 26-6. Dissolving zirconia electrode. (1) Zirconia bridge between (2) Pt,O_2 reference electrode and
(3) melt; (4) alumina electrode shaft; (5) alumina bolt; (6) zirconia grit; (7) thermocouple;
(8) four-bore alumina tube; (9) electrode head. T_r reference electrode temperature; T_m
temperature of measuring electrode in isothermal melt with T_m.

electrode is lowered as the melt dissolves away the lower end of the bridge. The arrangement is particularly suitable for streaming melts, which continuously remove traces of dissolved zirconia and thus eliminate errors caused by changes of melt composition and diffusion potentials. Since, however, the temperatures of reference electrode and melt are usually different, this electrode construction constitutes a nonisothermal cell. Thus, for practical application, either the thermoelectric voltage developed in the zirconia bridge must be taken into account [5–7] or the temperature of the reference electrode must be made equal to the temperature of the measuring electrode by automatic heating of the reference electrode compartment [23]. A platinum sleeve extending slightly over the lower end of the alumina tube and not in contact with either zirconia or melt protects the alumina electrode shaft from corrosion by condensing vapours from the melt [26].

26.4.2.4 Thermoelectric emfs of Oxide-Doped Zirconia

The "dissolving ZrO_2 reference electrode" (Figure 26-6) is represented by the cell scheme

$$Pt,O_2\ (r)\ (T_r)/ZrO_2/(T_m)\ melt,\ O_2\ (m),\ O^{2-}\ (m)\ (T_m)/Pt\ , \tag{III}$$

where T_r is the temperature of the reference Pt,O_2 electrode at the upper end of the zirconia bridge and T_m the temperature of the platinum measuring electrode in the melt, which is assumed to be uniform. The oxygen fugacity of the melt is given by

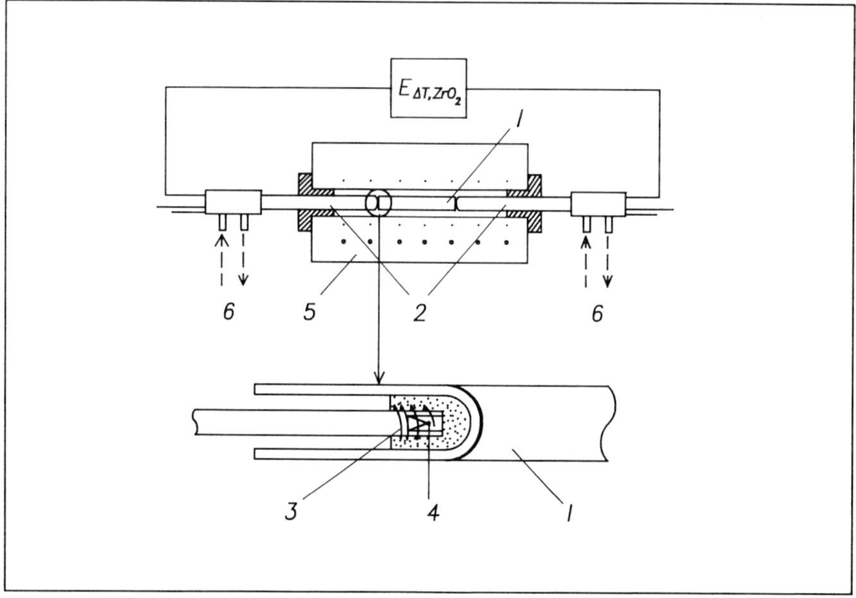

Figure 26-7. Experimental arrangement for measuring thermoelectric emfs of zirconia solid electrolytes. (1) Zirconia specimen; (2) zirconia tube electrodes; (3) Pt,O_2 reference electrode; (4) thermocouple; (5) temperature gradient furnace; (6) reference gases.

$$p^*_{O_2,m} = \exp\left\{\frac{4F}{RT_m}\left[E - E_{Th,ZrO_2}(T_r \to T_m)\right] + \frac{T_r}{T_m}\ln p_{O_2,r}\right\}, \qquad (26\text{-}20)$$

where E is the measured emf, and $E_{Th,ZrO_2}(T_r \to T_m)$ is the standard thermoelectric emf of the zirconia bridge between the reference and measuring temperature:

$$E_{Th,ZrO_2}(T_r \to T_m) = \int_{T_r}^{T_m}\left(\frac{d\varphi_{Th,ZrO_2}}{dT}\right)dT. \qquad (26\text{-}21)$$

φ_{Th,ZrO_2} is the thermoelectric potential of ZrO_2, and the integral in Equation 26-21 was defined as being the standard Seebeck coefficient and was measured for various oxide-doped zirconia ceramics in the set-up shown in Figure 26-7 [25]. The rod- or tube-shaped material to be investigated is contacted by two zirconia tube electrodes and subjected to certain, defined temperature profiles, which can be generated in the temperature gradient furnace. The maximum temperature gradient was 100 K/cm. The cell scheme of this arrangement is

$$\text{Pt}, O_2\,(T_1)/ZrO_2/(T_2)\text{Pt}, O_2 \qquad\qquad (IV)$$

and the measured thermoelectric emf by the equation

$$E_{\Delta T,ZrO_2} = \frac{RT_2}{4F}\ln p_{O_2}(T_2) - \frac{RT_1}{4F}\ln p_{O_2}(T_1) + \left[\varepsilon^0(T_2) - \varepsilon^0(T_1 +\right.$$

$$+ \frac{RT_1}{2F}\ln a_{O^{2-},ZrO_2}(T_1) - \frac{RT_2}{2F}\ln a_{O^{2-},ZrO_2}(T_2) +$$

$$\left. + \Delta\varepsilon_{Pt}(T_2 \to T_1) + \varepsilon_{ZrO_2}(T_1 \to T_2)\right], \qquad (26\text{-}22)$$

where $\varepsilon^0(T_1)$ and $\varepsilon^0(T_2)$ are the standard galvanic voltages at T_1 and T_2, respectively, $\Delta\varepsilon_{Pt}(T_2 \to T_1)$ is the thermoelectric voltage of platinum between T_2 and T_1, and $\varepsilon_{ZrO_2}(T_1 \to T_2)$ is the thermoelectric diffusion voltage of the doped ZrO_2 between T_1 and T_2. (Although usually presented in a different form in the literature, the thermoelectric emfs of cells (IV) and (VIII) (Section 4.4) are given by Equation (26-22) because of the clear presentation of the electrochemical quantities involved.) The sum of the terms in the square brackets in Equation (26-22), which are principally unknown, was defined as being the standard thermoelectric emf:

$$E_{Th,ZrO_2}(T_1 \to T_2) = \varepsilon^0(T_2) - \varepsilon^0(T_1) + \frac{RT_1}{2F}\ln a_{O^{2-},ZrO_2}(T_1) -$$

$$- \frac{RT_2}{2F}\ln a_{O^{2-},ZrO_2}(T_2) + \Delta\varepsilon_{Pt}(T_2 \to T_1) +$$

$$+ \varepsilon_{ZrO_2}(T_1 \to T_2), \qquad (26\text{-}23)$$

and is connected to the standard Seebeck coefficient as shown by Equation (26-21). We introduced the term "standard" into these definitions since $E_{\mathrm{Th,ZrO_2}}$ is directly measured by the arrangement (Figure 26-7) if both ZrO_2 electrodes contain 1 bar oxygen, and the quantity is thus referred to standard conditions:

$$E_{\Delta T,\mathrm{ZrO_2}} = E_{\mathrm{Th,ZrO_2}}(T_1 \rightarrow T_2) \quad \text{at } p_{O_2}(T_1) = p_{O_2}(T_2) = 1 \; .$$

Several zirconias doped with yttria in the concentration range 4–10 mol % were investigated [25]. At temperatures and oxygen partial pressures at which unit oxide transport number prevails [21], the standard Seebeck coefficient of the materials was found to be independent of temperature. This allows the application of Equation (26-24), a linear form of Equation (26-21):

$$E_{\mathrm{Th,ZrO_2}}(T_r \rightarrow T_m) = \frac{\mathrm{d}\varphi_{\mathrm{Th,ZrO_2}}}{\mathrm{d}T}(T_m - T_r) , \qquad (26\text{-}24)$$

for the calculation of the oxygen fugacity by Equation (26-20) [5, 7]. The dependence of the standard Seebeck coefficient on the yttria content of the zirconia is represented by

$$\frac{\mathrm{d}\varphi_{\mathrm{Th,ZrO_2}}}{\mathrm{d}T} = a\,c_{\mathrm{Y_2O_3}} + b , \qquad (26\text{-}25)$$

with $a = 4.643 \times 10^{-3}$ mV/(K mol%) and $b = -0.4949$ mV/K [25]. Equation (26-25) is in good agreement with a value reported by Fischer [26] and an equation given by Pizzini et al. [27]. The standard Seebeck coefficient of the most frequently applied zirconia, $(ZrO_2)_{0.9547}(Y_2O_3)_{0.0453}$, is (-0.4739 ± 0.0015) mV/K (700–1550 °C) [5]. The data must be determined with great accuracy because of their magnitude and since temperature differences between reference and measuring electrode can be large, eg, up to several hundred kelvin, in practical glass melting tanks.

24.4.2.5 Platinum Measuring Electrode

At first sight, the construction of the platinum measuring electrode does not seem to present any basic problems. It must be taken into account, however, that its surface contacts parts of the melt with different oxygen fugacities. In particular, the oxygen fugacity near the melt surface, in contact with the atmosphere, often differs considerably from that of the bulk of the melt. Since electrode components cannot be insulated satisfactorily from the melt because of the lack of sufficiently inert materials with low electrical conductivity, platinum measuring electrodes almost always form oxygen concentration cells, which are short-circuited by themselves. This results in two effects: (1) the electrode assumes a mixed potential, which can be different from the equilibrium potential of the part of the electrode contacting the bulk of the melt and to be measured, and (2) the oxygen fugacity of the melt contacting the lower region of the electrode changes owing to the short-circuit with the upper electrode part.

This effect cannot be eliminated completely. It is minimized to a large extent, however, by an electrode construction which is based on kinetic principles [5]. Figure 26-8 shows the basis

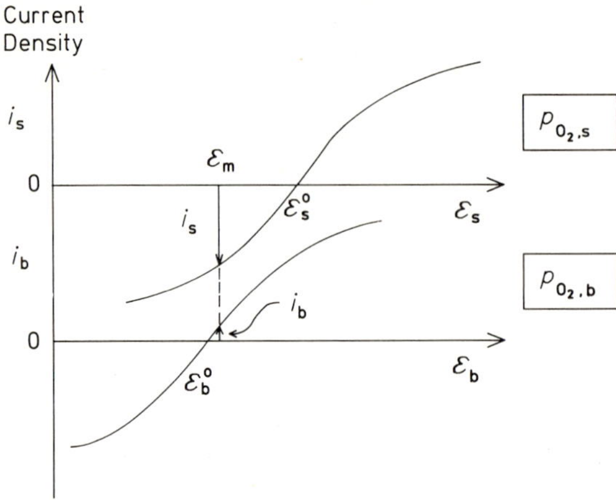

Figure 26-8.
Current density/potential curves of two platinum electrodes, in the melt near the surface ($i_s = f(\varepsilon_s)$) and in the bulk of the melt ($i_b = f(\varepsilon_b)$) at different oxygen fugacities, $p_{O_2,s} > p_{O_2,b}$. ε_s^0 and $\varepsilon_b^0 = $ equilibrium potentials; $\varepsilon_m = $ mixed potential; i_s and $i_b = $ current densities, caused by short-circuiting the electrodes if their surface areas are different, eg, $|i_b/i_s| = A_s/A_b \ll 1$. As a result, $(\varepsilon_s^0 - \varepsilon_m) > (\varepsilon_m - \varepsilon_b^0)$ and $\varepsilon_b^0 \simeq \varepsilon_m$.

with the simplifying assumption that the electrochemical cell representing the platinum measuring electrode consists of an electrode near the melt surface (denoted s) and an electrode in the bulk of the melt (denoted b) and is short-circuited by the intermediate electrode region, which behaves as an inert metal against the melt. In addition, the oxygen fugacity near the melt surface is assumed to be larger than that in the bulk, $p^*_{O_2,s} > p^*_{O_2,b}$. Correspondingly, the current density-potential curves of the electrodes (Figure 26-8) are at different positions on the potential scale and distinguished by different equilibrium potentials, ie, $\varepsilon_s^0 > \varepsilon_b^0$. The short-circuit causes currents with equal absolute magnitude, $|i_s A_s| = |i_b A_b|$, which means that the current densities through the electrodes are inversely proportional to their surface areas, $|i_s/i_b| = A_b/A_s$.

The ratio of the surface areas thus determines the relative magnitude of the current densities and, according to Figure 26-8, also the magnitude of the mixed potential, ε_m, of the cell relative to the equilibrium potentials of its upper and lower parts. Thus, for practical measuring electrodes, a ratio

$$A_b/A_s = |i_s/i_b| \gg 1 \tag{26-26}$$

is chosen, which results in the condition

$$(\varepsilon_s^0 - \varepsilon_m) \gg (\varepsilon_m - \varepsilon_b^0) \tag{26-27}$$

and thus in a mixed potential of the platinum measuring electrode which approximates the equilibrium potential of its lower part to be measured, $\varepsilon_b^0 \simeq \varepsilon_m$. A ratio $A_b/A_s = 20\text{--}60$, depending on the specific experimental conditions, has always been appropriate. Several electrode constructions with large lower surface areas are shown in Figure 26-9.

These electrode constructions also reduce the rate of the fugacity change at the lower part of the electrode caused by the short-circuit, but does not decrease the final magnitude of this effect. Complete elimination, however, is achieved by the relative motion of the melt and electrode, eg, by stirring the melt or in streaming melts as in continuously working industrial melting tanks.

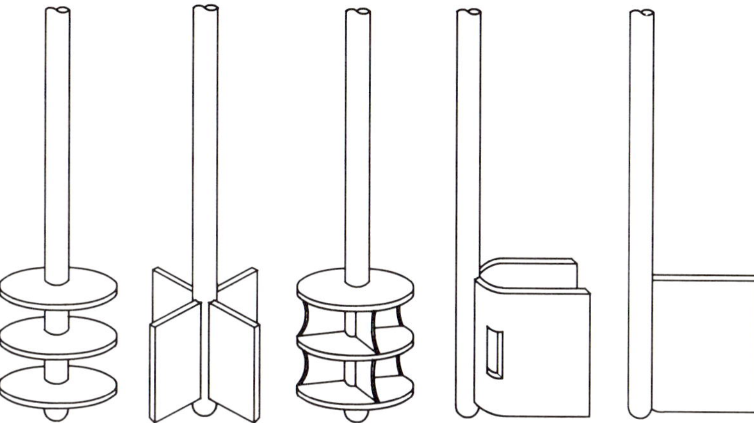

Figure 26-9. Various kinds of platinum measuring electrodes with large surface area of lower part, ie, in the bulk of the melt. Each kind has certain advantages and is applied in certain situations.

Platinum reference and measuring electrodes of the cells must be of pure platinum, although alloys, eg, platinum-rhodium and platinum-iridium, are generally preferred for glass-melting purposes because of their higher mechanical strength. Owing to the less noble character of rhodium and iridium compared with platinum, however, these metals are oxidized more easily and polarize the alloys. Figure 26-10 shows temperature- and time-dependent isothermal emfs of the cell

$$\text{Pt/melt/Pt,M,} \tag{V}$$

where M = 1% Ir, 5% Rh, or 10% Rh, respectively, and the melt is a sodium-calcium silicate containing 0.2 wt% Sb_2O_3 as fining agent [28]. The measurements demonstrate that the

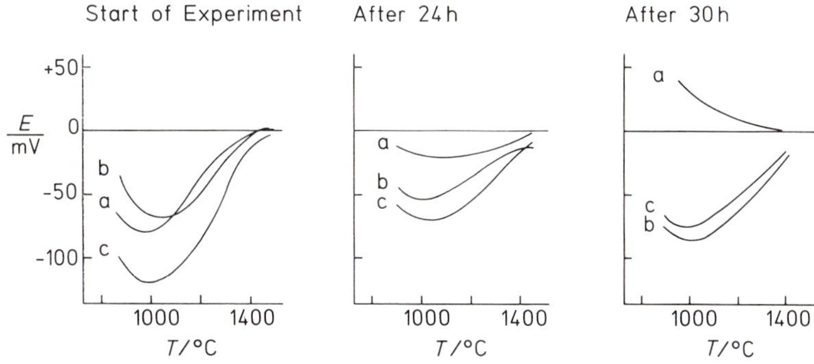

Figure 26-10. Time dependence of temperature-dependent emfs of the cell Pt/melt/Pt,M with M = 1% Ir (a), 5% Rh (b), and 10% Rh (c), showing irreversible behavior of the alloys and demonstrating that only pure platinum can be used for constructing measuring electrodes of oxygen fugacity cells. $dT/dt = -300$ K/h.

"alloying effect" is, in fact, fairly large and irreproducible, even after long periods, obviously because of individual, local, polarization processes at the metal surfaces.

26.4.3 Verification of Thermodynamically Correct Response of the Cells

It was essential for the application of the cell for scientific and technical purposes to confirm its correct and reversible response with respect to the intrinsic redox system, O_2/O^{2-}, of glass-forming melts. Since, however, melts are open systems with respect to dissolved gases and "standard melts" with defined oxygen contents are not available, the correct functioning was verified on a thermodynamic basis [5].

The measurements were conducted by applying the cell in oxidic melts containing small and equal molar concentrations of antimony(III) and antimony(V) (or arsenic(III) and arsenic(V)) oxide. The special arrangement used separated the melt essentially from the surrounding atmosphere and allowed electrochemical pumping of oxygen into, and out of, the melt by means of a platinum electrode and a zirconia electrode or the zirconia crucible, which contained the melt and served as the oxide ion-conducting wall [28]. The concentration ratio of the polyvalent ionic species could thus be well adjusted.

The measuring cell of the arrangement is represented by

$$\text{Pt}, O_2 \, (r)/ZrO_2/\text{melt}, \ Sb_2O_5 \, (m), \ Sb_2O_3 \, (m)/\text{Pt} \ . \tag{VI}$$

The cell reaction,

$$Sb_2O_5 \, (m) \ \rightleftharpoons \ Sb_2O_3 \, (m) + O_2 \, (r), \tag{26-28}$$

consists of an oxidation of oxide ions of the melt by antimony ions and the corresponding oxygen formation in the zirconia reference electrode. The cell entropy change is given by

$$\Delta S_{cell} = S^0_{Sb_2O_3,m} - S^0_{Sb_2O_5,m} + S^0_{O_2,r} - R \ln \frac{a_{Sb_2O_3,m}}{a_{Sb_2O_5,m}} -$$

$$- R \ln p_{O_2,r} - RT \frac{d}{dT} \left(\ln \frac{a_{Sb_2O_3,m}}{a_{Sb_2O_5,m}} \right) - RT \frac{d}{dT} (\ln p_{O_2,r}) , \tag{26-29}$$

where S^0 are the standard entropies and a are the activities of the compounds indicated. Since the reference partial pressure of oxygen is independent of temperature and the logarithmic terms containing activity ratios of antimony(III) and antimony(V) ions are approximately zero because of equal concentrations, $a_{Sb_2O_3}/a_{Sb_2O_5} \simeq c_{Sb_2O_3}/c_{Sb_2O_5} = 1$, the cell reaction entropy reduces to

$$\Delta S_{cell} = (S^0_{Sb_2O_3,m} - S^0_{Sb_2O_5,m}) + S^0_{O_2,r} - R \ln p_{O_2,r}, \tag{26-30}$$

which is further reduced to

$$\Delta S_{cell} \simeq S^0_{O_2,r} - R \ln p_{O_2,r} , \tag{26-31}$$

since the difference of the standard entropies of the dissolved antimony species [29] is considerably smaller than the standard entropy of oxygen [30] and can be neglected as a first approximation:

$$| S^0_{Sb_2O_3,m} - S^0_{Sb_2O_5,m} | < S^0_{O_2,r} .$$

(26-32)

The temperature coefficient of the emf of the cell (see Equation VI),

$$\frac{dE}{dT} = \frac{\Delta S_{cell}}{4F} \simeq \frac{1}{4F} (S^0_{O_2,r} - R \ln p_{O_2,r})$$

(26-33)

is thus expected to be approximately independent of temperature, eg, 0.59 mV/K mol at 1000 °C, 0.60 mV/K mol at 1250 °C, and 0.61 mV/K mol at 1500 °C at 1 bar oxygen partial pressure of the reference electrode and constant total oxygen content of the melt. In addition, the equation

$$\frac{d\,(dE/dT)}{d\,(\log p_{O_2,r})} = \frac{2.303\,R}{4F} = 0.0499 \text{ mV/K}$$

(26-34)

is expected to hold.

The measurements were conducted dynamically at constant rates of change of temperature, which guaranteed isothermal conditions throughout the cell ("quasi-isothermal measurements") and negligible oxygen exchange with the contacting atmosphere ("quasi-closed system") [5] in addition to the closed cell arrangement [28]. Figure 26-11 as an example shows

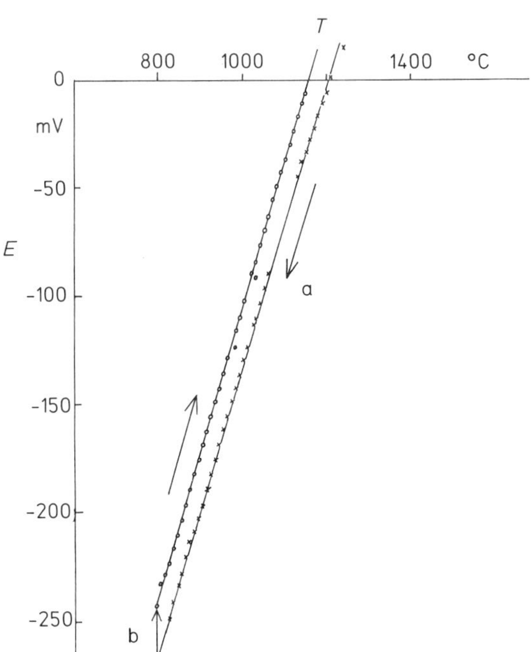

Figure 26-11.
Isothermal emf of cell (VI) in a sodium borate melt as a function of temperature (dynamic measurements) showing validity of Equation (26-33) and thermodynamically correct functioning of the cell with respect to the inherent oxygen/oxide system of the melt. (b) Effect of internal short circuit of platinum measuring electrode during a 10 h halt of the dynamic measurement after (a).

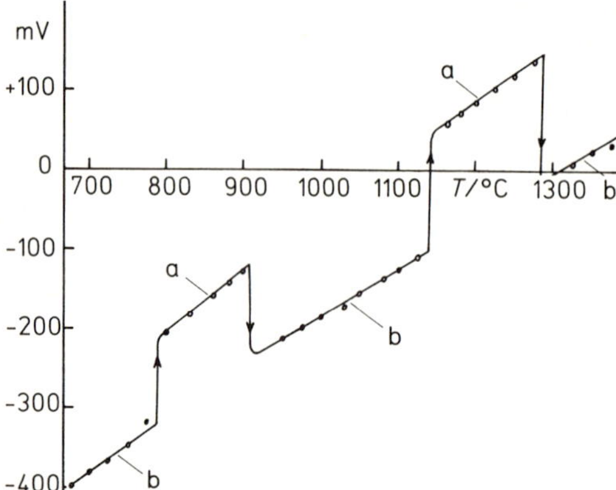

Figure 26-12.
Isothermal emf of cell (VI) as a function of temperature (dynamic measurement) at two different oxygen partial pressures of the reference electrode, resulting in $dE/dT = 0.70$ mV/K at 10^{-2} bar (a) and $dE/dT = 0.60$ mV/K at 1 bar oxygen (b), in agreement with Equation (26-33). The result d $(dE/dT)/$d $(\log p_{O_2}) = -0.05$ mV/K additionally confirms the thermodynamic reversibility of the zirconia reference electrode (Equation (26-34)); see also Figure 26-11.

that the experiments agree with the theoretical expectation. The measurements conducted in the unstirred melt were interrupted for 10 h at 800 °C and show a slight formation of oxygen at the lower part of the Pt measuring electrode owing to its internal short-circuit discussed in the preceding section, after which, however, an almost unchanged $dE/dT = 0.63$ mV/K is found. Some measurements were extended down to 10^{-11} bar oxygen fugacity and still showed a nearly constant temperature coefficient of the emf [5]. As the activity ratio of antimony(III) and antimony(V) ions is nearly constant during these experiments owing to the low concentration of physically dissolved oxygen, $c_{O_2} \ll (c_{Sb_2O_3} + c_{Sb_2O_5})$, the cell thus shows a thermodynamically correct response with respect to the intrinsic redox system of the melts. Additional proof of the correct functioning of the zirconia reference electrode is exhibited by Figure 26-12, which shows excellent agreement of the measurements with Equation (26-34).

26.4.4   Cells in Nonisothermal Glass-Forming Melts

The measurement of oxygen fugacities in glass melting tanks often requires the cell to be applied in nonisothermal melts either because of technical conditions of the melting unit or since, for economic reasons, one zirconia electrode is to be employed as the reference electrode for several platinum measuring electrodes at different locations. Since dissolving zirconia electrodes are usually applied under technical conditions, the measuring cell contains two media, ie, zirconia and melt, which develop thermoelectric voltages. Figure 26-13 shows such an arrangement. The cell scheme is

$$\text{Pt, O}_2 \text{ (r) } (T_r)/(T_r) \text{ ZrO}_2 \text{ } (T_c)/(T_c) \text{ melt, O}_2 \text{ (m), O}^{2-} \text{ (m) } (T_m)/\text{Pt} \tag{VII}$$

and the oxygen fugacity at the temperature T_m of the measuring electrode is calculated by

$$p^*_{O_2,m}(T_m) = \exp\left\{ \frac{4F}{RT_m}\left[E - E_{Th,\,ZrO_2}(T_r \to T_c) - E_{Th,\,m}(T_c \to T_m) \right] + \right.$$

$$\left. + \frac{T_r}{T_m} \ln p_{O_2,\,r}(T_r) \right\}, \tag{26-35}$$

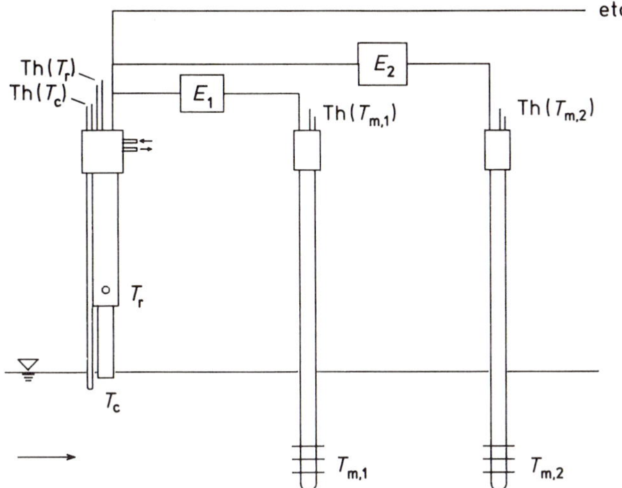

Figure 26-13. Arrangement of cells for measuring oxygen fugacities in a streaming nonisothermal melt by referring to only one nonisothermal dissolving zirconia reference electrode (see Equation 26-35). E_1 and E_2 = measured emfs; Th = thermocouples measuring indicated temperatures; T_r = temperature of reference electrode; T_c = temperature at contact ZrO_2/melt; $T_{m,1}$ and $T_{m,2}$ = temperatures of platinum measuring electrodes.

where T_r and T_c are the temperatures of the reference electrode and at the contact of zirconia bridge and melt, respectively, E is the measured emf, and $E_{Th,ZrO_2}(T_r \rightarrow T_c)$ and $E_{Th,m}$ $(T_c \rightarrow T_m)$ are the standard thermoelectric emf of zirconia between T_r and T_c, as given by Equation (26-21), and of the melt between T_c and T_m, as defined by

$$E_{Th,m}(T_c \rightarrow T_m) = \int_{T_c}^{T_m} \left(\frac{d\varphi_{Th,m}}{dT} \right) dT , \qquad (26\text{-}36)$$

respectively.

The quantity $(d\varphi_{Th,m}/dT)$ in the integral of Equation (26-36) is defined standard Seebeck coefficient and is a characteristic property of a glass-forming melt. It is determined by means of zirconia microelectrodes (Figure 26-14), which eliminate temperature-dependent redox potentials inherently included in the emf if platinum electrodes were applied to these measurements [12]. The cell scheme for measuring standard Seebeck coefficients according to Figure 26-14 is

$$\text{Pt, } O_2(T_1)/ZrO_2(T_1)/\text{melt, } O^{2-}(m)/ZrO_2(T_2)/\text{Pt, } O_2(T_2) \qquad \text{(VIII)}$$

and the thermoelectric emf measured, $E_{\Delta T,m}$, is given by Equation (26-22), in which, however, the oxide activities of zirconia, a_{O^{2-},ZrO_2}, are replaced with those of the melt, $a_{O^{2-},m}$, and the thermoelectric diffusion voltage of zirconia, $\varepsilon_{ZrO_2}(T_1 \rightarrow T_2)$, is replaced with that of the melt, $\varepsilon_m(T_1 \rightarrow T_2)$. At 1 bar oxygen partial pressure in both zirconia electrodes, the standard thermoelectric emf is directly measured by the arrangement (Figure 26-14)

$$E_{\Delta T,\mathrm{m}} = E_{\mathrm{Th,m}}\,(T_1 \rightarrow T_2) \quad \text{at } p_{O_2}(T_1) = p_{O_2}(T_2) = 1$$

corresponding to the measurements on zirconia (Figure 26-7).

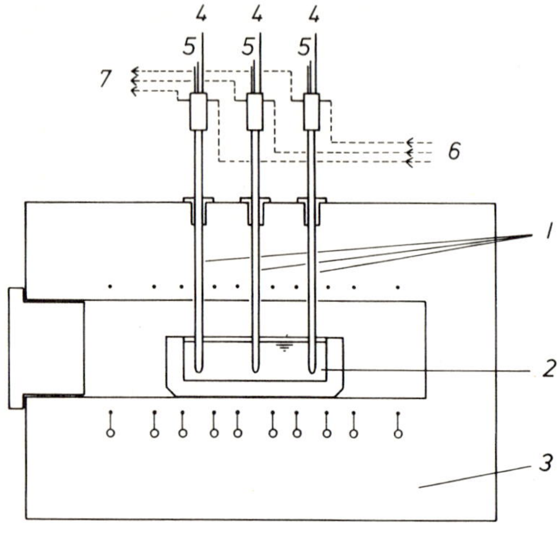

Figure 26-14.
Experimental arrangement for measuring thermoelectric emfs of glass-forming melts by means of zirconia microelectrodes. (1) Zirconia electrodes; (2) melt; (3) temperature gradient furnace; (4) leads from Pt,O$_2$ reference electrodes; (5) leads from thermocouples; (6) reference gas inlet; (7) outlet.

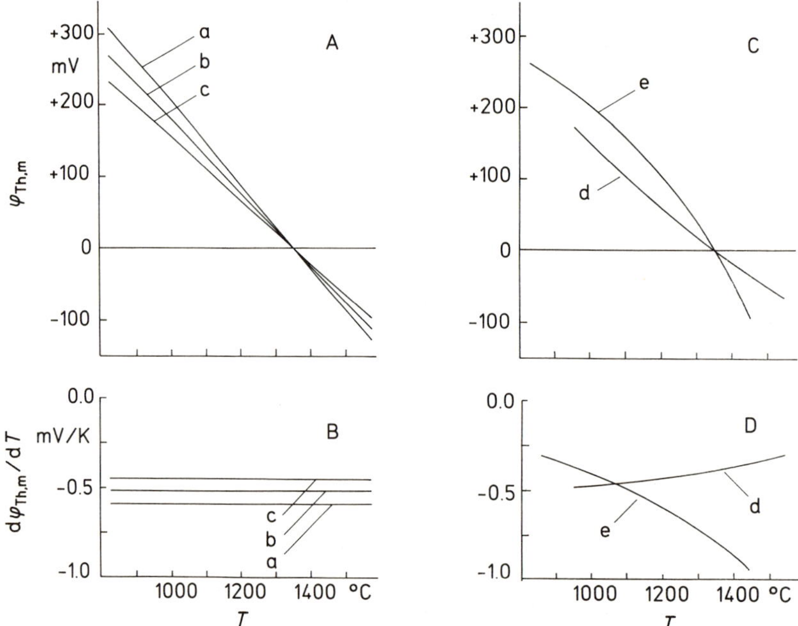

Figure 26-15. Relative potentials of a ZrO$_2$ electrode with 1 bar oxygen partial pressure in glass-forming melts with temperature-independent (A, B) and temperature-dependent (C, D) standard Seebeck coefficients. Reference potential: $\varphi_r = 0$. (a) Fiolax klar; (b) (Na$_2$O)$_{0.078}$ (K$_2$O)$_{0.078}$ (CaO)$_{0.107}$ (SiO$_2$)$_{0.737}$; (c) (Na$_2$O)$_{0.156}$ (CaO)$_{0.107}$ (SiO$_2$)$_{0.737}$; (d) BK7; (e) phosphate-based optical glass.

Standard Seebeck coefficients of oxidic glass-forming melts were found to be between -0.3 and -1.0 mV/K in the temperature range 800–1600 °C and to be dependent on or, in some cases, independent of temperature. Examples are given in Figure 26-15. Melts with temperature-independent standard Seebeck coefficients allow the application of linear equations such as Equation (26-24) for calculating standard thermoelectric emfs of the melt for an application in Equation (26-35). For melts with temperature-dependent standard Seebeck coefficients, however, the thermoelectric emf of the melt, $E_{Th,m}(T_c \rightarrow T_m)$, according to Equation (26-36) depends on the temperature difference, $(T_c - T_m)$, in addition to the absolute magnitude of the temperatures, T_c and T_m (see Figure 26-15), and are expressed, eg, by polynomials. The continuous calculation of the oxygen fugacity of a technical melt according to Equation (26-35) can thus only be accomplished by computer application.

26.5 Applications

26.5.1 Investigation of Redox Fining

The various kinds of electrochemical cells discussed are applied for production and research purposes, and some examples may illustrate the usefullness and the scientific values of these sensors. Figure 26-16 shows measurements which were conducted during the study of a short-time redox fining process [7]. During the 30-min fining period of a lead-containing silicate glass melt with 0.3 wt% As_2O_3, the large equilibrium constant at 1280 °C (Equation (26-9)) generated oxygen fugacities above 1 bar and thus caused the removal of gaseous impurities by bubble formation. Subsequently, the temperature was lowered to the values indicated in

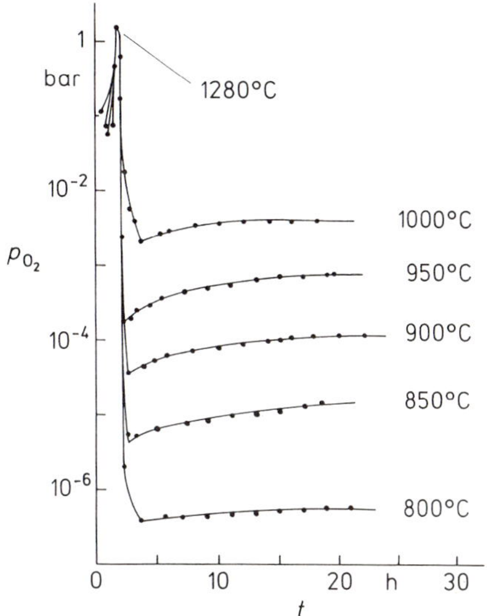

Figure 26-16.
Short-time fining process, with Sb_2O_3 as fining agent, consisting of fining period (30 min, 1280 °C) and subsequent temperature-dependent redissolution period of oxygen blisters.

order to achieve redissolution of the remaining oxygen blisters. The oxygen fugacities correspond to the smaller equilibrium constants, and their slow increase during this resorption period is determined by the different diffusion coefficients of oxygen in the melts.

26.5.2 Determination of Thermodynamic Standard Data of Redox Equilibria

In many cases, the determination of relative oxygen fugacities (Section 5.1) on the basis of measurements indicated by Figure 26-11 yield sufficient information for the processes to be studied. As stated in Section 3, however, a knowledge of the quantitative relationship between

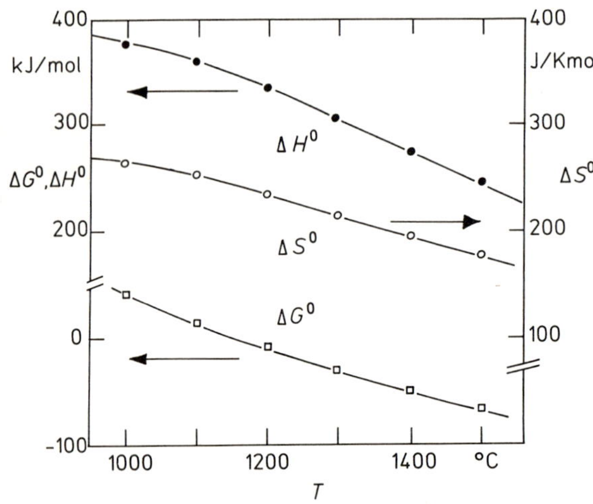

Figure 26-17.
Temperature-dependent thermodynamic standard data of antimony(III)/antimony(V) redox equilibrium (Equation (26-28)), in an oxidic glass-forming silicate melt; see also Figure 26-18.

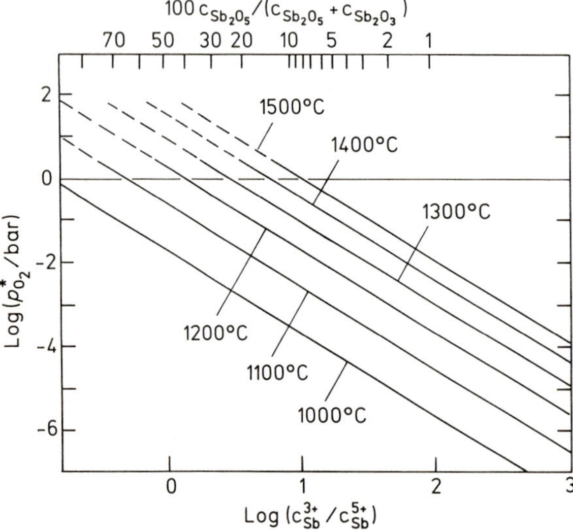

Figure 26-18.
Temperature-dependent oxygen fugacitiy as a function of antimony(III)/antimony(V) ratio in an oxidic glass-forming silicate melt; see also Figure 26-17.

oxygen fugacity and concentrations of the redox components concerned is often essential. An example is the investigation of the thermodynamics of the redox equilibrium of antimony (Equation (26-9)), and thus of the corresponding redox fining process. It was carried out in the closed cell arrangement quoted in Section 4.3 and additionally employed Mössbauer spectroscopy to obtain the concentrations of the polyvalent species [28, 31]. Figure 26-17 shows standard enthalpy, ΔH^0, entropy, ΔS^0, and free enthalpy, ΔG^0, as functions of the temperature in an alkali metal-alkaline earth metal silicate glass melt containing 0.2 mol% Sb_2O_3, and Figure 26-18 shows temperature-dependent oxygen fugacities as functions of the concentration ratio of the polyvalent species and thus presents the quantitative basis of the Sb^{3+}/Sb^{5+} redox fining process.

26.5.3   Elucidation of the Mechanism of a Spontaneous Heterogeneous Reaction

In addition to measuring oxygen fugacities in melts, the zirconia electrode is frequently applied as a reference electrode for other investigations, eg, for studying redox processes in melts and in refractories in contact with melts. Figure 26-19 shows an arrangement by which the spontaneous oxygen bubble formation at the interface between oxidic melts and new zirconium silicate (ZS) refractories was studied [32]. This reaction has long been known and is of high economic interest since it can impair and even prevent glass production in newly constructed glass melting tanks for long periods.

The zirconia electrode served as a reference electrode not only for the platinum electrode measuring the oxygen fugacity of the melt, but also for the platinum contact of the ZS crucible during the reaction between ZS and melt and was also short-circuited with the ZS contact in order to change the course of the reaction. The study resulted in an electrochromic-type

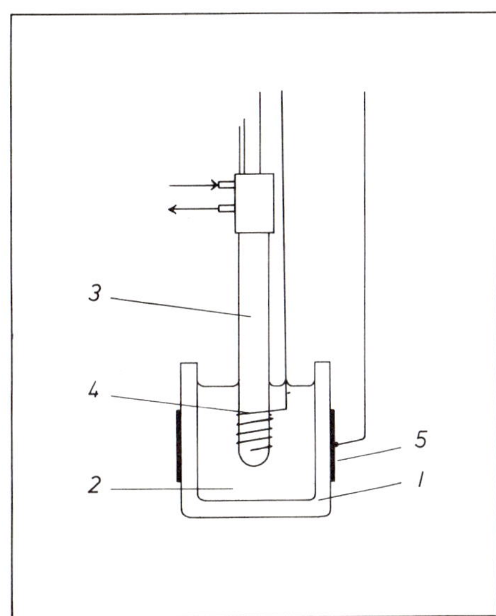

Figure 26-19.
Experimental arrangement for studying a heterogeneous reaction between zirconium silicate (ZS) refractory (1) and oxidic melt (2). The zirconia electrode (3) is used as a reference electrode for the platinum electrode (4) measuring the oxygen fugacity of the melt, for the ZS crucible (1), and for a short-circuit with ZS, via platinum contact (5).

electrochemical reaction mechanism consisting of an oxidation of oxide ions of the melt by an internal reduction of polyvalent impurities of the ZS and the simultaneous diffusion of alkali metal ions of the melt into the ZS, by which the charge imbalance caused by the heterogeneous redox process is compensated [32].

26.5.4   Study of Metal Electrodes in Nonisothermal Glass-Forming Melts

The final example is concerned with standard Seebeck coefficients of glass melts, which were accessible after zirconia electrodes had been developed [12]. Continuously working glass melting units are characterized by nonisothermal operation, and metals, eg, platinum-type metals, contacting the melt and often short-circuited are subject to electrode reactions, eg, generation and consumption of oxygen, which can indirectly impair the production of the melters. Thermoelectric emfs of such nonisothermal cells,

$$Pt\ (T_1)/\text{melt}/Pt\ (T_2)\ , \tag{IX}$$

cannot be measured with high precision because of insufficient experimental control of uniform oxygen content of the melt, but are obtained by a combination of standard thermoelectric emfs of glass melts, $E_{Th,m}\ (T_1 \to T_2)$, and temperature-dependent emfs, Equation (26-15) of cell (I), according to the equation

$$E_{\Delta T,\ Pt}\ (T_1 \to T_2) = E_{Th,m}\ (T_1 \to T_2) + [E\ (T_2) - E\ (T_1)]\ . \tag{26-37}$$

Figure 26-20 shows three examples. The cells contained melts with 0.2 wt% Sb_2O_3 and a constant total oxygen content and are thus represented by cell scheme (VI) under conditions

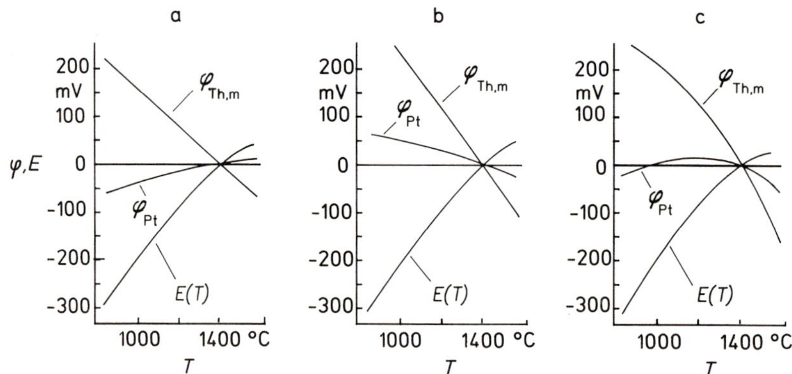

Figure 26-20. Practical significance of thermoelectric potentials. The relative potentials, φ_{Pt}, of a platinum electrode in melts satisfying Equation (26-33) were obtained from standard thermoelectric potentials, $\varphi_{Th,m}$, of a zirconia electrode and temperature-dependent emfs, E, of cell (VI) according to Equation (26-37). The temperature dependence of φ_{Pt} is determined by the standard Seebeck coefficients of the melts and is positive (a), negative (b), and, depending on the temperature, both negative and positive (c).

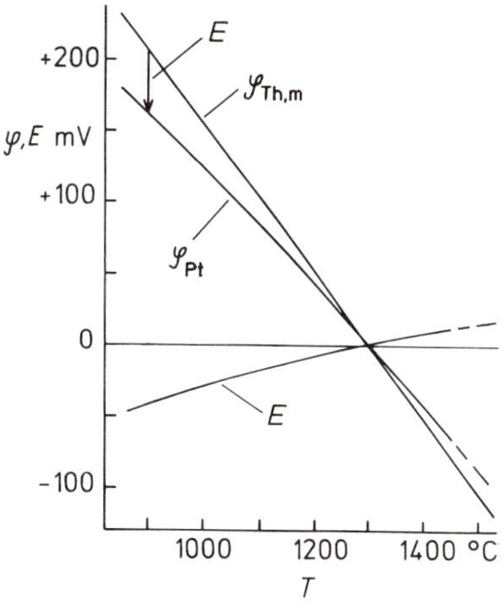

Figure 26-21.
Temperature-dependent relative potential of zirconia reference electrode, $\varphi_{Th,m}$, with 1 bar oxygen partial pressure, and of platinum measuring electrode, φ_{Pt}, and temperature-dependent emf, E, of cell (I), in an oxidic glass-forming melt with constant total oxygen content and containing no fining agent. Short-circuiting platinum electrodes at different temperatures leads to strong oxygen formation at the (negative) high-temperature electrode and to a large decrease in the reboil temperature. Reference potential, $\varphi = 0$: potential of the platinum electrode at 1 bar oxygen fugacity.

which satisfy the application of Equation (26-33). The thermoelectric emf of cell (IX) can either be positive, ie, oxygen is formed at the low-temperature electrode on short-circuiting, (Figure 26-20a), or negative, with oxygen formation at the high-temperature electrode (Figure 26-20b), or, depending on the temperatures chosen, positive and/or negative (Figure 26-20c) [12]. Which of these cases prevails is determined by the magnitude and temperature dependence of the standard Seebeck coefficient.

Melts causing oxygen formation at the high-temperature electrode are critical during glass production, since the large oxygen content generated due to the short-circuit decreases the reboil temperature, ie, the temperature at which the oxygen fugacity exceeds 1 bar according to equilibrium Equation (26-9), and causes spontaneous bubble formation. The fundamental knowledge gained during these studies explains also the difficulty encountered when bubble-free glasses without any redox fining agent are to be melted in platinum crucibles (Figure 26-21), since the extremely isothermal conditions necessary can only be established with great experimental care, especially with larger crucibles [12].

26.6   References

[1] Besson, J., Deportes, C., Darcy, M., *Compt. Rend. Acad. Sci.,* **251** (1961) 1630–1632.
[2] Plumat, E., Toussaint, F., Boffe, M., *J. Am. Ceram. Soc.,* **49** (1966) 551–558.
[3] Frey, T., Schaeffer, H. A., Baucke, F. G. K., *Glastechn. Ber.,* **53** (1980) 116–123.
[4] Schaeffer, H. A., Frey, T., Löh, I., Baucke, F. G. K., *J. Non-Cryst. Solids,* **49** (1982) 179–188.
[5] Baucke, F. G. K., *Glastechn. Ber.,* **56 K** (1983) Vol. 1, 307–312.
[6] Lenhart, A., Schaeffer, H. A., *Glastechn. Ber.,* **58** (1985) 139–147.
[7] Baucke, F. G. K., *Glastechn. Ber.,* **61** (1988) 87–90.
[8] Baucke, F. G. K., Röth, G., *German Patent P 31 09 454;* 1. 10. 1987.
[9] Schott Glaswerke, Mainz, *German Patent GBM 85 13 976;* 1. 8. 1985.
[10] Baucke, F. G. K., Röth, G., Werner, R.-D., *German Patent 38 11 865;* 24. 5. 1989.
[11] Baucke, F. G. K., Frank, W., Röth, G., *German Patent P 3028270;* 14. 8. 1986.
[12] Baucke, F. G. K., Mücke, K., *J. Non-Cryst. Solids,* **84** (1986) 174–182.

[13] Uhlmann, D. R., Kreidl, N. J. (eds.), *Glass: Science and Technology;* Orlando: Academic Press 1984.

[14] Kieffer, J., Borchardt, G., *Glastechn. Ber.,* **62** (1989) 337–344.

[15] Campbell, J. H., et al., *Elimination of Platinum Inclusions in Phosphate Laser Glasses;* Lawrence Livermore National Laboratory, California, UCRL 53932, Distribution Category UC-712, Livermore, California, 1989.

[16] Gliemeroth, G., Eichhorn, U., Hölzel, E., *Glastechn. Ber.,* **59** (1981) 162–174.

[17] Kitaigorodski, J. J., *Technologie des Glases;* Berlin: VEB-Verlag Technik, 1959, pp. 414–415.

[18] Stroud, J. S., *J. Amer. Cer. Soc.* **34** (1971) 401–406.

[19] Bamford, C. R., *Colour Generation and Control in Glass, Glass Science and Technology, 2;* Amsterdam: Elsevier, 1977, pp. 35–38.

[20] Cable, M., in: *Glass: Science and Technology,* Uhlmann, D. R. and Kreidl, N. J. (eds.); Orlando: Academic Press, 1984, Vol. 2, Processing I, pp. 1–44.

[21] Fischer, W. A., Janke, D., *Metallurgische Elektrochemie, Stahleisen;* Berlin: Springer, 1975, pp. 102–104.

[22] Baucke, F. G. K., Frey, Th., Schaeffer, H. A., *German Patent P 2908368;* 3. 3. 1979.

[23] Baucke, F. G. K., Röth, G., Werner, R.-D., *German Patent 3811864;* 8. 2. 1990.

[24] Baucke, F. G. K., Röth, G., Werner, R.-D., *German Patent 3811915;* 22. 3. 1990.

[25] Veith, J. A., *Diploma Thesis,* F. H. Rheinland-Pfalz, Bingen, 1983.

[26] Fischer, W., *Z. Naturforsch.,* **22a** (1967) 1575–1581.

[27] Pizzini, S., Riccardi, C., Wagner, V., Sinistri, C., *Z. Naturforsch.,* **25a** (1970) 559–565.

[28] Stahlberg, B., PhD Thesis, Münster, Mainz; 1987, 113–117.

[29] Coughlin, J. P., *Contribution to the Data of Theoretical Metallurgy, XII., Heats and Free Energies of Formation of Inorganic Oxides,* Bulletin 542, Bureau of Mines, Washington: United States Governement Printing Office, 1954, pp. 7–10.

[30] Lewis, G. N., Randall, M., *Thermodynamics,* 2nd edn.; New York: McGraw-Hill, 1961, p. 672.

[31] Stahlberg, B., Mosel, B. D., Müller-Warmuth, W., Baucke, F. G. K., *Glastechn. Ber.,* **61** (1988) 335–340.

[32] Baucke, F. G. K., Röth, G., *Glastechn. Ber.,* **61** (1988) 109–118.

Cumulated List of Symbols and Abbreviations of Volumes 2 and 3

The following list contains the symbols most frequently used in Volumes 2 and 3. To avoid redundancy, subscripts are only noted in exceptional cases. References to chapters (where the quantities are explained in more detail) are only given for symbols with special meanings or in cases of uncommon use. Chapters 1 to 13 are located in Volume 2, chapters 14 to 26 in Volume 3.

Symbol	Designation	Chapter
a	activity, sensitivity	6
	area	11, 15
	calibration coefficient	17
	chemical activity	8
	pre-exponential factor	11
	radius of microelectrode	5
a_i	activity of species i	
a_w	cross-sectional area of wire	11
a_x	mole fractional activity	25
A	area	
	absorbance	17
	aperture	13
	pre-exponential factor for particular oxidation	6
	surface	9
A_k	vector of a_{ik} of normalized sensor responses	6
$A(f_c)$	loop gain	13
A_{eff}	catalytically active area	6
b	channel width	10
	calibration coefficient	17
	distance between electrodes	4
b_A, b_B	adsorption equilibrium constant of species A, B	11
B	Biot number	11
B	material component in one of the electrode compartments	25
B_λ	interfering absorbance of background of dye and measuring instrument	17
Bi	Biot number	14
c	concentration	
C	capacitance	
	bulk gas concentration	11
	concentration	16
	cosubstrates	14
	elasticity of the crystal	13
	total concentration	17
$C(V)$	function of voltage	10
C_f	mass sensitivity	4

Symbol	Designation	Chapter
C_P	molar heat capacity at constant pressure	
C_s	function of the metal structure	10
C_V	molar heat capacity at constant volume	
Cl	clorinity of sea water	7
d	distance of electrodes	7
	diameter of pores	20
	ion-size parameter	17
	thickness	
d	orbital number	14
d_i	insulator thickness	4, 10
D	diffusion coefficient	
	Goos-Hänchen shift	18
D_eff	effective diffusion coefficient	14
D_K	Knudsen diffusion coefficient	8
D_MFC	dilution range	16
e, e_0	elementary charge	7
emf	electromotive force	
E	activation energy	11
	electric field strength	
	open-circuit voltage, electromotive force (*emf*)	5
	potential	7
	voltage	1, 8
E	functional or calibration matrix	6
ΔE	activation energy	6
E_a	activation energy	9
$E_\mathrm{a}, E_\mathrm{i}$	active (inactive) enzyme columns	14
$E_\mathrm{Ag/AgCl}$	reference electrode potential	10
E_C	conduction band edge energy	
E_CE	voltage of the counter electrode	7
E_d	molar activation energy of diffusion	16
E_F	Fermi energy	
E_g	band gap of the semiconductor	10
E_i	energy	4
E_j	liquid junction potential	10
E_pol	polarization voltage	7
E_V	valence band edge energy	
E_WE	effective voltage at the working electrode	7
f	frequency	
	activation function	6
f_E	enzyme loading factor	14
f_i	activity coefficient	7
F	Faraday constant	
	fluorescence intensity	17, 18
	electrical mobility	8
	molar flow rate	16

Symbol	Designation	Chapter
δF	change in the frequency of the crystal	14
F_e, F_i	number of unoccupied adsorption sites on the surface,	
	at the metal/insulator interface	10
g	geometric factor	11
g_m	transconductance	20
G	amplification factor	14
	conductance	
	total free enthalpy	4
ΔG	Gibb's free energy	
G_0	conductance under reference conditions	6
h	height	15
	Thiele modulus	11
h^+	defect electrons	4
H	enthalpy	4, 6
ΔH	heat of combustion, heat of adsorption	11
ΔH_i	oxidation heat of component i	4
ΔH_v	heat of vaporization	16
i	electric current	
$i(t)$	time dependent current	6
i_0	exchange current density	8
i_D	limiting current	14
I	electric current	
	ionic strength	7, 17
	intensity of light	4
I_D	diffusion-controlled limiting current	7, 14
I_h	hole current	8
I_i	current of the carriers i	8
I_l	plateau current of diffusion-limited regime	8
j	current density	5
J	flux	8
k	constant	
	Boltzmann constant	8
	attenuation factor	17
k_1, k_2	kinetic parameters	14
k_B	Boltzmann constant	
k_{ij}	selectivity coefficients	7
k_w	thermal conductivity of platinum wire	11
K	dissociation constant	7, 17
	equilibrium constant	20
	mass transfer coefficient	11
	ratio of molar heat capacities	16
K, K_{a1}, K_{a2}	function of activity and differently occupied	
	surface sites	10

Symbol	Designation	Chapter
K'	capacity ratio of a compound	15
K'_a	concentration constant	17
K_1, K_2, K_3, K_4	calibration coefficients	17
K_a	thermodynamic dissociation constant	17
K_d	quenching constant	18
K_g	thermal conductivity of the gas surrounding the	11
K_i	accelerating force on ions	7
	characteristic parameters	9
	equilibrium constants	25
K_{ij}	selectivity coefficients	1, 5
K_L	solubility product	7
K_M	Michaelis constant	14
K_m, K_1, K_2	net conductance of mixture, -of components 1 or 2	11
K_{xy}	equilibrium constant of the surface ion exchange	8
l	distance	
	length	
ℓ	optical path length	17
l_i	distance of electron-hole recombination from	8
	electrode II	8
L	length	
	inductance	
	mass of crystals	13
	membrane thickness	14
L_D	Debye-length	9
L_k	number of different partial pressures	6
m	characteristic parameter of defects	4
	electrical bridge constant	11
	intrinsic cell sensitivity	6
	mass	
	number of fragments upon dissociation	11
M	molecular weight	
	molecular weight of water	20
M_i	metal interstitial	4
M_i	molar weight of component i	16
n	counting number	
	exponent in the range 1–2	11
n_c	number of moles of gas in volume V_0	6
n_e	number of absorbed hydrogen atoms	10
n_i	reaction orders	4
	characteristic parameters	4, 9
n_i	number of hydrogen atoms at the inner interface	10
n_s	electron concentration at semiconductor surface	10

Symbol	Designation	Chapter
N	doping density in the semiconductor	10
	frequency constant	13
	internal hydrogencarbonate concentration	17
	number of interfering functions	17
ΔN	excess concentration	4
N_A	Avogadro number	
	bulk concentration of holes (acceptor concentration)	10
	concentration of electrons at the semiconductor/ insulator-interface	4
N_i	number of measuring signals	1
N_s	surface site density	10
N_{th}	theoretical plate number	15
NA	numerical aperture	
O_L	lattice oxygen	4
p	dipole moment	
	partial pressure	
	period	13
	total pressure	
	pyroelectric coefficient	11
p_i	analytical information	4
	partial pressure of component i	1, 4
	amount of species	15
p_R	reference pressure	8
p_t	tabular pressure	25
pa	arterial partial pressure	23
P	electrical power	
	partial pressure	10, 11
	product concentration	14
	statistical probability	1
	total pressure	
	water vapor pressure	20
P	analytical information	6
ΔP	excess concentration	4
$\Delta P(t)$	net flux of heat into the pyroelectric	11
P^d	product concentration at the electrode surface	14
P_3	product	14
P_i	partial pressure of component i	16
P_L	sum of heat losses	11
P_m	membrane permeability for diffusion transport	14
P_M	permeability	4
P_r	heat loss by radiation	11
P_S	permeability of the solution layer	14
q	elementary charge	
	specific humidity	20
q_i, q_j	transported charge due to ions i, j	5

Symbol	Designation	Chapter
Q	absolute humidity	20
	heat flow, heat production rate	16
	heat of adsorption	11
	quality factor	13
	redox ratio	25
Q_{ad}, Q_{react}	heat of adsorption, heat of reaction	4
$Q_{electric}$	electric power	16
Q_i	equivalent insulator charge	4, 10
r	catalytic activity	6
	mixing ratio	20
	radius	
	reaction rate	4, 11
r_d	rate of bulk diffusion	11
r_K	Kelvin radius	20
R	gas constant	
	losses in the crystal	13
	reflectance	17
	residence time	15
	resistance	
	resolution of two compounds	15
R_a	actual rate	11
R_B	resistor	20
R_d	rate of diffusion to the element	11
R_e	reagent phase	17
R_e	reaction rate on the bead in absence of external mass transfer control	11
R_G	gas constant per mole	6
R_i	partial resistance	8
R_L	load resistor	20
s^0	bulk substrate concentration	14
s_i	best estimate for a standard deviation	1
$sol(a_i)$	solution of activity	10
S	area of diffusion hole	8
	entropy	4
	salinity of seawater	7
	selectivity	15
	slope of potentiometric electrodes	7
	substrate concentration	14
S_0	sticking coefficient	4
S_i	sensitivity	1
S_{ij}	sensitivity	6
S_k	similarity	6
S_{MX}	solubility product of MX	8
SE	intermediate complex	14

Symbol	Designation	Chapter
t	time	
t	orbital number	14
t_e	average electronic transference number	8
$t_{1/2}$	time to reach half the steady state current	14
t_{90}	90 % level of response time	11
t_K	transference number of the K^{z+} carriers	8
$t_{\text{steady state}}$	time to reach a steady state current	14
tcp	transcutaneous partial pressure	23
T	temperature	
$\tilde{u}$	electrical mobility	8
u_i	mobility of ion i	4, 5
	flow rate	6
U	voltage	
	inner energy	4
	electrode potential	25
$U(\varphi_B)$	signal function	25
v	AC voltage	
	velocity	
	volume	
	maximal rate of enzyme reaction	14
V	voltage	
	DC voltage	10, 20
	volume	
$\dot{V}$	standard volumetric flow rate	16
V_0	volume of 1 mole of ideal gas, standard conditions	6, 16
V_M	metal vacancy	4
V_O	oxygen vacancy	
V_R	phase velocity	13
V_{T0}	threshold voltage of the transistor	10
w	chromatographic peak width	15
	flowrate	15
w_i	migration speed of ions	7
w_{ik}	statistical weight	6
W	reduced electro-chemical potential	8
	work funktion	10
W_d	width of depletion layer	10
W_{el}	electrical work	5
x	number	
	distance from the source	10
	species variable	11
	distance from the catalyst surface	11
	mole fraction	11, 25
	stoichiometric coefficient	25
x'	measuring signal	1, 4, 6

Symbol	Designation	Chapter
x_0'	reference signal	1
x_i	measuring signals	15
X	impedance	7
X	measuring signal, sensor signal	6
X_v	mole fraction of water vapor	20
y	amplitude of output signal	6
	mole fraction	16
	stoichiometric coefficient	25
$\tilde{Y}$	complex admittance	9
z_i	charge number	10
z_K	electrical mobility	8
z_r	number of electrons in electrochemical reactions	25
Z	number of electrons per val	8
	partition function	4
$\tilde{Z}$	complex impedance	9
a	a value	20
	concentration coefficient of a nozzle	16
	partition coefficient	14
	reaction order	11
	Seebeck coefficient	11
	transfer coefficient	5
a_i	corrected composition qantity of reactant i	25
	real potential of species i	10
a_{ik}	normalized sensor response	6
β	dimensionless pH-sensitive parameter	10
	reaction order	11
β_{ik}	exponential non-linearity factor	6
γ	activity coefficient	8, 17
	partial sensitivity	1
	reaction order	11
	surface tension of water	20
γ^*	normalization factor of sensor array	6
γ, γ'	proportionality constants	11
γ_{ik}	partial sensitivity	6
δ	partial charge	4
	reaction order	11
	thickness	
Δ	confidence range	1
Δ_k^*	Euclidean distance	6

Symbol	Designation	Chapter
ε	absorption yield	17
	dielectric constant	7, 3
	electrode potential	5
	emissivity of the sensing element	11
	error	16
	error, random variable	6
	molar absorption	15
	optical constants	4
	relative permittivity	9
$\boldsymbol{\varepsilon}$	error matrix	6
$(\varepsilon)_\lambda$	molar absorption coefficient	17
ε_i	dielectric permittivity of insulator	4, 10
ε_s	dielectric permitivity of semiconductor	4, 10
η	overpotential	7
	viscosity	5
ϑ	coverage	4, 11
	fractional coverage	20
	temperature	
κ	electrolytic conductivity	
λ	constant	11
	mobility	7
	wavelength	
Λ	molar conductivity	7
μ	chemical potential	
	mobility of electrons in the channel	10
$\tilde{\mu}$	electrochemical potential	5, 8
μ_i	statisitcal average	1
ν	frequency	
ν_i	stoichiometric coefficient of reactant i	25
Ξ	selectivity	6
π	orbital number	14
Π	fluorescence yield	17
ρ	density	
σ	conductivity	
	density	16
	diffusity of leak aperture	8
	force constant	11
	magnitude of induced mobile charge	10
	Stephan-Boltzmann constant	11

Symbol	Designation	Chapter
σ^0	conductivity	8
σ_i	standard deviation	1
σ_{ik}	dispersion	6
σ_K	partial conductivity of species K^{z+}	8
τ	transit (or delay) time	13
τ_s	time constant of sensor	6
φ	electrical potential	
	electrostatic potential, Galvani potential	8
	phase angle between current and voltage	9
φ_A	phase shift	13
φ_B	voltage drop	4, 10
	volume concentration of component B	25
Φ	work function	9
χ	electron affinity	4, 9, 10
Ψ_0	potential difference	10
Ψ_i	partial specificity	6
ω	angular frequency	
	frequency of AC pumping current	8
	rotation rate	14
Ω	collision integral	11

Abbreviation	Explanation
a. c.	alternating current
AAS	atomic absorption spectroscopy
AC	alternating current
ACHE	acetylcholinesterase
AES	Auger electron spectroscopy
	atomic emission spectroscopy
AFS	atomic fluorescence spectroscopy
AGC	automatic gain control
AIM	adsorption isotherm measurements
AIS	atom inelastic scattering
ALAT	alanine aminotransferase
ALE	atomic layer epitaxy
AOD	alcohol oxidase
APS	appearance potential spectroscopy
APTES	3-aminopropyltrienthoxysilane
ARD	acoustic resonance densitometry
ARIES	angular resolved ion and electron spectroscopy
ASAT	aspartate aminotransferase
ASEA	Allmaenna Svenska Elektriska Aktiebolaget

Abbreviation	Explanation
ASIA	atomizer, source, inductively coupled plasma in atomic fluorescence spectroscopy
ATR	attenuated total reflection
BAW	bulk acoustic waves
BCG	Bromocresol Green
BGM	blood gas machine
BLM	bilayer lipid membrane
BMFT	Bundesministerium für Forschung und Technologie
BOD	biochemical oxygen demand
	biological oxygen demand
BTB	Bromothymol Blue
CAB	cellulose acetate butylate
CADI	computer-assisted dispersive infrared
cAMP	cyclic adenosine monophosphate
CARS	coherent anti-Stokes Raman spectroscopy
CCC	counter current chromatography
CCM	cubic centimeter per minute
CD	calibration data
	circular dichroism
CDS	corona discharge spectroscopy
CE	counter electrode
CEA	Commissariat á l'Energie Atomique
CFS	constant final state spectroscopy
CHEMFET	chemically sensitive field effect transistor
Chl	chlorophyll
CI	chemical ionization
CIR	cylindrical internal reflection
CIS	constant initial state spectroscopy
CM	conductance measurements
COMAS	concentration-modulated absorption spectroscopy
CP/MAS	cross polarization/magic angle spinning
CPAC	Center for Process Analysis and Control
CPB	cardiopulmonary bypass
CPD	contact potential difference measurements
CSN/ICP	conductive solids nebulizer/inductively coupled plasma
CSP	chiral stationary phase
CSSD	chemically sensitive semiconductor devices
CTA	cellulose triacetate
CTC	chlortetracycline
CV	calorific value
CVD	chemical vapor deposition
CWE	coated wire electrode
d.c.	direct current
DAC	diamond anvil cell
DC	direct current

Abbreviation	Explanation
DCCC	droplet counter current chromatography
DCI	direct current ionization
DCP	direct current plasma
DCPAES	direct current plasma atomic emission spectroscopy
DIP	2,6-dichlorophenolindophenol
DM	diffusion measurements
DMS	dynamic mass spectrometer
DNA	desoxyribonucleinacid
DR	diffuse reflectance
DRIFTS	diffuse reflectance infrared Fourier transform spectroscopy
DTA	differential thermal analysis
EBD	electron beam deposition
EBIC	electron beam induced current
EC	enzyme catalog
	electrochemical sensors
ECG	electrocardiogram
EDS	energy dispersive spectroscopy
EDXRF	electron diffraction X-ray fluorescence
EDXS	energy dispersive X-ray spectroscopy
EE	enzyme electrode
EEC	European Economic Community
EELS	electron energy loss spectroscopy
EG(FET)	extended gate (field effect transistor)
EI	electron ionization
EIE	easily ionized element
EIEIO	easily ionized element interface observation
ELISA	enzyme-linked immunosorbent assay
ELL	ellipsometry
EMA	electron microprobe analysis
EMF	electromotive force
EMTC	external mass transfer control
ENDOR	electron nuclear double beam resonance
EnFET	enzyme-modified field effect transistor
EPR	electron paramagnetic resonance
ER	enzyme reactor
ES	electron spectroscopy
ESA	electrostatic analyzer
ESCA	electron spectroscopy for chemical analysis
ESD	electron stimulated desorption
ESIE	electron-stimulated ion emission
ESPRIT	European Strategic Program for Research in Information Technology
ESR	electron spin resonance
ETA	electrothermal atomization
$EtCo_2$	end-tidal partial pressure
EXAFS	extended X-ray absorption fine structure

Abbreviation	Explanation
FAB-MS	fast-atom bombardment mass spectrometry
FAD	flavine adenine dinucleotide
FD	field desorption
FDIR	fast dispersive infrared
FDM	field desorption microscopy
FEC	field effect of conductance
FEM	field effect microscopy
FER	field effect of reflectance
FES	field emission spectroscopy
	flame emission spectrometry
FET	field effect transistor
FIA	flow-injection analysis
FIAP	field ionisation atom probe
FIIA	flow injection immunoanalysis
FIM	field ion microscopy
FIMS	field ion mass spectrometry
FIR	far infrared
FITC	fluorescein isothiocyanat
FOCS	fiberoptic chemical sensors
FOS	fiberoptic sensors
FOTDR	frequency optical time domain reflectometry
FRC	functional residual capacity
FTIR	Fourier transform infrared
	frustrated total internal reflection
FTMS	Fourier transform mass spectrometry
FTR	frustrated total reflection
GASFET	gas sensitive field effect transistor
GC	gas chromatography
GCIR	gas chromatography infrared
GCMS	gas chromatography mass spectrometry
GCNMR	gas chromatography nuclear magnetic resonance
GDH	glucose dehydrogenase
GDMS	glow discharge mass spectrometry
gel	gelantine
GFAAS	graphite furnace atomic absorption spectroscopy
GL	gluconolactone
GOD	glucose oxidase
GPC	gel permeation chromatography
GPMAS	gas phase molecular absorption spectroscopy
HAM	heat of adsorption measurements
Hb	hemoglobin
HDC	hydrodynamic chromatography
HE	Hall effect
HEED	high energy electron diffraction
HID	hydrogen-induced drift
HIXSE	heavy ion induced X-ray satellite emission

Abbreviation	Explanation
HK	hexokinase
hmds	hexamethyldisiloxane
HOL	holography
HPLC	high performance liquid chromatography
HPTLC	high perfomance thin layer chromatography
HPTS	1-hydroxypyrene-3,6,8-trisulfonic acid
	hydroxypyrene trisulfonate
HREELS	high resolution electron energy loss spectrometer
HROC	high resolution gas chromatography
HUP	hydrogen uranyl phosphate tetrahydrate
	uranophosphoric acid
IC	integrated circuit
	ion chromatography
ICAP	inductively coupled argon plasma
ICB	ion cluster beam deposition
ICD	ion controlled diode
ICP	inductively coupled plasma
ICPAES	inductively coupled plasma atomic emission spectroscopy
ICPES	inductively coupled plasma emission spectroscopy
ICPMS	inductively coupled plasma mass spectrometry
ICRS	ion cyclotron resonance spectroscopy
IDT	interdigital transducer
IETS	inelastic electron tunneling spectroscopy
IEX	ion excited X-ray fluorescence
IGFET	insulated-gate field effect transistor
IgG	immuno-gamma-globulin
IID	ion impact desorption
ImFET	immuno-sensing field effect transistor
IMPA	ion microprobe analysis
IMXA	ion microprobe for X-ray analysis
INEPT	insensitive nuclei enhanced by polariz. transfer
INS	ion neutralization spectroscopy
IPS	inverse photoelectron spectroscopy
IR	infrared
IRE	internal reflection element
IRS	internal reflection spectroscopy
ISE	ion-selective electrode
ISFET	ion-sensitive field effect transistor
ISM	solvent polymeric membrane
ISS	ion scattering spectroscopy
LAMMA	laser microprobe mass analysis
LASER	light amplification by stimulated electromagnetic
	radiation
LB	Langmuir-Blodgett
LC	liquid chromatography
LCIR	liquid chromatography infrared

Abbreviation	Explanation
LCMS	liquid chromatography mass spectrometry
LDH	lactate dehydrogenase
LED	light emitting diode
LEED	low energy electron diffraction
LEL	lower explosive limit
LETI	Laboratoire d'Electronique et des Technologies de l'Information
LIDAR	light detection and ranging
LIMA	laser ionization mass analysis
LIMS	laser ionization mass spetrometry
	laboratory information managment system
LM	light microscope
	liter per minute
LMO	lactate monoxigenase
LOD	lactate oxidase
LPCVD	low-pressure chemical vapor deposition
LRS	laser Raman spectroscopy
MAK	maximal tolerable concentration in the ambient air which for a daily exposition of 8 h is without any influence on the health of a person
MBAS	molecular beam atom scattering
MBE	molecular beam epitaxy
MBT	molecular beam techniques
MCD	magnetic circular dichroism
MCP	methyl-accepting chemotaxis protein
MFC	mass flow controller
MIKE	mass selection followed by ion kinetic energy analysis
	mass-analyzed ion kinetic energy
MIP	microwave induced plasma
MIR	multiple internal reflection
	mid infrared
MISCAP	metal-insulator-semiconductor capacitor
MISFET	metal-insulator-semiconductor field effect transistor
MOS	metal oxide semiconductor
MOSCAP	metal oxide semiconductor capacitor
MOSFET	metal oxide semiconductor field effect transistor
MPS	modulated photoconductivity spectroscopy
MRI	magnetic resonance imaging
MS	mass spectrometry
MTX	methotrexate
NAD	nicotine adenine dinucleotide
NADH	nicotinamide adenine dinucleotide
	reduced nicotine adenine dinucleotide
NCI	nitrogen chemical ionization
NEC	National Electric Code

Abbreviation	Explanation
NHE	normal hydrogen electrode
NIR	near infrared
	near infrared spectroscopy
NIRA	near infrared reflectance analysis
NIS	neutron inelastic scattering
NMR	nuclear magnetic resonance
NQR	nuclear quadrupole resonance
NTIS	National Technical Information Service
OAS	optoacoustic spectroscopy
OES	optical emission spectroscopy
ORD	optical rational dispersion
ORP	oxidation reduced potential
OS	optical spectroscopy
	oxygen saturation
OSCA	Optical Sensor Collaborative Association
OTDR	optical time domain reflectometry
OxyHb	oxyhemoglobin
PAH	polynuclear aromatic hydrocarbons
PARUPS	polarization and angle resolved ultraviolet photoelectron spectroscopy
PAS	photoacoustical spectroscopy
PASCA	positron annihilation spectroscopy for chemical analysis
PBA	pyrenebutyric acid
PC	photoconductivity
Pc	phatalocyanines
PCS	plastic cladding silica
PD	photodesorption
PDMS	plasma desorption mass spectrometry
PDS	photodischarge spectroscopy
PEM	photoelastic modulator
PEP	phosphoenolpyruvate
PES	photoelectron spectroscopy
PIGME	particle-induced gamma ion emission
PIM	patient interface module
PIS	Penning ionization spectroscopy
PIXE	proton/particle-induced X-ray emission
PK	pyruvate kinase
PLP	pyridoxal phosphate
PM	permeation measurements
PNMR	proton nuclear magnetic resonance
POTDR	polarization optical time domain reflectometry
PQQ	pyrroloquinolinequinones
PT	paper tape
PTFE	polytetrafluoroethylene
PUR	polyurethane
PVA	polyvinyl alcohol

Abbreviation	Explanation
PVD	physical vapor deposition
PVS	photovoltage spectroscopy
r.h.	relativ humidity
RBS	Rutherford backscattering
RC	reaction coil
REFET	reference field effect transistor
RFF	remote fiber fluorimetric technique
RGA	residual gas analysis
RH	relative humidity
RHEED	reflection high energy electron diffraction
RI	refractive index
RIA	radioimmunoassay
ROA	Raman optical activity
RQ	respiratory quotient
RRS	resonance Raman spectroscopy
RT	response time
RTP	room temperature phosphorescence
RTPL	room temperature phosphorescence in liquids
s.i.c.	solid ionic conductor
SAM	scanning Auger microscopy
	scanning acoustic microscopy
SAW	surface acoustic waves
SAX	selected areas X-ray photo-electron spectroscopy
SCCM	standard cubic centimeter per minute
SCE	saturated calomel electrode
SDS	surface discharge spectroscopy
SEM	scanning electron microscopy
SERC	Science and Engineering Research Council
SERS	surface enhanced Raman spectroscopy
SES	spin echo spectroscopy
	secondary electron spectroscopy
SFC	supercritical fluid chromatography
SGFET	suspended gate field effect transistor
SIM	scanning ion microscopy
SIMS	secondary ion mass spectrometry
SIS	silicon-insulator-silicon
SLM	standard liter per minute
SNMS	secondary neutral mass spectrometry
	sputtered neutral mass spectrometry
SNR	signal-to-noise ratio
SOS	silicon-on-sapphire
SS	stain-less steel
	solid state sensors
SSIMS	scanning mode secondary ion mass spectrometry
SSMS	spark source mass spectrometer

Abbreviation	Explanation
STAT	slotted tube atom trap
STM	scanning tunneling microscope
TAB	tape automated bonding
TB	Thymol Blue
TDS	thermal desorption spectroscopy
TEM	transmission electron microscopy
TIMS	thermal ionization mass spectrometry
TL	thermoluminescence
TLC	thin layer chromatography
TLV	threshold-limited value
TMOS	ultra-thin gate metal oxide semiconductor
TOF	time-of-flight mass spectrometer
TOSFET	ion-sensitive Ta_2O_5-based field effect transistor
TPD	temperature-programmed desorption
UEL	upper explosive limit
UHV	ultra high vacuum
UPS	ultraviolet photoelectron spectroscopy
UV	ultraviolet
UV-VIS	ultraviolet-visible spectroscopy
VUV	vacuum ultraviolet
W	waste
WDM	wavelength multiplexing
WDS	wavelength dispersive spectroscopy
WE	working electrode
XAES	X-ray induced Auger electron spectroscopy
XANES	X-ray absorption near edge structure
XPS	X-ray photoelectron spectroscopy
XRD	X-ray diffraction
XRF	X-ray fluorescence
YSZ	yttria-stabilized zirconia

Cumulated Index of Volumes 2 and 3

Please note: numbers in bold type refer to the volumes

HOW TO GET A JOB IN DALLAS/FORT WORTH

CITRIN, RICHARD
CAMDEN

SURREY BOOKS
230 East Ohio Street
Suite 120
Chicago, Illinois 60611

...LLAS/FORT WORTH

...c., 230 E. Ohio St., Suite 120, Chicago, IL 60611.

...ks, Inc. All rights reserved, including the right to reproduce this book or portions thereof in any form, including any information storage and retrieval system, except for the inclusion of brief quotations in a review.

This book is manufactured in the United States of America.

4th Edition. 1 2 3 4 5

Library of Congress Cataloging-in-Publication data:

Citrin, Richard.
      How to get a job in Dallas/Fort Worth / by Richard Citrin and Thomas
      M. Camden.— 4th ed.
      420 p. cm.
      Previous eds. by Thomas M. Camden and Nancy Bishop.
      Includes bibliographical references and indexes.
      ISBN 0-940625-43-1 (pbk.): $15.95
1. Job hunting—Texas—Dallas. 2. Job Hunting—Texas—Fort Worth. 3. Job vacancies—Texas—Dallas. 4. Job vacancies—Texas—Fort Worth. 5. Professions—Texas—Dallas. 6. Professions—Texas—Fort Worth. 7. Occupations—Texas—Dallas. 8. Occupations—Texas—Fort Worth. 9. Dallas (Tex.)—Industries—Directories. 10. Fort Worth (Tex.)—Industries—Directories.
I. Camden, Thomas M., 1938-   . II. Camden, Thomas M., 1938-   How to get a job in Dallas/Fort Worth. III. Title.
HF5382.75.U62T42   1992   91-45308
650.14'09764'2812—dc20   CIP

AVAILABLE TITLES IN THIS SERIES — $15.95 *(Pacific Rim $17.95)*

How To Get a Job in Atlanta by Diane C. Thomas, Bill Osher, Ph.D., and
    Thomas M. Camden.

How To Get a Job in Greater Boston by Paul S. Tanklefsky and
    Thomas M. Camden.

How To Get a Job in Chicago by Thomas M. Camden and Susan Schwartz.

How To Get a Job in Dallas/Fort Worth by Richard S. Citrin, Ph.D., and
    Thomas M. Camden.

How To Get a Job in Europe by Robert Sanborn, Ed.D.

How To Get a Job in Houston by Thomas M. Camden and Robert Sanborn.

How To Get a Job in The New York Metropolitan Area by
    Thomas M. Camden and Susan Fleming-Holland.

How To Get a Job in the Pacific Rim by Robert Sanborn, Ed.D., and Anderson
    Brandao.

How To Get a Job in The San Francisco Bay Area by Thomas M. Camden and
    Evelyn Jean Pine.

How To Get a Job in Seattle/Portland by Thomas M. Camden and
    Sara Steinberg.

How To Get a Job in Southern California by Thomas M. Camden and
    Jonathan Palmer.

How To Get a Job in Washington, DC, by Thomas M. Camden and
    Karen Tracy Polk.

Single copies may be ordered directly from the publisher. Send check or money order plus $2.50 per book for postage and handling to Surrey Books at the above address. For quantity discounts, please contact the publisher.

Editorial production by Bookcrafters, Inc., Chicago.
Cover design by Hughes Design, Chicago.
Typesetting by On Track Graphics, Inc., Chicago.
"How To Get a Job Series" is distributed to the trade by Publishers Group West.

ii